Student Solutions Man...

M000274575

Contemporary Precalculus
A Graphing Approach

FIFTH EDITION

Thomas W. Hungerford
Saint Louis University

Douglas J. Shaw
University of Northern Iowa

Prepared by

Jerrett Dumouchel
Florida Community College at Jacksonville

Amanda Nunley
Florida Community College at Jacksonville

BROOKS/COLE
CENGAGE Learning™

Australia • Brazil • Japan • Korea • Mexico • Singapore • Spain • United Kingdom • United States

ISBN-13: 978-0-495-55399-1
ISBN-10: 0-495-55399-9

Brooks/Cole
10 Davis Drive
Belmont, CA 94002-3098
USA

Cengage Learning is a leading provider of customized learning solutions with office locations around the globe, including Singapore, the United Kingdom, Australia, Mexico, Brazil, and Japan. Locate your local office at: **international.cengage.com/region**

Cengage Learning products are represented in Canada by Nelson Education, Ltd.

For your course and learning solutions, visit
academic.cengage.com

Purchase any of our products at your local college store or at our preferred online store
www.ichapters.com

Printed in the United States of America
1 2 3 4 5 6 7 11 10 09 08

Table of Contents

Chapter 1 Basics

Chapter 2 Graphs and Technology

Chapter 3 Functions and Graphs

Chapter 4 Polynomial and Rational Functions

Chapter 5 Exponential and Logarithmic Functions

Chapter 6 Trigonometric Functions

Chapter 7 Trigonometric Identities and Equations

Chapter 8 Triangle Trigonometry

Chapter 9 Applications of Trigonometry

Chapter 10 Analytic Geometry

Chapter 11 Systems of Equations

Chapter 12 Discrete Algebra

Chapter 13 Limits and Continuity

Appendices

Chapter 1
Basics

1.1 The Real Number System

1.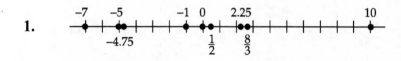

3. Since $b < 0$, therefore b is a negative real number. Since we want $-b$, it will be $-(-b)$, or a positive. Final answer is positive.

5. Since b is a negative real number, c is a positive real number, and d is a negative real number, therefore negative times a positive times a negative is a positive. Final answer is positive.

7. b times c will produce a negative real number. b times d produces a positive real number. Then, negative minus a positive will be a negative. Final answer is negative.

9. The calculator yields:

$$\frac{189}{37} = 5.108108108 \qquad \frac{4587}{691} = 6.638205499 \qquad \sqrt{47} = 6.8556546$$

$$\sqrt{27} = 5.196152423 \qquad \frac{2040}{523} = 3.900573614$$

Thus the order of the numbers from smallest to largest is:

$$\frac{2040}{523}, \frac{189}{37}, \sqrt{27}, \frac{4587}{691}, 6.735, \sqrt{47}$$

11. $-4 > -8$

13. $\pi < 100$

15. $z \geq -4$

17. $d \leq 7$

19. $z \geq -17$

21. $5 > -3$

23. $>$

25. 3

27. 5

29. 0

31. $b + c = a$

33. a lies to the right of b.

35. $a < b$

37.

39.

41.

43. $[5, 10]$

1

45. $[-9, \infty)$

47. $(-3, 14)$

49. $6,506,000,000 = 6.506 \times 1,000,000,000 = 6.506 \times 10^9$ [Decimal point is moved 9 places to the left and 10 is raised to the power 9.]

51. $5,910,000,000,000 = 5.91 \times 1,000,000,000,000 = 5.91 \times 10^{12}$ [Decimal point is moved 12 places to the left and 10 is raised to the power 12.]

53. $.000000002 = 2 \times \dfrac{1}{1,000,000,000} = 2 \times 10^{-9} \text{m}$ [Decimal point is moved 9 places to the right and 10 is raised to -9.]

55. $1.50 \times 10^{11} = 1.50 \times 100,000,000,000 = 150,000,000,000$

57. $1.6726 \times 10^{-19} = 1.6726 \times \dfrac{1}{10,000,000,000,000,000,000}$
$= .00000000000000000016726$

59. a. Debt:
   8365 billion $= 8365 \times 10^9 = 8.365 \times 10^3 \times 10^9 = 8.365 \times 10^{12}$ dollars
   Population:
   298.4 million $= 298.4 \times 10^6 = 2.984 \times 10^2 \times 10^6 = 2.984 \times 10^8$ persons
   b. $\dfrac{\text{Debt}}{\text{Population}} = \dfrac{8.365 \times 10^{12}}{2.984 \times 10^8} = 2.803284 \times 10^4 = \$28,032.84$ per person.

61. $\sqrt{2}\sqrt{8} = \sqrt{2 \cdot 8} = \sqrt{16} = 4$

63. $\sqrt{\dfrac{3}{5}} \cdot \sqrt{\dfrac{12}{5}} = \sqrt{\dfrac{3 \cdot 12}{5 \cdot 5}} = \sqrt{\dfrac{36}{25}} = \dfrac{6}{5}$

65. $\sqrt{6} + \sqrt{2}\left(\sqrt{2} - \sqrt{3}\right) = \sqrt{6} + \sqrt{4} - \sqrt{6} = \sqrt{4} = 2$

67. $\sqrt{u^4} = \sqrt{u^2 \cdot u^2} = \sqrt{u^2}\sqrt{u^2}$ [since u^2 is always positive] $= |u| \cdot |u|$ [by square roots of squares property] $= u^2$

69. $|3 - 14| = |-11| = 11$

71. $3 - |2 - 5| = 3 - |-3| = 3 - 3 = 0$

73. $\left|(-13)^2\right| = |169| = 169$

75. Since $\pi - \sqrt{2}$ is a positive number, $\left|\pi - \sqrt{2}\right| = \pi - \sqrt{2}$.

77. $|3-\pi|+3=-(3-\pi)+3=-3+\pi+3=\pi$ **79.** Since $|-2|=2$ and $|-5|=5$, $|-2|<|-5|$. $<$

81. Since $|3|=3$ and $-|4|=-4$, $|3|>-|4|$. $>$ **83.** Since $|-1|=1$, $-7<|-1|$. $<$

85. $|4-(-3)|=|7|=7$

87. $\left|\dfrac{15}{2}-(-7)\right|=\left|\dfrac{15}{2}+7\right|=\left|\dfrac{29}{2}\right|=\dfrac{29}{2}=14.5$

89. $|3-\pi|=-(3-\pi)=\pi-3$ **91.** $|\sqrt{3}-\sqrt{2}|=\sqrt{3}-\sqrt{2}$

93. $T=2\pi\sqrt{\dfrac{l}{g}}=2\pi\sqrt{\dfrac{4}{32.2}}\approx2.21$ seconds

95. For a person with height 5 ft 8 in, the difference between weight x and 143 lbs must be less than or equal to 21. Hence $|x-143|\le21$. For a person with height 6 ft 0 in, the difference between the weight x and 163 lbs must be less than or equal to 26. Hence $|x-163|\le26$.

97. $|(-11)-4|°=15°$ **99.** $|(-46)-(-16)|°=30°$

101. Since $t^2\ge0$, $|t^2|=t^2$.

103. If $b\ge3$, $b-3\ge0$, hence $|b-3|=b-3$.

105. If $c<d$, $c-d<0$, hence $|c-d|=-(c-d)=d-c$.

107. Since $|u-v|=|v-u|$, $|u-v|-|v-u|=|u-v|-|u-v|=0$.

109. Since $(c-d)^2\ge0$, $|(c-d)^2|=(c-d)^2=c^2-2cd+d^2$.

111. $|x-5|<4$ **113.** $|x+4|\le17$ **115.** $|c|<|b|$

117. The distance from x to 3 is less than 2.   **119.** x is at most 3 units from -7.

121. (a) — iii
(b) — i
(c) — ii
(d) — v
(e) — iv

123. The distance of x from 0 is equal to 1. Thus $x=1$ or $x=-1$.

125. The distance of x from 2 is equal to 1. Thus $x=2+1=3$ or $x=2-1=1$.

127. The distance of x from $-\pi$ is equal to 4. Thus $x = -\pi + 4$ or $x = -\pi - 4$.

129. x is between -7 and 7. Thus, $-7 < x < 7$.

131. The distance of x from 5 is less than 2. Thus x is between $5 - 2$ and $5 + 2$. $3 < x < 7$.

133. The distance of x from -2 is greater than or equal to 3. Thus $x \geq -2 + 3$ or $x \leq -2 - 3$. $x \geq 1$ or $x \leq -5$.

135. If, and only if, none of the three numbers a, b, c is different from 0, then $|a| = 0, |b| = 0, |c| = 0$, and $|a| + |b| + |c| = 0$. Otherwise $|a| + |b| + |c| > 0$.

1.1A Decimal Representation of Real Numbers

1. $.777\ldots$

3. $.8181\ldots$

5. $3.142857142857\ldots$

7.
$$100d = 37.3737\ldots$$
$$d = .3737\ldots$$
$$99d = 37$$
$$d = \frac{37}{99}$$

9.
$$10,000d = 766342.4242\ldots$$
$$100d = 7663.4242\ldots$$
$$9900d = 758,679$$
$$d = \frac{758,679}{9900} = \frac{252,893}{3300}$$

11.
$$1000d = 135.135135\ldots$$
$$d = .135135\ldots$$
$$999d = 135$$
$$d = \frac{135}{999} = \frac{5}{37}$$

13.
$$10,000d = 523127.2727\ldots$$
$$100d = 5231.2727\ldots$$
$$9900d = 517,896$$
$$d = \frac{517,896}{9900} = \frac{14,386}{275}$$

15. No

17. Yes

19. No

21. Yes

23. Let $d = .7499\ldots$
$$1000d = 749.99\ldots$$
$$100d = 74.99\ldots$$
$$900d = 675$$
$$d = \frac{675}{900} = \frac{3}{4}$$

Let $d = .7500\ldots$
$$1000d = 750.00\ldots$$
$$100d = 75.00\ldots$$
$$900d = 675$$
$$d = \frac{675}{900} = \frac{3}{4}$$

25. .0588235294117647 0588235294117647...

27. .0344827586206896551724137931 0344827586206896551724137931...

29. 6.021276595744680851063829787234042553191489 3617 02127659574
4680851063829787234042553191489 3617...

31. **a.** $.058823529411 = \dfrac{58823529411}{100000000000}$

b. $.0588235294117 = \dfrac{588235294117}{1000000000000}$

c. $.058823529411724 = \dfrac{58823529411724}{100000000000000}$

d. $.0588235294117985 = \dfrac{588235294117985}{10000000000000000}$

All of these numbers are approximations of $\dfrac{1}{17}$, but not equal to $\dfrac{1}{17}$.

1.2 Solving Equations Algebraically

1. $3x + 2 = 26$
$3x = 24$
$x = 8$

3. $3x + 2 = 9x + 7$
$-6x + 2 = 7$
$-6x = 5$
$x = -\dfrac{5}{6}$

5. $\dfrac{3y}{4} - 6 = y + 2$
$3y - 24 = 4y + 8$
$-y - 24 = 8$
$-y = 32$
$y = -32$

7. $x = 3y - 5$
$x + 5 = 3y$
$\dfrac{x + 5}{3} = y$
$y = \dfrac{x + 5}{3}$

9. $A = \dfrac{h}{2}(b + c)$
$\dfrac{2A}{h} = b + c$
$\dfrac{2A}{h} - c = b$
$b = \dfrac{2A}{h} - c$
$b = \dfrac{2A - hc}{h}, h \neq 0$

11. $V = \dfrac{\pi d^2 h}{4}$

$4V = \pi d^2 h$

$\dfrac{4V}{\pi d^2} = h$

$h = \dfrac{4V}{\pi d^2}, d \neq 0$

13. $x^2 - 8x + 15 = 0$

$(x-3)(x-5) = 0$

$x - 3 = 0 \ \text{ or } \ x - 5 = 0$

$x = 3 \ \text{ or } \ \ \ x = 5$

15. $x^2 - 5x = 14$

$x^2 - 5x - 14 = 0$

$(x+2)(x-7) = 0$

$x + 2 = 0 \ \text{ or } \ x - 7 = 0$

$x = -2 \ \text{ or } \ \ \ x = 7$

17. $2y^2 + 5y - 3 = 0$

$(2y-1)(y+3) = 0$

$2y - 1 = 0 \ \text{ or } \ y + 3 = 0$

$y = \dfrac{1}{2} \ \text{ or } \ \ \ y = -3$

19. $4t^2 + 9t + 2 = 0$

$(4t+1)(t+2) = 0$

$4t + 1 = 0 \ \text{ or } \ t + 2 = 0$

$t = -\dfrac{1}{4} \ \text{ or } \ \ t = -2$

21. $3u^2 - 4u = 4$

$3u^2 - 4u - 4 = 0$

$(3u+2)(u-2) = 0$

$3u + 2 = 0 \ \text{ or } \ u - 2 = 0$

$u = -\dfrac{2}{3} \ \text{ or } \ \ u = 2$

23. $x^2 - 2x = 12$

$x^2 - 2x + 1 = 13$

$(x-1)^2 = 13$

$x - 1 = \pm\sqrt{13}$

$x = 1 \pm \sqrt{13}$

25. $x^2 - x - 1 = 0$

$x^2 - x = 1$

$x^2 - x + \dfrac{1}{4} = \dfrac{5}{4}$

$\left(x - \dfrac{1}{2}\right)^2 = \dfrac{5}{4}$

$x - \dfrac{1}{2} = \pm\sqrt{\dfrac{5}{4}}$

$x = \dfrac{1}{2} \pm \dfrac{\sqrt{5}}{2}$

$x = \dfrac{1+\sqrt{5}}{2} \ \text{ or } \ \dfrac{1-\sqrt{5}}{2}$

27. $a = 1, b = 4, c = 1 \quad b^2 - 4ac = 4^2 - 4(1)(1) = 12.$

Two real solutions.

29. $9x^2 = 12x + 1$
$9x^2 - 12x - 1 = 0$
$a = 9, b = -12, c = -1$
$b^2 - 4ac = (-12)^2 - 4(9)(-1) = 180$
Two real solutions.

31. $25t^2 + 49 = 70t$
$25t^2 - 70t + 49 = 0$
$a = 25, b = -70, c = 49$
$b^2 - 4ac = (-70)^2 - 4(25)(49) = 0$
One real solution.

33. $x^2 - 4x + 1 = 0$
$a = 1, b = -4, c = 1$
$$x = \frac{-(-4) \pm \sqrt{(-4)^2 - 4(1)(1)}}{2(1)}$$
$$x = \frac{4 \pm \sqrt{12}}{2}$$
$$x = \frac{4 \pm 2\sqrt{3}}{2}$$
$$x = 2 \pm \sqrt{3}$$

35. $x^2 + 6x + 7 = 0$
$a = 1, b = 6, c = 7$
$$x = \frac{-6 \pm \sqrt{6^2 - 4(1)(7)}}{2(1)}$$
$$x = \frac{-6 \pm \sqrt{8}}{2}$$
$$x = \frac{-6 \pm 2\sqrt{2}}{2}$$
$$x = -3 \pm \sqrt{2}$$

37. $x^2 + 6 = 2x$
$x^2 - 2x + 6 = 0$
$a = 1, b = -2, c = 6$
$b^2 - 4ac = (-2)^2 - 4(1)(6) = -20$
No real solutions.

39. $4x^2 + 4x = 7$
$4x^2 + 4x - 7 = 0$
$a = 4, b = 4, c = -7$
$$x = \frac{-(4) \pm \sqrt{(4)^2 - 4(4)(-7)}}{2(4)}$$
$$x = \frac{-4 \pm \sqrt{128}}{8}$$
$$x = \frac{-4 \pm 8\sqrt{2}}{8}$$
$$x = \frac{-1 \pm 2\sqrt{2}}{2}$$
$$x = \frac{-1}{2} \pm \sqrt{2}$$

41. $4x^2 - 8x + 1 = 0$
$a = 4, b = -8, c = 1$

$$x = \frac{-(-8) \pm \sqrt{(-8)^2 - 4(4)(1)}}{2(4)}$$

$$x = \frac{8 \pm \sqrt{48}}{8}$$

$$x = \frac{8 \pm 4\sqrt{3}}{8}$$

$$x = \frac{2 \pm \sqrt{3}}{2}$$

43. $x^2 + 9x + 18 = 0$
$(x+3)(x+6) = 0$
$x + 3 = 0$ or $x + 6 = 0$
$x = -3$ or $x = -6$

45. $4x(x+1) = 1$
$4x^2 + 4x = 1$
$4x^2 + 4x - 1 = 0$
$a = 4, b = 4, c = -1$

$$x = \frac{-4 \pm \sqrt{4^2 - 4(4)(-1)}}{2(4)} = \frac{-4 \pm \sqrt{32}}{8}$$

$$x = \frac{-4 \pm 4\sqrt{2}}{8} = \frac{-1 \pm \sqrt{2}}{2}$$

47. $2x^2 = 7x + 15$
$2x^2 - 7x - 15 = 0$
$(2x + 3)(x - 5) = 0$
$2x + 3 = 0$ or $x - 5 = 0$
$$x = \frac{-3}{2} \text{ or } x = 5$$

49. $t^2 + 4t + 13 = 0$
$a = 1, b = 4, c = 13$
$b^2 - 4ac = 4^2 - 4(1)(13) = -36$
No real solutions

51. $\dfrac{7x^2}{3} = \dfrac{2x}{3} - 1$
$7x^2 = 2x - 3$
$7x^2 - 2x + 3 = 0$
$a = 7, b = -2, c = 3$
$b^2 - 4ac = (-2)^2 - 4(7)(3) = -80$
No real solutions

53. $4.42x^2 - 10.14x + 3.79 = 0$

$$x = \frac{-(-10.14) \pm \sqrt{(-10.14)^2 - 4(4.42)(3.79)}}{2(4.42)} \approx 1.824, 0.470$$

55. $3x^2 - 82.74x + 570.4923 = 0$

$$x = \frac{-(-82.74) \pm \sqrt{(-82.74)^2 - 4(3)(570.4923)}}{2(3)} \approx 13.79$$

57. To solve $y^4 - 7y^2 + 6 = 0$, let $u = y^2$,

$y^4 - 7y^2 + 6 = 0$

$u^2 - 7u + 6 = 0$

$(u-1)(u-6) = 0$

$u - 1 = 0$ or $u - 6 = 0$

$u = 1$ $u = 6$

Since $u = y^2$, we have the equivalent

statements: $y^2 = 1$ $y^2 = 6$

$y = \pm 1$ or $y = \pm\sqrt{6}$

59. To solve $x^4 - 2x^2 - 35 = 0$, let $u = x^2$,

$x^4 - 2x^2 - 35 = 0$

$u^2 - 2u - 35 = 0$

$(u-7)(u+5) = 0$

$u - 7 = 0$ or $u + 5 = 0$

$u = 7$ $u = -5$

Since $u = x^2$, we have the equivalent

statements:

$x^2 = 7$ or $x^2 = -5$ (no real solution)

$x = \pm\sqrt{7}$

61. To solve $2y^4 - 9y^2 + 4 = 0$, let $u = y^2$.

$2y^4 - 9y^2 + 4 = 0$

$2u^2 - 9u + 4 = 0$

$(u-4)(2u-1) = 0$

$u - 4 = 0$ or $2u - 1 = 0$

$u = 4$ $u = \dfrac{1}{2}$

Since $u = y^2$. we have the equivalent

statements:

$y^2 = 4$ or $y^2 = \dfrac{1}{2}$

$y = \pm\sqrt{\dfrac{1}{2}}$

$y = \pm 2$ or $y = \dfrac{\pm 1}{\sqrt{2}}$

63. To solve $10x^4 + 3x^2 = 1$, let $u = x^2$.

$10x^4 + 3x^2 = 1$

$10u^2 + 3u = 1$

$10u^2 + 3u - 1 = 0$

$(5u-1)(2u+1) = 0$

$5u - 1 = 0$ or $2u + 1 = 0$

$u = \dfrac{1}{5}$ $u = -\dfrac{1}{2}$

Since $u = x^2$, we have the equivalent

statements:

$x^2 = \dfrac{1}{5}$ or $x^2 = -\dfrac{1}{2}$ (no real solution)

$x = \pm\sqrt{\dfrac{1}{5}}$

$x = \pm\dfrac{\sqrt{5}}{5}$

65.

$\dfrac{1}{2t} - \dfrac{2}{5t} = \dfrac{1}{10t} - 1$

$10t \cdot \dfrac{1}{2t} - 10t \cdot \dfrac{2}{5t} = 10t \cdot \dfrac{1}{10t} - 10t$

$5 - 4 = 1 - 10t$

$0 = -10t$

$t = 0$

Impossible. No solution.

67. $\dfrac{2x-7}{x+4} = \dfrac{5}{x+4} - 2$

$2x - 7 = 5 - 2(x+4)$

$2x - 7 = 5 - 2x - 8$

$4x = 4$

$x = 1$

69.
$$25x + \frac{4}{x} = 20$$
$$25x^2 + 4 = 20x$$
$$25x^2 - 20x + 4 = 0$$
$$(5x - 2)^2 = 0 \Rightarrow x = \frac{2}{5}$$

71.
$$\frac{2}{x^2} - \frac{5}{x} = 4$$
$$x^2 \cdot \frac{2}{x^2} - x^2 \cdot \frac{5}{x} = 4x^2 \quad x \neq 0$$
$$2 - 5x = 4x^2$$
$$0 = 4x^2 + 5x - 2$$
$$x = \frac{-5 \pm \sqrt{5^2 - 4(4)(-2)}}{2(4)}$$
$$x = \frac{-5 \pm \sqrt{57}}{8}$$

73.
$$\frac{4x^2 + 5}{3x^2 + 5x - 2} = \frac{4}{3x - 1} - \frac{3}{x + 2} \quad x \neq -2, \ x \neq \frac{1}{3}$$
$$(3x - 1)(x + 2)\frac{4x^2 + 5}{(3x - 1)(x + 2)} = (3x - 1)(x + 2)\frac{4}{3x - 1} - (3x - 1)(x + 2)\frac{3}{x + 2}$$
$$4x^2 + 5 = (x + 2)4 - (3x - 1)3$$
$$4x^2 + 5 = 4x + 8 - 9x + 3$$
$$4x^2 + 5 = -5x + 11$$
$$4x^2 + 5x - 6 = 0$$
$$(4x - 3)(x + 2) = 0$$
$$4x - 3 = 0 \ \text{ or } \ x + 2 = 0$$
$$x = \frac{3}{4} \ \text{ or } \ x = -2 \ (\text{not allowed})$$

75.
$$12.62 = .79x + 3.93$$
$$12.62 - 3.93 = .79x$$
$$8.69 = .79x$$
$$11 = x$$
$$2000 + 11 = 2011$$

77.
$$1794.25 = 73.04x + 625.6$$
$$1794.25 - 625.6 = 73.04x$$
$$1168.65 = 73.04x$$
$$16 = x$$
$$1990 + 16 = 2006$$

79. $Y = C + I + G + (X - M)$

$Y = 120 + .9Y + 140 + 150 + \left[60 - (20 + .2Y)\right]$

$Y = 120 + .9Y + 290 + 60 - 20 - .2Y$

$Y = 450 + .7Y$

$.3Y = 450$

$Y = 1500$

81. In this case v_0 is 0 and h_0 is 640, so that the height equation is

$h = -16t^2 + 640$
We must solve

$0 = -16t^2 + 640$

$16t^2 = 640$

$t^2 = 40$

$t = \pm\sqrt{40}$

$t = \pm 2\sqrt{10} = \pm 6.32$

It takes about 6.32 seconds

83. In this case v_0 is 800 and h_0 is 0, so that the height equation is $h = -16t^2 + 800t$

a. Solve $3200 = -16t^2 + 800t$.

$16t^2 - 800t + 3200 = 0$

$t^2 - 50t + 200 = 0$

$t = \dfrac{50 \pm \sqrt{(-50)^2 - 4(1)(200)}}{2(1)}$

$t = \dfrac{50 \pm \sqrt{1700}}{2}$

$t = 4.38,\ 45.61$

It takes about 4.38 seconds for the rocket to first reach 3200 feet.

b. Solve $0 = -16t^2 + 800t$.

$0 = t^2 - 50t$

$0 = t(t - 50)$

$t = 0 \quad \text{or} \quad t = 50$

50 seconds for the rocket to reach the ground again.

85. **a.** At sea level, evaluate a for $h = 0$; $a = 2116.1$ pounds per square foot.
    For the top of Mt. Everest, evaluate

$$a = .8315h^2 - 73.93h + 2116.1 \quad \text{for} \quad h = 29.035$$

$$a = .8315(29.035)^2 - 73.93(29.035) + 2116.1$$

$a = $ It is about 670.5 pounds per square foot.

   b. Solve $1223.43 = .8315h^2 - 73.93h + 2116.1$

$$0 = .8315h^2 - 73.93h + 892.67$$

$$h = \frac{-(-73.93) \pm \sqrt{(-73.93)^2 - 4(.8315)(892.67)}}{2(.8315)}$$

$$h = 74.502 \quad \text{or} \quad 14.409$$

It is about 14,400 feet high.

87. Solve $804.2 = .08x^2 - 13.08x + 927$.

$$0 = .08x^2 - 13.08x + 122.8$$

$$x = \frac{-(-13.08) \pm \sqrt{(-13.08)^2 - 4(.08)(122.8)}}{2(.08)} = \frac{13.08 \pm 11.48}{2(.08)}$$

$$x = 153.5 \quad \text{or} \quad 10$$

Only the answer 10 represents a past year: 1960.

89. **a.** $4240 = 6.82x^2 - 1.55x + 666.8 \Rightarrow 6.82x^2 - 1.55x - 3573.2 = 0$
    $a = 6.82, b = -1.55, c = -3573.2$

$$x = \frac{-(-1.55) \pm \sqrt{(-1.55)^2 - 4(6.82)(-3573.2)}}{2(6.82)} = \frac{1.55 \pm \sqrt{97479.2985}}{13.64}$$

$x = 23 \text{ or } -22.8$
1980 − 22.8 will be earlier than 1980, so we will not use that.
$1980 + 23 = 2003$

   b. $5600 = 6.82x^2 - 1.55x + 666.8 \Rightarrow 6.82x^2 - 1.55x - 4933.2 = 0$
    $a = 6.82, b = -1.55, c = -4933.2$

$$x = \frac{-(-1.55) \pm \sqrt{(-1.55)^2 - 4(6.82)(-4933.2)}}{2(6.82)} = \frac{1.55 \pm \sqrt{134580.0985}}{13.64}$$

$x = 27 \text{ or } -26.7$
1980 − 22.8 will be earlier than 1980, so we will not use that.
$1980 + 27 = 2007$

   c. $7000 = 6.82x^2 - 1.55x + 666.8 \Rightarrow 6.82x^2 - 1.55x - 6333.2 = 0$
    $a = 6.82, b = -1.55, c = -6333.2$

$$x = \frac{-(-1.55) \pm \sqrt{(-1.55)^2 - 4(6.82)(-6333.2)}}{2(6.82)} = \frac{1.55 \pm \sqrt{172767.2935}}{13.64}$$

$x = 30 \text{ or } -30$
1980 − 30 will be earlier than 1980, so we will not use that.
$1980 + 30 = 2010$. Since the result was 30.5, we could say mid 2010.

91. a. $1 = .0031x^2 - .291x + 7.1$

$0 = .0031x^2 - .291x + 6.1$

$a = .0031, b = -.291, c = 6.1$

$$x = \frac{(-.291) \pm \sqrt{(-.291)^2 - 4(.0031)(6.1)}}{2(.0031)} = \frac{(-.291) \pm \sqrt{.009041}}{.0062}$$

$x = 62.27 \text{ or } 31.6$

The ages are about 32 and 62.

b. Since we need the rate to be three times greater, we will set $D = 3$.

$3 = .0031x^2 - .291x + 7.1$

$0 = .0031x^2 - .291x + 4.1$

$a = .0031, b = -.291, c = 4.1$

$$x = \frac{(-.291) \pm \sqrt{(-.291)^2 - 4(.0031)(4.1)}}{2(.0031)}$$

$$x = \frac{(-.291) \pm \sqrt{.033841}}{.0062}$$

$x = 76.6 \text{ or } 17.3$

The ages are about 17 and 77.

93. a. $\dfrac{18x}{100-x} = 50$

$18x = 50(100-x)$

$18x = 5000 - 50x$

$68x = 5000$

$x = 73.529$

73.53%

b. $\dfrac{18x}{100-x} = 100$

$18x = 100(100-x)$

$18x = 10000 - 100x$

$118x = 10000$

$x = 84.745$

84.75%

c. $\dfrac{18x}{100-x} = 200$

$18x = 200(100-x)$

$18x = 20000 - 200x$

$218x = 20000$

$x = 91.743$

91.74%

95. This is a quadratic equation with $a = 1, b = k, c = 25$. It will have exactly one real solution only if $b^2 - 4ac = 0$. Thus

$k^2 - 4(1)(25) = 0$

$k^2 - 100 = 0$

$k^2 = 100$

$k = \pm 10$

$k = -10 \text{ or } 10$

97. This is a quadratic equation with $a = k, b = 8, c = 1$. It will have exactly one real solution only if $b^2 - 4ac = 0$. Thus

$8^2 - 4(k)(1) = 0$

$64 - 4k = 0$

$k = 16$

99. Yes, because 25 is a perfect square.

101. No, because 72 is not a perfect square.

103. Since 4 is a solution of the equation, $4^2 - 5 \cdot 4 + k = 0$, hence $k - 4 = 0$ and $k = 4$. Also $1^2 - 5 \cdot 1 + k = 0$, which also yields $k = 4$.

105. a. $x^2 + 5x + 2 = 0$

$$x = \frac{-(5) \pm \sqrt{(5)^2 - 4(1)(2)}}{2(1)} = \frac{-5 \pm \sqrt{17}}{2}$$

   b. Answers will vary
   c. $x \approx -.3068$ or $.6959$ or -1.1183 or 1.3959

1.2.A Absolute Value Equations

1. The equation is equivalent to
$$2x + 3 = 9 \quad \text{or} \quad -(2x + 3) = 9$$

$$2x = 6 \qquad\qquad -2x - 3 = 9$$

$$x = 3 \qquad\qquad -2x = 12$$

$$\qquad\qquad\qquad x = -6$$

Checking each possible solution, we see
$|2 \cdot 3 + 3| = 9$ and $|2 \cdot (-6) + 3| = |-9| = 9$.
Thus both 3 and –6 are solutions.

3. The equation is equivalent to
$$6x - 9 = 0$$

$$x = \frac{9}{6} = \frac{3}{2}$$

Checking, we see $\left| 6\left(\frac{3}{2}\right) - 9 \right| = 0$.

$\frac{3}{2}$ is the solution.

5. The equation is equivalent to
$$2x + 3 = 4x - 1 \quad \text{or} \quad -(2x + 3) = 4x - 1$$

$$-2x = -4 \qquad\qquad -2x - 3 = 4x - 1$$

$$x = 2 \qquad\qquad -6x = 2$$

$$\qquad\qquad\qquad x = -\frac{1}{3}$$

Checking, we see that 2 is a solution
because $|2 \cdot 2 + 3| = 7$ and $4 \cdot 2 - 1 = 7$.
However, $-\frac{1}{3}$ is not a solution because
$$\left| 2\left(-\frac{1}{3}\right) + 3 \right| = \frac{7}{3} \quad \text{but} \quad 4\left(-\frac{1}{3}\right) - 1 = -\frac{7}{3}.$$

7. The equation is equivalent to
$$x - 3 = x \quad \text{or} \quad -(x - 3) = x$$

$$-3 = 0 \qquad\qquad 3 = 2x$$

no solution $\qquad\qquad x = \frac{3}{2}$

Checking, we see that $\frac{3}{2}$ is a solution
because $\left| \frac{3}{2} - 3 \right| = \frac{3}{2}$.

9. The equation is equivalent to

$$x^2 + 4x - 1 = 4 \quad \text{or} \quad -(x^2 + 4x - 1) = 4$$

$$x^2 + 4x - 5 = 0 \qquad\qquad x^2 + 4x + 3 = 0$$

$$(x+5)(x-1) = 0 \qquad\qquad (x+3)(x+1) = 0$$

$$x = -5 \text{ or } x = 1 \qquad\qquad x = -3 \text{ or } x = -1$$

All four of these numbers are solutions of the original equation.

11. The equation is equivalent to

$$x^2 - 5x + 1 = 3 \qquad\qquad \text{or} \qquad\qquad -(x^2 - 5x + 1) = 3$$

$$x^2 - 5x - 2 = 0 \qquad\qquad\qquad\qquad x^2 - 5x + 4 = 0$$

$$x = \frac{-(-5) \pm \sqrt{(-5)^2 - 4(1)(-2)}}{2(1)} \qquad\qquad (x-1)(x-4) = 0$$

$$x = \frac{5 \pm \sqrt{33}}{2} \qquad\qquad\qquad\qquad x = 1 \text{ or } x = 4$$

$$x = \frac{5 + \sqrt{33}}{2} \text{ or } x = \frac{5 - \sqrt{33}}{2}$$

All four of these numbers are solutions of the original equation.

13. Substituting the given values for $\bar{p}$ and n, we obtain

$$|CL - .02| = 3\sqrt{\frac{.02(1 - .02)}{200}}$$

$$|CL - .02| = .0297$$

This equation is equivalent to

$$CL - .02 = .0297 \text{ or } -(CL - .02) = .0297$$

$$-CL + .02 = .0297$$

$$CL = .0497$$

$$CL = -.0097$$

The upper control limit is .0497 and the lower control limit is −.0097 (which may be interpreted as 0 depending on the situation).

1.2.B Variation

1. Area varies directly as the square of the radius: the constant of variation is π.

3. Area varies jointly as the length and the width; the constant of variation is 1.

5. Volume varies jointly as the square of the radius and the height; the constant of variation is $\pi/3$.

7. $a = k/b$ **9.** $z = kxyw$ **11.** $d = k\sqrt{h}$

13. $v = ku$

$8 = k(2)$

$k = \dfrac{8}{2} = 4$

15. $v = \dfrac{k}{u}$

$8 = \dfrac{k}{2}$

$k = 16$

17. $t = krs$

$24 = k(2)(3)$

$k = \dfrac{24}{6} = 4$

19. $w = kxy^2$

$96 = k \cdot 3 \cdot 4^2$

$k = \dfrac{96}{3 \cdot 4^2} = 2$

21. $T = \dfrac{kpv^3}{u^2}$

$24 = \dfrac{k \cdot 3 \cdot 2^3}{4^2}$

$k = \dfrac{24 \cdot 4^2}{3 \cdot 2^3} = 16$

23. $r = kt$

$6 = k(3)$

$k = 2$

Therefore,

the variation equation is $r = 2t$.

When $t = 2$,

$r = 2(2) = 4$.

25. $b = \dfrac{k}{x}$

$9 = \dfrac{k}{3}$

$k = 27$

Therefore, the variation equation is $b = \dfrac{27}{x}$.

When $x = 12$, $b = \dfrac{27}{12} = \dfrac{9}{4}$.

27. $w = k(u + v^2)$

$200 = k(1 + 7^2)$

$200 = 50k$

$k = 4$

Therefore,

the variation equation

is $w = 4(u + v^2)$.

When $w = 300$ and

$v = 5$,

$300 = 4(u + 5^2)$

$300 = 4(u + 25)$

$75 = u + 25$

$u = 50$

29. $r = \dfrac{k}{st}$

$12 = \dfrac{k}{3 \cdot 1}$

$k = 36$

Therefore,

the variation equation

is $r = \dfrac{36}{st}$.

When $s = 6$ and $t = 2$,

$r = \dfrac{36}{6 \cdot 2}$

$r = 3$

31. Let $t = $ tax, $x = $ AGI,

Then,

$t = kx$

$936 = k(24,000)$

$k = \dfrac{936}{24,000}$

Therefore,

the variation equation

is $t = \dfrac{936}{24,000}x$.

When $x = 39,000$,

$t = \dfrac{936}{24,000}(39,000)$

$t = \$1521$

33.
$$d = kW$$
$$15.75 = k(7)$$
$$k = 2.25$$
Therefore, the variation equation is
$d = 2.25W$. When $d = 27$,
$$27 = 2.25W$$
$$W = \frac{27}{2.25}$$
$$W = 12 \text{ pounds}$$

35. Let P = pressure, V = volume. Then
$$P = \frac{k}{V}$$
$$50 = \frac{k}{200}$$
$$k = 10,000$$
Therefore, the variation equation is
$P = \frac{10,000}{V}$. When $V = 125$,
$$P = \frac{10,000}{125}$$
$$P = 80 \text{ kilograms per sq centimeter}$$

37. Let d = distance, t = time. Then
$$d = kt^2$$
$$100 = k(2.5)^2$$
$$k = 16$$
Therefore, the variation equation is
$d = 16t^2$. When $t = 5$,
$$d = 16(5)^2$$
$$d = 400 \text{ feet}$$

39. Let W = weight, h = height, r = radius. Then
$$W = khr^2$$
$$250 = k(20)(5)^2$$
$$k = \frac{250}{20 \cdot 5^2}$$
$$k = \frac{1}{2}$$
Therefore, the variation equation is
$W = \frac{1}{2}hr^2$. When $W = 960$ and $r = 8$,
$$960 = \frac{1}{2}h \cdot 8^2$$
$$h = \frac{2 \cdot 960}{8^2}$$
$$h = 30 \text{ inches}$$

41. Let F = force, W = weight, v = speed.

$$F = \frac{kWv^2}{r}$$

$$1500 = \frac{k(1000)50^2}{200}$$

$$k = \frac{(1500)(200)}{1000(50)^2}$$

$$k = .12$$

Therefore, the variation equation is

$$F = \frac{.12\,Wv^2}{r}.$$

When $W = 1000$, $v = 100$, $r = 320$

$$F = \frac{.12(1000)100^2}{320}$$

$$F = 3750 \text{ kilograms}$$

43. Let L = the maximum safe load. Then

$$L = \frac{kwh^2}{l}$$

$$1000 = \frac{k \cdot 4 \cdot 2^2}{6}$$

$$k = \frac{6000}{4 \cdot 2^2}$$

$$k = 375$$

Therefore, the variation equation is

$$L = \frac{375wh^2}{l}.$$

a. When $w = 4$, $h = 4$, $l = 10$,

$$L = \frac{375 \cdot 4 \cdot 4^2}{10}$$

$$L = 2400 \text{ pounds}$$

b. When $w = 4$, $h = 4$, $L = 6000$

$$6000 = \frac{375 \cdot 4 \cdot 4^2}{l}$$

$$6000l = 24,000$$

$$l = 4$$

No more than 4 feet

1.3 The Coordinate Plane

1. $A(-3,3)$; $B(-1.5,3)$; $C(-2.5,0)$; $D(-1.5,-3)$; $E(0,2)$; $F(0,0)$; $G(2,0)$; $H(3,1)$; $I(3,-1)$

3. $(-6,3)$

5. $\left(4, \frac{1}{2} \times 4\right) = (4,2)$

7.

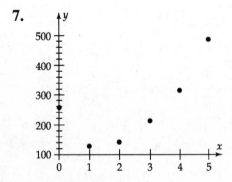
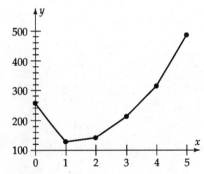

9. **a.** Since x is positive and y is negative, the point lies in quadrant IV.
  b. Since y is negative, the point lies in quadrant III or IV.

11. **a.**

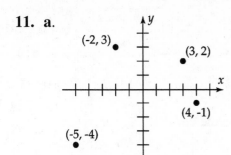

b.

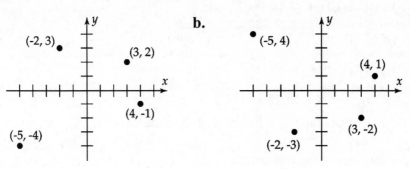

c. The new points are found by reflecting the original points in the x-axis.

13.   distance $= \sqrt{(x_1 - x_2)^2 + (y_1 - y_2)^2}$

$= \sqrt{(-3-2)^2 + (5-(-7))^2}$

$= \sqrt{(-5)^2 + 12^2}$

$= \sqrt{25 + 144} = \sqrt{169} = 13$

The midpoint is

$\left(\dfrac{x_1 + x_2}{2}, \dfrac{y_1 + y_2}{2} \right) = \left(\dfrac{-3+2}{2}, \dfrac{5+(-7)}{2} \right)$

$= \left(-\dfrac{1}{2}, -1 \right)$

15.   distance $= \sqrt{(x_1 - x_2)^2 + (y_1 - y_2)^2}$

$= \sqrt{(-2-(-1))^2 + (5-2)^2}$

$= \sqrt{(-1)^2 + 3^2}$

$= \sqrt{1 + 9} = \sqrt{10}$

The midpoint is

$\left(\dfrac{x_1 + x_2}{2}, \dfrac{y_1 + y_2}{2} \right) = \left(\dfrac{-2+(-1)}{2}, \dfrac{5+2}{2} \right)$

$= \left(\dfrac{-3}{2}, \dfrac{7}{2} \right)$

17.   distance $= \sqrt{(x_1 - x_2)^2 + (y_1 - y_2)^2}$

$= \sqrt{\left(\sqrt{2} - \sqrt{3} \right)^2 + (1-2)^2}$

$= \sqrt{2 - 2\sqrt{2}\sqrt{3} + 3 + 1}$

$= \sqrt{6 - 2\sqrt{6}} \approx 1.05$

The midpoint is

$\left(\dfrac{x_1 + x_2}{2}, \dfrac{y_1 + y_2}{2} \right) = \left(\dfrac{\sqrt{2} + \sqrt{3}}{2}, \dfrac{1+2}{2} \right)$

$= \left(\dfrac{\sqrt{2} + \sqrt{3}}{2}, \dfrac{3}{2} \right)$

19.   distance $= \sqrt{(x_1 - x_2)^2 + (y_1 - y_2)^2}$

$= \sqrt{(a-b)^2 + (b-a)^2}$

$= \sqrt{2(a-b)^2}$

$= |a-b|\sqrt{2}$

The midpoint is

$\left(\dfrac{x_1 + x_2}{2}, \dfrac{y_1 + y_2}{2} \right) = \left(\dfrac{a+b}{2}, \dfrac{b+a}{2} \right)$

$= \left(\dfrac{a+b}{2}, \dfrac{a+b}{2} \right)$

21. We will have to get the distance of each one from the origin

$$\sqrt{(4-0)^2+(4.2-0)^2}=\sqrt{16+17.64}=\sqrt{33.64}$$

$$\sqrt{(-3.5-0)^2+(4.6-0)^2}=\sqrt{12.25+21.16}=\sqrt{33.41}$$

$$\sqrt{(-3-0)^2+(-5-0)^2}=\sqrt{9+25}=\sqrt{34}$$

$$\sqrt{(2-0)^2+(-5.5-0)^2}=\sqrt{4+30.25}=\sqrt{34.25}$$

So, the point (-3.5,4.6) has the least distance, so it is the closest.

23. The perimeter is the sum of the distances:
between $(0,0)$ and $(4,0)=4$

between $(4,0)$ and $(2,2)=\sqrt{(4-2)^2+(0-2)^2}=\sqrt{4+4}=\sqrt{8}=2\sqrt{2}$

between $(2,2)$ and $(4,5)=\sqrt{(2-4)^2+(2-5)^2}=\sqrt{4+9}=\sqrt{13}$

between $(4,5)$ and $(0,5)=4$

between $(0,5)$ and $(0,0)=5$

Thus, the perimeter $=4+2\sqrt{2}+\sqrt{13}+4+5=13+2\sqrt{2}+\sqrt{13}$.

25. The area of the rectangle formed by $(0,0),(4,0),(4,5)$, and $(0,5)$ is $4(5)=20$ square units. The triangle formed by $(4,0),(2,2)$, and $(4,5)$ is $\frac{1}{2}(5)(2)=5$ square units. Thus, the area of the shaded region is $20-5=15$ square units.

27. The lengths of the sides of the triangle are the distances:
between $(0,0)$ and $(1,1)=\sqrt{(0-1)^2+(0-1)^2}=\sqrt{1+1}=\sqrt{2}$

between $(1,1)$ and $(2,-2)=\sqrt{(1-2)^2+(1-(-2))^2}=\sqrt{1+9}=\sqrt{10}$

between $(2,-2)$ and $(0,0)=\sqrt{(2-0)^2+(-2-0)^2}=\sqrt{4+4}=\sqrt{8}$

Since $\left(\sqrt{2}\right)^2+\left(\sqrt{8}\right)^2=\left(\sqrt{10}\right)^2$, the triangle is a right triangle. The length of the hypotenuse is $\sqrt{10}$ units.

29. The lengths of the sides of the triangle are the distances:
between $(1,4)$ and $(5,2)=\sqrt{(1-5)^2+(4-2)^2}=\sqrt{16+4}=\sqrt{20}$

between $(1,4)$ and $(3,-2)=\sqrt{(1-3)^2+(4-(-2))^2}=\sqrt{4+36}=\sqrt{40}$

between $(5,2)$ and $(3,-2)=\sqrt{(5-3)^2+(2-(-2))^2}=\sqrt{4+16}=\sqrt{20}$

Since $\left(\sqrt{20}\right)^2+\left(\sqrt{20}\right)^2=\left(\sqrt{40}\right)^2$, the triangle is a right triangle. The length of the hypotenuse is $2\sqrt{10}$ units.

31. a. The quarterback has coordinates $(20,10)$. The receiver has coordinates $\left(48\tfrac{1}{3},45\right)$. The length of the pass can be calculated from the distance formula:

$$\text{distance} = \sqrt{\left(x_1 - x_2\right)^2 + \left(y_1 - y_2\right)^2} = \sqrt{\left(20 - 48\tfrac{1}{3}\right)^2 + \left(10 - 45\right)^2} \approx 45 \text{ yds}$$

(The pass would be recorded as the change in y coordinate only, however, or 35 yds.)

b. The midpoint is

$$\left(\frac{x_1 + x_2}{2}, \frac{y_1 + y_2}{2}\right) = \left(\frac{20 + 48\tfrac{1}{3}}{2}, \frac{10 + 45}{2}\right) = \left(34\tfrac{1}{6}, 27\tfrac{1}{2}\right).$$

33. a.

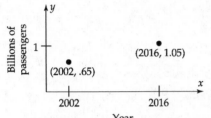

b. $\left(\dfrac{2002 + 2016}{2}, \dfrac{.65 + 1.05}{2}\right) = \left(2009, .85\right)$

c. If linear growth is assumed, the midpoint suggests that there will be 850 million passengers in 2009.

35. Since $3(2) - (-1) - 5 = 6 + 1 - 5 = 2 \neq 0, (2,-1)$ does not satisfy the equation $3x - y - 5 = 0$ and the point is not on the graph.

37. Since $3 \cdot 2 + 6 = 6 + 6 = 12, (6,2)$ satisfies the equation $3y + x = 12$ and the point is on the graph.

39. Since $(1-2)^2 + (-4+5)^2 = (-1)^2 + (1)^2 = 2 \neq 4, (1,-4)$ does not satisfy the equation $(x-2)^2 + (y+5)^2 = 4$ and the point is not on the graph.

41. To find the x-intercepts, set $y = 0$ and solve for x.
$$x^2 - 6x + 5 = 0 \Rightarrow (x-1)(x-5) = 0 \Rightarrow x = 1 \text{ or } x = 5$$
To find the y-intercepts, set $x = 0$ and solve for y.
$$y + 5 = 0 \Rightarrow y = -5$$

43. To find the x-intercepts, set $y = 0$ and solve for x.
$$(x-2)^2 = 9 \Rightarrow x - 2 = \pm 3$$
$$x - 2 = 3 \text{ or } x - 2 = -3$$
$$x = 5 \qquad x = -1$$
To find the y-intercepts, set $x = 0$ and solve for y.
$$(-2)^2 + y^2 = 9 \Rightarrow y^2 = 5 \Rightarrow y = \pm\sqrt{5} \Rightarrow y = \sqrt{5} \text{ or } y = -\sqrt{5}$$

45. To find the x-intercepts, set $y = 0$ and solve for x.

$9x^2 + 90x = 0$

$9x(x + 10) = 0$

$9x = 0$ or $x + 10 = 0$

$x = 0$ $x = -10$

To find the y-intercepts, set $x = 0$ and solve for y.

$16y^2 - 128y = 0$

$16y(y - 8) = 0$

$16y = 0$ or $y - 8 = 0$

$y = 0$ $y = 8$

47. a. Approximately 7.3 million. **b.** Approximately 2007; approximately 7.8 million.
c. 2000 — 2003, 2011 —.

49. a. Under 56   **b.** Age 56   **c.** Retiring at age 60: about \$30,000; retiring at age 65: about \$35,000

51. B   **53.** C

55. $\left(x - (-3)\right)^2 + (y - 4)^2 = 2^2$ **57.** $x^2 + y^2 = \left(\sqrt{3}\right)^2$

$(x + 3)^2 + (y - 4)^2 = 4$ $x^2 + y^2 = 3$

59.

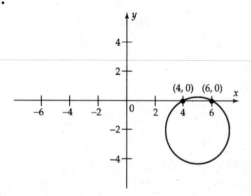

61.

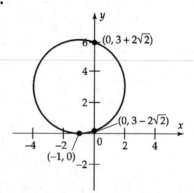

63.
$$x^2 + y^2 + 8x - 6y - 15 = 0$$
$$\left(x^2 + 8x\right) + \left(y^2 - 6y\right) = 15$$
$$\left(x^2 + 8x + 16\right) + \left(y^2 - 6y + 9\right) = 15 + 16 + 9$$
$$(x+4)^2 + (y-3)^2 = 40$$
Center: $(-4, 3)$, Radius: $\sqrt{40} = 2\sqrt{10}$

65.
$$x^2 + y^2 + 6x - 4y - 15 = 0$$
$$\left(x^2 + 6x\right) + \left(y^2 - 4y\right) = 15$$
$$\left(x^2 + 6x + 9\right) + \left(y^2 - 4y + 4\right) = 15 + 9 + 4$$
$$(x+3)^2 + (y-2)^2 = 28$$
Center: $(-3, 2)$, Radius: $\sqrt{28} = 2\sqrt{7}$

67.
$$x^2 + y^2 + 25x + 10y = -12$$
$$\left(x^2 + 25x\right) + \left(y^2 + 10y\right) = -12$$
$$\left(x^2 + 25x + 625/4\right) + \left(y^2 + 10y + 25\right) = -12 + 625/4 + 25$$
$$(x+25/2)^2 + (y+5)^2 = 677/4$$
Center: $\left(-\dfrac{25}{2}, -5\right)$, Radius: $\dfrac{\sqrt{677}}{2}$

69. The radius of the circle from $(1,3)$ is 2

    a. $\sqrt{(1-2.2)^2 + (3-4.6)^2} = \sqrt{1.44 + 2.56} = \sqrt{4} = 2$ on

    b. $\sqrt{(1-(-.2))^2 + (3-4.7)^2} = \sqrt{1.44 + 2.89} = \sqrt{4.33} \approx 2.08$ outside

    c. $\sqrt{(1-(-.1))^2 + (3-1.4)^2} = \sqrt{1.21 + 2.56} = \sqrt{3.77} \approx 1.94$ inside

    d. $\sqrt{(1-2.6)^2 + (3-4.3)^2} = \sqrt{2.56 + 1.69} = \sqrt{4.25} \approx 2.06$ outside

    e. $\sqrt{(1-(-.6))^2 + (3-1.8)^2} = \sqrt{2.56 + 1.44} = \sqrt{4} = 2$ on

71. The radius of the circle is the distance from $(3,3)$ to $(0,0)$.
$$\sqrt{(3-0)^2 + (3-0)^2} = \sqrt{9+9} = \sqrt{18} = 3\sqrt{2}$$
The equation of the circle with center $(3, 3)$ and radius $3\sqrt{2} = \sqrt{18}$ is
$$(x-3)^2 + (y-3)^2 = \left(\sqrt{18}\right)^2$$
$$(x-3)^2 + (y-3)^2 = 18$$

73. The radius of the circle is the distance from $(1,2)$ to $(3,0)$.
The equation of the circle with center $(1,2)$ and radius $\sqrt{8}$ is

$$(x-1)^2 + (y-2)^2 = \left(\sqrt{8}\right)^2$$
$$(x-1)^2 + (y-2)^2 = 8$$

75. The radius of the circle is the distance from $(-5,4)$ to the x-axis, i.e. 4.
The equation of the circle with center $(-5,4)$ and radius 4 is

$$(x-(-5))^2 + (y-4)^2 = 4^2$$
$$(x+5)^2 + (y-4)^2 = 16$$

77. The center of the circle is the midpoint of the diameter.

$$\left(\frac{3+1}{2}, \frac{3+(-1)}{2}\right) = (2,1)$$

The radius of the circle is the distance from the center $(2,1)$ to $(3,3)$.

$$\sqrt{(2-3)^2 + (1-3)^2} = \sqrt{1+4} = \sqrt{5}$$

The equation of the circle with center $(2,1)$ and radius $\sqrt{5}$ is

$$(x-2)^2 + (y-1)^2 = \left(\sqrt{5}\right)^2$$
$$x^2 - 4x + 4 + y^2 - 2y + 1 = 5$$
$$x^2 + y^2 - 4x - 2y = 0$$

79. Interchange the x-coordinates of the given points to obtain $(2,1)$ and $(-3,-4)$. These points will be the endpoints of the other diagonal.

81. Assume $k > d$. If (c,d) and (c,k) are adjoining vertices, then the square has length of side $k-d$. Then two squares could be formed with two sides of this length perpendicular to the given line segment. One would lie $k-d$ units on one side of the given segment, with vertices $(c+(k-d),d)$ and $(c+(k-d),k)$. The other would lie $|d-k|$ units on the other side, with vertices $(c-(k-d),d)$ and $(c-(k-d),k)$. The third square is formed if (c,d) and (c,k) are opposite vertices. Then the midpoint of this diagonal would be $\left(c, \frac{d+k}{2}\right)$, and a second diagonal of equal length, perpendicular to the given diagonal, would have endpoints $\left(c-\frac{k-d}{2}, \frac{d+k}{2}\right)$ and $\left(c+\frac{k-d}{2}, \frac{d+k}{2}\right)$.

83. Using the hint, we have: the distance from $(x,0)$ to $(3,4)$ is 5. Thus

$$\sqrt{(x-3)^2 + (0-4)^2} = 5 \Rightarrow (x-3)^2 + 16 = 25 \Rightarrow (x-3)^2 = 9$$
$$x-3 = 3 \quad \text{or} \quad x-3 = -3$$
$$x = 6 \qquad\qquad x = 0$$

The points are $(6,0)$ and $(0,0)$.

85. The distance from $(3, y)$ to $(-2, -5)$ is 6. Thus

$$\sqrt{(3-(-2))^2 + (y-(-5))^2} = 6 \Rightarrow \sqrt{25 + (y+5)^2} = 6$$

$$25 + (y+5)^2 = 36 \Rightarrow (y+5)^2 = 11$$

$$y + 5 = \sqrt{11} \quad \text{or} \quad y + 5 = -\sqrt{11}$$

$$y = -5 + \sqrt{11} \qquad\qquad y = -5 - \sqrt{11}$$

The points are $\left(3, -5 + \sqrt{11}\right)$ and $\left(3, -5 - \sqrt{11}\right)$.

87. The distance from $(0,0)$ to $(3,2)$ equals the distance from $(x,0)$ to $(3,2)$. Thus

$$\sqrt{(0-3)^2 + (0-2)^2} = \sqrt{(x-3)^2 + (0-2)^2} \Rightarrow \sqrt{9+4} = \sqrt{(x-3)^2 + 4}$$

$$13 = (x-3)^2 + 4 \Rightarrow (x-3)^2 = 9$$

$$x - 3 = 3 \text{ or } x - 3 = -3$$

$$x = 6 \qquad\qquad x = 0$$

$x = 6$, since $x = 0$ is already a vertex.

89. From the figure, the coordinates of M are $\left(\frac{0+s}{2}, \frac{r+0}{2}\right) = \left(\frac{s}{2}, \frac{r}{2}\right)$.

Then the distances from M to the three vertices are given by

$$\sqrt{\left(0 - \frac{s}{2}\right)^2 + \left(0 - \frac{r}{2}\right)^2} = \sqrt{\left(\frac{s}{2}\right)^2 + \left(\frac{r}{2}\right)^2} = \sqrt{\frac{s^2 + r^2}{4}}$$

$$\sqrt{\left(s - \frac{s}{2}\right)^2 + \left(0 - \frac{r}{2}\right)^2} = \sqrt{\left(\frac{s}{2}\right)^2 + \left(\frac{r}{2}\right)^2} = \sqrt{\frac{s^2 + r^2}{4}}$$

$$\sqrt{\left(0 - \frac{s}{2}\right)^2 + \left(r - \frac{r}{2}\right)^2} = \sqrt{\left(\frac{s}{2}\right)^2 + \left(\frac{r}{2}\right)^2} = \sqrt{\frac{s^2 + r^2}{4}}$$

Thus the three distances are equal.

91. Sketch a figure.

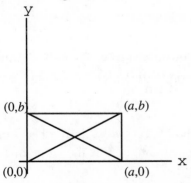

The diagonals have lengths $\sqrt{(0-a)^2 + (b-0)^2} = \sqrt{a^2 + b^2}$ and

$\sqrt{(0-a)^2 + (0-b)^2} = \sqrt{a^2 + b^2}$. Thus the diagonals have the same length.

93. The circles have centers at $(k,0)$ on the x-axis and radius $|k|$, therefore, they are tangent to the y-axis at the origin.

95. The midpoint of the line segment joining (c,d) and $(-c,-d)$ is given by
$$\left(\frac{x_1+x_2}{2},\frac{y_1+y_2}{2}\right)=\left(\frac{c+(-c)}{2},\frac{d+(-d)}{2}\right)=(0,0)$$
that is, the origin. Therefore, (c,d) and $(-c,-d)$ lie on a straight line through the origin and equidistant from the origin. Since the x- and y-coordinates both have opposite signs, the points are on opposite sides of the origin.

1.4 Lines

1. a. C   **b.** B   **c.** B   **d.** D

3. Slope $=\dfrac{y_2-y_1}{x_2-x_1}=\dfrac{7-2}{3-1}=\dfrac{5}{2}$

5. Slope $=\dfrac{y_2-y_1}{x_2-x_1}=\dfrac{2-0}{\frac{3}{4}-\frac{1}{4}}=\dfrac{2}{\frac{1}{2}}=4$

7. Slope $=\dfrac{y_2-y_1}{x_2-x_1}$

$-2=\dfrac{4-t}{9-0}$

$-18=4-t$

$t=22$

9. Slope $=\dfrac{y_2-y_1}{x_2-x_1}$

$-2=\dfrac{(-3t+7)-5}{6-(t+1)}$

$-2=\dfrac{-3t+2}{5-t}$

$-10+2t=-3t+2$

$t=\dfrac{12}{5}$

11. Sketch a figure:

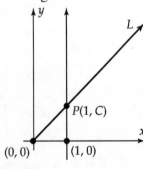

Slope of $L=\dfrac{C-0}{1-0}=C$

From the figure, the slope m of line L is given by
$$\frac{C-0}{1-0}=m$$
Thus $C=m$, and the second coordinate of P is the slope of L.

13. d   **15.** a   **17.** $y = 4x + 5$ **19.** $y = -2.3x + 1.5$

21. $y = \dfrac{-2}{3}x + 2$ **23.** $y = \dfrac{3}{4}x - 3$

25. Solve $2x - y + 5 = 0$ for y in terms of x.

$$2x - y + 5 = 0$$
$$2x + 5 = y$$
slope: 2, y-intercept: 5

27. Solve for y in terms of x.

$$3(x - 2) + y = 7 - 6(y + 4)$$
$$3x - 6 + y = 7 - 6y - 24$$
$$7y = -3x - 11$$
$$y = -\frac{3}{7}x - \frac{11}{7}$$
slope: $-\frac{3}{7}$, y-intercept: $-\frac{11}{7}$

29. Substitute 1 for m and $(4,7)$ for (x_1, y_1) in the point-slope equation.

$$y - y_1 = m(x - x_1)$$
$$y - 7 = 1(x - 4)$$
$$y - 7 = x - 4$$
$$y = x + 3$$

31. Substitute -1 for m and $(6,2)$ for (x_1, y_1) in the point-slope equation.

$$y - y_1 = m(x - x_1)$$
$$y - 2 = (-1)(x - 6)$$
$$y - 2 = -x + 6$$
$$y = -x + 8$$

33. The slope of the line is given by

$$m = \frac{y_2 - y_1}{x_2 - x_1} = \frac{-2 - (-5)}{-3 - 0} = -1$$

Use the slope and one of the points to find the equation of the line.

$$y - y_1 = m(x - x_1)$$
$$y - (-2) = -1(x - (-3))$$
$$y + 2 = -x - 3$$
$$y = -x - 5$$

35. The slope of the line is given by

$$m = \frac{y_2 - y_1}{x_2 - x_1} = \frac{3 - \frac{3}{5}}{\frac{1}{5} - \frac{6}{5}} = -\frac{12}{5}$$

Use the slope and one of the points to find the equation of the line.

$$y - y_1 = m(x - x_1)$$
$$y - \frac{3}{5} = -\frac{12}{5}\left(x - \frac{6}{5}\right)$$
$$y - \frac{3}{5} = -\frac{12}{5}x + \frac{72}{25}$$
$$y = -\frac{12}{5}x + \frac{87}{25}$$

37.

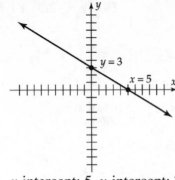

x-intercept: 5 y-intercept: 3

39.

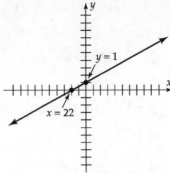

x-intercept: –2 y-intercept: 1

41.

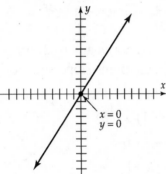

x-intercept: 0 y-intercept: 0

43. Slope of line $PQ = \dfrac{-1-5}{-1-2} = 2$

Slope of line $RS = \dfrac{1-2}{6-4} = -\dfrac{1}{2}$

Since $2\left(-\frac{1}{2}\right) = -1$, line PQ is perpendicular to line RS.

45. Slope of line $PQ = \dfrac{-1-\frac{1}{3}}{1-(-3)} = -\dfrac{1}{3}$

Slope of line $RS = \dfrac{-\frac{2}{3}-0}{4-2} = -\dfrac{1}{3}$

Since these slopes are equal, line PQ is parallel to line RS.

47. Write both equations in the slope-intercept form.

$2x + y - 2 = 0 \qquad\qquad 4x + 2y + 18 = 0$

$\qquad\quad y = -2x + 2 \qquad\qquad\quad y = -2x - 9$

Both lines have slope –2, hence they are parallel.

49. Write both equations in the slope-intercept form.

$y = 2x + 4 \qquad .5x + y = -3$

$\qquad\qquad\qquad y = -.5x - 3$

Since $2(-.5) = -1$, the lines are perpendicular.

51. Slope of the line containing $(9,6)$ and $(-1,2) = \dfrac{2-6}{-1-9} = \dfrac{2}{5}$

Slope of the line containing $(-1,2)$ and $(1,-3) = \dfrac{-3-2}{1-(-1)} = -\dfrac{5}{2}$

Slope of the line containing $(9,6)$ and $(1,-3) = \dfrac{-3-6}{1-9} = \dfrac{9}{8}$

Since $\left(\frac{2}{5}\right)\left(-\frac{5}{2}\right) = -1$, the first two lines are perpendicular, and the points are vertices of a right triangle.

53. Find the slope by $\dfrac{7-3}{3-1} = 2.$

Then the slope that we want is $m = -\dfrac{1}{2}.$

The midpoint is $\left(\dfrac{1+3}{2}, \dfrac{7+3}{2}\right) = (2,5).$

$$y - y_1 = m(x - x_1)$$

$$y - 5 = -\dfrac{1}{2}(x-2)$$

$$y - 5 = -\dfrac{1}{2}x + 1$$

$$y = -\dfrac{1}{2}x + 6$$

55. Find the slope by $\dfrac{7-(-3)}{4-2} = 5.$

Then the slope that we want is $m = -\dfrac{1}{5}.$

The midpoint is $\left(\dfrac{2+4}{2}, \dfrac{-3+7}{2}\right) = (3,2)$

$$y - y_1 = m(x - x_1)$$

$$y - 2 = -\dfrac{1}{5}(x-3)$$

$$y - 2 = -\dfrac{1}{5}x + \dfrac{3}{5}$$

$$y = -\dfrac{1}{5}x + \dfrac{13}{5}$$

57. Substitute 3 for m and $(-2,1)$ for (x_1, y_1) in the point-slope form.

$$y - y_1 = m(x - x_1)$$

$$y - 1 = 3(x - (-2))$$

$$y - 1 = 3x + 6$$

$$y = 3x + 7$$

59. First find the slope of the given line.

$$3x - 2y = 5$$

$$y = \dfrac{3}{2}x - \dfrac{5}{2}$$

The slope is $\frac{3}{2}$. This is also the slope of the required line.

Now substitute $\frac{3}{2}$ for m and $(2,3)$ for (x_1, y_1) in the point-slope equation.

$$y - y_1 = m(x - x_1)$$

$$y - 3 = \dfrac{3}{2}(x-2)$$

$$y - 3 = \dfrac{3}{2}x - 3$$

$$y = \dfrac{3}{2}x$$

61. The line passes through $(5,0)$ and $(0,-5)$. Hence its slope is given by
$$\frac{y_2 - y_1}{x_2 - x_1} = \frac{-5 - 0}{0 - 5} = 1$$
Substitute 1 for m and -5 for b in the slope-intercept equation.
$$y = mx + b$$
$$y = 1x + (-5)$$
$$y = x - 5$$

63. First find the slope of the line through $(0,1)$ and $(2,3)$.
$$\frac{y_2 - y_1}{x_2 - x_1} = \frac{3 - 1}{2 - 0} = 1$$
Therefore, the slope of the required perpendicular line is -1.
Now substitute -1 for m and $(-1,3)$ for (x_1, y_1) in the point-slope equation.
$$y - y_1 = m(x - x_1)$$
$$y - 3 = -1(x - (-1))$$
$$y - 3 = -x - 1$$
$$y = -x + 2$$

65. $(3,-2)$ is on the line $kx - 2y + 7 = 0$ if its coordinates satisfy the equation. Substituting, we obtain
$$3k - 2(-2) + 7 = 0$$
$$3k + 11 = 0$$
$$k = -\frac{11}{3}$$

67. Since C is $(0,0)$ and P is $(3,4)$, the slope of radius CP is given by $\dfrac{y_2 - y_1}{x_2 - x_1} = \dfrac{4 - 0}{3 - 0}$

$= \dfrac{4}{3}$. Therefore the slope of the tangent line is $-\dfrac{3}{4}$. The equation of the tangent line

is found by substituting $-\dfrac{3}{4}$ for m and $(3,4)$ for (x_1, y_1) in the point-slope form.

$$y - y_1 = m(x - x_1) \Rightarrow y - 4 = -\frac{3}{4}(x - 3) \Rightarrow y - 4 = -\frac{3}{4}x + \frac{9}{4} \Rightarrow y = -\frac{3}{4}x + \frac{25}{4}$$

69. Since C is $(1,3)$ and P is $(2,5)$, the slope of radius CP is given by
$$\frac{y_2 - y_1}{x_2 - x_1} = \frac{5 - 3}{2 - 1} = 2$$
Therefore the slope of the tangent line is $-\frac{1}{2}$. The equation of the tangent line is found by substituting $-\frac{1}{2}$ for m and $(2,5)$ for (x_1, y_1) in the point-slope form.

$$y - y_1 = m(x - x_1) \Rightarrow y - 5 = -\frac{1}{2}(x - 2) \Rightarrow y - 5 = -\frac{1}{2}x + 1 \Rightarrow y = -\frac{1}{2}x + 6$$

71. Put the equations of the lines into slope-intercept form.

$$Ax + By + C = 0 \qquad\qquad Ax + By + D = 0$$

$$By = -Ax - C \qquad\qquad By = -Ax - D$$

$$y = -\frac{A}{B}x - \frac{C}{B} \qquad\qquad y = -\frac{A}{B}x - \frac{D}{B}$$

Thus, both lines have the slope $-A/B$. Therefore the lines are parallel.

73. a. $(0, 60),\ (5, 66)$

b. $m = \dfrac{66 - 60}{5 - 0} = 1.2$

$$y - 60 = 1.2(x - 0)$$

$$y = 1.2x + 60$$

c. $y = 1.2(4) + 60 = 64.8\,\text{million}$

d. $72 = 1.2x + 60$

$$12 = 1.2x \Rightarrow x = 10 \Rightarrow 2010$$

75. a. $y = .03x$

b.
$$y^2 + \left(\frac{y^2}{.03}\right)^2 = 5280^2$$

$$y^2 + \frac{y^2}{.0009} = 27{,}878{,}400 \Rightarrow .0009\,y^2 + y^2 = 25{,}090.56$$

$$1.0009\,y^2 = 25{,}090.56 \Rightarrow y^2 = \frac{25{,}090.56}{1.0009}$$

$$y^2 \approx 25067.9988011$$

$$y \approx \sqrt{25067.9988011}\ \text{(negative value not considered)}$$

$$y \approx 158.33\ \text{ft}$$

77. a. Use the data points $(0, 212)$ and $(1100, 210)$.

The slope is given by $\dfrac{y_2 - y_1}{x_2 - x_1} = \dfrac{210 - 212}{1100 - 0} = -\dfrac{1}{550}$.

Substitute $-\frac{1}{550}$ for m and 212 for b in $y = -\dfrac{1}{550}x + 212$.

b. Substitute 550 for x.

$$y = -\frac{1}{550}(550) + 212 = 211°$$

c. Substitute 1300 for x.

$$y = -\frac{1}{550}(1300) + 212 = 209.6°$$

d. Substitute 3120 for x.

$$y = -\frac{1}{550}(3120) + 212 = 206.3°$$

e. Substitute 6900 for x.

$$y = -\frac{1}{550}(6900) + 212 = 199.5°$$

79. a. The points are (0, 3.2) and (10, 5.5)

$$m = \frac{5.5 - 3.2}{10 - 0} = .23$$

$$y - 3.2 = .23(x - 0)$$

$$y = .23x + 3.2$$

b. $y = .23(20) + 3.2 = 7.8$ million

c. $10.1 = .23x + 3.2$

 $6.9 = .23x$

 $x = 30$

 2020

81. a. The average cost is,

$$y = \frac{2.75x + 26,000}{x}$$

b. Plug in 3 for y.

$$3 = \frac{2.75x + 26,000}{x}$$

$$3x = 2.75x + 26,000$$

$$.25x = 26,000$$

$$x = 104,000$$

83. a. Since profit=revenue – cost, revenue = profit + cost. Thus,

$r = p + c$

$$r = (.6x - 14.5) + (.8x + 14.5)$$

$$r = 1.4x$$

b. Set $r = c$ and solve for x.

$1.4x = .8x + 14.5$

$.6x = 14.5$

$$x = \frac{145}{6} \text{ thousand items}$$

$x = 24,167$ items

85. a. Slope $= \dfrac{0 - 60}{30 - 0} = -2$; y-intercept $= 60$

 $y = mx + b$

 $y = -2x + 60$

c. Slope $= \dfrac{0 - 160}{40 - 0} = -4$; y-intercept $= 160$

 $y = mx + b$

 $y = -4x + 160$

b. Slope $= \dfrac{0 - 80}{40 - 0} = -2$; y-intercept $= 80$

 $y = mx + b$

 $y = -2x + 80$

87. a. Use the slope formula,

 Since $m = \dfrac{212 - 32}{100 - 0} = \dfrac{9}{5}$, $F - 32 = \dfrac{9}{5}(C - 0) \Rightarrow F = \dfrac{9}{5}C + 32.$

b. Solve the previous equation for C.

$$F = \frac{9}{5}C + 32 \Rightarrow F - 32 = \frac{9}{5}C \Rightarrow \frac{5}{9}(F - 32) = C \Rightarrow C = \frac{5}{9}(F - 32)$$

c. Replace F with C

$$C = \frac{9}{5}C + 32 \Rightarrow 5C = 9C + 160 \Rightarrow -4C = 160 \Rightarrow C = -40$$

Temperature is $-40°$ (Celsius)

89. a. The rate of emptying is given by the opposite of the slope.

During the first two minutes, the slope $= \dfrac{y_2 - y_1}{x_2 - x_1} = \dfrac{50 - 75}{2 - 0} = -12.5$

Emptying at 12.5 gallons per minute.

During the next three minutes, the slope $= \dfrac{y_2 - y_1}{x_2 - x_1} = \dfrac{25 - 50}{5 - 2} = -8.33$

Emptying at 8.33 gallons per minute.

During the last minute, the slope $= \dfrac{y_2 - y_1}{x_2 - x_1} = \dfrac{0 - 25}{6 - 5} = -25$

Emptying at 25 gallons per minute.

b.

Since the slope is given as -10, substitute -10 for m and 75 for b in the slope-intercept form to obtain $y = -10x + 75$.

91. a. Using the points (0, 15350) and (4, 9910)

b. It is the slope, $1360 per year.

c. $y = -1360(6) + 15{,}350$
$= \$7190$

$m = \dfrac{9910 - 15350}{4 - 0} = -1360$

$y - 15{,}350 = -1360(x - 0)$

$y = -1360x + 15{,}350$

93. Using the hint, if (x_1, y_1) satisfied both equations, then

$y_1 = mx_1 + b$

$y_1 = mx_1 + c$

$mx_1 + b = mx_1 + c$

$b = c$

Since $b \neq c$, this is a contradiction, and no point can lie on both lines, that is, the lines are parallel.

95. Sketch a figure:

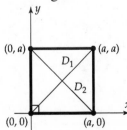

The slopes of the diagonals are given by

$\dfrac{a - 0}{a - 0} = 1$ and $\dfrac{0 - a}{a - 0} = -1$.

Since $1(-1) = -1$, the diagonals are perpendicular.

Chapter 1 Review Exercises

1. **a.** > **b.** < **c.** < **d.** > **e.** =

3. **a.** $-10 < y < 0$ **b.** $0 \le x \le 10$

5. **a.** $x > -8$, thus $(-8, \infty)$
   b. $x \le 5$, thus $(-\infty, 5]$

7. **a.** $12{,}320{,}000{,}000{,}000{,}000 = 1.232 \times 10{,}000{,}000{,}000{,}000{,}000 = 1.232 \times 10^{16}$
   b. $.0000000000789 = 7.89 \times \dfrac{1}{100{,}000{,}000{,}000} = 7.89 \times 10^{-11}$

9. **a.** $|x - (-7)| < 3$ or $|x + 7| < 3$ **b.** $|y| > |x - 3|$

11. The equation is equivalent to
    $$x - 5 = 3 \quad \text{or} \quad -(x-5) = 3$$
    $$x = 8 \qquad\qquad x = 2$$
    Checking each possible solution, we see
    $$|8 - 5| = 3 \text{ and } |2 - 5| = |-3| = 3$$
    Thus both 8 and 2 are solutions.

13. The equation is equivalent to
    $$x + 3 = \frac{5}{2} \quad \text{or} \quad -(x+3) = \frac{5}{2}$$
    $$x = -\frac{1}{2} \qquad\qquad x = -\frac{11}{2}$$
    Checking each possible solution, we see
    $$\left|-\frac{1}{2} + 3\right| = \frac{5}{2} \text{ and } \left|-\frac{11}{2} + 3\right| = \left|-\frac{5}{2}\right| = \frac{5}{2}$$
    Thus both $-\frac{1}{2}$ and $-\frac{11}{2}$ are solutions.

15. The inequality is equivalent to
    $$x + 2 > 2 \quad \text{or} \quad -(x+2) > 2$$
    $$-x - 2 > 2$$
    $$-x > 4$$
    $$x > 0 \quad \text{or} \qquad x < -4$$

17. **a.** $|\pi - 7| = -(\pi - 7) = 7 - \pi$ **b.** $|\sqrt{23} - \sqrt{3}| = \sqrt{23} - \sqrt{3}$

19. Let $d = .2828\ldots$
$100d = 28.2828\ldots$

$$\underline{d = .2828\ldots}$$
$99d = 28$

$$d = \frac{28}{99}$$

21. $2\left(\dfrac{x}{5} + 7\right) - 3x = \dfrac{x+2}{5} - 4$

$$\frac{2x}{5} + 14 - 3x = \frac{x+2}{5} - 4$$

$$2x + 70 - 15x = x + 2 - 20$$

$$70 - 13x = x - 18$$

$$-14x = -88$$

$$x = \frac{44}{7}$$

$$\frac{44}{7}$$

23. Since $b^2 - 4ac = (-2)^2 - 4(3)(5) = -56$, the equation has no real solutions.

25. $\quad 5z^2 + 6z = 7$
$\quad 5z^2 + 6z - 7 = 0$

$$z = \frac{-6 \pm \sqrt{6^2 - 4(5)(-7)}}{2(5)}$$

$$z = \frac{-6 \pm \sqrt{176}}{10} = \frac{-3 \pm 2\sqrt{11}}{5}$$

$$\frac{-3 + 2\sqrt{11}}{5}, \frac{-3 - 2\sqrt{11}}{5}$$

27. Let $x =$ width, $\frac{3}{4}x =$ height, $d =$ diagonal, $A =$ area. Then,

$$A = x\left(\frac{3}{4}x\right) = \frac{3}{4}x^2$$

$$d = \sqrt{x^2 + \left(\frac{3}{4}x\right)^2} = \sqrt{x^2 + \frac{9}{16}x^2} = \sqrt{\frac{25}{16}x^2} = \frac{5}{4}x, \text{ since the width is positive.}$$

Thus, $x = \dfrac{4}{5}d$ and $A = \dfrac{3}{4}\left(\dfrac{4}{5}d\right)^2 = \dfrac{3}{4}\left(\dfrac{16}{25}d^2\right) = \dfrac{12}{25}d^2$.

If $d_1 = 14$, $A_1 = \dfrac{12}{25} \cdot 14^2$. If $d_2 = 21$, $A_2 = \dfrac{12}{25} \cdot 21^2$.

$$\frac{A_2}{A_1} = \left(\frac{12}{25} \cdot 21^2\right) \div \left(\frac{12}{25} \cdot 14^2\right)$$

$$\frac{A_2}{A_1} = 21^2 \div 14^2$$

$$A_2 = \frac{9}{4}A_1 = 2.25A_1 \quad \text{This implies 2.25 times as large.}$$

29. $20x^2 + 12 = 31x$
$20x^2 - 31x + 12 = 0$

$b^2 - 4ac = (-31)^2 - 4(20)(12) = 1$
Since $b^2 - 4ac$ is positive, the
equation has two real solutions.

31.
$$\frac{3}{x} + \frac{5}{x+2} = 2$$

$$x(x+2) \cdot \frac{3}{x} + x(x+2) \cdot \frac{5}{x+2} = 2x(x+2) \quad x \neq 0, -2$$

$$3(x+2) + 5x = 2x^2 + 4x$$

$$3x + 6 + 5x = 2x^2 + 4x$$

$$8x + 6 = 2x^2 + 4x$$

$$0 = 2x^2 - 4x - 6$$

$$0 = 2(x^2 - 2x - 3)$$

$$0 = 2(x-3)(x+1)$$

$$x - 3 = 0 \text{ or } x + 1 = 0$$

$$x = 3 \text{ or } x = -1$$

33. $y = kx$
$36 = k(12)$
$k = 3$
Therefore, the variation equation is
$y = 3x$. When $x = 2$, $y = 3(2) = 6$.

35. $T = \dfrac{kR^2}{S}$

$.6 = \dfrac{k(3)^2}{15}$

$9 = k(3)^2$

$k = 1$

37. To solve $x^4 - 11x^2 + 18 = 0$, let
$u = x^2$.
$x^4 - 11x^2 + 18 = 0$
$u^2 - 11u + 18 = 0$
$(u-2)(u-9) = 0$

$\qquad u - 2 = 0 \quad \text{or} \quad u - 9 = 0$

$\qquad\qquad u = 2 \qquad\qquad u = 9$

Since $u = x^2$, we have the equivalent
statements: $x^2 = 2 \qquad x^2 = 9$

$\qquad\qquad x = \pm\sqrt{2} \quad x = \pm 3$

$\sqrt{2}, -\sqrt{2}, 3, -3$

39. The equation is equivalent to
$3x - 1 = 4 \quad \text{or} \quad -(3x-1) = 4$

$\qquad 3x = 5 \qquad\qquad -3x = 3$

$\qquad x = \dfrac{5}{3} \qquad\qquad x = -1$

Checking each possible solution, we
see $\left| 3(\frac{5}{3}) - 1 \right| = |5 - 1| = 4$ and
$|3(-1) - 1| = |-4| = 4$.
Thus, both $\frac{5}{3}$ and -1 are solutions.

41. distance $= \sqrt{(x_1 - x_2)^2 + (y_1 - y_2)^2}$

$= \sqrt{(1-4)^2 + (-2-5)^2}$

$= \sqrt{(-3)^2 + (-7)^2}$

$= \sqrt{9 + 49} = \sqrt{58}$

43. distance $= \sqrt{(x_1 - x_2)^2 + (y_1 - y_2)^2}$

$= \sqrt{(c - (c-d))^2 + (d - (c+d))^2}$

$= \sqrt{(c-c+d)^2 + (d-c-d)^2}$

$= \sqrt{d^2 + (-c)^2} = \sqrt{c^2 + d^2}$

45. The midpoint is

$$\left(\frac{x_1 + x_2}{2}, \frac{y_1 + y_2}{2} \right) = \left(\frac{c + 2d - c}{2}, \frac{d + c + d}{2} \right) = \left(\frac{2d}{2}, \frac{c + 2d}{2} \right) = \left(d, \frac{c + 2d}{2} \right)$$

47. a. The radius of the circle is the distance from $(2, -3)$ to $(1, 1)$.

$$\sqrt{(2-1)^2 + (-3-1)^2} = \sqrt{1^2 + (-4)^2} = \sqrt{17}$$

b. The equation of the circle with center $(2, -3)$ and radius $\sqrt{17}$ is

$$(x-2)^2 + (y-(-3))^2 = (\sqrt{17})^2$$

$$(x-2)^2 + (y+3)^2 = 17$$

49.

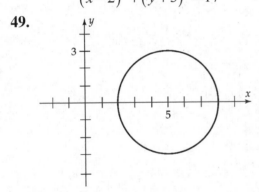

51. (b) and (d)   **53. (c)**

55.

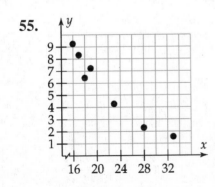 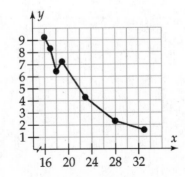

57. Rewrite the equation in slope-intercept form.

$$y = x - \frac{x-2}{5} + \frac{3}{5}$$

$$y = x - \frac{1}{5}x + \frac{2}{5} + \frac{3}{5}$$

$$y = \frac{4}{5}x + 1$$

a. The y-intercept is 1. **b.** The slope is $\frac{4}{5}$.

59. Substitute 3 for m and $(2,-1)$ for (x_1, y_1) in the point-slope form.

$$y - (-1) = 3(x-2)$$

$$y + 1 = 3x - 6$$

$$y = 3x - 7$$

61. The line $2y - x = 5$, in slope-intercept form, becomes $y = \frac{1}{2}x + \frac{5}{2}$. Its slope is $\frac{1}{2}$. The slope of any perpendicular line is -2. Thus the required line has equation $y = -2x + b$. Since $(0,1)$ satisfies the equation, $1 = -2(0) + b$ and $b = 1$. Hence $y = -2x + 1$ is the required equation.

63. The line through $(1,3)$ and $(-4,2)$, has slope given by

$$\frac{y_2 - y_1}{x_2 - x_1} = \frac{2-3}{-4-1} = \frac{1}{5}$$

This is the slope of the required line.
Substitute $\frac{1}{5}$ for m and $(-4,5)$ for (x_1, y_1) in the point-slope form.

$$y - 5 = \frac{1}{5}\left(x - (-4)\right)$$

$$y - 5 = \frac{1}{5}(x + 4)$$

$$y - 5 = \frac{1}{5}x + \frac{4}{5}$$

$$y = \frac{1}{5}x + \frac{29}{5} \Rightarrow 5y = x + 29 \Rightarrow x - 5y = -29$$

65. Since slope $= \dfrac{\text{change in } y}{\text{change in } x}$ we can

write $\dfrac{1}{5} = \dfrac{5000}{\text{change in } x}$

change in x = 25,000 feet

67. False. Since $x = 5y + 6$, we have $x - 6 = 5y$ or $y = \dfrac{1}{5}x - \dfrac{6}{5}$. The y-intercept is $-\dfrac{6}{5}$.

69. False. Since $3x + 4y = 12$, we have $4y = -3x + 12$ or $y = -\dfrac{3}{4}x + 3$. The slope of this line is $-\dfrac{3}{4}$. Since $4x + 3y = 12$, we have $3y = -4x + 12$ or $y = -\dfrac{4}{3}x + 4$. The slope of this line is $-\dfrac{4}{3}$. The product of the slopes is $\left(-\dfrac{3}{4}\right)\left(-\dfrac{4}{3}\right) = 1 \neq -1$.

71. False. As x-values increase, y-values decrease. This indicates a negative slope.

73. False. The y-intercept is 1, which is positive.

75. (d)   **77.** (e)

79. (c) To find the equation, we use (0,2) and (3,0)
$$m = \frac{0-2}{3-0} = \frac{-2}{3} \Rightarrow y - 0 = \frac{-2}{3}(x-3) \Rightarrow y = \frac{-2}{3}x + \frac{6}{3}$$
$$3y = -2x + 6 \Rightarrow 2x + 3y = 6$$

81. Use the datapoints (0,74.7) and (30,77.8).
a. The slope is given by
$$\frac{y_2 - y_1}{x_2 - x_1} = \frac{77.8 - 74.7}{20 - 0} = .155$$
Substitute .155 for m and 74.7 for b in the slope-intercept form.
$y = mx + b$
$y = .155x + 74.7$

b. 1990 corresponds to $x = 5$.
$y = .155(5) + 74.7$
$y = 75.5$ years.

c. Set $y = 80$ and solve for x.
$80 = .155x + 74.7$
$5.3 = .155x$
$x = 34$
This corresponds to the year 2019.

83. Graph (c) has slope +75.   **85.** Graph (d) has slope +20.

Chapter 1 Test

1. a.

b. $(-\infty, -2]$

2. $1.1x = .7x + 19.6$

$\quad .4x = 19.6$

$\quad x = 49$

3. a. $.8315(26.504)^2 - 73.93(26.504) + 2116.1 = 740.76$ pounds per square ft

b. $1238.41 = .8315h^2 - 73.93h + 2116.1$

$\quad 0 = .8315h^2 - 73.93h + 877.69$

$\quad h = 14.112$

$\quad 14,112\ ft$

4. a. 5.622×10^{12} **b.** 2.811×10^8 **c.** Divide

$$\frac{5.622 \times 10^{12}}{2.811 \times 10^8} = 2.0 \times 10^4 = \$20,000$$

5. Multiply by the common denominator

$$\frac{1}{3t} - \frac{3}{4t} = \frac{1}{12t} + 1$$

$\quad -5 = 1 + 12t$

$\quad -6 = 12t$

$\quad t = \dfrac{-1}{2}$

6. $2x^2 + 13x - 7 = 0$

$\quad (2x - 1)(x + 7) = 0$

$\quad x = \dfrac{1}{2} \qquad x = -7$

7. $100n = 14.141414...$

$\quad n = .1414...$

$\quad 99n = 14$

$\quad n = \dfrac{14}{99}$

8. $2E = h(b + c)$

$\quad 2E = bh + hc$

$\quad 2E - hc = bh$

$\quad b = \dfrac{2E - hc}{h}$

9. $4x^2 - 6x - 5 = 0$

$\quad x = \dfrac{6 \pm \sqrt{36 + 80}}{8}$

$\quad x = \dfrac{6 \pm \sqrt{116}}{8}$

$\quad x = \dfrac{6 \pm 2\sqrt{29}}{8}$

$\quad x = \dfrac{3 \pm \sqrt{29}}{4}$

10. a. All numbers that are less than 6 units from 16 on the number line.
  b. $|c-16|>5$

11.   We want $(-k)^2-4(1)(16)=0$. Thus, $k^2-64=0 \Rightarrow k^2=64 \Rightarrow k=8$ or -8.

12. a. $(y-1)^2$ **b.** x^2-1

13. $|4x-5|=12$

   $4x-5=12$ or $-(4x-5)=12$

   $4x=17$ $-4x=7$

   $x=\dfrac{17}{4}$ $x=\dfrac{-7}{4}$

14.   Given the center, we have $(x+1)^2+(y+6)^2=r^2$

   $r=\sqrt{(-7-(-1))^2+(-5-(-6))^2}=\sqrt{36+1}=\sqrt{37}$

   Therefore, we have $(x+1)^2+(y+6)^2=37$

15. Neither. One line has a slope of -4 and the other line has a slope of -2.

16. a. Using the midpoint formula,

   $\left(\dfrac{-1+2}{2},\dfrac{3+(-1)}{2}\right)=\left(\dfrac{1}{2},1\right)$

b. Using the distance formula,

   $\sqrt{(-1-2)^2+(3-(-1))^2}=\sqrt{9+16}=\sqrt{25}=5$

17. Using the slope formula,

   $\dfrac{-8-(-6)}{5-\sqrt{5}}=\dfrac{-2}{5-\sqrt{5}}$ or $\dfrac{2}{\sqrt{5}-5}$

18. y-intercept
   $-y-2=0$
   $y=-2$

   x-intercept
   $3x^2+x-2=0$
   $(3x-2)(x+1)$
   $x=\dfrac{2}{3}$ $x=-1$

19. a. Using the slope formula,
   $m=\dfrac{19309-13359}{2004-1990}=\$425\,a\,year$

b. Using x=0 for 1990
   $y=425(10)+13,359=\$17,609\,for\,2000$

c. Using x=0 for 1990
   $y=425(20)+13,359=\$21,859\,for\,2010$

20. a. $(1,1)$

b. Using the slope formula, $m = \dfrac{3-1}{2-1} = 2$.

Since the line we want is perpendicular, our slope needs to be $\frac{-1}{2}$.

$$y - 3 = \frac{-1}{2}(x-2)$$

$$y - 3 = \frac{-1}{2}x + 1$$

$$y = \frac{-1}{2}x + 4$$

21. $3x^2 + 3y^2 + 6x + 24 = 24y$

$x^2 + y^2 + 2x + 8 = 8y$

$(x^2 + 2x) + (y^2 - 8y) = -8$

$(x^2 + 2x + 1) + (y^2 - 8y + 16) = 9$

$(x+1)^2 + (y-4)^2 = 9$

center : $(-1, 4)$ radius : 3

22.

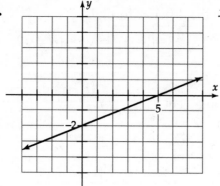

x-intercept 5 and y-intercept -2

23. The perimeter is the sum of the distances:

between $(0,0)$ and $(4,0) = 4$

between $(4,0)$ and $(2,2) = \sqrt{(4-2)^2 + (0-2)^2} = \sqrt{4+4} = \sqrt{8}$

between $(2,2)$ and $(4,4) = \sqrt{(2-4)^2 + (2-4)^2} = \sqrt{4+4} = \sqrt{8}$

between $(4,4)$ and $(0,4) = 4$

between $(0,4)$ and $(0,0) = 4$

Thus, the perimeter $= 4 + \sqrt{8} + \sqrt{8} + 4 + 4 = 12 + 2\sqrt{8}$.

24. a. Using the slope formula,

$$\frac{232.3 - 412.1}{23 - 0} = -7.82$$

$$y - 412.1 = -7.82(x - 0)$$

$$y = -7.82x + 412.1$$

b. $y = -7.82(21) + 412.1 = 247.88$ per 100,000

c. $y = -7.82(29) + 412.1 = 185.32$ per 100,000

d. No. It predicts a negative death rate in 2033.

25. Since it is inversely proportional,

$$r = \frac{k}{t}$$

$$9 = \frac{k}{3}$$

$$k = 27$$

$$r = \frac{27}{t}$$

$$r = \frac{27}{12}$$

$$r = \frac{9}{4}$$

26. Find the slope of each of the lines.

$$\overline{PQ} = \frac{1 - \frac{3}{2}}{1 - 0} = \frac{-1}{2}$$

$$\overline{RS} = \frac{5 - 4}{7 - 6} = 1$$

Neither

Chapter 2
Graphs and Technology

2.1 Graphs

1. Sketch:

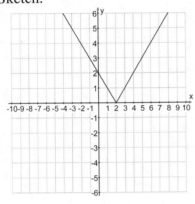

Calculator:

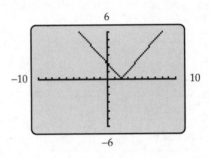

3. Sketch:

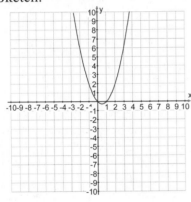

Calculator:

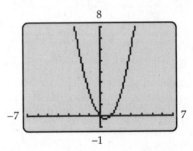

5. Sketch:

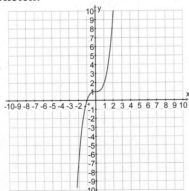

Calculator:

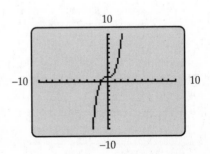

7. Graph $y = .5x - 3$

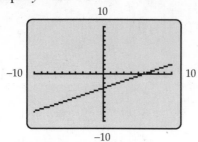

9.

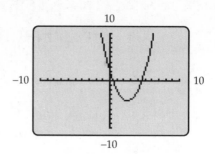

11.

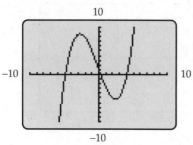

13. a.

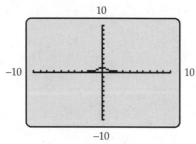

b. Since the pixels are approximately .32 units high, by the time the calculator gets to $x = 2$ the y values are so small that it is using the same pixel to draw the curve as well as the x-axis.

c. Use the window $-10 \leq x \leq 10, -1 \leq y \leq 1$.

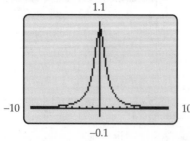

15. Solve $x^2 + y^2 = 9$ for y to obtain two solutions: $y = \sqrt{9 - x^2}$ and $y = -\sqrt{9 - x^2}$.

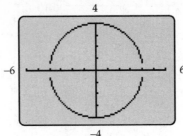

17. Solve to obtain two solutions: $y = \sqrt{\dfrac{48 - 3x^2}{2}}$ and $y = -\sqrt{\dfrac{48 - 3x^2}{2}}$.

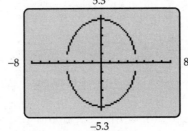

19. Solve to obtain two solutions: $y = -2 + \sqrt{25 - (x - 4)^2}$ and $y = -2 - \sqrt{25 - (x - 4)^2}$.

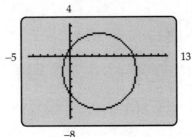

21. Solve to obtain two solutions: $y = \sqrt{\dfrac{4x^2 - 36}{9}}$ and $y = -\sqrt{\dfrac{4x^2 - 36}{9}}$.

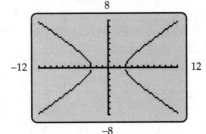

23. Solve to obtain two solutions: $y = \sqrt{\dfrac{45 - 9x^2}{5}}$ and $y = -\sqrt{\dfrac{45 - 9x^2}{5}}$.

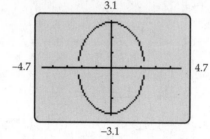

25. Answers vary, but most calculators are correct to 6-8 decimal places.

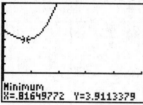

27. Graph the equation in the window specified and use the maximum and minimum routines.

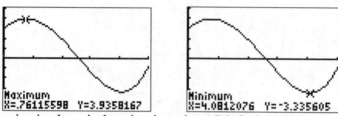

The highest point in the window is given by (.7612, 3.9358) and the lowest point is given by (4.0812,-3.3356) (disregarding calculator round-off error).

29. It will be late 2003 and about 5.17%.

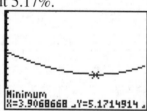

31. Of all the choices given, **c** is the best.

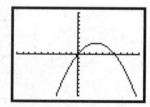

33. Of all the choices given, **d** is the best.

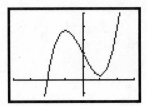

35. Of all the choices given, **d** is the best, although a window which extends further in the negative x direction would be even better. **e**

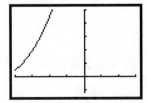

37. The window shown is $-5 \leq x \leq 5$, $-100 \leq y \leq 100$ (scale 10 shown below).

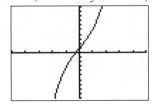

39. The window shown is $-10 \leq x \leq 10$, $-2 \leq y \leq 20$. Note that $\sqrt{x^2} - x = 0$ if $x \geq 0$; $\sqrt{x^2} - x = -2x$ elsewhere.

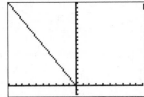

41. The window shown is $-6 \leq x \leq 12, -100 \leq y \leq 250$.

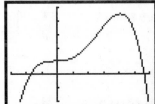

43.

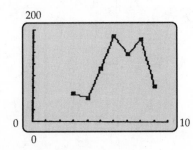

45.

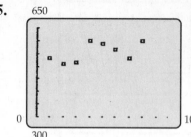 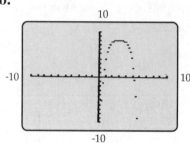

47. a.   **b.**

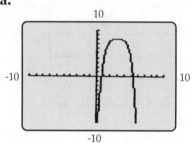 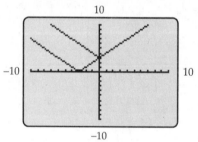

c. Because the *vertical* distance between points on the graph is very small at the top of the graph, the graphed pixels make a solid line on the screen. When the vertical distance between adjacent pixels is larger, more "space" shows and the individual points appear isolated.

49. The two equations are not the same. $y = |x+3|$ and $y = |x| + 3$ are graphed below.

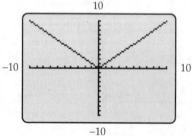

51. The two equations are the same and have the same graph:

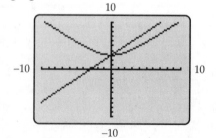

53. The two equations are not the same. $y = \sqrt{x^2 + 9}$ and $y = x + 3$ are graphed below.

55. a. The two equations given have the same graph, hence the factorization can be considered confirmed.

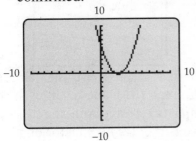

b. The two equations have different graphs, as is shown here in the window. Thus the expressions are not equivalent.

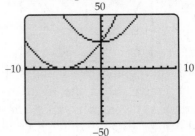

57. The two expressions give rise to the same graph, hence the statement can be considered possibly true.

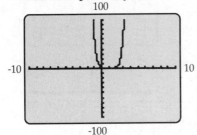

59. Of all the choices, **(c)** is the best, as it shows the rocket ascending, then descending to earth. The window $0 \le x \le 10, 0 \le y \le 250$ is shown.

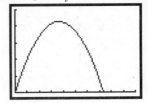

61. The window shown is $0 \le x \le 40$, $0 \le y \le 110,000$.

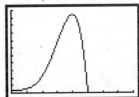

63. The window shown is $0 \le x \le 48$, $0 \le y \le 40$.

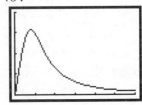

65. a. Values are approximately 12.3 and 24.2, corresponding to 1917 and 1929.
  b. Use the maximum routine to find a high point at $x = 20$, corresponding to 1925. There were approximately 104,000 deer at this time.

67. a. Use the maximum routine to find a high point at approximately $x = 6.3$ hours.
  b. The value is approximately $x = 18.7$ hours.

69. The window shown is $-4 \leq x \leq 4$, $-5 \leq y \leq 15$.

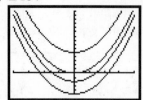

All four graphs have the same shape, but graph **(b)** is the result of shifting graph **(a)** vertically up 5 units, **(c)** of shifting vertically down 5 units, and **(d)** of shifting down 2 units.

71. The window shown is $0 \leq x \leq 10$, $0 \leq y \leq 10$.

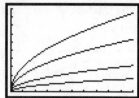

All four graphs have the same shape, but graph **(b)** is the result of stretching graph **(a)** away from the x-axis by a factor or 2, **(c)** by a factor of 3, and **(d)** by a factor of 1/2.

73. The window shown is $-5 \leq x \leq 5$, $-5 \leq y \leq 5$.

The graphs are reflections of each other in the line $y = x$.

75. The   window   shown   is $-20 \leq x \leq 20, -20 \leq y \leq 20.$

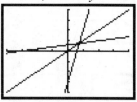

The graphs are reflections of each other in the line $y = x$.

2.2 Solving Equations Graphically and Numerically

1.   Graph $y = x^5 + 5$ and $y = 3x^4 + x$ in the same viewing window to find the points of intersection; these will be the solutions of the equation. First use a window such as $-4 \leq x \leq 4, -5 \leq y \leq 15$. This shows two intersections. A second window such as $2 \leq x \leq 5, 10 \leq y \leq 500$ shows a third intersection. 3 solutions.

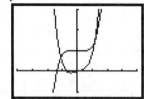   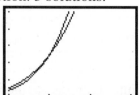

3.   Graph $y = x^7 - 10x^5 + 15x + 10$ in the standard window. This suggests that there are three x-intercepts, that is, three solutions of the equation. A larger window such as $-20 \leq x \leq 20, -10 \leq y \leq 10$ does not indicate any more solutions.

   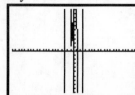

5. Graph $y = x^4 + 500x^2 - 8000x$ and $y = 16x^3 - 32000$ in the same viewing window to find the points of intersection. The window $-5 \le x \le 20, -50{,}000 \le y \le 50{,}000$ is shown. 2 solutions.

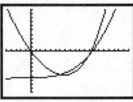

7. $x = -2.426453$ **9.** $x = -1.4526$ **11.** $x = -1.4751$

13. $x = 1.1921235$ **15.** $x = -1.379414$ **17.** $x = 1.3289$

19. $x = -2.1149$

21. Graph $y = 2x^3 - 4x^2 + x - 3$ and use the root (zero) routine to find the solution $x = 2.1016863$. The window $-2 \le x \le 3, -10 \le y \le 10$ is shown.

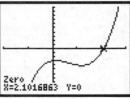

23. Graph $y = x^5 - 6x + 6$ and use the root (zero) routine to find the solution $x = -1.752119$. The window $-3 \le x \le 3, -5 \le y \le 15$ is shown.

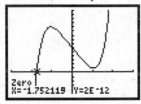

25. Graph $y = 10x^5 - 3x^2 + x - 6$ and use the root (zero) routine to find the solution $x = .95054589$. The window $-1 \le x \le 2, -20 \le y \le 10$ is shown.

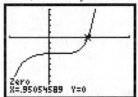

27. Rewrite the equation by multiplying both sides by 12 to obtain $24x - 6x^2 - x^4 = 0$. Graph $y = 24x - 6x^2 - x^4$ in the window $-1 \le x \le 3, -10 \le y \le 20$ to see two intercepts. One is obviously 0; the other is found by the root (zero) routine to be $x = 2.2073898$.

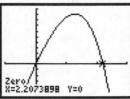

29. Graph $y = \dfrac{5x}{x^2 + 1} - 2x + 3$ and use the root (zero) routine to find the solution $x = 2.3901454$. The window $-3 \le x \le 3, -2 \le y \le 5$ is shown.

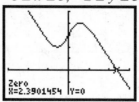

31. Graph $y = |x^2 - 4|$ and $y = 3x^2 - 2x + 1$ in the same window $-5 \le x \le 5, 0 \le y \le 10$ and use the intersection routine to locate the two solutions at $x = -.6513878$ and $x = 1.1513878$.

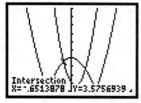 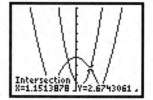

33. Graph $y = \sqrt{x^2 + 3}$ and $y = \sqrt{x - 2} + 5$ in the same window $0 \le x \le 10, 0 \le y \le 10$ and use the intersection routine to locate the only solution at $x = 7.033393$.

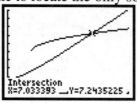

35. Graph $y = 3x^3 - 2x^2 + 3x - 2$ in the window $0 \le x \le 1, -2 \le y \le 2$ and use the root (zero) routine to find the solution approximated as $x = .66666667$. This suggests that the solution could be exactly $x = \frac{2}{3}$. Substituting into the original equation confirms this.

37. Graph $y = 12x^4 - x^3 - 12x^2 + 25x - 2$ in the window $0 \le x \le 1, -5 \le y \le 5$ and use the root (zero) routine to find the solution approximated as $x = .083333333$. This suggests that the solution could be exactly $x = \frac{1}{12}$. Substituting into the original equation confirms this.

39. Graph $y = 4x^4 - 13x^2 + 3$ in the window $1 \le x \le 2, -10 \le y \le 10$ and use the root (zero) routine to find the solution approximated as $x = 1.7320508$. This suggests that the solution could possibly be exactly $x = \sqrt{3}$. Substituting into the original equation confirms this.

41. Graph $y = 10^x - \frac{1}{4}x$ and $y = 28$ in the same window $-120 \le x \le 3, -10 \le y \le 40$ and use the intersection routine to locate the solutions at $x = 1.4528$ and $x = -112$.

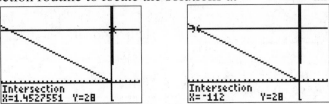

In a different window, the first solution can more readily be identified.

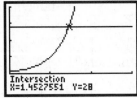

43. Graph $y = x + \sin\left(\frac{1}{2}x\right)$ and $y = 4$ in the standard window and use the intersection routine to obtain $x = 3.00242$.

45. Graph $y = 5\ln x + x^3 - x^2$ and $y = 5$ in the standard window and use the intersection routine to obtain $x = 1.7388$.

47. Graph $y = \sqrt{180115x^2 + 2863851x + 11383876}$ and $y = 7000$ in the same window $0 \le x \le 25, 0 \le y \le 8000$ and use the intersection routine to locate the solution that corresponds to the year 2009.

49. a. Graph the given function and the line $y = 944$ in the same window and use the intersection routine to locate the only solution at $x = 8.8$. This corresponds to late 2003.

b. Graph the given function and the line $y = 1,100$ in the same window $0 \le x \le 25, 0 \le y \le 1500$ and use the intersection routine to locate the only solution at $x = 15.6$. This corresponds to mid 2010.

51. Graph the given function and the line $y = 12$ in the same window and use the intersection routine to locate the only solution at $x = 96$. This corresponds to 96 ft.

53. Graph the given function and the line $y = 1400$ in the same window and use the intersection routine to locate the only solution at $x = 12.78$. This corresponds to 2012.

2.3 Applications of Equations

1.

English Language	Mathematical Language
Score on fourth exam	x
Sum of scores on four exams	$88 + 62 + 79 + x$
Average of scores on four exams	$\dfrac{88 + 62 + 79 + x}{4}$

3.

English Language	Mathematical Language
The width and length of a rectangle	x and y
The perimeter is 45 centimeters	$2x + 2y = 45$
The area is 112.5 square centimeters	$xy = 112.5$

5.

English Language	Mathematical Language
Old salary	x
8% pay raise	$.08x$
New salary	$x + .08x$

$$x + .08x = 2619$$

7.

English Language	Mathematical Language
Area before decrease	$\pi\left(\dfrac{16}{2}\right)^2 = \pi(8)^2 = 64\pi$
Decrease radius	x
Radius after decrease	$8 - x$
Area after decrease	$\pi(8-x)^2$

$$\pi(8-x)^2 = 64\pi - 48\pi \Rightarrow \pi(8-x)^2 = 16\pi$$

9. Let $x =$ amount invested at 12%, $1100 - x =$ amount invested at 6%
Return $= .12x + .06(1100 - x) + .11(550)$

$$12x + .06(1100 - x) + .11(550) = .09(1650)$$

$$.06x + 126.5 = 148.5$$

$$.06x = 22$$

$$x = \$366.67 \text{ at } 12\% \text{ and}$$

$$\$1100 - \$366.67 = \$733.33 \text{ @ } 6\%$$

11. Let $x =$ amount drained

Then $.4(8-x) =$ amount of antifreeze in drained radiator

$x =$ amount of antifreeze added

$.6(8) =$ amount of antifreeze required

$$.4(8-x)+x = .6(8)$$
$$3.2+.6x = 4.8$$
$$.6x = 1.6$$
$$x = 2\tfrac{2}{3} \text{ quarts}$$

13. Use distance = rate $\times$ time

distance going $= 2.5(360+r)$

distance returning $= 3.5(360-r)$

Since the distances are equal,
$$2.5(360+r) = 3.5(360-r)$$
$$900 + 2.5r = 1260 - 3.5r$$
$$6r = 360$$
$$r = 60 \text{ mph}$$

15. Solve the equation in the example.
$$x + \frac{1683}{x} = 82.25$$
$$x^2 - 82.25x + 1683 = 0$$
$$x = \frac{82.25 \pm \sqrt{(82.25)^2 - 4(1683)}}{2(1)}$$
$$x = \frac{82.25 \pm 5.75}{2}$$
$$x = 44 \text{ and } 38.25$$

17. Let $x =$ rate of southbound car

$50 =$ rate of northbound car

$3x =$ distance of southbound car

$3(50) =$ distance of northbound car

$$3x + 3(50) = 345$$
$$3x = 195$$
$$x = 65 \text{ miles per hour}$$

19. Before: After:

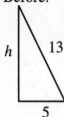

Before: $h^2 = 13^2 - 5^2$, hence $h = 12$

After: $y^2 = 13^2 - 8^2$, hence $y = \sqrt{105} \approx 10.2$

$h - y = 12 - 10.2 = 1.8$ feet.

21.   Let $x =$ width of walk

$12 =$ radius of pool

$x + 12 =$ outer radius of walk

Then, Area of walk $= \pi(x+12)^2 - \pi(12)^2$

$$52\pi = \pi(x+12)^2 - \pi(12)^2$$

$$52 = x^2 + 24x + 144 - 144$$

$$0 = x^2 + 24x - 52$$

$$0 = (x-2)(x+26)$$

$$x = 2 \ \text{ or } \ x = -26$$

Only the positive answer makes sense: 2 meters.

23. Let $16t^2 =$ distance fallen

$1100(3 - t) =$ distance covered by sound

Then $16t^2 = 1100(3 - t)$

$$16t^2 + 1100t - 3300 = 0$$

$$x = \frac{-1100 \pm \sqrt{(1100)^2 - 4(16)(-3300)}}{2(16)}$$

$$x = \frac{-1100 \pm 1192.14}{32}$$

$x = -71.6 \ \text{ or } \ 2.88$ sec

Only the positive answer makes sense. Then $d = 16t^2 = 16(2.88)^2 = 132.7$ feet.

25. Let x = Red's speed
  $x - 6$ = Wolf's speed

$$\frac{432}{x} = \text{Red's time and } \frac{432}{x-6} = \text{Wolf's time}$$

$$\frac{432}{x-6} - \frac{432}{x} = 1$$

$$432x - 432(x-6) = x(x-6)$$

Then
$$0 = x^2 - 6x - 2592$$

$$0 = (x-54)(x+48)$$

$$x = 54 \text{ or } x = -48$$

Only the positive answer makes sense: Red drives 54 mph, Wolf drives 48 mph.

27. In Example 10, the dimensions of the box were found to be x, $22 - 2x$, and $30 - 2x$ and the equation was found to be $(30 - 2x)(22 - 2x)x = 1000$ which simplifies to be

$4x^3 - 104x^2 + 660x - 1000 = 0$.   To have one dimension greater than 18, look for a solution with x less than 6, then $30 - 2x$ is greater than 18. The window $0 \le x \le 11, -10 \le y \le 10$ shown in the text yields the solution $x = 2.234$ inches.

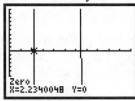

Zero
X=2.2340048 Y=0

Thus, we have a 2.234 in.$\times$2.234 in. square.

29. Use the given formula with $h = 5$, $S = 100$. Thus, $100 = \pi r \sqrt{r^2 + 25}$.
A calculator graph in the window $0 \le x \le 10$, $0 \le y \le 150$:

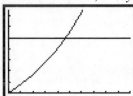

shows the root at 4.658 inches.

31. Use the given formula with $h = \frac{1}{3} r$, volume $= 180$.

$$180 = \pi r^2 \left(\frac{1}{3} r\right)/3 \Rightarrow 180 = \frac{\pi r^3}{9}$$

$$r = \sqrt[3]{\frac{1620}{\pi}} = 8.02 \text{ inches}$$

33. Label the text figure:

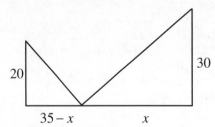

20

30

$35 - x$ x

From the Pythagorean Theorem, the length of the rope is given by

$$L = \sqrt{20^2 + (35 - x)^2} + \sqrt{30^2 + x^2}$$

Solve $63 = \sqrt{20^2 + (35 - x)^2} + \sqrt{30^2 + x^2}$. Graph the left and right sides of the equation in the window $0 \leq x \leq 35, 50 \leq y \leq 70$ and apply the intersection routine to obtain $x = 11.47$ feet and $x = 29.91$ feet.

35. Redraw the text figure:

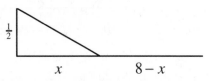

$\frac{1}{2}$

x $8 - x$

Cost (in thousands) = Underwater Cost (in thousands) + Land Cost (in thousands)

$$72 = 12\sqrt{x^2 + \left(\tfrac{1}{2}\right)^2} + 8(8 - x)$$

Solve by graphing the left and right sides of the equation in the window $0 \leq x \leq 8, 0 \leq y \leq 100$ and apply the intersection routine to obtain $x = 1.795$ miles. Then $8 - x = 8 - 1.795 = 6.205$ miles from the substation.

37. Let r be the speed of Ray Harroun, $r + 83$ be the speed of Dan Wheldon, $t - 3.53$ be the time of Ray Harroun, t be the time of Dan Wheldon. Since they both traveled 500 miles and $d = rt$, we have $500 = rt$ and $500 = (r + 83)(t - 3.53)$. Solving for t in the first equation, we have $t = \dfrac{500}{r}$. Substituting this into the second equation, we have $500 = (r + 83)\left(\dfrac{500}{r} - 3.53\right)$. Solve by graphing $y = (x + 83)\left(\dfrac{500}{x} - 3.53\right)$ and $y = 500$ in the same window and apply the intersection routine. Using the window $0 \leq x \leq 100$ (scale 5 shown below) $0 \leq y \leq 1000$ (scale 100 shown below) we one solution. It is $x = 74.6$.

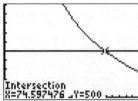

Intersection
X=74.597476   Y=500

This solution represents the speed of Ray Harroun. Since we seek the speed of Dan Wheldon, the answer is $74.6 + 83 = 157.6$ mph.

39. The dimensions of the base of the box are given by x = width and $2x$ = length. The height is found from length $\times$ width $\times$ height = Volume $\Rightarrow (2x)xh = 38.72$ $\Rightarrow h = \frac{19.36}{x^2}$. The cost is given by Cost = area of base $\times$ 12 + area of sides $\times$ 8

$\Rightarrow 538.56 = 2x^2 \times 12 + 2\left(\frac{19.36}{x^2}\right)x \times 8 + 2\left(\frac{19.36}{x^2}\right)2x \times 8$ which simplifies to

$538.56 = \frac{929.28}{x} + 24x^2$. A calculator graph of $y = \frac{929.28}{x} + 24x^2$ and $y = 538.56$ in the window $0 \le x \le 5$, $400 \le y \le 700$ (scale 100 shown below):

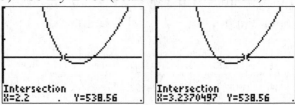

shows two roots, one at $x = 2.2$ and one at $x = 3.2$. Only the first leads to an acceptable height of 4 feet and the dimensions are 2.2 by 4.4 by 4 feet.

2.4 Optimization Applications

1. Graph $y = 2x^3 - 3x^2 - 12x + 1$ and use the maximum routine to find the high point at $(-1, 8)$. The window $-3 \le x \le 3, -20 \le y \le 10$ is shown.

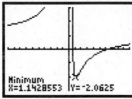

3. Graph $y = \frac{4}{x^2} - \frac{7}{x} + 1$ and use the minimum routine to find the low point at $(1.1428553, -2.0625)$. The window $-10 \le x \le 10, -3 \le y \le 3$ is shown.

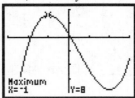

5. Graph $y = \frac{x^2(x+1)^3}{(x-2)^2(x-4)^2}$ and use the maximum routine to find the high point at $(-.3409026, .00032224)$. The window $-1 \le x \le 0, 0 \le y \le .0004$ is shown.

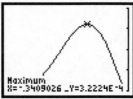

7. **a.** In the window $-2 \leq x \leq 0, -1 \leq y \leq 5$ the high point is at $(-1,4)$.
   b. In the window $-2 \leq x \leq 2, -1 \leq y \leq 5$ there is a high point at $(-1,4)$ and another at the endpoint $(2,4)$ of the interval.
   c. In the window $-2 \leq x \leq 3, -1 \leq y \leq 25$ the high point occurs at the right endpoint of the interval $(3,20)$.

9. Graph the given function in the window $0 \leq x \leq 100$ (scale 5), $0 \leq y \leq 100$ (scale 5) and use the maximum function to find $x = 52.86$ which corresponds to 52.86 mph.

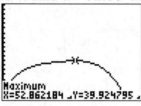

11. The length function from Ex 33 in Section 2.3 is given by $L = \sqrt{20^2 + (35-x)^2} + \sqrt{30^2 + x^2}$. Graphing this in the window $0 \leq x \leq 30$, $50 \leq y \leq 70$ and use the minimum function to find $x = 21.0$ which corresponds to 21 ft.

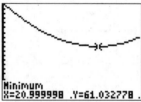

13. Let x be the length of the sides perpendicular to the river. The side that parallels the river would be $1800 - 2x$. The area would be given by $A = x(1800 - 2x)$ $= -2x^2 + 1800x$. Graphing this in the window $0 \leq x \leq 1000$ (scale 100), $0 \leq y \leq 500,000$ (scale 10,000) and use the maximum function to find $x = 450.0$. Thus, the dimensions should be 450 ft $\times$ 900 ft.

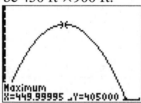

15. Let x be the measurement on the east/west sides and y be the measurement on the north/south sides. The given a that the area is 24,200 sq ft, we have $xy = 24,200$ or $y = \dfrac{24,200}{x}$. The cost would be given by $C = 2(10x) + 2(20y) = 20x + 40y$ $= 20x + 40\left(\dfrac{24,200}{x}\right) = 20x + \dfrac{968,000}{x}$. Graphing this in the window $0 \le x \le 1000$ (scale 50), $5000 \le y \le 15,000$ (scale 1000) and use the minimum function to find the value assigned to y is the minimum cost, $8800.

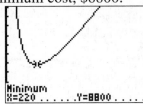

17. The dimensions of the base of the box are given by x = width = length. Then the height is found from
$$\text{length} \times \text{width} \times \text{height} = \text{Volume}$$
$$xxh = 20,000$$
$$h = \frac{20,000}{x^2}$$
The total surface area is given by
$$\text{Area} = \text{Area of base} + 4 \times \text{Area of one side}$$
$$y = x^2 + 4 \cdot x \cdot \frac{20,000}{x^2}$$
$$y = x^2 + \frac{80,000}{x}$$
Graph this in the window $0 \le x \le 100, 0 \le y \le 20,000$ and use the minimum routine to find $x = 34.2$ cm and $h = 20,000/x^2 = 17.1$ cm   Therefore, we have 34.2 cm × 34.2 × cm × 17.1 cm.

19. As in the example, the dimensions of the box are given by x = height, $20 - 2x$ = length = width. Thus, $\text{Volume} = \text{length} \times \text{width} \times \text{height} \Rightarrow y = (20 - 2x)(20 - 2x)x$.
   a. Graph $y = (20 - 2x)(20 - 2x)x$ and $y = 550$ in the window $4 \le x \le 10$, $0 \le y \le 700$ and use the intersection routine to find $x = 4.4267$ inches. Thus, we have 4.4267 by 4.4267 in.
   b. Graph $y = (20 - 2x)(20 - 2x)x$ in the window $0 \le x \le 10, 0 \le y \le 700$ and use the maximum routine to find $x = 3.33$ inches. Thus we have 10/3 by 10/3 in.

21. The average cost per unit is given by
$$y = \frac{c(x)}{x} = \frac{.13x^3 - 70x^2 + 10,000x}{x}$$
$$y = .13x^2 - 70x + 10,000$$

a. Graph $y = .13x^2 - 70x + 10,000$ and $y = 1100$ in the window $0 \leq x \leq 400$, $0 \leq y \leq 5000$ and use the intersection routine to find $x = 206$. The other intersection shows x greater than 300 units and must be discarded.

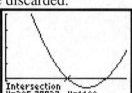

b. Graph $y = .13x^2 - 70x + 10,000$ in the window $0 \leq x \leq 400$, $0 \leq y \leq 5000$ and use the minimum routine to find $x = 269$ units, which leads to an average cost of \$577 per unit.

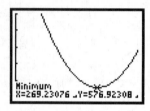

23. Use Revenue $= 142x$, Cost $= x^3 - 8x^2 + 20x + 40$. Then
$$\text{Profit} = \text{Revenue} - \text{Cost}$$
$$y = 142x - \left(x^3 - 8x^2 + 20x + 40 \right)$$
$$y = -x^3 + 8x^2 + 122x - 40$$

a. Graph this equation in the window $0 \leq x \leq 6$, $-100 \leq y \leq 1000$. The high point occurs at the right hand endpoint $x = 6$, so 600 bookmarks should be made.

b. Graph this equation in the window $0 \leq x \leq 16$, $-100 \leq y \leq 1000$ and use the maximum routine to find $x = 9.5788127$, so 958 bookmarks should be made.

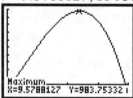

25. The figure of the garden consists of a rectangle of width 10, length $2x$ plus a triangle with base $2x$, height h. Since $h^2 + x^2 = 10^2$, we can write $h = \sqrt{100 - x^2}$. Then
$$\text{Area} = \text{area of rectangle} + \text{area of triangle}$$
$$y = 10(2x) + \tfrac{1}{2}(2x)\sqrt{100 - x^2} = 20x + x\sqrt{100 - x^2}$$

Graph this equation in the window $0 \leq x \leq 10$, $0 \leq y \leq 300$ and use the maximum routine to find $x = 9.306$ feet, then area$= y = 220.18$ square feet.

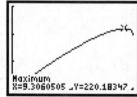

27. Any point on the graph of $y = 5 - x^2$ must have coordinates $(x, 5 - x^2)$. Then the distance from $(x, 5 - x^2)$ to $(0,1)$ is given by

$$y = \sqrt{(x-0)^2 + ((5-x^2)-1)^2}$$

$$y = \sqrt{x^2 + (4-x^2)^2}$$

$$y = \sqrt{x^4 - 7x^2 + 16}$$

Graph this equation in the window $0 \le x \le 5, 0 \le y \le 10$ and use the minimum routine to find $x = 1.871$.

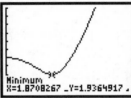

Hence the coordinates of the required point are given by $x = 1.871$, $y = 5 - x^2 = 5 - (1.871)^2 = 1.5$.

29. Let x = number of times ladders are ordered per year. Then

$$\text{Cost} = \text{cost of ordering} + \text{cost of storage}$$

$$y = 20x + 10\left(\frac{300}{x}\right)$$

$$y = 20x + \frac{3000}{x}$$

Graph this equation in the window $0 \le x \le 24, 0 \le y \le 3000$ and use the minimum routine to find $x = 12.2$. Since x must be an integer, the answer is 12 times per year.

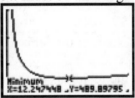

2.5 Linear Models

1. **a.** Here is a table of the residuals for each model.

$y = x$

Data Point (x, p)	Model Point (x, y)	Residual $p - y$	Squared Residual $(p - y)^2$
(1,2)	(1,1)	1	1
(2,2)	(2,2)	0	0
(3,3)	(3,3)	0	0
(4,3)	(4,4)	-1	1
(5,5)	(5,5)	0	0

$y = .5x + 1.5$

Data Point (x, p)	Model Point (x, y)	Residual $p - y$	Squared Residual $(p - y)^2$
(1,2)	(1,2)	0	0
(2,2)	(2,2.5)	-.5	.25
(3,3)	(3,3)	0	0
(4,3)	(4,3.5)	-.5	.25
(5,5)	(5,4)	-1	1

The sum of the residuals of the first model is 0. The sum of the residuals of the second model is 0.

b. The sum of the squares of the residuals in the first model is 2. The sum of the squares of the residuals in the second model is 1.5.

c. Clearly the second model is the better fit.

3. **a.** Here is a table of the residuals for each model.

$y = 4.9x + 170$

Data Point (x, p)	Model Point (x, y)	Residual $p - y$	Squared Residual $(p - y)^2$
(0,171.3)	(0,170)	+1.3	1.69
(2,179.8)	(2,179.8)	0	0
(4,188)	(4,189.6)	−1.6	2.56
(6,201.5)	(6,199.4)	+2.1	4.41

$y = 5x + 171$

Data Point (x, p)	Model Point (x, y)	Residual $p - y$	Squared Residual $(p - y)^2$
(0,171.3)	(0,171)	+.3	.09
(2,179.8)	(2,181)	−1.2	1.44
(4,188)	(4,191)	−3	9
(6,201.5)	(6,201)	+.5	.25

The sum of the residuals of the first model is 1.8. The sum of the residuals of the second model is -3.4

b. The sum of the squares of the residuals in the first model is 8.66. The sum of the squares of the residuals in the second model is 10.78.

c. Clearly the first model is the better fit.

5. Here is a table of the residuals for the model.

$y = .71x + 13.05$

Data Point (x, p)	Model Point (x, y)	Residual $p - y$	Squared Residual $(p - y)^2$
(0,13)	(0,13.05)	-.05	.0025
(1,14)	(1,13.76)	+.24	.0576
(2,14)	(2,14.47)	- .47	.2209
(3,15)	(3,15.18)	-.18	.0324
(4,17)	(4,15.89)	+1.11	1.2321
(5,16)	(5,16.6)	-.6	.36

The sum of the squares of the residuals is 1.9055. The other models are 2.8 and 2.75.

7.   There seems to be a negative correlation between the variables.

9.   There seems to be a negative correlation between the variables.

11. a. The data appears to be linear.
   b. There appears to be a positive correlation.

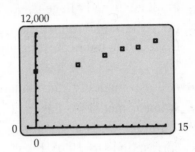

13. a. The data does not appear to be linear.

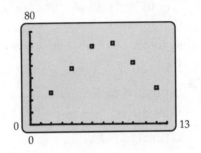

15. a. The data appears to be linear.
   b. There appears to be a negative correlation.

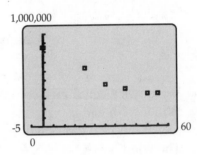

17. a. A linear model for this data is $y = 14.9x + 2822$.
   b. Evaluating this model, we have the following:

   For 150 sq ft: $y = 14.9(150) + 2822 = 5057$ BTUs

   For 280 sq ft: $y = 14.9(280) + 2822 = 6994$ BTUs

   For 420 sq ft: $y = 14.9(420) + 2822 = 9080$ BTUs

   c. For 235 sq ft: $y = 14.9(235) + 2822 = 6325.5$ BTUs; Choose the 6500 BTU model

19. a. $(6, 1.8), (7, 2.3), (8, 2.5), (9, 3.1), (10, 3.9), (11, 3.8), (12, 4.0), (13, 4.4), (14, 4.8), (15, 5.1)$
   b. A linear model for this data is $y = .3594x - .2036$.
   c. For 2010, $x = 20$: $.3594(20) - .2036 = 6.9844$ which yields \$6.98 billion estimated funds spent on research and development.

21. **a.** A linear model for the median weekly earnings of full-time workers 25 and older who attended college but did not graduate is $y = 9.9429x + 600.8095$. A linear model for the median weekly earnings of full-time workers 25 and older who did graduate from college is $y = 20.2286x + 900.4286$.

   b. Four the four groups, the coefficient of x represents the yearly rate of increase. Rounded to the nearest penny we have $9.63, $16.29, $9.94, and $20.23.

23. **a.** A linear model for this data is $y = .2179x + 52.2805$.

   b. A linear model for this data is $y = .1447x + 65.4246$.

   c. Solving $.2179x + 52.2805 = .1447x + 65.4246$ we have $x \approx 179.56$. Since $x = 70$ corresponds to 1970, the year in which the life expectancies would be the same would be $1900 + 179 = 2079$.

25. **a.** A linear model for this data is $y = .03119x + .5635$.

   b. For 2012, $x = 12$: $.03119(12) + .5635 = .93778$ which yields .938 billion estimated passengers.
   For 2016, $x = 16$: $.03119(12) + .5635 = 1.06254$ which yields 1.06 billion estimated passengers.

Chapter 2 Review Exercises

1. **a.** Windows **a** and **d** give complete graphs.
   b. Windows **b** and **c** make the interesting part of the graph too close to the x-axis to see. Window **e** shows nothing sensible at all.
   c. The window $-4 \le x \le 6, -10 \le y \le 10$ (shown) would be a better choice.

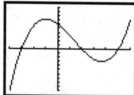

3. **a.** No window gives a complete graph.
   b. Windows **a** and **b** do not show the maxima and minima. Window **c** does not show the minima. Window **e** shows nothing at all, and window **d** is just a random stab.
   c. The window $-7 \le x \le 11, -1000 \le y \le 500$ (scale 100) would be a better choice.

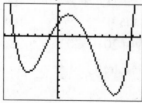

5. a. Windows **b** and **c** give a complete graph.
   b. Window **a** does not show the maxima and minima. Window **d** squeezes everything uselessly. Window **e** shows nothing at all.
   c. The window $-10 \le x \le 10, -150 \le y \le 150$ (scale 50) would be a better choice.

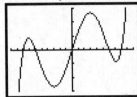

7.

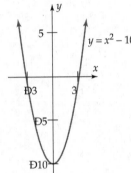

This shows all intercepts and the minimum.

9.

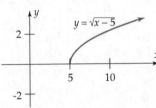

This shows all intercepts.

11.

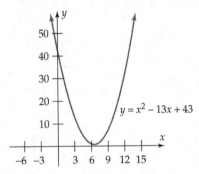

13.

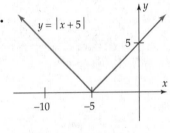

15. Graph $y = x^3 + 2x^2$ and $y = 11x + 6$ in the window $0 \le x \le 5, -10 \le y \le 60$ and use the intersection routine to obtain $x = 2.7644$.

17. Graph $y = x^4 + x^3 - 10x^2$ and $y = 8x + 16$ in the window $0 \le x \le 5, -20 \le y \le 60$ and use the intersection routine to obtain $x = 3.2678$.

19. The equation is equivalent to $x^3 + 2x^2 - 3x + 4 = 0$ as long as the denominator is defined, that is, $x \ne -5, 3$. Graph this equation in the window $-6 \le x \le 3$, $-10 \le y \le 11$ and use the root (zero) routine to obtain $x = -3.2843$.

21. The equation is equivalent to $x^3 + 2x^2 - 3x - 5 = 0$. Graph this equation in the window $0 \le x \le 3, -10 \le y \le 10$ and use the root (zero) routine to obtain $x = 1.6511$.

23. Let $x =$ amount of gold

Then $1 - x =$ amount of silver

$600x =$ value of gold

$50(1-x) =$ value of silver

$600x + 50(1-x) = 200$

$600x + 50 - 50x = 200$

$550x = 150$

$x = 3/11$ ounce of gold

$1 - x = 8/11$ ounce of silver

25. Let $x =$ time working

$\dfrac{x}{5} =$ Karen's rate

$\dfrac{x}{4} =$ Claire's rate

$\dfrac{x}{5} + \dfrac{x}{4} = 1$

$4x + 5x = 20$

$x = \frac{20}{9} = 2\frac{2}{9}$ hrs

27. Let $x =$ length of smaller piece

$4x =$ length of bigger piece

$x + 4x = 12$

$x = 2.4$

$4x = 9.6$ feet

29. Let $x =$ original length

$x + 2 =$ new length

$2x =$ new width

$x^2 =$ original area

$2x(x+2) =$ new area

$2x(x+2) = 3x^2$

$2x^2 + 4x = 3x^2$

$0 = x^2 - 4x$

$x = 0$ (unreasonable) or $x = 4$ feet

31. Average cost $= \dfrac{\text{cost}}{x} = \dfrac{600x^2 + 600x}{x^2 + 1} \div x = \dfrac{600x + 600}{x^2 + 1}$

$\dfrac{600x + 600}{x^2 + 1} = 25$

$600x + 600 = 25x^2 + 25$

$0 = 25x^2 - 600x - 575$

$x = 12 \pm \sqrt{167}$

Only the positive answer makes sense, so $x = 12 + \sqrt{167} = 25$ caseloads

33. Let $x =$ one of the two sides

$z =$ one of the three sides

$2x + 3z = 120$

$z = 40 - \frac{2}{3}x$

Area $= xz = x\left(40 - \frac{2}{3}x\right)$

Graph $y = x\left(40 - \frac{2}{3}x\right)$; use the maximum routine to find $x = 30$ yards, $z = 20$ yards.

35. Width of rectangle $= 2x$, length $= 9 - x^2$, hence area $= 2x\left(9 - x^2\right) = 18x - 2x^3$. Graph this equation and use the maximum routine to find $x = 1.7321$. This suggests that the exact value is $x = \sqrt{3}$, which can be proven.

37. a.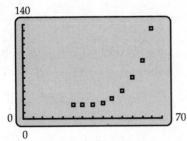

    b. The data does not appear to be linear.

39. a. A linear model for this data is $y = 1.5868x + .2473$.

    b. For 56 in., $y = 1.5868(56) + .2473 = 89.1081$.   Thus, we have approximately 89 years old.

    For 68 in., $y = 1.5868(68) + .2473 = 108.1497$. Thus, we have approximately 108 years old.

    c. Solving $151 = 1.5868x + .2473$ we have $x \approx 95.0042$ in. circumference. Thus, the diameter is approximately $95.0042 / \pi \approx 30.1408$ in. or about 2.5 ft.

41. a. A linear model for this data is $y = 14.8998x + 59.0163$.

    b. Evaluating this model, we have the following:

    For 1999, $x = 9$: $14.8998(9) + 59.0163 = 193.1145$.   Thus, have approximately $193.1 billion.

    For 2006, $x = 16$: $14.8998(16) + 59.0163 = 297.4131$. Thus, have approximately $297.4 billion.

    c. Solving $372 = 14.8998x + 59.0163$ we have $x \approx 21.0059$. Thus, in 1990 + 21 = 2011.

Chapter 2 Test

1. The standard viewing window clearly shows the parts of the graph near the origin.

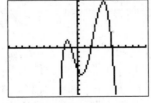

To see the rest of the graph a larger window is needed, such as $-4 \le x \le 14$, and $-212 \le y \le 10$ (scale 10).

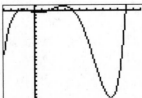

2.   Graph $y = \dfrac{7x^2}{x^2 + 24x}$ and $y = x$ in the same window. $-20 \le x \le 50$ (scale 5), and $-20 \le y \le 20$ (scale 2) is shown below. Using the intersection feature, the solution to the equation is $x = -17$. Although it appears that there is a solution of $x = 0$, origin is not a point on the graph of $y = \dfrac{7x^2}{x^2 + 24x}$.

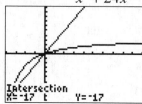

3.   The graph the function. The window $-10 \le x \le 20$, and $-20 \le y \le 100$ (scale 20) shows the maximum. Using the maximum routine we find the highest point is approximately (8.1510, 95.8884).

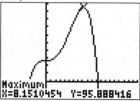

4.   a. When $k = 0$, we graph $y = .3x^5 - 2x^3 + x$ in the window $-5 \le x \le 5$, and $-5 \le y \le 5$ to see that the graph has 5 x-intercepts, which implies 5 solutions.

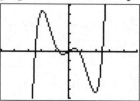

   b. To determine where the graph would have one solution, we need to either shift vertically up or down. To determine by how much, we need to locate the relative maximum and minimum points.

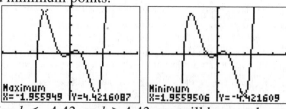

   This implies for $k \le -4.43$ or $k \ge 4.43$ we will have only one real solution in that the graph will intersect the **x**-axis in only one place.

5. Solve to obtain two solutions: $y = \sqrt{\dfrac{144-x^2}{4}}$ and $y = -\sqrt{\dfrac{144-x^2}{4}}$. A window such as $-12 \le x \le 12$, and $-8 \le y \le 8$ will show the graph clearly.

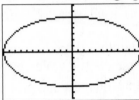

Window choices will vary for different models of graphing calculators.

6. Graph $y = \dfrac{x^3 - 2x^2 - 4x + 8}{x^2 + x - 6}$ in a window such as $-10 \le x \le 10$, and $-20 \le y \le 10$.
Using the zero feature, the solution to the equation is $x = 2$.

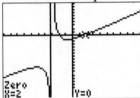

At $x = -2$, $y = \dfrac{x^3 - 2x^2 - 4x + 8}{x^2 + x - 6}$ is not defined.

7. **a.** A window such as $0 \le x \le 48$ (scale 2), and $0 \le y \le 30$ (scale 2) is appropriate.

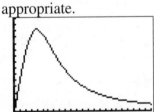

b. Using the maximum routine we find the highest point is approximately (7.937, 26.457) implying a result of 7.937 hours and 26.457 mg per liter.

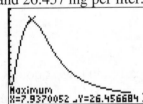

8. Graph $y = \sqrt{x^4 - x^2 + 2x + 1}$ in a window such as $-2 \le x \le 1$, and $-5 \le y \le 5$. Zoom in about the x-intercepts to find $x \approx -1.3865$ or $-.4258$.

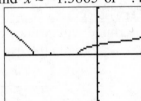

9. If x is the length of the strip, then one side measures $35 - x$ and the other measures $40 - x$. Since the product of the two side lengths is the area, the equation would be $(35 - x)(40 - x) = 785$.

10. If x be the width of the page and y be the length. Since the area of the printed part is 40 sq in., we have $40 = \left[x - (1.7 + .4)\right]\left[y - (2 \cdot .4)\right]$. which implies $y = \dfrac{40}{x - 2.1} + .8$.

Since we want to minimize the total area, we have $A = xy = x\left(\dfrac{40}{x - 2.1} + .8\right)$. Since the width of the page cannot be more than 7.4 inches, an appropriate window would be $0 \le x \le 8$, and $0 \le y \le 100$ (scale 10). We can see that the minimum of the total area would be making the width of the page as large as possible. Thus, the width would be 7.4 in. and a length of the page would be $\dfrac{40}{7.4 - 2.1} + .8 \approx 8.347$ in.

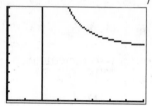

11. Let x be the amount of fluid drained and replaced with pure antifreeze. Since we have 8 quarts of fluid, we get the following equation, $.30(8 - x) + 1.00x = .40(8)$.

Using the graphing calculator we can consider the equation $.30(8 - x) + x = 3.2$. Graphing $y = .30(8 - x) + x$ and $y = 3.2$ in the window $0 \le x \le 8$, and $0 \le y \le 8$, we can use the intersection routine to find that 1.14 qt of fluid should be drained and replaced.

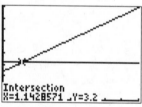

12. a. Clearly the lowest point is $(1, 1.1)$ for this window.

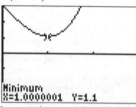

b. In this case, we need to determine with values at the endpoints. Using the trace feature and assigning the $x = -2$ and .99, we see that the lowest point is at $(-2, 1.1)$.

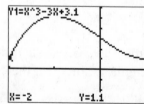
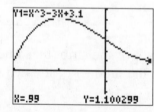

13. Let $x, x+1,$ and $x+2$ be the consecutive integers. We need to solve the equation $x(x+1)(x+2) = 21,924$. Graphing $y = x(x+1)(x+2)$ and $y = 21,924$ in the window $0 \le x \le 30$, and $0 \le y \le 25,000$ (scale 5000), we can use the intersection routine to find that the first consecutive integer is 27. Thus, the dimensions are 27 cm by 28 cm by 29 cm.

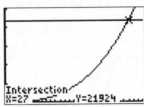

14. The graph the function. The window $0 \le x \le 70$ (scale 5), and $0 \le y \le 1000$ (scale 100) shows the maximum. Using the maximum routine we find the highest point is is when $x \approx 42.75$. Thus, the year would be 1942.

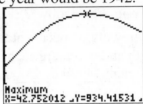

15. Let x be the measurement of the width. $2x$ is the measurement of the length and h is the height. Given a volume of 42.32 cubic feet, we have $x(2x)h = 42.32$ or

$$h = \frac{42.32}{2x^2} = \frac{21.16}{x^2},$$ where $h \ge 3$. The equation relating the dimensions to the cost to build this box would be as follows.

$$13(2x^2) + 2(9)(2x)h + 2(9)xh = 634.34 \Rightarrow 26x^2 + 36xh + 18xh = 634.34$$

$$26x^2 + 54xh = 634.34$$

Substituting for h, we have the following.

$$26x^2 + 54x\left(\frac{21.16}{x^2}\right) = 634.34 \Rightarrow 26x^2 + \frac{1142.64}{x} = 634.34$$

Graph $y = 26x^2 + \dfrac{1142.64}{x}$ and $y = 634.34$ in the same window. Since $\dfrac{21.16}{x^2} \ge 3$,

we have $x^2 \le \dfrac{21.16}{3} \approx 7.05.$ Since $x \ge 0$, we know that $0 \le x \le 2.66$ since

$\sqrt{7.05} \approx 2.66.$ Thus, an appropriate window would be $0 \le x \le 3$, and $0 \le y \le 700$ (scale 100). Using the intersection feature, we have $x = 2.3$.

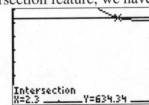

Thus, the dimensions are 2.3 ft by 4.6 ft by 4 ft high.

16. Let x be the height of the box. Each of the sides of the base measure $10 - 2x$. Since x is being cut from each end, we have the following restrictions on x.

$$0 < 10 - 2x < 6 \Rightarrow -10 < -2x < -4 \Rightarrow 2 < x < 5$$

  a. The equation that relates the measurements to the volume is $(10 - 2x)^2 x = 50$.

  Graph $y = (10 - 2x)^2 x$ and $y = 50$ in the same window. An appropriate window would be $2 \le x \le 5$, and $40 \le y \le 60$. Using the intersection feature, we have $x = 2.937$.

  Thus, the dimensions to be cut are 2.937 in. by 2.937 in.

  b. If we graph $y = (10 - 2x)^2 x$ in the window $2 \le x \le 5$, and $0 \le y \le 100$ (scale 10), we can see that the maximum volume occurs at the endpoint. Although $x < 2$ by the restriction that the length and width are each less than 6 in, rounded to three decimal places the measurement would be 2.000 in. by 2.000 in.

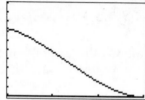

17. a.

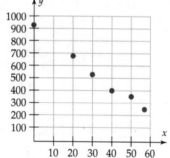

  b. Yes. It appears to be negative correlation.

  c. The linear regression model would be $y = -11.823x + 912.576$.

18. a. Here is a table of the residuals for the model.

$y = 4x + 131$

Data Point (x, p)	Model Point (x, y)	Residual $p - y$	Squared Residual $(p - y)^2$
(4,147.4)	(4,147)	.4	.16
(8,162.5)	(8,163)	-.5	.25
(10,171.3)	(10,171)	.3	.09
(12,179.8)	(12,179)	.8	.64

The sum of the residuals is 1.
The sum of the squares of the residuals is 1.14.

b. Here is a table of the residuals for the model.

$y = 4.1x + 131.5$

Data Point (x, p)	Model Point (x, y)	Residual $p - y$	Squared Residual $(p - y)^2$
(4,147.4)	(4,147.9)	-5	.25
(8,162.5)	(8,164.3)	-1.8	3.24
(10,171.3)	(10,172.5)	-1.2	1.44
(12,179.8)	(12,180.7)	-.9	.81

The sum of the residuals is -4.4.
The sum of the squares of the residuals is 5.74.

c. The model in part a is a better fit.

19. a. The linear regression equation would be $y = .3272x + 5.0259$.

b. Evaluating this model, we have the following:
For 2008, $x = 8$: $.3272(8) + 5.0259 = 7.6435$ which yields about 7.6 million barrels per day.
For 2018, $x = 18$: $.3272(18) + 5.0259 = 10.9155$ which yields about 10.9 million barrels per day.

20. a. The linear regression model would be $y = 19.2286x + 206.0952$.

b. Evaluating this model, we have the following:
For 2007, $x = 7$: $19.2286(7) + 206.0952 = 340.6954$ which yields about 340,700 vehicles.

c. About 19,200 cars per year.

Chapter 3
Functions and Graphs

3.1 Functions

1. This could be a table of values of a function, because each input determines one and only one output.

3. This could not be a table of values of a function, because two output values are associated with the input –5.

5. 6   **7.** –2   **9.** –17

11. This defines y as a function of x.

13. Since $x = \dfrac{y^2 - 1}{4}$, this defines x as a function of y.

15. Since $y = \dfrac{12 - 3x}{2}$ and $x = \dfrac{12 - 2y}{3}$ this defines both y as a function of x and x as a function of y.

17. Neither

19.

x	–2	–1.5	–1	–.5	0	0.5	1
y	–2	–3.25	–4	–4.25	–4	–3.25	–2
x	1.5	2	2.5	3	3.5	4	
y	–.25	2	4.75	8	11.75	16	

21.

79

23. $\$400 \;\text{—}\; \$0.021\,(400) = \$8.40$

$\$1{,}509 \;\text{—}\; \$0.021(1{,}509) = \$31.69$

$\$25{,}000 \;\text{—}\; \$693.75 + 0.048\,(25{,}000 - 25{,}000) = \$693.75$

$\$20{,}000 \;\text{—}\; \$262.50 + .0345\,(20{,}000 - 12{,}500) = \$521.25$

$\$12{,}500 \;\text{—}\; \$262.50 + .0345\,(12{,}500 - 12{,}500) = \$262.50$

$\$55{,}342 \;\text{—}\; \$693.75 + 0.048\,(55{,}342 - 25{,}000) = \$2150.17$

25. A function may assign the same output to many different inputs.

27. Postage is a function of weight, but weight is not a function of postage; for example, all letters less than one ounce use the same postage amount.

29. Since this would assign two numbers, –2 and 2, to the same input 4, it is not the rule of a function.

31. a. January 2000 — 8.5%.   January 2001 — 9.5%.   mid-2005 — 6%.
b. The prime rate was below 5% from late 2001 to late 2004.
c. The prime rate can be considered a function of time, but time cannot be considered a function of the prime rate, since there is more than one time associated with a particular prime rate.

33. a. $A = x^2$ **b.** Since $d^2 = x^2 + x^2 = 2x^2$, $A = \frac{d^2}{2}$.

35. $S = 2\pi r^2 + 2\pi rh$. Since $h = 2d = 4r$, subsitution yields
$S = 2\pi r^2 + 2\pi r\,(4r) = 2\pi r^2 + 8\pi r^2 = 10\pi r^2$

37. Cost = Fixed Cost + Variable Cost ; $C = 26{,}000 + 125x$

39. a. The TI-83 yields the following: **b.** The function derived in part **a** yields $\$40.9$ million in 2004.

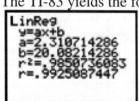

41. –2 — –2
  –1 — 0
  0 — $1\frac{1}{4}$
  1 — $2\frac{3}{4}$

43.   –2 — –1
   0 — 3
   1 — 2
   2.5 — –1
   –1.5 — 0

45. The graph yields: $f(-2) = 1$, $f(-1) = -2.9$, $f(0) = -1$, $f(1/2) = 1$, $f(1) = 1\frac{1}{2}$.

47. a. A calculator graph of $y = \cos(\ln x)$ in the standard window is shown.

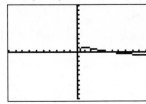

The largest possible domain is $(0, \infty)$ or $x > 0$.

b. The range of the function is $-1 \le y \le 1$, where $y = \cos(\ln x)$.

49. a. For non-negative numbers, the part of the number to the left of the decimal point (integer part) is the closest integer on the left side of the number.

b. For negative integers, the above statement also holds.

c. However, for negative real numbers that are not integers, for example, -1.5, the "integer part" will be 1 more than the greatest integer function, thus $\text{int}(-1.5) = -2$ while $\text{iPart}(-1.5) = -1$.

51. a.

X	Y_1
-5	-112
-11	-1306
8	499
7.2	361.85
-.44	3.7948

b. $10^3 - 2(10) + 3 = 983$

3.2 Functional Notation

1. Using $f(x) = \dfrac{x-3}{x^2+4}$, obtain the following table:

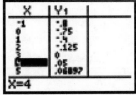

Thus, $f(-1) = -.8$, and so on.

3. $f(x) = \sqrt{x+3} - x + 1$
$f(0) = \sqrt{0+3} - 0 + 1 = \sqrt{3} + 1$

5. $f(x) = \sqrt{x+3} - x + 1$
$f(\sqrt{2}) = \sqrt{\sqrt{2}+3} - \sqrt{2} + 1$

7. $f(x) = \sqrt{x+3} - x + 1$

$f(-2) = \sqrt{-2+3} - (-2) + 1 = \sqrt{1} + 3 = 4$

9. $h(x) = x^2 + \dfrac{1}{x} + 2$

$h(-4) = (-4)^2 + \dfrac{1}{-4} + 2 = \dfrac{71}{4} = 17.75$

11. $h(x) = x^2 + \dfrac{1}{x} + 2$

$h(\pi + 1) = (\pi + 1)^2 + \dfrac{1}{\pi + 1} + 2$

$= \pi^2 + 2\pi + 3 + \dfrac{1}{\pi + 1}$

13. $h(x) = x^2 + \dfrac{1}{x} + 2$

$h(a + k) = (a + k)^2 + \dfrac{1}{a + k} + 2$

15. $h(x) = x^2 + \dfrac{1}{x} + 2$

$h(-x) = (-x)^2 + \dfrac{1}{(-x)} + 2 = x^2 - \dfrac{1}{x} + 2$

17. $h(x) = x^2 + \dfrac{1}{x} + 2$

$h(x - 3) = (x - 3)^2 + \dfrac{1}{x - 3} + 2 = x^2 - 6x + 9 + \dfrac{1}{x - 3} + 2 = x^2 - 6x + 11 + \dfrac{1}{x - 3}$

19. $g(t) = t^2 - 1$

$g(s + 1) = (s + 1)^2 - 1 = s^2 + 2s$

21. $g(t) = t^2 - 1$

$g(-t) = (-t)^2 - 1 = t^2 - 1$

23. $g(t) = t^2 - 1$

$g(3) = 3^2 - 1 = 8$

$f(8) = \sqrt{8 + 3} - 8 + 1 = \sqrt{11} - 7$

25. $f(x) = x + 1$

$\dfrac{f(x + h) - f(x)}{h} = \dfrac{(x + h + 1) - (x + 1)}{h} = \dfrac{h}{h} = 1$

27. $f(x) = 3x + 7$

$\dfrac{f(x + h) - f(x)}{h} = \dfrac{(3(x + h) + 7) - (3x + 7)}{h} = \dfrac{3h}{h} = 3$

29. $f(x) = x - x^2$

$\dfrac{f(x + h) - f(x)}{h} = \dfrac{\left((x + h) - (x + h)^2\right) - \left(x - x^2\right)}{h} = \dfrac{x + h - x^2 - 2xh - h^2 - x + x^2}{h}$

$= \dfrac{h - 2xh - h^2}{h} = 1 - 2x - h$

31. $f(x) = \sqrt{x}$

$\dfrac{f(x + h) - f(x)}{h} = \dfrac{\sqrt{x + h} - \sqrt{x}}{h} = \dfrac{1}{\sqrt{x + h} + \sqrt{x}}$

33. $f(x) = x^2 + 3$

$$\frac{f(x+h) - f(x)}{h} = \frac{\left((x+h)^2 + 3\right) - \left(x^2 + 3\right)}{h} = \frac{x^2 + 2xh + h^2 + 3 - x^2 - 3}{h}$$

$$= \frac{2xh + h^2}{h} = 2x + h$$

35. a. $f(x) = x^2$, $f(a) = a^2$, $f(b) = b^2$, $f(a+b) = (a+b)^2 = a^2 + 2ab + b^2$
Since $(a+b)^2 \neq a^2 + b^2$, $f(a+b) \neq f(a) + f(b)$.
b. $f(x) = 3x$, $f(a) = 3a$, $f(b) = 3b$, $f(a+b) = 3(a+b)$
Since $3(a+b) = 3a + 3b$, $f(a+b) = f(a) + f(b)$.
c. $f(x) = 5$, $f(a) = 5$, $f(b) = 5$, $f(a+b) = 5$
Since $10 \neq 5$, $f(a+b) \neq f(a) + f(b)$.

37. Since $2 = f(1)$, we have
$$f(x) = x^3 + cx^2 + 4x - 1$$
$$2 = 1^3 + c \cdot 1^2 + 4 \cdot 1 - 1$$
$$2 = 4 + c$$
$$c = -2$$

39. a. $-3 \leq x \leq 4$
b. $-2 \leq y \leq 3$
c. -2
d. $\frac{1}{2}$
e. 1
f. -1

41. This is the absolute value function.

43. a. $x \leq 20$
b. Since $-3 < 2$, $f(-3) = (-3)^2 + 2(-3) = 3$
c. Since $-1 < 2$, $f(-1) = (-1)^2 + 2(-1) = -1$
d. Since $2 \leq 2 \leq 20$, $f(2) = 3 \cdot 2 - 5 = 1$
e. Since $2 \leq 7/3 \leq 20$, $f(7/3) = 3(7/3) - 5 = 2$

45. All real numbers

47. All real numbers

49. $x \geq 0$

51. All real numbers except 0

53. All real numbers

55. If $u^2 - u - 6 = 0$, then $(u-3)(u+2) = 0$, thus $u = 3$ or $u = -2$.
Domain: all real numbers except 3 and –2.

57. We must have $9 - (x-9)^2 \geq 0$, thus $|x-9| \leq 3$ and $6 \leq x \leq 12$. Domain: $6 \leq x \leq 12$

59. $f(x) = x^2$ and $g(x) = x^4$ are two of many examples.

61. $g(x) = x^3$ is one of many examples.

63. $f(x) = g(x)$

$2x^2 + 13x - 14 = 8x - 2$

$2x^2 + 5x - 12 = 0$

$(2x - 3)(x + 4) = 0$

$x = \dfrac{3}{2}$ or $x = -4$

65. $f(x) = g(x)$

$2x^2 - x + 1 = x^2 - 4x + 4$

$x^2 + 3x - 3 = 0$

$x = \dfrac{-3 \pm \sqrt{3^2 - 4(1)(-3)}}{2} = \dfrac{-3 \pm \sqrt{21}}{2}$

67. $f(x) = 8 - 3x^2$

69. Let x = number of pounds sold

Cost = Fixed Cost + Variable Cost = $1800 + .5x$

Revenue = $1.2x$

Profit = Revenue – Cost = $P(x) = 1.2x - (1800 + .5x) = .7x - 1800$

Thus, the profit is a function of the number of pounds x.

71. **a.** Length = $38 + .72y$

b. Let $x = y + 50$, then $y = x - 50$. Hence length = $f(x) = 38 + .72(x - 50)$, that is $f(x) = .72x + 2$.

73. Since the distance is determined by the time, the distance is a function of the time.

$$d(t) = \begin{cases} 55t & \text{if } 0 \le t \le 2 \\ 20 + 45t & \text{if } t > 2 \end{cases}$$

75. **a.** Since girth = $2x + 2x = 4x$, $y + 4x = 108$, $y = 108 - 4x$.

b. Volume = length × width × height = $(108 - 4x)(x)(x) = 108x^2 - 4x^3$

77. Let h be the height of the box. Then $t^2 h = 10$, $h = \dfrac{10}{t^2}$. The total cost is given by

$$C = .85t^2 + 4(.50th) + 1.15t^2 = 2t^2 + 2t\left(\dfrac{10}{t^2}\right) = 2t^2 + \dfrac{20}{t}$$

79. **a.**

```
LinReg
y=ax+b
a=18.30434783
b=244.4347826
r2=.9182345515
r=.9582455591
■
```

b. $g(8) = 391$ stations; $g(11) = 446$ stations.

The first figure is higher than actual; the second, lower.

c. $g(10) = 427$ stations

d. $g(21) = 629$ stations

3.3 Graphs of Functions

1. This could be the graph of a function, since it passes the vertical line test. $f(3) = 0$.

3. This could not be the graph of a function, since it does not pass the vertical line test.

5.

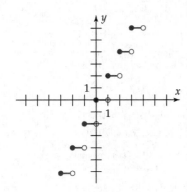

7.

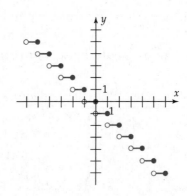

9.

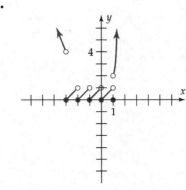

11. The graph fails the vertical line test, hence it cannot be the graph of a function. For example, $x = 3$ corresponds to both 5 and 9.

13. $[-3, 5]$

15. $[-3, 3)$

17. $g(2 + 1.5) = g(3.5) = 3.5$

19. $g(2) + 1.5 = 3 + 1.5 = 4.5$

21. $\alpha = 1$ and 5

23. $h(x) = \begin{cases} \frac{1}{2}x - 2 \text{ if } x \geq 0 \\ -\frac{1}{2}x - 2 \text{ if } x < 0 \end{cases}$

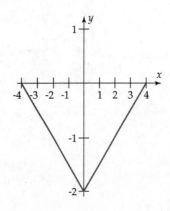

25. On the interval $[0,2]$, $0 \le x \le 2$ hence $|x| + |x-2| = x + [-(x-2)] = x - x + 2 = 2$.

27. Graph the function using the window $-4 \le x \le 4, -6 \le y \le 0$. The local maxima are at the endpoints $(\pm 4, 0)$; use the minimum routine (if necessary) to find the local minimum at the point $(0, -4)$.

29. Graph the function using the window $-3 \le x \le 3, -3 \le y \le 5$. Use the maximum routine to find the local maximum at the point $(-1, 3)$ and the minimum routine to find the local minimum at the point $(1, -1)$.

31. There is no local maximum or minimum because the graph is increasing $(-\infty, \infty)$.

33. Graph the function in the window $-10 \le x \le 5, -120 \le y \le 20$

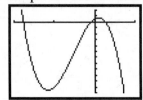
Using the minimum and maximum routines, we see that the function is decreasing on the intervals $(-\infty, -5.8)$ and $(.46, \infty)$. It is increasing on the interval $(-5.8, .46)$.

35. Graph the function in the window $-10 \le x \le 10, -10 \le y \le 10$

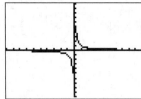
The function is decreasing $(-\infty, 0)$ and $(0, \infty)$.

37. a. $2x + 2z = 100$

b. From the above equation, $z = 50 - x$, hence area $= f(x) = xz = x(50 - x)$.

c. Graph the function in the window $0 \le x \le 50, 0 \le y \le 1000$ and use the maximum routine to find the local maximum at the point $(25, 625)$. Then the maximum area occurs when $x = 25$ inches, $z = 50 - x = 25$ inches.

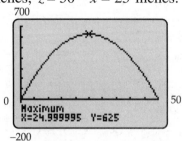

39. a. The surface area $S = \text{top} + \text{bottom} + \text{four sides} = 2x^2 + 4xh$.

   b. Since Volume = length × width × height, $867 = x^2h$, hence $h = 867/x^2$.

   c. Using parts a and b, we can write
   $$S = 2x^2 + 4xh = 2x^2 + 4x(867/x^2) = 2x^2 + 3468/x$$

   d. Graph the function in the window $0 \le x \le 25, 0 \le y \le 1000$ and use the minimum routine to find the local minimum at the point $(9.5354, 545.545)$. Therefore the minimum volume occurs when $x = 9.5354$ inches. Then the height
   $$h = 867/(9.5354)^2 = 9.5354 \text{ inches also.}$$

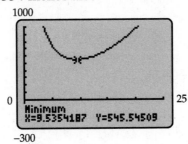

41. a — iv
     b — i
     c — v
     d — iii
     e — ii

43. A possible graph is given below. The domain is $x \ge 0$, since the oven will continue to exist for some indefinite time. The range is $50 \le y \le 350$.

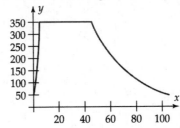

45. a. The point $(k, f(k))$ is on the function directly above k.

   b. The point $(-k, f(-k))$ is on the function directly above $-k$, which is on the x-axis opposite k.

   c. The point $(k, -f(k))$ is the mirror image of the point $(k, f(k))$ in the x-axis.

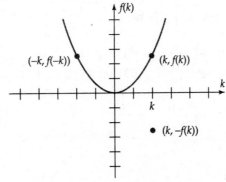

47. a. 9%; 16%; 8%
   b. lowest: 2003, highest: 1990.
   c. between 1980 and 1982; steepness of curve

49. a. $T(4 \cdot 3) = T(12) = 67°$ but $T(4) \cdot T(3) = 39 \cdot 42 = 1638$. False
   b. $T(4 \cdot 3) = T(12) = 67°$ but $4 \cdot T(3) = 4 \bullet 42° = 168°$. False
   c. $T(4+14) = T(18) = 74°$ but $T(4) + T(14) = 39° + 72° = 111°$. False

51.

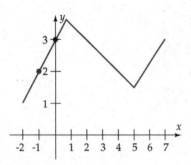

53. a.

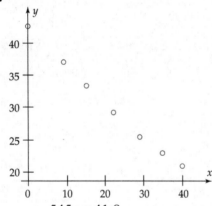

   b. $y = -.545x + 41.8$
   c. Evaluating this model, we have the following:
   For 1991, $x = 26$: $-.545(26) + 41.8 = 27.63$ which yields about 27.7% if you round up to the next tenth of a percent.
   For 2013, $x = 48$: $-.545(48) + 41.8 = 15.64$ which yields about 15.7% if you round up to the next tenth of a percent.

   d. Solving $15 = -.545x + 41.8$ we get $x \approx 49.17$. This implies the year 2014.
   e. It will disappear completely after 2042 according to this model.

55.

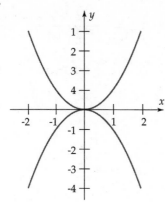

57. a. $(-\infty, 4]$

b. $[2, \infty)$

c.

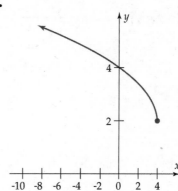

59. After 15 minutes she took a break then picked up her pace for 10 minutes. After 30 minutes she jogged back home at a constant rate for a total jog of 55 minutes.

3.3.A Parametric Graphing

1. Use the window
$-6 \le x \le 44$ (scale 4), $0 \le y \le 16$.

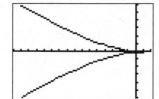

3. Use the window $-2 \le x \le 32$
(scale 2), $-60 \le y \le 60$ (scale 10)

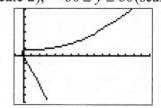

5. Use the window $-16 \le x \le 2$,
$-60 \le y \le 60$ (scale 10).

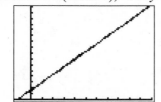

7. Graph $x = t^3 + 5t^2 - 4t - 5$, $y = t$ in
the window $-50 \le x \le 50$
(scale 10), $-7 \le y \le 3, -7 \le t \le 3$.

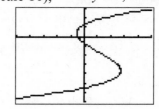

9. First, solve for x.

$$xy^2 + xy + x = y^3 - 2y^2 + 4 \Rightarrow x(y^2 + y + 1) = y^3 - 2y^2 + 4 \Rightarrow x = \frac{y^3 - 2y^2 + 4}{y^2 + y + 1}$$

Now graph $x = \dfrac{t^3 - 2t^2 + 4}{t^2 + t + 1}$, $y = t$ in the window $-8 \le x \le 5, -5 \le y \le 5, -5 \le t \le 5$.

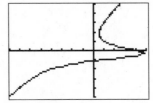

11. First, solve for x to obtain $x = \sqrt{y} - y^2 - 8$. Now graph $x = \sqrt{t} - t^2 - 8$, $y = t$ in the window $-10 \le x \le -7$, $0 \le y \le 2, 0 \le t \le 2$.

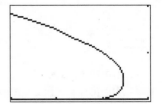

13. It crosses itself 6 times.

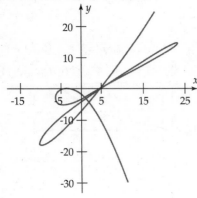

3.4 Graphs and Transformations

1. H   **3.** F   **5.** K   **7.** C

9.

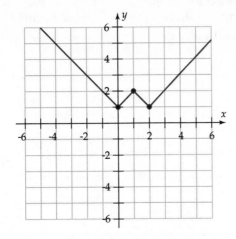

11.

t	$f(t)$	$g(t) = f(t) - 3$	$h(t) = 4f(-t)$	$i(x) = f(t-1) - 2$
−2	3	0	20	Not possible
−1	6	3	0	1
0	8	5	32	4
1	0	−3	24	6
2	5	2	12	−2

13.

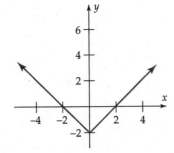

15.

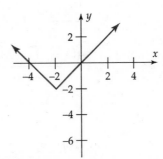

17. $-15 \le x \le 15, \ -12 \le y \le 10$

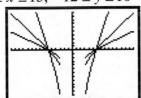

19. $-5 \le x \le 7, \ -10 \le y \le 10$

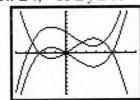

21.

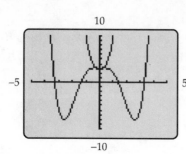

23. Shift the graph of f 2 units horizontally to the left, then 5 units vertically up.

25. Reflect across the x-axis, then stretch vertically by a factor of 2, then shift up 10 units.

27. $g(x) = -x^2 + x + 2$

29. $g(x) = -\frac{1}{2}\sqrt{-(x+3)}$

31. **a.** $g(x) = (x-1)^2 + 5 = x^2 - 2x + 1 + 5 = x^2 - 2x + 6$

   b. $f(x) = x^2 + 5$

$$\frac{f(x+h) - f(x)}{h} = \frac{\left((x+h)^2 + 5\right) - \left(x^2 + 5\right)}{h} = \frac{x^2 + 2xh + h^2 + 5 - x^2 - 5}{h}$$

$$= \frac{2xh + h^2}{h} = 2x + h$$

$$\frac{g(x+h) - g(x)}{h} = \frac{\left((x+h)^2 - 2(x+h) + 6\right) - \left(x^2 - 2x + 6\right)}{h}$$

$$= \frac{x^2 + 2xh + h^2 - 2x - 2h + 6 - x^2 + 2x - 6}{h}$$

$$= \frac{2xh + h^2 - 2h}{h}$$

$$= 2x + h - 2$$

   c. $2x + h - 2 = 2(x-1) + h = d(x-1)$

33.

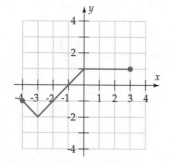

35.

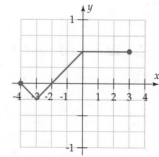

37.

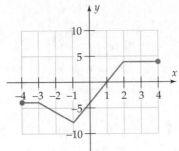

39.

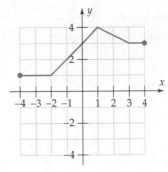

41.

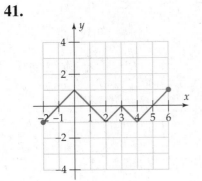

43.

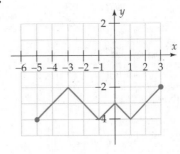

45.

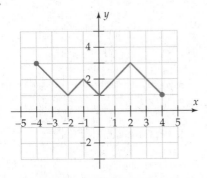

47. a. This shifts the graph vertically by 35 units up.

b. The graph stretches in the y-direction.

49.

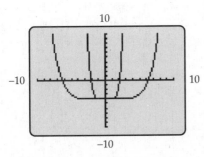

51. The graph of $f(cx), c > 1$, is the graph of $f(x)$ contracted toward the y-axis by a factor of c.

53.

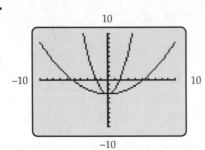

55.

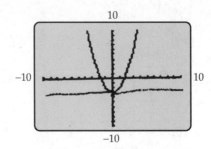

57.

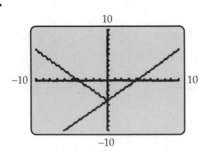

59.

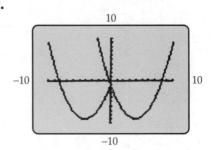

61. For positive x, $f(x)$ and $f(|x|)$ have the same graph; for negative x,
$f(|x|) = f(-x)$.

63.

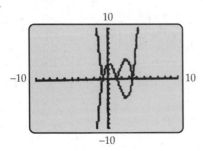

65.

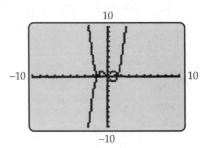

67.

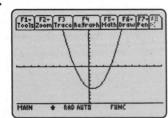

3.4.A Symmetry

1.

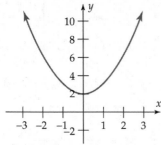

The graph has y-axis symmetry.

3.

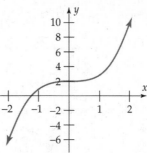

The graph has none of the three symmetries.

5. $f(-x) = 4(-x) = -4x = -f(x)$. Therefore, the function is odd.

7. $f(-x) = (-x)^2 - |-x| = x^2 - |x| = f(x)$. Therefore, the function is even.

9. $k(-t) = (-t)^4 - 6(-t)^2 + 5 = t^4 - 6t^2 + 5 = k(t)$. Therefore, the function is even.

11. $f(-x) = -x\left[(-x)^4 - (-x)^2\right] + 4(-x)$

$= -x(x^4 - x^2) - 4x = -\left[x(x^4 - x^2) + 4x\right] = -f(x)$.

Therefore, the function is odd.

13. $h(-x) = \sqrt{7 - 2(-x)^2} = \sqrt{7 - 2x^2} = h(x)$. Therefore, the function is even.

Neither even nor odd.

15. $g(-x) = \dfrac{(-x)^2 + 1}{(-x)^2 - 1} = \dfrac{x^2 + 1}{x^2 - 1} = g(x)$. Therefore, the function is even.

17. Replace y with $-y$:

$x^2 - 6x + (-y)^2 + 8 = 0$

$x^2 - 6x + y^2 + 8 = 0$

Equation is unchanged, graph has x-axis symmetry.

19. Replace y with $-y$:

$x^2 - 2x + (-y)^2 + 2(-y) = 2$

$x^2 - 2x + y^2 - 2y = 2$

Equation is changed, graph has no y-axis symmetry.

21. The graph has origin symmetry.

23. The graph has origin symmetry.

25. The graph has y-axis symmetry.

27. The graph has no symmetry.

29.

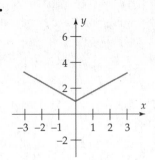

31.

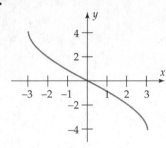

33. a.

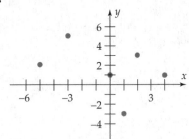

b.

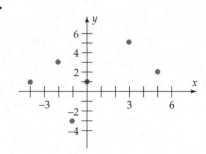

35. a.

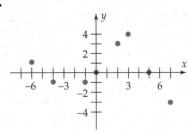

b.

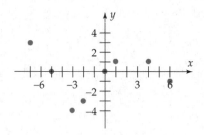

c. One possibility is shown below:

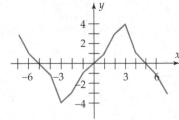

37. If replacing x by $-x$ does not change the graph, and replacing y by $-y$ does not change the graph, then doing both does not change the graph. Thus x-axis plus y-axis symmetry guarantees origin symmetry.

If replacing x by $-x$ does not change the graph, and replacing y by $-y$ DOES change the graph, then doing both must change the graph.

Thus y-axis symmetry and origin symmetry without x-axis symmetry is impossible.

The argument is analogous for x-axis and origin symmetry.

39. a. No, because as the advertising budget increases, so do the sales at which is an increasing function which is not a property of even functions.

b. Yes, because as the advertising budget increases, so do the sales which is an increasing function which is a property of odd functions.

3.5 Operations on Functions

1. $(f+g)(x) = f(x)+g(x) = (-3x+2)+x^3 = x^3 - 3x + 2$

$(f-g)(x) = f(x)-g(x) = (-3x+2)-x^3 = -x^3 - 3x + 2$

$(g-f)(x) = g(x)-f(x) = x^3 -(-3x+2)+ = x^3 + 3x - 2$

3. $(f+g)(x) = f(x)+g(x) = \dfrac{1}{x} + x^2 + 2x - 5$

$(f-g)(x) = f(x)-g(x) = \dfrac{1}{x} - (x^2 + 2x - 5) = \dfrac{1}{x} - x^2 - 2x + 5$

$(g-f)(x) = g(x)-f(x) = (x^2 + 2x - 5) - \dfrac{1}{x} = x^2 + 2x - 5 - \dfrac{1}{x}$

5. $(fg)(x) = f(x)g(x) = (-3x+2)x^3$

$\qquad\qquad\qquad = -3x^4 + 2x^3$

$\left(\dfrac{f}{g}\right)(x) = \dfrac{f(x)}{g(x)} = \dfrac{-3x+2}{x^3}$

$\left(\dfrac{g}{f}\right)(x) = \dfrac{g(x)}{f(x)} = \dfrac{x^3}{-3x+2}$

7. $(fg)(x) = f(x)g(x) = (x+5)(x-5)$

$\qquad\qquad\qquad = x^2 - 25$

$\left(\dfrac{f}{g}\right)(x) = \dfrac{f(x)}{g(x)} = \dfrac{x+5}{x-5}$

$\left(\dfrac{g}{f}\right)(x) = \dfrac{g(x)}{f(x)} = \dfrac{x-5}{x+5}$

9. The domain of f consists of all real numbers.

The domain of g consists of all real $x \neq 0$. $g(x)$ is never 0.

Therefore the domain of both fg and f/g is all real $x \neq 0$.

11. The domain of f is $-2 \leq x \leq 2$.

The domain of g is $x \geq -\frac{4}{3}$.

Therefore the domain of fg is $-\frac{4}{3} \leq x \leq 2$.

$g\left(-\frac{4}{3}\right) = 0$, therefore the domain of f/g is $-\frac{4}{3} < x \leq 2$.

13.
$$f(x)=1+x$$
$$f(0)=1+0=1$$
$$g(t)=t^2-t$$
$$g(f(0))=g(1)$$
$$=1^2-1=0$$

15.
$$f(x)=1+x$$
$$f(2)=1+2=3$$
$$g(t)=t^2-t$$
$$g(f(2)+3)=g(3+3)$$
$$=g(6)=6^2-6=30$$

17.
$$f(x)=3x-2$$
$$f(3)=3\cdot3-2$$
$$=7$$
$$g(x)=x^2$$
$$(g\circ f)(3)=g(f(3))$$
$$=g(7)$$
$$=7^2$$
$$=49$$

$$g(1)=1^2$$
$$=1$$
$$(f\circ g)(1)=f(g(1))$$
$$=f(1)$$
$$=3\cdot1-2$$
$$=1$$

$$f(0)=3\cdot0-2$$
$$=-2$$
$$(f\circ f)(0)=f(f(0))$$
$$=f(-2)$$
$$=3(-2)-2$$
$$=-8$$

19.
$$f(x)=x$$
$$f(3)=3$$
$$g(x)=-3$$
$$(g\circ f)(3)=g(f(3))$$
$$=g(3)=-3$$

$$g(1)=-3$$
$$(f\circ g)(1)=f(g(1))$$
$$=f(-3)$$
$$=-3$$

$$f(0)=0$$
$$(f\circ f)(0)=f(f(0))$$
$$=f(0)$$
$$=0$$

21.
$$f(x)=-3x+2$$
$$g(x)=x^3$$
$$(f\circ g)(x)=f(g(x))$$
$$=-3g(x)+2=-3x^3+2$$
Domain: all real numbers

$$(g\circ f)(x)=g(f(x))$$
$$=(f(x))^3$$
$$=(-3x+2)^3$$
Domain: all real numbers

23.
$$f(x) = \frac{1}{2x+1}$$
$$g(x) = x^2 - 1$$

$$(f \circ g)(x) = f(g(x)) = \frac{1}{2g(x)+1}$$

$$= \frac{1}{2(x^2-1)+1} = \frac{1}{2x^2-1}$$

Domain: all $x \neq \pm\sqrt{2}/2$

$$(g \circ f)(x) = g(f(x))$$
$$= (f(x))^2 - 1$$
$$= \left(\frac{1}{2x+1}\right)^2 - 1$$
$$= \frac{-4x^2-4x}{4x^2+4x+1}$$

Domain: all $x \neq -1/2$

25. $f(x) = x^3$

$$(ff)(x) = f(x)f(x) = x^3 \cdot x^3 = x^6$$

$$(f \circ f)(x) = f(f(x)) = (f(x))^3 = (x^3)^3 = x^9$$

27. $f(x) = \frac{1}{x}$

$$(ff)(x) = f(x)f(x) = \frac{1}{x} \cdot \frac{1}{x} = \frac{1}{x^2}$$

$$(f \circ f)(x) = f(f(x)) = \frac{1}{f(x)} = \frac{1}{1/x} = x$$

29. $(f \circ g)(x) = f(g(x))$
$$= 9g(x) + 8$$
$$= 9\left(\frac{x-8}{9}\right) + 8 = x - 8 + 8 = x$$

$$(g \circ f)(x) = g(f(x))$$
$$= \frac{f(x)-8}{9} = \frac{9x+8-8}{9} = \frac{9x}{9} = x$$

31. $(f \circ g)(x) = f(g(x))$
$$= \sqrt[3]{g(x)} + 2$$
$$= \sqrt[3]{(x-2)^3} + 2$$
$$= x - 2 + 2 = x$$

$$(g \circ f)(x) = g(f(x))$$
$$= (f(x)-2)^3$$
$$= (\sqrt[3]{x}+2-2)^3 = (\sqrt[3]{x})^3 = x$$

33.

x	$f(x)$	$g(x) = f(f(x))$
–4	–3	–1
–3	–1	1/2
–2	0	1
–1	1/2	5/4
0	1	3/2
1	3/2	2
2	1	3/2
3	–2	0
4	–2	0

35.

x	$(g \circ f)(x)$
1	4
2	2
3	5
4	4
5	4

37.

x	$(f \circ f)(x)$
2	3
3	3
4	5
5	1

39. Let $g(x) = x^2 + 2$, $h(t) = \sqrt[3]{t}$, then $(h \circ g)(x) = h(g(x)) = \sqrt[3]{g(x)} = \sqrt[3]{x^2 + 2}$.

41. Let $g(x) = 7x^3 - 10 + 17$, $f(t) = t^7$, then $(f \circ g)(x) = f(g(x)) = (7x^3 - 10x + 17)^7$.

43. Let $h(x) = x^2 + 2x$, $k(t) = t + 1$, then $(k \circ h)(x) = k(h(x)) = x^2 + 2x + 1$.

or

Let $h(x) = x + 2$, $k(t) = (t-1)^2$, then $(k \circ h)(x) = k(h(x)) = (h(x) - 1)^2 = x^2 + 2x + 1$.

45. $(f \circ g)(x) = (g(x))^3 = (\sqrt{x})^3$ Defined, domain $x \geq 0$

$(g \circ f)(x) = \sqrt{f(x)} = \sqrt{x^3}$ Defined, domain $x \geq 0$

47. $(f \circ g)(x) = \sqrt{g(x) + 10} = \sqrt{5x + 10}$ Defined, domain $x \geq -2$

$(g \circ f)(x) = 5f(x) = 5\sqrt{x + 10}$ Defined, domain $x \geq -10$

49. $f(x) = 2x^3 + 5x - 1$

a. $f(x^2) = 2(x^2)^3 + 5x^2 - 1 = 2x^6 + 5x^2 - 1$

b. $(f(x))^2 = (2x^3 + 5x - 1)^2 = 4x^6 + 20x^4 - 4x^3 + 25x^2 - 10x + 1$

c. The answers are not the same. In general, $f(x^2) \neq (f(x))^2$.

51. $(f \circ g)(x) = f(g(x)) = (x-2)^5 - (x-2)^3 - (x-2)$;

$(g \circ f)(x) = g(f(x)) = x^5 - x^3 - x - 2$

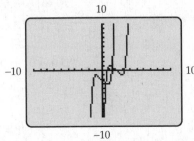

Graphing in the standard window shows that the two functions are not the same.

53. $f(x) = x + 3 \quad g(x) = x^2 + 1$

$(g \circ f)(x) = g(f(x)) = (f(x))^2 + 1 = (x+3)^2 + 1 = x^2 + 6x + 10$

$$\frac{(g \circ f)(x+h) - (g \circ f)(x)}{h} = \frac{\left((x+h)^2 + 6(x+h) + 10\right) - \left(x^2 + 6x + 10\right)}{h}$$

$$= \frac{x^2 + 2xh + h^2 + 6x + 6h + 10 - x^2 - 6x - 10}{h}$$

$$= \frac{2xh + h^2 + 6h}{h} = 2x + h + 6$$

55. $f(x) = x + 1 \quad g(x) = \dfrac{2}{x-1} \qquad (g \circ f)(x) = \dfrac{2}{f(x)-1} = \dfrac{2}{x+1-1} = \dfrac{2}{x}$

$$\frac{(g \circ f)(x+h) - (g \circ f)(x)}{h} = \frac{\dfrac{2}{x+h} - \dfrac{2}{x}}{h} \cdot \frac{x(x+h)}{x(x+h)}$$

$$= \frac{2x - 2(x+h)}{h(x)(x+h)}$$

$$= \frac{2x - 2x - 2h}{h(x)(x+h)}$$

$$= \frac{-2h}{h(x)(x+h)}$$

$$= \frac{-2}{x(x+h)}$$

57. a. $A = \pi\left(\dfrac{18}{2t+3}\right)^2$

After 1 day, $t = 1440$ min, $A = \pi\left(\dfrac{18}{2\cdot 1440+3}\right)^2 = 1.22\times 10^{-4}$ sq. in.

After 1 week, $t = 10{,}080$ min, $A = \pi\left(\dfrac{18}{2\cdot 10{,}080+3}\right)^2 = 2.5\times 10^{-6}$ sq. in.

After 1 month = 30 days, $t = 43{,}200$ min,

$$A = \pi\left(\dfrac{18}{2\cdot 43{,}200+3}\right)^2 = 1.36\times 10^{-7} \text{ sq. in.}$$

b. In this model, the puddle never totally evaporates. This is not realistic, however, the model may be valid over a certain time period.

59. a. $A = \dfrac{\pi}{4}d^2 = \dfrac{\pi}{4}\left(6-\dfrac{50}{t^2+10}\right)^2$

b. At $t = 0$, $A = \dfrac{\pi}{4}\left(6-\dfrac{50}{0^2+10}\right)^2 = .7854$ sq. in.

At $t = 8$, $A = \dfrac{\pi}{4}\left(6-\dfrac{50}{8^2+10}\right)^2 = 22.265$ sq. in.

When $A = 25$, we must solve $25 = \dfrac{\pi}{4}\left(6-\dfrac{50}{t^2+10}\right)^2$.

Using the calculator intersection routine, we obtain $t = 11.4$ weeks.

61. Since $r = 4t$ and $V = 4\pi r^3/3$, we have $V(t) = \dfrac{4\pi(4t)^3}{3} = \dfrac{256\pi t^3}{3}$

At $t = 4$, $V(4) = \dfrac{256\pi 4^3}{3} = 17{,}157$ cm^3

63. From similar triangles, we have

$$\dfrac{s}{6} = \dfrac{s+d}{15} \Rightarrow 15s = 6s+6d \Rightarrow 9s = 6d \Rightarrow s = 2d/3$$

Using distance = rate $\times$ time, $d = 5t$, hence
$$s(t) = 2(5t)/3 = 10t/3.$$

65. One example is $f(x) = |x|$. Another example is the function $f(x) = \dfrac{x-3}{x-2}, x \neq 1$.

Note that the domain of this function excludes 1 and 2. In fact, all functions of form $f(x) = \dfrac{x+a}{bx-2}$, where $ab = -3$, satisfy the condition, if the necessary values are excluded from the domain.

67. a. As the composition is applied you will stabilize at −.1708.

```
0²−.2
                -.2
Ans²−.2
               -.16
             -.1744
         -.16958464
       -.1712410499
```
```
-.1712410499
-.1706765028
-.1708695314
-.1708036032
-.1708261291
-.1708184336
-.1708210627
```

b. Yes.

```
1²−.2
                 .8
Ans²−.2
                .44
             -.0064
         -.19995904
       -.1600163823
```
```
-.1600163823
-.1743947574
-.1695864686
-.1712404297
-.1706767152
-.1708694589
-.170803628
```

c. Now we wind up oscillating between −.8873 and −.1127 (starting at either 0 or 1).

```
0²−.9
                -.9
Ans²−.9
               -.09
             -.8919
        -.10451439
       -.8890767423
```
```
-.8890767423
-.1095425463
-.8880004305
-.1114552354
-.8875777305
-.1122057723
-.8874098647
```
```
-.8874098647
-.1125037321
-.8873429103
-.1126225596
-.8873161591
-.1126700339
-.8873054635
```

```
1²−.9
                 .1
Ans²−.9
               -.89
             -.1079
        -.88835759
       -.1108207923
```
```
-.1108207923
-.887718752
-.1119554174
-.8874659845
-.1124041263
-.8873653124
-.1125828024
```
```
-.1125828024
-.8873251126
-.1126541445
-.8873090437
-.1126826609
-.8873026179
-.1126940642
```

d. We wind up cycling between four numbers this time.

```
0²−1.3
               -1.3
Ans²−1.3
                .39
            -1.1479
          .01767441
       -1.299687615
```
```
-1.299687615
 .3891878972
-1.148532781
 .0191275483
-1.299634137
 .3890488898
-1.148640961
```
```
-1.148640961
 .0193760581
-1.299624568
 .3890240187
-1.148660313
 .0194205143
-1.299622844
```

```
1²−1.3
                -.3
Ans²−1.3
              -1.21
              .1641
        -1.27307119
        .3207102548
```
```
 .3207102548
-1.197144932
 .1331559893
-1.282269483
 .3442150258
-1.181516016
 .0959800961
```
```
 .0959800961
-1.290787821
 .3661331992
-1.16594648
 .0594311952
-1.296467933
 .3808291014
```

3.6 Rates of Change

1.   average speed $= \dfrac{\text{distance traveled}}{\text{time interval}}$

   a. $\dfrac{d(10)-d(0)}{10-0} = \dfrac{140-0}{10-0} = 14$ ft per sec

   b. $\dfrac{d(20)-d(10)}{20-10} = \dfrac{680-140}{20-10} = 54$ ft per sec

   c. $\dfrac{d(30)-d(20)}{30-20} = \dfrac{1800-680}{30-20} = 112$ ft per sec

   d. $\dfrac{d(30)-d(15)}{30-15} = \dfrac{1800-400}{30-15} = 93.3$ ft per sec

3.   **a.** average rate of change $= \dfrac{f(1996)-f(1994)}{1996-1994} = \dfrac{44,043-24,134}{2} = 9954.5$

The number of cell phone users was increasing at the rate of 9,954,500,000 users per year.

   b. average rate of change $= \dfrac{f(2004)-f(1999)}{2004-1999} = \dfrac{182,140-86,047}{5} = 19,218.6$

The number of cell phone users was increasing at the rate of 19,218,600,000 users per year.

5.   **a.** average rate of change $= \dfrac{f(1985)-f(1980)}{1985-1980} = \dfrac{39,422-40,877}{5} = -291$

Decreasing at the rate of 291,000 per year.

   b. average rate of change $= \dfrac{f(1995)-f(1985)}{1995-1985} = \dfrac{44,840-39,422}{10} = 541.8$

Increasing at the rate of 541,800 per year.

   c. average rate of change $= \dfrac{f(2005)-f(1995)}{2005-1995} = \dfrac{48,375-44,840}{10} = 353.5$

Increasing at the rate of 353,500 per year.

   d. average rate of change $= \dfrac{f(2014)-f(2005)}{2014-2005} = \dfrac{49,993-48,375}{9} \approx 179.8$

Increasing at the rate of 179,800 per year.

   e. Fastest: 1985 - 1995. Slowest: 2005 - 2014.

7. a. average rate of change $= \dfrac{f(1999)-f(1995)}{1999-1995} = \dfrac{203.9-87.2}{4} = 29.18$

Increasing at the rate of 29.18 billion shares per year.

b. average rate of change $= \dfrac{f(2001)-f(1999)}{2001-1999} = \dfrac{307.5-203.9}{2} = 51.8$

Increasing at the rate of 51.8 billion shares per year.

c. average rate of change $= \dfrac{f(2005)-f(2001)}{2005-2001} = \dfrac{403.8-307.5}{4} = 24.08$

Increasing at the rate of 24.08 billion shares per year.

d. average rate of change $= \dfrac{f(2005)-f(1995)}{2005-1995} = \dfrac{403.8-87.2}{10} = 31.66$

Increasing at the rate of 31.66 billion shares per year.

e. 1999 - 2001

9. a. average rate of change $= \dfrac{f(20)-f(10)}{20-10} = \dfrac{100-50}{10} = 5$: \$5000 per page

b. average rate of change $= \dfrac{f(60)-f(20)}{60-20} = \dfrac{175-100}{40} = 1.875$: \$1875 per page

c. average rate of change $= \dfrac{f(100)-f(60)}{100-60} = \dfrac{200-175}{40} = .625$: \$625 per page

d. average rate of change $= \dfrac{f(100)-f(0)}{100-0} = \dfrac{200-25}{100} = 1.75$: \$1750 per page

e. The average rate of change between 70 and 80 pages is:

$\dfrac{f(80)-f(70)}{80-70} = \dfrac{190-180}{10} = 1$: \$1000 per page.

Therefore, it is only worthwhile to buy more than 70 pages of ads if the cost of a page is less than \$1000.

11. $\dfrac{f(2)-f(0)}{2-0} = \dfrac{11-3}{2} = 4$

13. $\dfrac{f(3)-f(-1)}{3-(-1)} = \dfrac{-18-10}{4} = -7$

15. $\dfrac{f(1.01)-f(1)}{1.01-1} = \dfrac{1.417921-1.414214}{.01} = \dfrac{.003707}{.01} = .371$

17. $\dfrac{f(6)-f(3)}{8-3} = \dfrac{5.0833-3}{5} = .417$

19. $\dfrac{f(x+h)-f(x)}{h} = \dfrac{(7(x+h)+2)-(7x+2)}{h} = 7$

21. $\dfrac{f(x+h)-f(x)}{h} = \dfrac{\left((x+h)^2 + 3(x+h)-1\right)-(x^2+3x-1)}{h}$

$$= \dfrac{x^2 + 2xh + h^2 + 3x + 3h - 1 - x^2 - 3x + 1}{h} = 2x + h + 3$$

23. $\dfrac{V(x+h)-V(x)}{h} = \dfrac{(x+h)^3 - x^3}{h} = \dfrac{x^3 + 3x^2h + 3xh^2 + h^3 - x^3}{h} = 3x^2 + 3xh + h^2$

25. $\dfrac{V(p+h)-V(p)}{h} = \dfrac{\dfrac{5}{p+h} - \dfrac{5}{p}}{h} = \dfrac{5p - 5p - 5h}{hp(p+h)} = \dfrac{-5}{p(p+h)}$

27. From Exercise 23, average rate of change between x and $x+h$ is $3x^2 + 3xh + h^2$

    a. Here $x=4$, $h=.1$, thus, average rate $= 3 \cdot 4^2 + 3 \cdot 4(.1) + .1^2 \approx 49.2$

    b. Here $x=4$, $h=.01$, thus, average rate $= 3 \cdot 4^2 + 3 \cdot 4(.01) + (.01)^2 \approx 48.1$

    c. Here $x=4$, $h=.001$, thus, average rate $= 3 \cdot 4^2 + 3 \cdot 4(.001) + (.001)^2 \approx 48$

    d. 48

29. From Exercise 25, average rate of change between p and $p+h$ is $-\dfrac{5}{p(p+h)}$.

    If h is set to 0, this becomes $-5/p^2$.

    When $p=50$, $\text{rate} = -\dfrac{5}{50^2} = -\dfrac{5}{2500} = -\dfrac{1}{500}$.

31. a. Since the amount of change is equal for all three years, the average rates of change over the equal intervals are equal.

    b. December 2008

33. a. $\dfrac{36-31}{6/12} = 10$

    b. Answers will vary depending on interpretation of the data from the graph. One possible answer is -15.

    c. Answers will vary depending on interpretation of the data from the graph. One possible answer is 5.

35. a. The linear equation that models this data is $y = 13.58x + 184.95$.

    b. The average rate of change is the slope of the line: 13.58.

    c. 1980-1990: $\dfrac{f(10)-f(0)}{10-0} = \dfrac{319-191}{10} = 12.8$;

        2000-2005: $\dfrac{f(25)-f(20)}{25-20} = \dfrac{538-454}{5} = 16.8$

    These rates are somewhat larger than the overall rate found in b.

    d. Solve $13.58x + 184.95 = 600$ to obtain $x = 30.56$, which would correspond to 2010.

3.7 Inverse Functions

1.

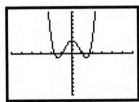

This fails the horizontal line test.
The function is not one-to-one.

3.

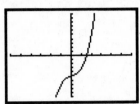

This passes the horizontal line test.
The function is one-to-one.

5.

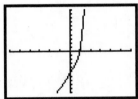

This passes the horizontal line test.
The function is one-to-one.

7.

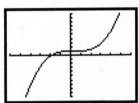

This fails the horizontal line test.
The function is not one-to-one.

9. Set $y = -x$

$$x = -y$$
$$g(y) = -y$$
$$g(x) = -x$$

11. Set $y = 5x - 4$

$$y + 4 = 5x$$
$$x = \frac{y+4}{5}$$
$$g(y) = \frac{y+4}{5}$$
$$g(x) = \frac{x+4}{5}$$

13. Set $y = 5 - 2x^3$

$$y - 5 = -2x^3$$
$$\frac{5-y}{2} = x^3$$
$$x = \sqrt[3]{\frac{5-y}{2}}$$
$$g(y) = \sqrt[3]{\frac{5-y}{2}}$$
$$g(x) = \sqrt[3]{\frac{5-x}{2}}$$

15. Set $y = \sqrt{4x - 7}$ $(y \geq 0)$

$$y^2 = 4x - 7$$
$$y^2 + 7 = 4x$$
$$x = \frac{y^2 + 7}{4}$$
$$g(y) = \frac{y^2 + 7}{4} \quad (y \geq 0)$$
$$g(x) = \frac{x^2 + 7}{4} \quad (x \geq 0)$$

17. Set $y = \dfrac{1}{x}$

$x = \dfrac{1}{y}$

$g(y) = \dfrac{1}{y}$

$g(x) = \dfrac{1}{x}$

19. Set $y = \dfrac{1}{2x+1}$

$2xy + y = 1$

$2xy = 1 - y$

$x = \dfrac{1-y}{2y}$

$g(y) = \dfrac{1-y}{2y}$

$g(x) = \dfrac{1-x}{2x}$

21. Set $y = \dfrac{x^3 - 1}{x^3 + 5}$

$x^3 y + 5y = x^3 - 1$

$5y + 1 = x^3 - x^3 y$

$5y + 1 = x^3(1 - y)$

$x^3 = \dfrac{5y+1}{1-y}$

$x = \sqrt[3]{\dfrac{5y+1}{1-y}}$

$g(y) = \sqrt[3]{\dfrac{-5y-1}{x-1}}$

$g(x) = \sqrt[3]{\dfrac{-5x-1}{x-1}}$

23. $g(f(x)) = g(x+1) = x + 1 - 1 = x$

$f(g(x)) = f(x-1) = x - 1 + 1 = x$

Thus g is the inverse of f.

25. $g(f(x)) = g\left(\dfrac{1}{x+1}\right) = \dfrac{1 - \dfrac{1}{x+1}}{\dfrac{1}{x+1}} = \dfrac{x+1-1}{1} = x$

$f(g(x)) = f\left(\dfrac{1-x}{x}\right) = \dfrac{1}{\dfrac{1-x}{x}+1} = \dfrac{x}{1-x+x} = x$

Thus g is the inverse of f.

27. $g(f(x)) = g(x^5) = \sqrt[5]{x^5} = x$

$f(g(x)) = f(\sqrt[5]{x}) = (\sqrt[5]{x})^5 = x$

Thus g is the inverse of f.

29. Calculate $f(f(x))$:

$$f(f(x)) = \frac{2f(x)+1}{3f(x)-2} = \frac{2 \cdot \dfrac{2x+1}{3x-2} + 1}{3 \cdot \dfrac{2x+1}{3x-2} - 2} = \frac{4x+2+3x-2}{6x+3-6x+4} = \frac{7x}{7} = x$$

Since $f(f(x)) = x$, f is the inverse of itself.

31. a. 48
   b. 2
   c. 5
   d. 12
   e. Not enough information
   f. Not enough information
   g. Not enough information

33.

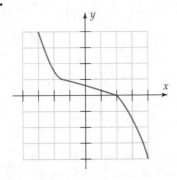

35. Sketch $x = \sqrt{3t-2}$, $y = t$, $\frac{2}{3} \le t \le 10$ in parametric mode:

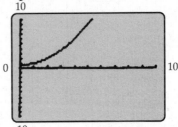

37. Sketch $x = \sqrt[3]{t+3}$, $y = t$, $-10 \le t \le 10$ in parametric mode:

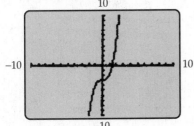

39. The inverse function is given by:

$$y = \begin{cases} -\sqrt{x+1} & x \geq -1 \\ -2(x+1) & x < -1 \end{cases}$$

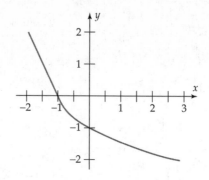

41. Let $h(x) = -x^2,\ x \geq 0$. Then set

$$y = -x^2,\ x \geq 0,\ y \leq 0$$

$$-y = x^2$$

$$x = \sqrt{-y}$$

$$g(y) = \sqrt{-y},\ y \leq 0$$

$$g(x) = \sqrt{-x},\ x \leq 0$$

is the inverse function.

43. Let $h(x) = \dfrac{1}{x^2+1},\ x \geq 0$. Then set

$$y = \frac{1}{x^2+1},\ x \geq 0,\ 0 < y \leq 1$$

$$x^2 + 1 = \frac{1}{y}$$

$$x^2 = \frac{1}{y} - 1$$

$$x = \sqrt{\frac{1}{y} - 1}$$

$$g(y) = \sqrt{\frac{1}{y} - 1},\ 0 < y \leq 1$$

$$g(x) = \sqrt{\frac{1}{x} - 1},\ 0 < x \leq 1$$

is the inverse function.

45. a.   Set $y = 3x + 2$.

$$3x = y - 2$$

$$x = \frac{y-2}{3}$$

$$f^{-1}(y) = \frac{y-2}{3}$$

$$f^{-1}(x) = \frac{x-2}{3}$$

b. $f^{-1}(1) = \dfrac{1-2}{3} = -\dfrac{1}{3}$

$$\frac{1}{f(1)} = \frac{1}{3(1)+2} = \frac{1}{5}$$

Therefore f^{-1} is not the same as $\dfrac{1}{f}$.

47. First note that f is one-to-one, since if $a \neq b,\ ma \neq mb$ and $ma + b \neq mb + b$. To find the inverse function, set

$$y = mx + b$$

$$y - b = mx$$

$$x = \frac{y-b}{m}$$

$$g(y) = \frac{y-b}{m}$$

$$g(x) = \frac{x-b}{m} \text{ is the inverse function.}$$

49. a. slope $= \dfrac{a-b}{b-a} = -1$

b. Since the slope of $y = x$ is 1, and $(-1)1 = -1$, the two lines are perpendicular.

c. Length of $PR = \sqrt{(c-a)^2 + (c-b)^2}$.

Length of $RQ = \sqrt{(b-c)^2 + (a-c)^2}$.

Clearly, these are equal, since $(b-c)^2 = (c-b)^2$ and $(c-a)^2 = (a-c)^2$.
Thus, using the result from part b, $y = x$ is perpendicular to PQ and bisects PQ.
$y = x$ is the perpendicular bisector of PQ and thus P and Q are symmetric with respect to $y = x$.

51. Let g be the inverse function of f. Then if $f(a) = f(b)$, $g(f(a)) = g(f(b))$.
Applying the round-trip properties, $a = b$. Thus f obeys the definition of a one-to-one function.

53. True. Rotating the graph of an increasing function over $y = x$ gives an increasing function.

Chapter 3 Review Exercises

1. a. $[-5/2] = [-2.5] = -3$ **b.** $[1755] = 1755$

c. $[18.7] + [-15.7] = 18 + (-16) = 2$ **d.** $[-7] - [7] = -7 - 7 = -14$

3.

x	0	1	2	-4	t	k	$b-1$	$1-b$	$6-2u$
$f(x)$	7	5	3	15	$7-2t$	$7-2k$	$9-2b$	$5+2b$	$4u-5$

5. Many examples are possible:

a. Let $f(x) = x+1, a=1, b=2$ **b.** Let $f(x) = x+1, a=1, b=2$

$f(a+b) = f(3) = 4$ $f(ab) = f(2) = 3$

$f(a) + f(b) = 2 + 3 = 5$ $f(a)f(b) = 2 \cdot 3 = 6$

7. $r \geq 4$

9. $h(x) = x^2 - 3x$
$h(t+2) = (t+2)^2 - 3(t+2) = t^2 + 4t + 4 - 3t - 6 = t^2 + t - 2$

11. $f(x) = 2x^3 + x + 1$

$f\left(\dfrac{x}{2}\right) = 2\left(\dfrac{x}{2}\right)^3 + \dfrac{x}{2} + 1 = \dfrac{x^3}{4} + \dfrac{x}{2} + 1 = \dfrac{x^3 + 2x + 4}{4}$

13. a. $f(t) = 50\sqrt{t}$

b. $g(t) = \pi r^2 = \pi\left(f(t)\right)^2 = \pi\left(50\sqrt{t}\right)^2 = 2500\pi t$

c. radius: $f(9) = 50\sqrt{9} = 150$ meters

area: $g(9) = 2500\pi(9) = 22{,}500\pi$ square meters

d. Solve $100{,}000 = 2500\pi t$

$$t = \frac{100{,}000}{2500\pi}$$

$$t \approx 12.7$$

After approximately 12.7 hours.

15.

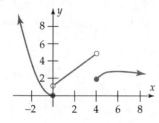

17. a fails the vertical line test. **b** is the graph of a function of x.

19. Graph the function using the window $-10 \le x \le 10, -1 \le y \le 10$ and use the minimum routine to find the local minimum at $x = -.5$. Then the function is decreasing on the interval $(-\infty, -.5)$ and increasing on the interval $(-.5, \infty)$.

21. Graph the function using the window $-8 \le x \le 3, -10 \le y \le 60$ and use the minimum routine to find the local minimum at $x = -.263$ and the maximum routine to find the local maximum at $x = -5.0704$. Then the function is increasing on the intervals $(-\infty, -5.0704)$ and $(-.263, \infty)$ and decreasing on the interval $(-5.0704, -.263)$.

23.

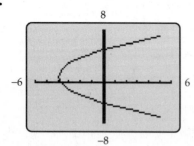

25. Here is one possibility:

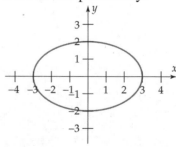

27. Replace x with $-x$:

$$\left(-x\right)^2 = y^2 + 2$$
$$x^2 = y^2 + 2$$

Equation unchanged;
graph has y-axis symmetry.

Replace y with $-y$:

$$x^2 = \left(-y\right)^2 + 2$$
$$x^2 = y^2 + 2$$

Equation unchanged;
graph has x-axis symmetry.

Since the graph has both x- and y-axis symmetry, it has origin symmetry.

29. $g(-x) = 9 - \left(-x\right)^2 = 9 - x^2 = g(x)$

Therefore, the function is even.

31. $h\left(-x\right) = 3\left(-x\right)^5 - \left(-x\right)\left(\left(-x\right)^4 - \left(-x\right)^2\right)$

$h\left(-x\right) = -3x^5 + x\left(x^4 - x^2\right) \neq h\left(x\right)$

Therefore, the function is not even.

$-h\left(x\right) = -\left(3x^5 - x\left(x^4 - x^2\right)\right) = -3x^5 + x\left(x^4 - x^2\right) = h\left(-x\right)$

Therefore, the function is odd.

33. Replace x with $-x$:

$$\left(-x\right)^2 + y^2 + 6y = -5$$
$$x^2 + y^2 + 6y = -5$$

Equation is unchanged;
graph has y-axis symmetry.

Replace y with $-y$:

$$x^2 + \left(-y\right)^2 + 6\left(-y\right) = -5$$
$$x^2 + y^2 - 6y = -5$$

Equation is changed;
graph has no x-axis symmetry.

Since the graph has y-axis symmetry but not x-axis symmetry, it cannot have origin symmetry.

35.

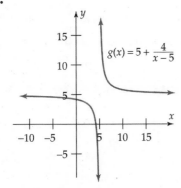

$g(x) = 5 + \dfrac{4}{x-5}$

37. $-3 \le y \le 4$

39. $f(3) < f(2)$, thus 2 is a possible answer.   Many correct answers are possible including any number from 2 to 3.5 or 4.5 to 6.

41. 1

43. $f(-1) = 0, f(1) = -3$
$f(-1) + f(1) = -3$

45. True. $3f(2) = -3$
$= -f(4)$

47. 4   **49.** $x \le 3$ **51.** $x < 3$

53. a. King Richard made 2 pitstops while Fireball Bob made 1.
   b. King Richard started out faster (greater slope).
   c. Fireball Bob won (finished at an earlier time).

55. Shrink vertically toward the x-axis by a factor of .25, then shift vertically 2 units up.

57. Shift horizontally 7 units to the right, then stretch vertically away from the x-axis by a factor of 3, then reflect in the x-axis, then shift vertically 2 units up.

59. e (only)

61. a. $f(2) = \dfrac{1}{2-1} = 1 \quad g(2) = \sqrt{2^2+5} = \sqrt{9} = 3.$ Hence $(f/g)(2) = \dfrac{f(2)}{g(2)} = \dfrac{1}{3}$

b. $(g/f)(x) = \dfrac{g(x)}{f(x)} = \dfrac{\sqrt{x^2+5}}{1/(x-1)} = (x-1)\sqrt{x^2+5} \quad (x \ne 1)$

c. $(fg)(c+1) = f(c+1)g(c+1) = \dfrac{1}{c+1-1}\sqrt{(c+1)^2+5} = \dfrac{\sqrt{(c+1)^2+5}}{c}, c \ne 0$

63.

x	-4	-3	-2	-1	0	1	2	3	4
$g(x)$	1	4	3	1	-1	-3	-2	-4	-3
$h(x) = g(g(x))$	-3	-3	-4	-3	1	4	3	1	4

65. Since $f(2) = \dfrac{1}{2+1} = \dfrac{1}{3}, \quad (g \circ f)(2) = g(f(2)) = g\left(\dfrac{1}{3}\right) = \left(\dfrac{1}{3}\right)^3 + 3 = \dfrac{82}{27}.$

67. Since $f(x-1) = \dfrac{1}{(x-1)+1} = \dfrac{1}{x}$, we have the following.

$(g \circ f)(x-1) = g(f(x-1)) = (f(x-1))^3 + 3 = \left(\dfrac{1}{x}\right)^3 + 3 = \dfrac{1}{x^3} + 3$

69. Since $g(1) = 1^3 + 3 = 4, \quad f(g(1)-1) = f(4-1) = f(3) = \dfrac{1}{3+1} = \dfrac{1}{4}.$

71. $(f \circ g)(x) = f(g(x)) = \dfrac{1}{g(x)} = \dfrac{1}{x^2-1}$

$(g \circ f)(x) = g(f(x)) = (f(x))^2 - 1 = \left(\dfrac{1}{x}\right)^2 - 1 = \dfrac{1}{x^2} - 1$

73. The domain of $f \circ g$ is those numbers in the domain of g for which $f\big(g(x)\big)$ is defined. The domain of g is $x \geq 0$. $f\big(g(x)\big)$ defined for all of these except 1. The domain of $f \circ g$ is $x \geq 0, x \neq 1$.

75. a. $g(-1) = \dfrac{(-1)^3 - (-1) + 1}{(-1) + 2} = \dfrac{1}{1} = 1$ $g(1) = \dfrac{(1)^3 - 1 + 1}{1 + 2} = \dfrac{1}{3}$

average rate of change $= \dfrac{g(1) - g(-1)}{1 - (-1)} = \dfrac{\frac{1}{3} - 1}{2} = -\dfrac{1}{3}$

b. $g(2) = \dfrac{2^3 - 2 + 1}{2 + 2} = \dfrac{7}{4}$ $g(0) = \dfrac{0^3 - 0 + 1}{0 + 2} = \dfrac{1}{2}$

average rate of change $= \dfrac{g(2) - g(0)}{2 - 0} = \dfrac{\frac{7}{4} - \frac{1}{2}}{2} = \dfrac{5}{8}$

77. $(f \circ g)(x) = 2g(x) + 1 = 2(3x - 2) + 1 = 6x - 3$

$(f \circ g)(5) = 6 \cdot 5 - 3 = 27$ $(f \circ g)(3) = 6 \cdot 3 - 3 = 15$

average rate of change $= \dfrac{(f \circ g)(5) - (f \circ g)(3)}{5 - 3} = \dfrac{27 - 15}{2} = 6$

79. $\dfrac{f(x + h) - f(x)}{h} = \dfrac{(3(x + h) + 4) - (3x + 4)}{h} = \dfrac{3x + 3h + 4 - 3x - 4}{h} = \dfrac{3h}{h} = 3$

81. $\dfrac{g(x + h) - g(x)}{h} = \dfrac{\left((x + h)^2 - 1\right) - (x^2 - 1)}{h} = \dfrac{x^2 + 2xh + h^2 - 1 - x^2 + 1}{h}$

$= \dfrac{2xh + h^2}{h} = 2x + h$

83. $P(4) = .2(4)^2 + .5(4) - 1 = 4.2$

a. $P(8) = .2(8)^2 + .5(8) - 1 = 15.8$

average rate of change $= \dfrac{P(8) - P(4)}{8 - 4} = \dfrac{15.8 - 4.2}{4} = 2.9$ or \$290 per ton.

b. $P(5) = .2(5)^2 + .5(5) - 1 = 6.5$

average rate of change $= \dfrac{P(5) - P(4)}{5 - 4} = \dfrac{6.5 - 4.2}{1} = 2.3$ or \$230 per ton.

c. $P(4.1) = .2(4.1)^2 + .5(4.1) - 1 = 4.412$

average rate of change $= \dfrac{P(4.1) - P(4)}{4.1 - 4} = \dfrac{4.412 - 4.2}{.1} = 2.12$ or \$212 per ton.

85. **a.** approximately 40 to 45
  b. approximately 25 to 35
  c. approximately 30 to 44

87. Let t = time, P = price

  a. $\dfrac{P(1998)-P(1985)}{1998-1985} = \dfrac{164,800-95,400}{13} \approx \$5338$ per year

  $\dfrac{P(2001)-P(1998)}{2001-1998} = \dfrac{194,500-164,800}{3} = \$9900$ per year

  $\dfrac{P(2001)-P(1985)}{2001-1985} = \dfrac{194,500-95,400}{16} = \$6194$ per year

  b. Four years of change at \$9900 per year = \$39,600. Adding this to the price in 2001 yields $\$194,500 + \$39,600 = \$234,100$ as the (projected) median price.

89. Set $y = \sqrt{5-x} + 7 \quad (y \geq 7)$

  $\sqrt{5-x} = y - 7$

  $5 - x = (y-7)^2$

  $x = 5 - (y-7)^2$

  $g(y) = 5 - (y-7)^2,$

  $g(x) = 5 - (x-7)^2, \quad x \geq 7$

  $g(x) = 5 - (x^2 - 14x + 49), \quad x \geq 7$

  $g(x) = 5 - x^2 + 14x - 49, \quad x \geq 7$

  $g(x) = -x^2 + 14x - 44, \quad x \geq 7$

91.

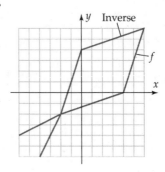

93. The graph of f passes the horizontal like test and hence has an inverse function. It is easy to verify either geometrically [by reflecting the graph of f in the line $y = x$] or algebraically [by calculating $f(f(x))$] that f is its own inverse function.

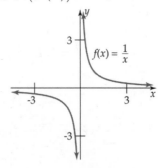

95. There is no inverse function because the graph of f fails the horizontal line test (use the viewing window with $-10 \le x \le 20$ and $-200 \le y \le 100$ (scale 10).

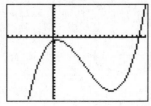

Chapter 3 Test

1. $A(r) = \pi r^2$

2.

$f(x) - 3$ $\dfrac{1}{3} f(x)$ $f(-x)$

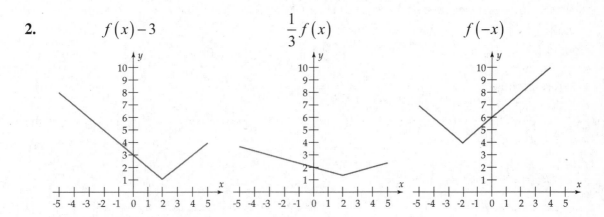

3. **a.** Domain: $-12 \le x \le 12$ or $\mathbb{R}$
    b. $-2 \le y < -1, 0 \le y \le 6$
    c. $g(0) = 6$

4. It is a function of x because every input has exactly one output. Answers invoking the vertical line test are also acceptable.

5.
$$\frac{f(x+h) - f(x)}{h} = \frac{\left(\dfrac{2}{x+h} + 3\right) - \left(\dfrac{2}{x} - 3\right)}{h} = \frac{\dfrac{2}{x+h} + 3 - \dfrac{2}{x} + 3}{h} = \frac{\dfrac{2}{x+h} - \dfrac{2}{x}}{h}$$

$$= \frac{\dfrac{2x}{x(x+h)} - \dfrac{2(x+h)}{x(x+h)}}{h} = \frac{2x - 2x - 2h}{hx(x+h)} = \frac{-2h}{hx(x+h)} = -\frac{2}{x(x+h)}$$

6.

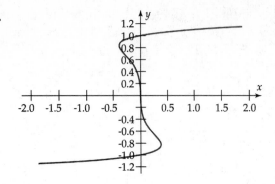

7.   a. $s(0) = 25;$ If the child has read no books, the score will be 25.

$s^{-1}(35) = 30;$ If the child wants to score 35, s/he must read 30 books.

b. Solving $45 = 15 + \sqrt{10b + 100}$, we get $b = 80$. They must read 80 books

8.   a. The origin
b. The x-axis
c. None of these

9. If you shift f up 7 units and to the left 5 units, the graph of f will transform into the graph of g.

10. $h(3) = -16 \cdot 3^2 + 500 = 356$

a. Since $h(3.1) = -16 \cdot 3.1^2 + 500 = 346.24,$ we have the following.

$$\frac{h(3.1) - h(3)}{3.1 - 3} = \frac{346.24 - 356}{.1} = -97.6\text{, thus we have 97.6 feet/second.}$$

b. Since $h(3.01) = -16 \cdot 3.01^2 + 500 = 355.0384,$ thus we have the following.

$$\frac{h(3.01) - h(3)}{3.01 - 3} = \frac{355.0384 - 356}{.01} = -96.16\text{, thus we have 96.2 feet/second.}$$

c. Estimated as 96 feet/second.

11. We would calculate the following.

$$\frac{\text{Volume in April } - \text{ Volume in January}}{\text{Number of months between April and January}} = \frac{108.24 - 110.15}{3} \approx -.63667$$

Thus, we have $-.63667$ million shares/month

12.

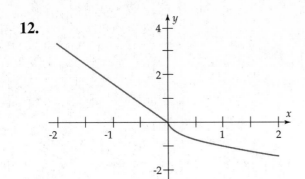

13. $(f \circ g)(x) = f(g(x))$

$\qquad = f(x^5 + 1)$

$\qquad = \sqrt[5]{(x^5 + 1) - 1}$

$\qquad = \sqrt[5]{x^5}$

$\qquad = x$

$(g \circ f)(x) = g(f(x))$

$\qquad = g(\sqrt[5]{x - 1})$

$\qquad = (\sqrt[5]{x - 1})^5 + 1$

$\qquad = x - 1 + 1$

$\qquad = x$

Chapter 4
Polynomial and Rational Functions

4.1 Quadratic Functions

1. The parabola opens upward and has vertex $(0,2)$. Graph **I.**

3. The parabola opens upward and has vertex $(2,0)$. Graph **K.**

5. The parabola opens upward and has vertex $(2,2)$. Graph **J.**

7. The parabola opens downward and has vertex $(-2,2)$. Graph **F.**

9. Vertex: $(5,2)$. Since $a=3$, opens upward.

11. Vertex: $(\sqrt{2},\pi)$. Since $a=-1$, opens downward.

13. Complete the square:
$$f(x)=2x^2-16x+29$$
$$=2(x^2-8x)+29$$
$$=2(x^2-8x+16)+29-2(16).$$
$$=2(x-4)^2+29-32$$
$$=2(x-4)^2-3$$
Vertex: $(4,-3)$. Since $a=2$, opens upward.

15. Complete the square:
$$h(x)=x^2-3x+1=x^2-3x+\frac{9}{4}+1-\frac{9}{4}=\left(x-\frac{3}{2}\right)^2+\frac{4}{4}-\frac{9}{4}=\left(x-\frac{3}{2}\right)^2-\frac{5}{4}$$
Vertex: $\left(\frac{3}{2},-\frac{5}{4}\right)$. Since $a=1$, opens upward.

17. Complete the square:
$$y=-4x^2+8x-1$$
$$=-4(x^2-2x)-1$$
$$=-4(x^2-2x+1)-1+4(1)$$
$$=-4(x-1)^2-1+4$$
$$=-4(x-1)^2+3$$
Vertex: $(1,3)$. Since $a=-4$, opens downward.

19. Vertex: $(0,3)$. Since $a = 2$, opens upward.

21. a. $\dfrac{f(x+h)-f(x)}{h} = \dfrac{\left(-3(x+h)^2 + (x+h)\right) - \left(-3x^2 + x\right)}{h}$

$= \dfrac{\left(-3\left(x^2 + 2xh + h^2\right) + (x+h)\right) - \left(-3x^2 + x\right)}{h}$

$= \dfrac{-3x^2 - 6xh - 3h^2 + x + h + 3x^2 - x}{h}$

$= \dfrac{-6xh - 3h^2 + h}{h} = -6x - 3h + 1$

b. Complete the square:

$f(x) = -3x^2 + x = -3\left(x^2 - \dfrac{1}{3}x\right) = -3\left(x^2 - \dfrac{1}{3}x + \dfrac{1}{36}\right) + 3\left(\dfrac{1}{36}\right) = -3\left(x - \dfrac{1}{6}\right)^2 + \dfrac{1}{12}$

Vertex: $\left(\dfrac{1}{6}, \dfrac{1}{12}\right)$

c. Substitute $\dfrac{1}{6}$ for x in $-6x - 3h + 1$ to obtain $-6\left(\dfrac{1}{6}\right) - 3h + 1 = -3h$.

23. a. $\dfrac{f(x+h)-f(x)}{h} = \dfrac{\left(-2(x+h)^2 + 2(x+h) - 1\right) - \left(-2x^2 + 2x - 1\right)}{h}$

$= \dfrac{-2x^2 - 4xh - 2h^2 + 2x + 2h - 1 + 2x^2 - 2x + 1}{h}$

$= \dfrac{-4xh - 2h^2 + 2h}{h} = -4x - 2h + 2$

b. Complete the square:

$f(x) = -2x^2 + 2x - 1$

$= -2\left(x^2 - x\right) - 1$

$= -2\left(x^2 - x + \dfrac{1}{4}\right) - 1 + 2\left(\dfrac{1}{4}\right)$

$= -2\left(x - \dfrac{1}{2}\right)^2 - \dfrac{2}{2} + \dfrac{1}{2}$

$= -2\left(x - \dfrac{1}{2}\right)^2 - \dfrac{1}{2}$

Vertex: $\left(\dfrac{1}{2}, -\dfrac{1}{2}\right)$

c. Substitute $\dfrac{1}{2}$ for x in $-4x - 2h + 2$ to obtain $-4\left(\dfrac{1}{2}\right) - 2h + 2 = -2h$.

25. Rule: $g(x) = 2x^2 - 3$ Vertex: $(0, -3)$

27. Since $\frac{1}{2}\left[h(x+5)-4\right]=f(x)=x^2$, we

can write: $h(x+5)-4=2x^2$

$h(x+5)=2x^2+4$

$h(x)=2(x-5)^2+4$

Vertex: $(5,4)$

29. Since the vertex is at $(0,0)$, the rule must be of form $f(x)=ax^2$.
Since $(2,12)$ is on the graph, $f(2)=12$, thus

$$12=a(2)^2$$

$$a=3$$

The rule is $f(x)=3x^2$.

31. Since the vertex is at $(3,4)$, the rule must be of form $f(x)=a(x-3)^2+4$.

Since $(-3,76)$ is on the graph, $f(-3)=76$, thus

$$76=a(-3-3)^2+4$$

$$72=36a$$

$$a=2$$

The rule is $f(x)=2(x-3)^2+4$ or $f(x)=2x^2-12x+22$.

33. Looking at the configuration of the points, you can see that $(1,4)$ is the vertex of the

parabola. Using the vertex, you can write the rule in the form $f(x)=a(x-1)^2+4$.
Either of the remaining points may be used to find a.

$$5=a(0-1)^2+4$$

$$5=a+4 \qquad \text{The rule is } f(x)=(x-1)^2+4.$$

$$1=a$$

35. Looking at the configuration of the points, you can see that $(0,6)$ is the vertex of the

parabola. Using the vertex, you can write the rule in the form $f(x)=ax^2+6$. You
may use either of the remaining points to find a.

$$7=a(-1)^2+6$$

$$7=a+6 \qquad \text{The rule is } f(x)=x^2+6.$$

$$1=a$$

37. Since the x-coordinate of the vertex satisfies $-b/(2a)=-b/(2(1))=-b/2=0, b=0$.

39. Since the x-coordinate of the vertex is $x = -b/2a$, we have

$$2 = -\frac{b}{2(1)} \Rightarrow b = -4.$$

Since $f(2) = 4$, we have

$$f(x) = x^2 - 4x + c \Rightarrow 4 = 2^2 - 4(2) + c \Rightarrow c = 8$$

41. Let x be the first number and y be the second number. Since the sum of the numbers must be 111 we have $x + y = 111$ or $y = 111 - x$. We would like to maximize the product of the numbers given by the following function.

$$f(x) = xy = x(111 - x) = 111x - x^2 = -x^2 + 111x$$

Since this function is a quadratic, the maximum occurs at the vertex.
The x-coordinate of the vertex is

$$x = \frac{-b}{2a} = \frac{-111}{2(-1)} = \frac{111}{2}$$

The second number y can be found by the following.

$$y = 111 - x = 111 - \frac{111}{2} = \frac{222}{2} - \frac{111}{2} = \frac{111}{2}$$

43. a. $p(50) = 1600(50) - 4(50)^2 - 50,000 = \$20,000$

$p(250) = 1600(250) - 4(250)^2 - 50,000 = \$100,000$

b. The maximum profit occurs at the vertex of the parabola that is the graph of $p(x)$. Since $p(x) = -4x^2 + 1600x - 50,000$ has vertex at

$x = -b/(2a) = -\frac{1600}{2(-4)} = 200,$ 200 units will maximize the profit. Then

$p(200) = 1600(200) - 4(200)^2 - 50,000 = \$110,000$.

c. The change from profit to loss occurs when $p(x) = 0$, thus we need to solve the equation $p(x) = 1600x - 4x^2 - 50,000$. We can do this with the quadratic formula.

$$x = \frac{-b \pm \sqrt{b^2 - 4ac}}{2a}$$

$$= \frac{-1600 \pm \sqrt{1600^2 - 4(-4)(-50000)}}{2(-4)}$$

$$= \frac{-1600 \pm \sqrt{2560000 - 800000}}{-8}$$

$$\approx 34.17 \text{ or } 365.83$$

The most units that we can sell and still make a profit is 365.

45. The maximum height occurs at the vertex of the parabola that is the graph of $p(x)$. Since the vertex is at $x = -b/2a = -\frac{.23}{2(-.0000167)} = 6886$ feet, the maximum height is given by $p(6886) = -.0000167(6886)^2 + .23(6886) + 50 = 842$ feet. The shell hits the water when $p(x) = 0$. Solve

$$-.0000167x^2 + .23x + 50 = 0$$

$$x = \frac{-.23 \pm \sqrt{(.23)^2 - 4(-.0000167)(50)}}{2(-.0000167)} = \frac{-.23 \pm .2371497}{-.0000334}$$

$$x = -214.06 \quad \text{or} \quad 13986.52$$

The shell hits the water 13,987 feet, or 2.65 miles, away.

47. a. $B(30) = .01(30)^2 + .7(30) = 30$ meters; $B(100) = .01(100)^2 + .7(100) = 170$ meters
b. Substitute 60 for $B(s)$ and solve for s.

$$60 = .01s^2 + .7s$$

$$0 = .01s^2 + .7s - 60$$

$$s = \frac{-.7 \pm \sqrt{(.7)^2 - 4(.01)(-60)}}{2(.01)}$$

$$s = -120 \quad \text{or} \quad 50$$

50 kilometers per hour

49. Substitute 80 for v_0 and 96 for h_0.
$$h = -16t^2 + 80t + 96$$
The maximum height occurs when
$$t = -b/(2a) = -\frac{80}{2(-16)} = 2.5 \text{ sec.}$$

At that time
$$h = -16(2.5)^2 + 80(2.5) + 96 = 196 \text{ ft.}$$

The ball reaches the maximum height of 196ft 2.5 seconds after being thrown.

51. Substitute 11 for v_0 and 5 for h_0.
$$h = -16t^2 + 11t + 5$$
The maximum height occurs when
$$t = -b/(2a) = -\frac{11}{2(-16)} = .34 \text{ sec.}$$

At that time
$$h = -16(.34)^2 + 11(.34) + 5 = 6.9 \text{ ft.}$$

53. Set $h = 30 - b$. The area is given by $A = \frac{1}{2}bh = \frac{1}{2}b(30 - b) = -\frac{1}{2}b^2 + 15b$.

This will be maximum when $b = -\frac{15}{2\left(-\frac{1}{2}\right)} = 15$.

Thus $b = 15$, $30 - b = h = 15$, for maximum area.

55. Set x = length perpendicular to river and y = length parallel to river. Then
$$2x + y = 200 \Rightarrow y = 200 - 2x \Rightarrow A = xy \Rightarrow A = x(200 - 2x) \Rightarrow A = -2x^2 + 200x$$

This will be maximum when $x = -\frac{200}{2(-2)} = 50$ feet.

Then $y = 200 - 2(50) = 100$ feet.

57. Set x = length and y = width. Then
$$2x + 3y = 130$$
$$y = \frac{130 - 2x}{3}$$
$$A = xy = x\left(\frac{130 - 2x}{3}\right) = -\frac{2}{3}x^2 + \frac{130}{3}x$$
This will be maximum when $x = -\frac{130}{2} \div 2\left(-\frac{2}{3}\right) = 32.5$ feet.

Then $A = -\frac{2}{3}(32.5)^2 + \frac{130}{3}(32.5) = 704.2$ square feet.

59. a. Since the vertex is at the origin, the equation has form $y = ax^2$. Since the end of the board is half of 8 feet, or 48 inches horizontally from the vertex, and y is 2 inches there, we have $2 = a(48)^2$, $a = 1/1152$; $y = (1/1152)x^2$.

 b. Substitute 1 for y and solve. $1 = \dfrac{1}{1152}x^2$. Thus $x = \sqrt{1152} \approx 34$ inches.

61. Let x = the number of 50-cent decreases. Then
   price per bowl $= 5 - .5x$

   number of bowls $= 120 + 20x$

   Income $= (120 + 20x)(5 - .5x)$
$$= -10x^2 + 40x + 600$$
This will be maximum when $x = -\frac{40}{2(-10)} = 2$.

The price should be $5 - .5(2) = \$4.00$.

63. Let x = the number of 25-cent decreases. Then
   price per ticket $= 10 - .25x$

   attendance $= 500 + 30x$

   Income $= (500 + 30x)(10 - .25x) = -7.5x^2 + 175x + 5000$

This will be maximum when $x = -\frac{175}{2(-7.5)} = \frac{35}{3}$. The price will be $10 - .25\left(\frac{35}{3}\right) = \$7.08$.

65. We would like to minimize the area inside the figure. The formula for the area inside the track is $A = \pi r^2 + 2rx$. Before we can minimize it, we must get the formula for the area in terms of one variable. Using the fact that the perimeter must be 200 meters, we get the equation $2\pi r + 2x = 200$. Solve for x in the perimeter equation and you have $x = 100 - \pi r$. Substituting this into the area formula, we get $A = \pi r^2 + 2r(100 - \pi r) = \pi r^2 + 200r - 2\pi r^2 = -\pi r^2 + 200r$. Looking at the graph of the area function, we see that minimum area occurs when $r = 0$ making the area 0.

4.2 Polynomial Functions

1. Write the expression in descending order as $x^3 + 1$.
This is a polynomial with leading coefficient 1, constant term 1, and degree 3.

3. Write the expression as $(x-1)(x^2+1) = x^3 - x^2 + x - 1$.
This is a polynomial with leading coefficient 1, constant term −1, and degree 3.

5. Write the expression as $(x-\sqrt{3})(x-\sqrt{3}) = x^2 - 3$.
This is a polynomial with leading coefficient 1, constant term −3, and degree 2.

7. This is not a polynomial.

9. This is not a polynomial.

11.
$$
\begin{array}{r}
3x^3 - 3x^2 + 11x - 17 \\
x+1\overline{)3x^4 \qquad\quad +8x^2 - 6x + 1} \\
\underline{3x^4 + 3x^3} \qquad\qquad\qquad \\
-3x^3 + 8x^2 - 6x + 1 \\
\underline{-3x^3 - 3x^2} \qquad\qquad \\
11x^2 - 6x + 1 \\
\underline{11x^2 + 11x} \qquad \\
-17x + 1 \\
\underline{-17x - 17} \\
18
\end{array}
$$

Quotient: $3x^3 - 3x^2 + 11x - 17$ Remainder: 18

Check: $(3x^3 - 3x^2 + 11x - 17)(x+1) + 18 =$
$$3x^4 + 8x^2 - 6x + 1$$

13.

$$\begin{array}{r}
x^2+2x-6 \\
x^3+1\overline{\smash{\big)}x^5+2x^4-6x^3+x^2-5x+1} \\
\underline{x^5+x^2} \\
2x^4-6x^3-5x+1 \\
\underline{2x^4+2x} \\
-6x^3-7x+1 \\
\underline{-6x^3-6} \\
-7x+7
\end{array}$$

Quotient: x^2+2x-6 Remainder: $-7x+7$

Check: $\left(x^2+2x-6\right)\left(x^3+1\right)+\left(-7x+7\right)=x^5+2x^4-6x^3+x^2-5x+1$

15.

$$\begin{array}{r}
x^2+2x+3 \\
2x^3+x^2-7x+4\overline{\smash{\big)}2x^5+5x^4+x^3-7x^2-13x+12} \\
\underline{2x^5+x^4-7x^3+4x^2} \\
4x^4+8x^3-11x^2-13x \\
\underline{4x^4+2x^3-14x^2+8x} \\
6x^3+3x^2-21x+12 \\
\underline{6x^3+3x^2-21x+12} \\
0
\end{array}$$

Quotient: x^2+2x+3 Remainder: 0

Check: $\left(x^2+2x+3\right)\left(2x^3+x^2-7x+4\right)+(0)=2x^5+5x^4+x^3-7x^2-13x+12$

17.

$$\begin{array}{r}
5x^2+5x+5 \\
x^2-x+1\overline{\smash{\big)}5x^4+5x^2+5} \\
\underline{5x^4-5x^3+5x^2} \\
5x^3+5 \\
\underline{5x^3-5x^2+5x} \\
5x^2-5x+5 \\
\underline{5x^2-5x+5} \\
0
\end{array}$$

Quotient: $5x^2+5x+5$ Remainder: 0

Check: $\left(5x^2+5x+5\right)\left(x^2-x+1\right)+0=5x^4+5x^2+5$

19. Divide:

$$
\begin{array}{r}
x-3 \\
x^2+5x-1{\overline{\smash{\big)}\,x^3+2x^2-5x-6}} \\
\underline{x^3+5x^2-x} \\
-3x^2-4x-6 \\
\underline{-3x^2-15x+3} \\
11x-9
\end{array}
$$

Since the remainder is non-zero, the first polynomial is not a factor of the second.

21. Divide:

$$
\begin{array}{r}
x^2-1 \\
x^2+3x-1{\overline{\smash{\big)}\,x^4+3x^3-2x^2-3x+1}} \\
\underline{x^4+3x^3-x^2} \\
-x^2-3x+1 \\
\underline{-x^2-3x+1} \\
0
\end{array}
$$

Since the remainder is 0, the first polynomial is a factor of the second.

23. Substitute:

$g(2)=2^4+6\cdot 2^3-2^2-30\cdot 2=0.$ 2 is a root.

$g(3)=3^4+6\cdot 3^3-3^2-30\cdot 3=144.$ 3 is not a root.

$g(-5)=(-5)^4+6\cdot(-5)^3-(-5)^2-30\cdot(-5)=0.$ -5 is a root.

$g(1)=(1)^4+6(1)^3-(1)^2-30(1)=-24.$ 1 is not a root.

25. Substitute: $h\left(2\sqrt{2}\right)=\left(2\sqrt{2}\right)^3+\left(2\sqrt{2}\right)^2-8\left(2\sqrt{2}\right)-8=0.$ $2\sqrt{2}$ is a root.

$h\left(\sqrt{2}\right)=\left(\sqrt{2}\right)^3+\left(\sqrt{2}\right)^2-8\left(\sqrt{2}\right)-8=-6\sqrt{2}-6.$ $\sqrt{2}$ is not a root.

$h\left(-\sqrt{2}\right)=\left(-\sqrt{2}\right)^3+\left(-\sqrt{2}\right)^2-8\left(-\sqrt{2}\right)-8=6\sqrt{2}-6.$ $-\sqrt{2}$ is not a root.

$h(1)=(1)^3+(1)^2-8(1)-8=-14.$ 1 is not a root.

$h(-1)=(-1)^3+(-1)^2-8(-1)-8=0.$ -1 is a root.

27. Substitute:

$l(3)=3(3+3)^{27}=3(6)^{27}.$ 3 is not a root.

$l(-3)=-3(-3+3)^{27}=-3(0)^{27}=0.$ -3 is a root.

$l(0)=0(0+3)^{27}=0(3)^{27}=0.$ 0 is a root.

29. Apply the Remainder Theorem with $c = 1$.
The remainder is $f(1) = 1^{10} + 1^8 = 2$.

31. Apply the Remainder Theorem with $c = -3/2$. The remainder is

$$f(-3/2) = 3(-3/2)^4 - 6(-3/2)^3 + 2(-3/2) - 1 = {}^{503}\!/_{16} = 31.4375.$$

33. Apply the Remainder Theorem with $c = -2$. The remainder is
$$f(-2) = (-2)^3 - 2(-2)^2 + 8(-2) - 4 = -36.$$

35. Apply the Remainder Theorem with $c = 10$. The remainder is
$$f(10) = 2(10)^5 - \sqrt{3}(10)^4 + (10)^3 - \sqrt{3}(10)^2 + \sqrt{3}(10) - 100 = 183,424.$$

37. Apply the Remainder Theorem with $c = 20$. The remainder is
$$f(20) = 2\pi(20)^5 - 3\pi(20)^4 + 2\pi(20)^3 - 8\pi(20) - 8\pi = 5,935,832\pi.$$

39. $h(x)$ is a factor of $f(x)$ if 1 is a root of $f(x)$. Evaluating $f(x)$ at 1 shows that
$f(1) = 1^5 + 1 = 2$. Therefore, 1 is not a root and $h(x)$ is not a factor.

41. $h(x)$ is a factor of $f(x)$ if -3 is a root of $f(x)$. Evaluating $f(x)$ at -3 shows that
$$f(-3) = (-3)^3 - 3(-3)^2 - 4(-3) - 12 = -54.$$
Therefore, -3 is not a root and $h(x)$ is not a factor.

43. $h(x)$ is a factor of $f(x)$ if 1 is a root of $f(x)$. Evaluating $f(x)$ at 1 shows that
$f(1) = 14 \cdot 1^{99} - 65 \cdot 1^{56} + 51 = 0$. Therefore, 1 is a root and $h(x)$ is a factor.

45. $h(x)$ is a factor of $f(x)$ if $\sqrt{2}$ is a root of $f(x)$. Evaluating $f(x)$ at $\sqrt{2}$ shows that
$$f(\sqrt{2}) = 3(\sqrt{2})^3 - 4(\sqrt{2})^2 - 6(\sqrt{2}) + 8 = 0.$$
Therefore, $\sqrt{2}$ is a root and $h(x)$ is a factor.

47. After some trial and error, or a calculator plot, we find that -4 is a root, since
$f(-4) = 6(-4)^3 - 7(-4)^2 - 89(-4) + 140 = 0$. By the Factor Theorem, $x + 4$ is a
factor. Division by $x + 4$ yields
$$f(x) = (x+4)(6x^2 - 31x + 35)$$
$$f(x) = (x+4)(2x-7)(3x-5)$$

49. After some trial and error, or a calculator plot, we find that 3 is a root, since $f(3) = 0$. By the Factor Theorem, $x - 3$ is a factor. Division by $x - 3$ yields
$$f(x) = (x-3)(4x^3 + 16x^2 + 13x + 3) = (x-3)g(x)$$
Similarly, –3 is a root, since $g(-3) = 4(-3)^3 + 16(-3)^2 + 13(-3) + 3 = 0$
By the Factor Theorem, $x - 3$ is a factor. Division by $x - 3$ yields
$$f(x) = (x-3)(x+3)(4x^2 + 4x + 1)$$
$$f(x) = (x-3)(x+3)(2x+1)^2$$

51. The graph indicates that –2, –1, 1, 2, and 3 are roots. By the Factor Theorem,
$$f(x) = (x+2)(x+1)(x-1)(x-2)(x-3)$$
$$f(x) = x^5 - 3x^4 - 5x^3 + 15x^2 + 4x - 12$$

53. The graph indicates that –1, 0, 1, 2, and 3 are roots. By the Factor Theorem,
$$f(x) = (x+1)x(x-1)(x-2)(x-3)$$
$$f(x) = x^5 - 5x^4 + 5x^3 + 5x^2 - 6x$$

55. $f(x) = (x-1)(x-7)(x+4) = x^3 - 4x^2 - 25x + 28$

57. $f(x) = (x-1)(x+1)$

59. There are many possible polynomials, such as $f(x) = (x-1)^4(x-2)(x-\pi)$ or $f(x) = (x-1)^2(x-2)^2(x-\pi)^2$, and so on.

61. Let $f(x) = a(x-8)(x-5)x$. Then
$$25 = f(10) = a(10-8)(10-5)10$$
$$25 = 100a$$
$$a = .25$$
$$f(x) = .25(x-8)(x-5)x$$

63. Since $x - 2$ is a factor, 2 is a root. Thus $(2)^3 + 3(2)^2 + k(2) - 2 = 0$
$$2k + 18 = 0$$
$$k = -9$$

65. Since $x - 1$ is a factor, 1 is a root. Thus $k^2 \cdot 1^4 - 2k \cdot 1^2 + 1 = 0$
$$k^2 - 2k + 1 = 0$$
$$(k-1)^2 = 0$$
$$k = 1$$

67. If $x - c$ were a factor, by the Factor Theorem, c would be a root. Then $c^4 + c^2 + 1 = 0$. This equation has discriminant given by $1^2 - 4(1)(1) = -3$, hence it has no real roots. Thus $x - c$ cannot be a factor for any real c.

69. a. There are many examples. One example might be $c = 2$ and $n = 3$. $x + 2$ is not a factor of $x^3 - 2^3$ since -2 is not a root because $(-2)^3 - 2^3 = -8 - 8 = -16 \neq 0$.

  b. $x + c$ is a factor of $x^n + c^n$ if n is odd, because $-c$ is a root, which follows because $(-c)^n + c^n = -c^n + c^n = 0$.

71. The difference quotient is calculated in the following way.

$$\frac{f(x+h) - f(x)}{h} = \frac{\left((x+h)^2 + k(x+h)\right) - \left(x^2 + kx\right)}{h}$$

$$= \frac{\left((x^2 + 2xh + h^2) + kx + kh\right) - x^2 - kx}{h}$$

$$= \frac{x^2 + 2xh + h^2 + kx + kh - x^2 - kx}{h}$$

$$= \frac{2xh + h^2 + kh}{h}$$

$$= 2x + h + k$$

If the difference quotient is equal to $2x + 5 + h$, by equating coefficients we find that $k = 5$.

73. $(x+3)^2 = x^2 + 6x + 9$

Use long division to reduce the polynomial.

$$
\begin{array}{r}
x^2 - 4x - 76 \\
x^2 + 6x + 9 \overline{)\; x^4 + 2x^3 - 91x^2 - 492x - 684} \\
\underline{x^4 + 6x^3 + 9x^2} \\
-4x^3 - 100x^2 - 492x \\
\underline{-4x^3 - 24x^2\; - 36x} \\
-76x^2 - 456x - 684 \\
\underline{-76x^2 - 456x - 684} \\
0
\end{array}
$$

Find the remaining zeros by the quadratic formula.

$x^2 - 4x - 76 = 0$

$$x = \frac{-(-4) \pm \sqrt{(-4)^2 - 4(1)(-76)}}{2(1)}$$

$$= \frac{4 \pm \sqrt{16 + 304}}{2}$$

$$= \frac{4 \pm \sqrt{320}}{2}$$

$$= \frac{4 \pm 8\sqrt{5}}{2}$$

$$= 2 \pm 4\sqrt{5}$$

Roots: -3, $2 - 4\sqrt{5}$, and $2 + 4\sqrt{5}$.

4.2.A Synthetic Division

1.
$$\underline{2}|\;3\;\;-8\;\;\;\;0\;\;\;\;9\;\;\;\;8$$
$$\phantom{\underline{2}|\;3\;\;}6\;\;-4\;\;-8\;\;\;\;2$$
$$\overline{3\;\;-2\;\;-4\;\;\;\;1\;\;\underline{|10}}$$

Quotient: $3x^3 - 2x^2 - 4x + 1$

Remainder: 10

3.
$$\underline{-3}|\;2\;\;\;\;7\;\;\;\;0\;\;\;-2\;\;\;-8$$
$$\phantom{\underline{-3}|\;2\;\;\;}-6\;\;-3\;\;\;\;9\;\;\;-21$$
$$\overline{2\;\;\;\;1\;\;-3\;\;\;\;7\;\;\underline{|-29}}$$

Quotient: $2x^3 + x^2 - 3x + 7$

Remainder: -29

5.
$$\underline{7}|\;5\;\;\;\;0\;\;\;-3\;\;\;\;-4\;\;\;\;\;\;6$$
$$\phantom{\underline{7}|\;5\;}35\;\;245\;\;1694\;\;11830$$
$$\overline{5\;\;35\;\;242\;\;1690\;\;\underline{|11836}}$$

Quotient: $5x^3 + 35x^2 + 242x + 1690$

Remainder: 11836

7.
$$\underline{1}|\;1\;\;\;0\;\;\;0\;\;\;0\;\;\;0\;\;\;0\;\;\;-1$$
$$\phantom{\underline{1}|\;1\;\;\;}1\;\;\;1\;\;\;1\;\;\;1\;\;\;1\;\;\;1$$
$$\overline{1\;\;\;1\;\;\;1\;\;\;1\;\;\;1\;\;\;1\;\;\underline{|0}}$$

Quotient: $x^5 + x^4 + x^3 + x^2 + x + 1$

Remainder: 0

9.
$$\tfrac{1}{4}|\;3\;\;\;\;0\;\;\;-2\;\;\;\;0\;\;\;\;2$$
$$\phantom{\tfrac{1}{4}|\;3\;}\tfrac{3}{4}\;\;\;\tfrac{3}{16}\;\;-\tfrac{29}{64}\;\;-\tfrac{29}{256}$$
$$\overline{3\;\;\;\tfrac{3}{4}\;\;\;\tfrac{29}{16}\;\;-\tfrac{29}{64}\;\;\underline{|\tfrac{483}{256}}}$$

Quotient: $3x^3 + \dfrac{3}{4}x^2 - \dfrac{29}{16}x - \dfrac{29}{64}$

Remainder: $\dfrac{483}{256}$

11.
$$\sqrt{2}|\;1\;\;\;-1\;\;\;\;\;-2\;\;\;\;\;\;2$$
$$\phantom{\sqrt{2}|\;1\;}\sqrt{2}\;\;\;-\sqrt{2}+2\;\;-2$$
$$\overline{1\;\;-1+\sqrt{2}\;\;-\sqrt{2}\;\;\;\;\underline{|0}}$$

Quotient: $x^2 + \left(-1 + \sqrt{2}\right)x - \sqrt{2}$

Remainder: 0

13.
$$\underline{-4}|\;1\;\;\;\;3\;\;\;-34\;\;\;-120$$
$$\phantom{\underline{-4}|\;1\;\;}-4\;\;\;\;4\;\;\;\;\;120$$
$$\overline{1\;\;-1\;\;-30\;\;\;\;\underline{|0}}$$

$$x^3 + 3x^2 - 34x - 120 = (x+4)(x^2 - x - 30)$$

15.
$$\tfrac{1}{2}|\;2\;\;\;-7\;\;\;15\;\;\;-6\;\;\;-10\;\;\;\;5$$
$$\phantom{\tfrac{1}{2}|\;2\;\;\;}1\;\;\;-3\;\;\;\;6\;\;\;\;\;0\;\;\;-5$$
$$\overline{2\;\;-6\;\;\;12\;\;\;\;0\;\;\;-10\;\;\underline{|0}}$$

$$2x^5 - 7x^4 + 15x^3 - 16x^2 - 10x + 5 = \left(x - \frac{1}{2}\right)\left(2x^4 - 6x^3 + 12x^2 - 10\right)$$

17.

$$\begin{array}{r} 3.12\rfloor \quad 1 \quad -5.27 \quad 10.708 \quad -10.23 \\ 3.12 \quad -6.708 \quad 12.48 \\ \hline 1 \quad -2.15 \quad \quad 4 \quad \lfloor 2.25 \end{array}$$

Quotient: $x^2 - 2.15x + 4$ or $x^2 - \dfrac{43}{20}x + 4$

Remainder: 2.25

19.

$$\begin{array}{r} -2\rfloor \quad 1 \quad 0 \quad c \quad 4 \\ -2 \quad 4 \quad -2c-8 \\ \hline 1 \quad -2 \quad c+4 \quad \lfloor -2c-4 \end{array}$$

Since the remainder is given

as 4, we must have: $-2c - 4 = 4$

$$-2c = 8$$
$$c = -4$$

21. **a.**

$$\begin{array}{r} 3\rfloor \quad 1 \quad -2 \quad -1 \quad 0 \quad 3 \quad 1 \\ 3 \quad 3 \quad 6 \quad 18 \quad 63 \\ \hline 1 \quad 1 \quad 2 \quad 6 \quad 21 \quad \lfloor 64 \end{array}$$

$x^4 + x^3 + 2x^2 + 6x + 21$; remainder 64

b.

$$\begin{array}{r} a\rfloor \quad 1 \quad -2 \quad -1 \quad 0 \quad 3 \quad 1 \\ a \quad a^2-2a \quad a^3-2a^2-a \quad a^4-2a^3-a^2 \quad a^5-2a^4-a^3+3a \\ \hline 1 \quad a-2 \quad a^2-2a-1 \quad a^3-2a^2-a \quad a^4-2a^3-a^2+3 \quad \lfloor a^5-2a^4-a^3+3a+1 \end{array}$$

Since,
$$a > 3$$
$$a - 2 > 3 - 2$$
$$a - 2 > 1$$

Thus $a-2$, (the second coefficient) is positive .

Since $a^2 - 2a - 1 = a(a-2) - 1$ by the above $a > 3$ and $a - 2 > 1$ which implies $a(a-2) > 1$ and $a(a-2) - 1 > 0$ Thus $a^2 - 2a - 1$, (the third coefficient) is positive. The rest of the coefficients and remainder are positive since you are multiplying a positive number and adding positive numbers.

c. By showing that the remainder is positive for all real numbers greater than 3, we know that for all numbers greater than 3 the remainder is non-zero. This implies that there are no roots greater than 3.

4.3 Real Roots of Polynomials

1. Using the Rational Root Test, the possible rational roots are

    $$\frac{\pm 1}{\pm 1}, \frac{\pm 3}{\pm 1}.$$

    Eliminating duplicates we have possible rational roots of $1, -1, 3, -3$.
    Graph the function in the window $-4 \le x \le 4, -10 \le y \le 10$. The
    graph shows the numbers on our list that are possible are $-1, 1, 3$.
    We use the table feature to evaluate $f(x)$ at these values. The table
    shows that $1, -1$, and 3 are the only rational roots.

    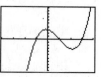
    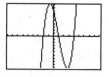

3. Using the Rational Root Test, the possible rational roots are

    $$\frac{\pm 1}{\pm 1}, \frac{\pm 2}{\pm 1}, \frac{\pm 4}{\pm 1}, \frac{\pm 8}{\pm 1}.$$

    Eliminating duplicates we have possible rational roots of
    $1, -1, 2, -2, 4, -4, 8, -8$.

    Graph the function in the window $-9 \le x \le 9, -10 \le y \le 10$. $-2, 1, 4$
    are intercepts. Use the table feature to verify this.

    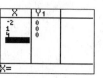

5. Using the Rational Root Test, the possible rational roots are
    $$\frac{\pm 1}{\pm 1}, \frac{\pm 2}{\pm 1}, \frac{\pm 3}{\pm 1}, \frac{\pm 6}{\pm 1}, \frac{\pm 1}{\pm 2}, \frac{\pm 2}{\pm 2}, \frac{\pm 3}{\pm 2}, \frac{\pm 6}{\pm 2}, \frac{\pm 1}{\pm 3}, \frac{\pm 2}{\pm 3}, \frac{\pm 3}{\pm 3}, \frac{\pm 6}{\pm 3}, \frac{\pm 1}{\pm 6}, \frac{\pm 2}{\pm 6}, \frac{\pm 3}{\pm 6}, \frac{\pm 6}{\pm 6}$$
    Eliminating duplicates we have possible rational roots of
    $1, -1, 2, -2, 3, -3, 6, -6, \dfrac{1}{2}, -\dfrac{1}{2}, \dfrac{3}{2}, -\dfrac{3}{2}, \dfrac{1}{3}, -\dfrac{1}{3}, \dfrac{2}{3}, -\dfrac{2}{3}, \dfrac{1}{6}, -\dfrac{1}{6}$.

    Graph the function in the window $-7 \le x \le 7, -5 \le y \le 5$. It seems
    that one root is 3 but the other roots are unclear. We can use the
    table feature to see that one of the roots is -.6667 which is the same
    as -2/3 and the other is -.5. The rational roots are
    $3, -1/6$, and $-1/2$.

    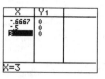

7. $f(x) = 2x^5 - 3x^4 - 11x^3 + 6x^2$

$= x^2(2x^3 - 3x^2 - 11x + 6) = x^2 g(x)$

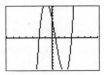

0 is a rational root of f.

Using the Rational Root Test, the possible rational roots are

$$\frac{\pm 1}{\pm 1}, \frac{\pm 2}{\pm 1}, \frac{\pm 3}{\pm 1}, \frac{\pm 6}{\pm 1}, \frac{\pm 1}{\pm 2}, \frac{\pm 2}{\pm 2}, \frac{\pm 3}{\pm 2}, \frac{\pm 6}{\pm 2}.$$

Eliminating duplicates we have possible rational roots of

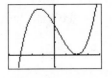

$1, -1, 2, -2, 3, -3, 6, -6, \dfrac{1}{2}, -\dfrac{1}{2}, \dfrac{3}{2}, -\dfrac{3}{2}.$

Graph $g(x)$ in the window $-7 \le x \le 7, -10 \le y \le 10$. The graph

seems to have roots of -2, 1/2 and 3. We verify this with the table

feature. The rational roots of $f(x)$ are $-2, 0, \dfrac{1}{2}, 3.$

.

9. $12 f(x) = g(x) = x^3 - x^2 - 8x + 12$. The roots of $f(x)$ are the same

as the roots of $g(x)$.

Using the Rational Root Test, the possible rational roots are

$$\frac{\pm 1}{\pm 1}, \frac{\pm 2}{\pm 1}, \frac{\pm 3}{\pm 1}, \frac{\pm 4}{\pm 1}, \frac{\pm 6}{\pm 1}, \frac{\pm 12}{\pm 1}.$$

Eliminating duplicates we have possible rational roots of
$1, -1, 2, -2, 3, -3, 4, -4, 6, -6, 12, -12.$

Graph $g(x)$ in the window $-4 \le x \le 4, \ -5 \le y \le 20$. The graph

has intercepts -3 and 2. These are the rational roots. Use the table
feature to verify this.

11. $6\left(\dfrac{1}{3}x^3 - \dfrac{5}{6}x^2 - \dfrac{1}{6}x + 1\right) = g(x) = 2x^3 - 5x^2 - x + 6$

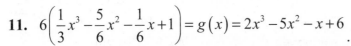

The roots of the original polynomial are the same as the roots of $g(x)$.

Using the Rational Root Test, the possible rational roots are

$$\frac{\pm 1}{\pm 1}, \frac{\pm 2}{\pm 1}, \frac{\pm 3}{\pm 1}, \frac{\pm 6}{\pm 1}, \frac{\pm 1}{\pm 2}, \frac{\pm 2}{\pm 2}, \frac{\pm 3}{\pm 2}, \frac{\pm 6}{\pm 2}.$$

Eliminating duplicates we have possible rational roots of

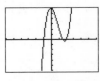

$1, -1, 2, -2, 3, -3, 6, -6, \dfrac{1}{2}, -\dfrac{1}{2}, \dfrac{3}{2}, -\dfrac{3}{2}.$

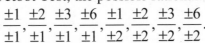

Graph g in the window $-6 \le x \le 6, \ -5 \le y \le 5$. Using the table we can see

that the rational roots of the polynomial are $-1, \dfrac{3}{2}$, and 2.

13. $10\left(.1x^3 - 1.9x + 3\right) = g(x) = x^3 - 19x + 30$. The roots of the original

polynomial are the same as the roots of $g(x)$.

Using the Rational Root Test, the possible rational roots are
$$\frac{\pm 1}{\pm 1}, \frac{\pm 2}{\pm 1}, \frac{\pm 3}{\pm 1}, \frac{\pm 5}{\pm 1}, \frac{\pm 6}{\pm 1}, \frac{\pm 15}{\pm 1}, \frac{\pm 30}{\pm 1}.$$
Eliminating duplicates we have possible rational roots of
$1, -1, 2, -2, 3, -3, 5, -5, 6, -6, 15, -15, 30, -30$.

Graph $g(x)$ in the window $-6 \le x \le 6$, $-10 \le y \le 30$. Using the graph we
see that there are intercepts at -5, 2, and 3. Use the table to verify this.

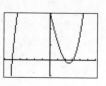

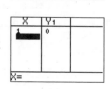

15. Using the Rational Root Test, the possible rational roots are
$$\frac{\pm 1}{\pm 1}.$$
Eliminating duplicates we have possible rational roots of $1, -1$.

Graph $g(x)$ in the window $-1 \le x \le 3$, $-3 \le y \le 3$. Using the graph we
see that there may be an intercept at 1. Use the table to verify this.

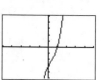

17. Using the Rational Root Test, the possible rational roots are
$$\frac{\pm 1}{\pm 1}, \frac{\pm 3}{\pm 1}, \frac{\pm 1}{\pm 2}, \frac{\pm 3}{\pm 2}.$$
Eliminating duplicates we have possible rational roots of
$1, -1, 3, -3, \dfrac{1}{2}, -\dfrac{1}{2}, \dfrac{3}{2}, -\dfrac{3}{2}.$

Graph the function in the window $-5 \le x \le 5, -5 \le y \le 5$ and observe that
1 is a root, hence $x - 1$ is a factor. Synthetically divide by $x - 1$

$$
\begin{array}{r|rrrr}
1 & 2 & -2 & 3 & -3 \\
  &   & 2 & 0 & 3 \\
\hline
  & 2 & 0 & 3 & 0
\end{array}
$$

From this we obtain $2x^2 + 3$, which has no further

rational roots. $(x-1)\left(2x^2 + 3\right)$

19. Note that the function can be factored into $x^3\left(x^3-4x^2+3x-12\right)$ The roots

of $x^3-4x^2+3x-12$ are the same as the original function. Using the Rational Root Test, the possible rational roots are

$$\frac{\pm1}{\pm1},\frac{\pm2}{\pm1},\frac{\pm3}{\pm1},\frac{\pm4}{\pm1},\frac{\pm6}{\pm1},\frac{\pm12}{\pm1}.$$

Eliminating duplicates we have possible rational roots of $1,-1,2,-2,3,-3,4,-4,6,-6,12,-12.$ Graph the function in the window $-3\le x\le6,-25\le y\le20$ and see that 4 is a root, hence $x-4$ is a factor.

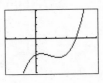

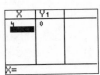

$$\begin{array}{r|rrrr} 4 & 1 & -4 & 3 & -12 \\ & & 4 & 0 & 12 \\ \hline & 1 & 0 & 3 & \underline{|0} \end{array}$$

Synthetically divide by $x-4$ From this we obtain

x^2+3 which has no further rational roots. $x^3(x-4)(x^2+3)$.

21. Using the Rational Root Test, the possible rational roots are

$$\frac{\pm1}{\pm1},\frac{\pm2}{\pm1},\frac{\pm4}{\pm1},\frac{\pm7}{\pm1},\frac{\pm8}{\pm1},\frac{\pm14}{\pm1},\frac{\pm28}{\pm1},\frac{\pm56}{\pm1}.$$

Eliminating duplicates we have possible rational roots of $1,-1,2,-2,4,-4,7,-7,8,-8,14,-14,28,-28,56,-56.$ Graph the function in the window $-6\le x\le6,-.5\le y\le.5$ and observe: the graph is tangent to the x-axis at 2, hence 2 is a root and $x-2$ is a factor twice. Synthetically divide by $(x-2)^2$

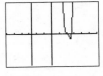

$$\begin{array}{r|rrrrr} 2 & 1 & -6 & 5 & 34 & -84 & 56 \\ & & 2 & -8 & -6 & 56 & -56 \\ \hline & 1 & -4 & -3 & 28 & -28 & \underline{|0} \end{array} \qquad \begin{array}{r|rrrrr} 2 & 1 & -4 & -3 & 28 & -28 \\ & & 2 & -4 & -14 & 28 \\ \hline & 1 & -2 & -7 & 14 & \underline{|0} \end{array}$$

$$\begin{array}{r|rrrr} 2 & 1 & -2 & -7 & 14 \\ & & 2 & 0 & -14 \\ \hline & 1 & 0 & -7 & \underline{|0} \end{array}$$

From this we obtain x^2-7 which has no further rational roots.

$(x-2)^3(x^2-7)$.

23. Synthetic division by $x-2$ yields a bottom row which is all positive, hence 2 is an upper bound. Synthetic division by $x+5$ yields a bottom row which alternates in sign, hence -5 is a lower bound.

25. Synthetic division by $x-5$ yields a bottom row which is all positive, hence 5 is an upper bound (no upper bound less than 3). Synthetic division by $x+1$ yields a bottom row which alternates in sign, hence -1 is a lower bound.

27. Following the hint, analyze $-f(x) = x^5 + 5x^4 - 9x^3 - 18x^2 + 68x - 176$, which has the same roots as f. Synthetic division by $x - 4$ yields a bottom row which is all positive, hence 4 is an upper bound for the roots of $-f$ and f. Synthetic division by $x + 8$ yields a bottom row which alternates in sign, hence -8 is a lower bound for the roots of $-f$ and f.

29. Using the Rational Root Test, the possible rational roots are
$$\frac{\pm 1}{\pm 1}, \frac{\pm 2}{\pm 1}, \frac{\pm 3}{\pm 1}, \frac{\pm 6}{\pm 1}, \frac{\pm 1}{\pm 2}, \frac{\pm 2}{\pm 2}, \frac{\pm 3}{\pm 2}, \frac{\pm 6}{\pm 2}.$$
Eliminating duplicates we have possible rational roots of

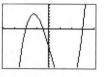

$1, -1, 2, -2, 3, -3, 6, -6, \dfrac{1}{2}, -\dfrac{1}{2}, \dfrac{3}{2}, -\dfrac{3}{2}$. Graph the function in the window $-4 \le x \le 4, -13 \le y \le 8$ and observe that the roots are $-\frac{1}{2}, 3$, and -2. Use the table feature to verify. These are the only roots since a polynomial of degree 3 has at most 3 roots.

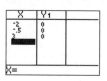

31. Using the Rational Root Test, the possible rational roots are
$$\frac{\pm 1}{\pm 1}, \frac{\pm 2}{\pm 1}, \frac{\pm 1}{\pm 2}, \frac{\pm 2}{\pm 2}, \frac{\pm 1}{\pm 3}, \frac{\pm 2}{\pm 3}, \frac{\pm 1}{\pm 6}, \frac{\pm 2}{\pm 6}.$$
Eliminating duplicates we have possible rational roots of

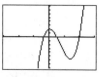

$1, -1, 2, -2, \dfrac{1}{2}, -\dfrac{1}{2}, \dfrac{1}{3}, -\dfrac{1}{3}, \dfrac{2}{3}, -\dfrac{2}{3}, \dfrac{1}{6}, -\dfrac{1}{6}.$ Graph the function in the window $-3 \le x \le 3, -8 \le y \le 8$ and observe that the roots are $-\frac{1}{3}, , \frac{1}{2}$, and 2. Use the table feature to verify. These are the only roots since a polynomial of degree 3 has at most 3 roots.

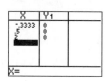

33. Using the Rational Root Test, the possible rational roots are
$$\frac{\pm 1}{\pm 1}, \frac{\pm 2}{\pm 1}, \frac{\pm 3}{\pm 1}, \frac{\pm 4}{\pm 1}, \frac{\pm 6}{\pm 1}, \frac{\pm 12}{\pm 1}.$$
Eliminating duplicates we have possible rational roots of
$1, -1, 2, -2, 3, -3, 4, -4, 6, -6, 12, -12$. Graph the function in the window $-6 \le x \le 6, -20 \le y \le 20$ and observe that there is a repeated root at 2 and it is not clear what the other roots are. Reducing the polynomial by the factor of $x - 2$ twice, will result in a quadratic.

$$
\begin{array}{r|rrrrr}
2 & 1 & 1 & -19 & 32 & -12 \\
  &   & 2 & 6 & -26 & 12 \\
\hline
  & 1 & 3 & -13 & 6 & \underline{0}
\end{array}
\qquad
\begin{array}{r|rrrr}
2 & 1 & 3 & -13 & 6 \\
  &   & 2 & 10 & -6 \\
\hline
  & 1 & 5 & -3 & \underline{0}
\end{array}
$$

We can find the roots of the quadratic using the quadratic formula.
$$x^2 + 5x - 3 = 0 \Rightarrow x = \frac{-5 \pm \sqrt{5^2 - 4(1)(-3)}}{2(1)} \Rightarrow x = \frac{-5 \pm \sqrt{37}}{2}$$

The real roots of the polynomial are 2 and $\frac{-5 \pm \sqrt{37}}{2}$.

35. Using the Rational Root Test, the possible rational roots are

$$\frac{\pm1}{\pm1}, \frac{\pm2}{\pm1}, \frac{\pm3}{\pm1}, \frac{\pm6}{\pm1}, \frac{\pm1}{\pm2}, \frac{\pm2}{\pm2}, \frac{\pm3}{\pm2}, \frac{\pm6}{\pm2}.$$

Eliminating duplicates we have possible rational roots of

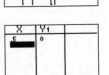

$1, -1, 2, -2, 3, -3, 6, -6, \dfrac{1}{2}, -\dfrac{1}{2}, \dfrac{3}{2}, -\dfrac{3}{2}$. Graph the function in the window

$-3 \le x \le 3, -3 \le y \le 3$. Five roots are seen, of which the only rational

root is $\frac{1}{2}$. Use synthetic division to reduce the polynomial.

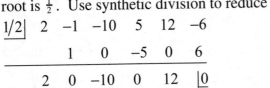

The remaining polynomial can be factored

$2x^4 - 10x^2 - 12 = 2(x^2 - 3)(x^2 - 2)$. The remaining roots are $\pm\sqrt{3}$ and

$\pm\sqrt{2}$.

37. Using the Rational Root Test, the possible rational roots are

$$\frac{\pm1}{\pm1}, \frac{\pm2}{\pm1}, \frac{\pm3}{\pm1}, \frac{\pm4}{\pm1}, \frac{\pm6}{\pm1}, \frac{\pm9}{\pm1}, \frac{\pm12}{\pm1}, \frac{\pm18}{\pm1}, \frac{\pm36}{\pm1}.$$

Eliminating duplicates we have possible rational roots of

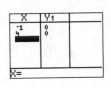

$1, -1, 2, -2, 3, -3, 4, -4, 6, -6, 9, -9, 12, -12, 18, -18, 36, -36$. Graph the

function in the window $-8 \le x \le 8, -10 \le y \le 10$. Four roots are seen,

of which the only rational roots are -1 and 4. Use synthetic division to

reduce the polynomial.

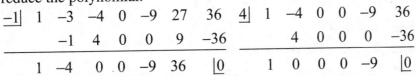

The remaining polynomial can be factored

$x^4 - 9 = (x^2 - 3)(x^2 + 3)$. The remaining real roots are $\pm\sqrt{3}$.

39. Using the Rational Root Test, the possible rational roots are

$$\frac{\pm 1}{\pm 1}, \frac{\pm 2}{\pm 1}, \frac{\pm 5}{\pm 1}, \frac{\pm 10}{\pm 1}, \frac{\pm 25}{\pm 1}, \frac{\pm 50}{\pm 1}.$$

Eliminating duplicates we have possible rational roots of $1, -1, 2, -2, 5, -5, 10, -10, 25, -25, 50, -50$. First graph the function in the window $-60 \le x \le 60$, $-10,000 \le y \le 10,000$ and observe that 50 is a root, but -50 is not. Synthetically divide by $x-50$ to obtain $x^3 + 2x^2 - x - 1$.

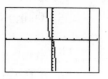

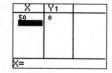

```
50 | 1   -48   -101   49    50
   |      50    100   -50   -50
   ----------------------------
     1    2     -1    -1   |0
```

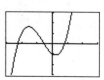

Then graph this in the window $-3 \le x \le 3, -3 \le y \le 3$.
Use the root (zero) routine to locate the roots at $-2.24698, -.5549581, .80193774$, as well as 50.

41. a. By the Rational Roots Theorem, the only possible rational roots of $x^2 - 2$ are ± 1 and ± 2. Since $\sqrt{2}$ is a root of the polynomial, and it is not one of those four numbers, it is not rational.

b. By the Rational Roots Theorem, the only possible rational roots of $x^2 - 3$ are ± 1 and ± 3. Since $\sqrt{3}$ is a root of the polynomial, and it is not one of those four numbers, it is not rational.

c. The only possible rational roots of $x^2 - 4$ are $\pm 1, \pm 2,$ and ± 4. We try all of them and find that 2 and -2 both are roots of the polynomial.

43. a. In 2000, $x = 5$, the murder rate was

$$f(5) = -.0002724 \cdot 5^5 + .005237 \cdot 5^4 - .03027 \cdot 5^3 + .1069 \cdot 5^2 - .9062 \cdot 5 + 0.9003$$

$$= 5.78 \text{ per } 100,000$$

In 2003, $x = 8$, hence the murder rate was

$$f(8) = -.0002724 \cdot 8^5 + .005237 \cdot 8^4 - .03027 \cdot 8^3 + .1069 \cdot 8^2 - .9062 \cdot 8 + 9.003$$

$$= 5.62 \text{ per } 100,000$$

b. Graph the function and the horizontal line $y = 7$ in the window $0 \le x \le 11$, $0 \le y \le 15$ and use the intersection routine to find $x = 2.67$, that is in the middle of 1997.

c. By visual inspection of the graph the highest murder rate was in 1995.
d. The lowest murder rate was in 2004.
e. Use the minimum and maximum features of the calculator to find that the murder rate was increasing from the middle of 2001 to 2004.

45. Since height $= x$, length $= 36 - 4x$, width $= 12 - 2x$ we can write:
volume $= V(x) = x(36 - 4x)(12 - 2x) = 8x^3 - 120x^2 + 432x$. Graph this function and the horizontal line $y = 448$ in the window $0 \le x \le 3, 0 \le y \le 600$ and use the intersection routine to find the only root less than 2.5 exactly at $x = 2$.

47. a. At $t = 0$, $F(0) = 6$ degrees per day. At $t = 11$, $F(11) = 6.6435$ degrees per day.
  b. Graph the function and the horizontal line $y = 4$ in the window $0 \le x \le 12$, $-10 \le y \le 30$ and use the intersection routine to find $t = 2.0330$ and $t = 10.7069$.
  c. Graph the function and the horizontal line $y = -3$ (the temperature is decreasing, hence the rate is negative) in the same window and use the intersection routine to find $t = 5.0768$ and $t = 9.6126$.
  d. Graph the function and use the minimum routine to find $t = 7.6813$.

49. a. The carrying capacity is an upper bound on the population of bunnies, while the threshold population is a lower bound.
  b. The population is increasing.
  c. The population is decreasing.
  d. The population is decreasing.
  e. $k\left(-x^3 + x^2(T + C) - CTx\right) = -kx\left(x^2 - x(T + C) + CT\right)$
$$= -kx(x - C)(x - T) \text{ or } kx(x - T)(C - x)$$
  f. $-kx(x - C)(x - T) = 0$
  $x = 0$ or $x = C$ or $x = T$

4.4 Graphs of Polynomial Functions

1. This could be a graph of a polynomial function (of a very large degree).

3. This is discontinuous and could not be the graph of a polynomial function.

5. This is discontinuous and could not be the graph of a polynomial function.

7. This could be the graph of a polynomial function. Since the ends go in opposite directions, the polynomial has odd degree. Since there are 3 intercepts and 2 local extrema, it must have degree at least 3 or 5..

9. Not the graph of a polynomial function, since the domain is not all real numbers.

11. This could be the graph of a polynomial function. Since the ends go in opposite directions, the polynomial has odd degree. Since there are 5 intercepts and 4 local extrema, it must have degree at least 5.

13. In a very large window like $-50 \le x \le 50$, $-100,000 \le y \le 1,000,000$ the graph looks like the graph of $y = x^4$.

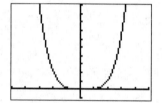

15. Roots –2, 1, 3; each has (odd) multiplicity 1.

17. Root –2 has odd multiplicity, root –1 has odd multiplicity, and root 2 has even multiplicity (2 or higher).

19. Graph (e) since the graph in (e) is a straight line.

21. Graph (f) since the graph has three roots.

23. Graph (c) since the graph has four roots and is down facing.

25.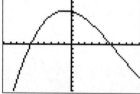

Since this is a polynomial function of degree 3, the complete graph must have another x intercept.

27.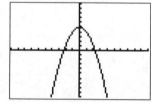

Since this is a polynomial function of degree 4 with positive leading coefficient, both of the ends should point up; here they are pointing down.

29. $-5 \le x \le 5, -50 \le y \le 30$

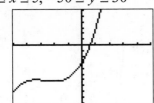

31. $-1 \le x \le 5, -5 \le y \le 5$

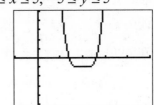

33. $-33 \le x \le -2, -50,000 \le y \le 260,000$ then $-2 \le x \le 3, -20 \le y \le 30$

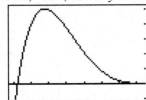

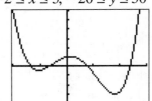

35. $-90 \le x \le 120, -15,000 \le y \le 5000$

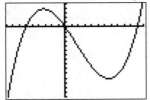

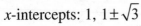

37. a. The graph of a cubic polynomial can have no more than two local extrema. If only one, the ends would go in the same direction. Hence it can have two or none.

   b. When the end behavior and the number of extrema are both accounted for, these four shapes are the only possible ones.

39. a. The ends go in opposite directions, so the degree is odd.

   b. Since if x is large and positive, so is $f(x)$, the leading coefficient is positive.

   c. Root -2 has multiplicity 1, root 0 has multiplicity at least 2, root 4 has multiplicity 1, root 6 has multiplicity 1.

   d. Adding the multiplicities from **c**, the polynomial must have degree at least 5.

41. Since the degree is odd and the leading coefficient is negative, the graph should decrease as x increases. Since there are 3 roots, the shape matches (d).

43. Use the window $-4 \le x \le 6, -50 \le y \le 50$.

   x-intercepts: $1, 1 \pm \sqrt{3}$

   local maximum: $(2,2)$

   local minimum: $(0,-2)$

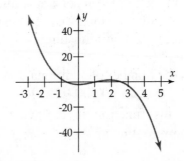

45. Use the window $0 \le x \le 4, 0 \le y \le 3$.

x-intercepts: 2, 3

local minimum: $\left(\dfrac{11}{4}, -\dfrac{27}{256} \right)$

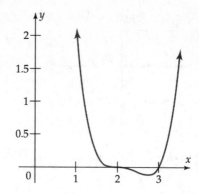

47. Use the window $-2 \le x \le 2, -1 \le y \le 3$.

x-intercepts: $\dfrac{\pm 1 \pm \sqrt{5}}{2}, 0$

local maxima: $(-1.30, 1.58), (.345, .227)$

local minima: $(-.345, -.227), (1.30, -1.58)$

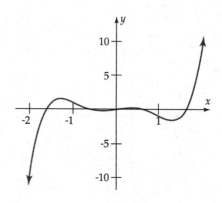

49. a.

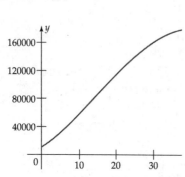

b.

$$\frac{f(15) - f(0)}{15 - 0} = \frac{86,250 - 12,000}{15} = 4950$$

$$\frac{f(40) - f(15)}{40 - 15} = \frac{180,000 - 86,250}{25} = 3750$$

Sales are increasing faster in the first interval.

c. The inflection point separates intervals on which the rate of increase is increasing from those on which it is decreasing. When the rate of increase is decreasing, the returns are diminishing.

51. Let x = the number of additional trees planted per acre. Then

y = yield = $(22 + x)(500 - 15x)$

Graph this function in the window $0 \le x \le 10, 10,000 \le y \le 12,000$ and use the maximum routine to obtain $x = 5.67$, that is, 6 additional trees per acre.

53. Let r be the radius and h be the height of the can. Then $\pi r^2 h = 355$, $h = 355/\pi r^2$. The total surface area is given by $S = \pi r^2 + 3\pi r^2 + 2\pi r(355/\pi r^2) = 4\pi r^2 + 710/r$. Graph this function in the window $0 \le x \le 8$, $0 \le y \le 900$ and use the minimum routine to obtain $r = 3.046$, $h = 12.18$ cm.

55. $p(x)$ is a good approximation for $f(x)$ when $-3 < x < 3$.

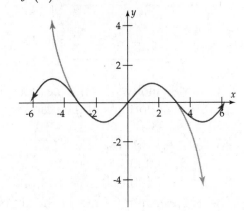

57. $p(x)$ is a good approximation for $f(x)$ when $-3 \le x \le 2.25$.

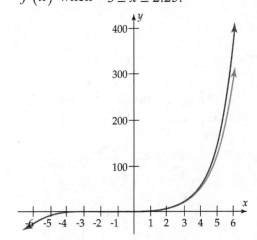

59. a.

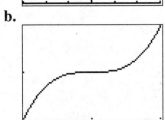

If the graph had the horizontal portion appearing in the window, then the equation $g(x) = 4$ would have infinite solutions, which is impossible since a polynomial equation of degree 3 can have at most 3 solutions.

b.

When graphed in the window shown, which is $1 \le x \le 3$, $3.99 \le y \le 4.01$, the graph has a much smaller incorrect horizontal portion.

c. If the graph of a polynomial of degree n had a horizontal portion appearing in the window, it would be a portion also of the line $y = k$. Then the equation $f(x) = k$ would have infinite solutions, which is impossible since a polynomial equation of degree n can have at most n solutions.

61. a. The general shape of the graph would be something like this, with an intercept and a minimum at $x = 2$.

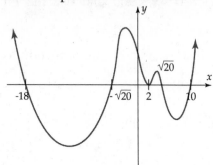

b. The graph does not resemble the sketch. The intercepts $\pm\sqrt{20}$ and 10 are shown. However, the intercept 2 is not shown.

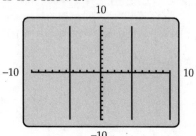

c. In this graph, the intercepts $\pm\sqrt{20}$, 10, and -18. However, 2 is still not shown.

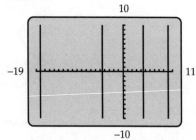

d. Different windows would be needed for different portions of the graph.

$-20 \leq x \leq -3, -5 \times 10^6 \leq y \leq 10^6$
$-3 \leq x \leq 2, -5000 \leq y \leq 60,000$
$1 \leq x \leq 5, -5000 \leq y \leq 5000$
$5 \leq x \leq 11, -100,000 \leq y \leq 100,000$

The Table feature of the TI-83 can be very useful in estimating the size of the y-values.

4.4.A Polynomial Models

1. A curve that would fit this data well would have one local maximum and one local minimum and therefore be a polynomial of degree 3. A cubic model seems best.

3. A curve that would fit this data well would have one local maximum and therefore be a polynomial of degree 2. A quadratic model seems best.

5. **a.** Entering the data into the TI-83 and applying the cubic regression routine yields the following $f(x) = 1.239x^3 - 26.221x^2 + 19.339x + 4944.881$

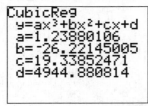

   b. Evaluating the function in **(a)** gives:

   In 1998, $f(8) = 1.239(8)^3 - 26.221(8)^2 + 19.339(8) + 4944.881 = 4055.7$

   In 2001, $f(11) = 1.239(11)^3 - 26.221(11)^2 + 19.339(11) + 4944.881 = 3633.7$

   c. Very accurate.
   d. The model predicts a large surge in crime after 2005. The data does not support this conclusion. So the model should not be used to predict the crime rate after 2005.

7. **a.** Entering the data into the TI-83 yields the following:

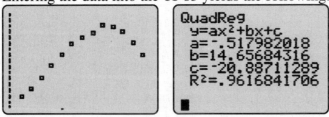

   b. Evaluating the function in **(a)** gives:
   $(x = 12)$ 80.4°, 9 A.M. $(x = 9)$ 69.1°, 2 P.M. $(x = 14)$ 82.8°

9. **a.** A scatter plot for the data is as follows.

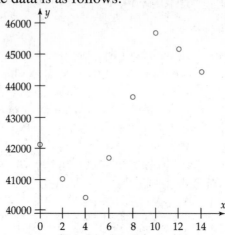

b. Since the data points suggest a curve that is concave up on the left and concave down on the right, a cubic polynomial would provide a more reasonable model then the quadratic model.

```
CubicReg
 y=ax³+bx²+cx+d
 a=-14.84659091
 b=321.8468615
 c=-1444.238095
 d=42266.86364
 R²=.9675257055
```

$$f(x) = -14.847x^3 + 321.847x^2 - 1444.238x + 42266.864$$

c. $f(22) = -14.847(22)^3 + 321.847(22)^2 - 1444.238(22) + 42266.864 = \$8181$

d. The model predicts that a dropoff in the median income that began in 2002 will continue, and get worse and worse. The reasonability of the model depends on your opinion of this prediction.

11. a. The scatter plot for the data is as follows.

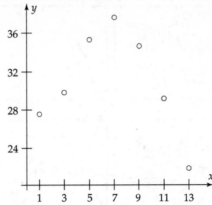

b. Quadratic, cubic, and quartic models are shown and graphed.

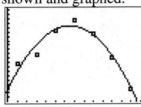

$$f(x) = -.340x^2 + 4.410x + 22.092$$

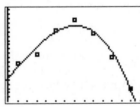

$$f(x) = -.0156x^3 - .0115x^2 + 2.55x + 24.39$$

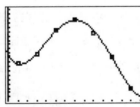

$$f(x) = .00746x^4 - .224x^3 + 1.896x^2 - 3.684x + 29.531$$

c. The quartic model fits the data best, but given the trend in smoking, the cubic model is probably best for the future.

13. a. Cubic and quartic models are shown and graphed:

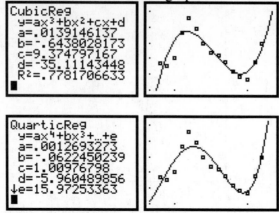

b. Both models seem to rise disturbingly steeply and only a pessimist would expect them to be accurate very far into the future.

4.5 Rational Functions

1. $f(x)$ is defined for all real numbers except when $3x - 4 = 0$, or $x = 4/3$.

Domain: $(-\infty, 4/3) \cup (4/3, \infty)$ or $x \neq 4/3$

3. $h(x)$ is defined for all real numbers except when $x^2 + 9 = 0$. Since the solutions to this equation are imaginary numbers, the domain $h(x)$ is all real numbers.

Domain: $(-\infty, \infty)$

5 $j(x)$ is defined for all real number except when $\pi x^3 + \pi x^2 - 9\pi x - 9\pi = 0$.

$$\pi x^3 + \pi x^2 - 9\pi x - 9\pi = 0$$

$$\pi \left(x^3 + x^2 - 9x - 9 \right) = 0$$

$$\pi \left[x^2 (x+1) - 9(x+1) \right] = 0$$

$$\pi (x+1) \left(x^2 - 9 \right) = 0$$

$$\pi (x+1)(x-3)(x+3) = 0$$

$$x = -1, \;\; x = 3, \;\; x = -3$$

Domain: $(-\infty, -3) \cup (-3, -1) \cup (-1, 3) \cup (3, \infty)$

7. Since there are vertical asymptotes at $x = 3$ and $x = -3$, the factors $(x-3)(x+3)$ must be in the denominator of the rational function. Since there is a horizontal asymptote at $y = 2$, the degree of the numerator must be the same as the degree of the denominator and the ratio of the leading coefficients must be 2. One example with these properties is $f(x) = \dfrac{2x^2}{x^2 - 9}$

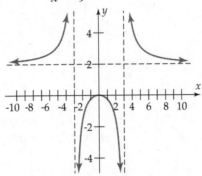

9. Since the function is even, if we have vertical asymptotes at $x = 1$ and $x = 3$ we will also have vertical asymptotes at $x = -1$ and $x = -3$. Since there is a horizontal asymptote at $y = -1$, the degree of the numerator and the degree of the denominator must be the same and the ratio of the leading coefficients must be -1. $\dfrac{-x^4}{(x+3)(x+1)(x-1)(x-3)}$ would satisfy these properties. In order to satisfy the property of passing through the point $(0,4)$ we will add 36 to the numerator.

$$f(x) = \frac{-x^4 + 36}{x^4 - 10x^2 + 9}$$

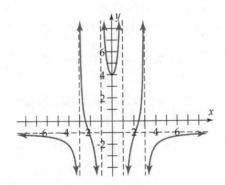

11. The graph of f has vertical asymptote $x = 2$. **F.**

13. The graph of h has vertical asymptote $x = -1$. **A.**

15. $f(x) = \dfrac{x^3 + 6x^2 + 11x + 6}{x^3 - x} = \dfrac{(x+3)(x+2)(x+1)}{x(x-1)(x+1)}$

Vertical asymptotes: $x = 0$ and $x = 1$

Hole: $(-1, 1)$

17. $f(x) = \dfrac{x^2 + 8x + 2}{x^2 + 7x + 2}$ if $x = \dfrac{-7 \pm \sqrt{7^2 - 4(1)(2)}}{2(1)} = \dfrac{-7 \pm \sqrt{41}}{2}$

Vertical asymptotes: $x = \dfrac{-7 \pm \sqrt{41}}{2}$

19. Since the degree of the numerator is less than the degree of the denominator, there is a horizontal asymptote at $y = 0$. Any viewing window will work. To see the graph look in the window $-4 < x < 4;\ \ -.5 < y < .5$

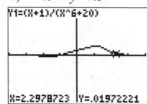

21. Since the degree of the numerator is more than the degree of the denominator, there is no horizontal asymptote.

23. Since the degree of the numerator is the same as the degree of the denominator there is a horizontal asymptote at the ratio of the leading coefficients. The horizontal asymptote is at $y = 2/3$. The viewing window $-25 < x < 25;\ \ -5 < y < 5$ shows the graph within .1 of the asymptote.

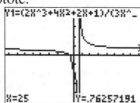

25. $f(x) = \dfrac{1}{x-2}$

vertical asymptote: $x = 2$
horizontal asymptote: $y = 0$
holes: none

$f(0) = \dfrac{1}{0-2} = -\dfrac{1}{2}$ y-int: $\left(0, -\dfrac{1}{2}\right)$

$f(x) = 0$ has no solution, so there is no x-intercept.

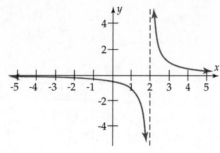

27. $f(x) = \dfrac{2x}{x+1}$

vertical asymptote: $x = -1$
horizontal asymptote: $y = 2$
holes: none

$f(0) = \dfrac{0}{0+1} = 0$ y-intercept $(0,0)$

$(0,0)$ This is also an x-intercept.

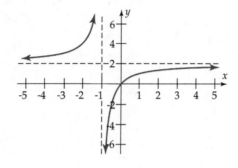

29. $f(x) = \dfrac{x}{x(x-2)(x-3)}$

vertical asymptotes: $x = 2$; $x = 3$
horizontal asymptote: $y = 0$

holes: $\left(0, \dfrac{1}{6}\right)$

$f(0) =$ undefined; No y-intercept

$f(x) = 0$ has no solution, so there is no x-intercept.

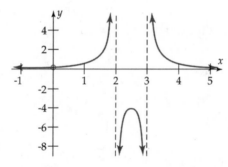

31. $f(x) = \dfrac{2}{x^2+1}$

vertical asymptote: none
horizontal asymptote: $y = 0$
holes: none

$f(0) = \dfrac{2}{0^2+1} = 2$ y-intercept $(0,2)$

$f(x) = 0$ has no solution, so there is no x-intercept.

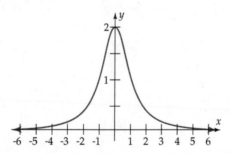

33. $f(x) = \dfrac{(x^2+6x+5)(x+5)}{(x+5)^3(x-1)} = \dfrac{(x+1)(x+5)^2}{(x+5)^3(x-1)}$

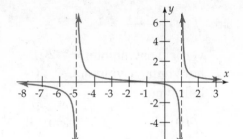

$\qquad = \dfrac{x+1}{(x+5)(x-1)}, x \neq -5$

vertical asymptotes: $x = -5$, $x = 1$

horizontal asymptote: $y = 0$

holes: none

$f(0) = \dfrac{(0^2+6(0)+5)(0+5)}{(0+5)^3(0-1)} = \dfrac{5^2}{-5^3} = -\dfrac{1}{5}$

y-intercept: $\left(0, -\dfrac{1}{5}\right)$

x-intercept: $(-1, 0)$

35. $f(x) = \dfrac{2x^3+3x^2-3x-2}{x^3+x^2-4x-4} = \dfrac{(x-1)(2x+1)(x+2)}{(x+1)(x-2)(x+2)}$

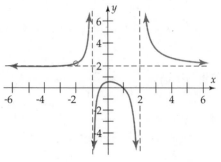

vertical asymptotes: $x = -1$, $x = 2$

horizontal asymptote: $y = 2$

hole: $\left(-2, \dfrac{9}{4}\right)$

$f(0) = \dfrac{2(0)^3+3(0)^2-3(0)-2}{0^3+0^2-4(0)-4} = \dfrac{-2}{-4} = \dfrac{1}{2}$

y-intercept: $\left(0, \dfrac{1}{2}\right)$

x-intercepts: $(1, 0)$, $(-1/2, 0)$

37. $f(x) = \dfrac{x(x-1)(x+5)}{(x-2)(x+2)(x+3)(x-3)}$

holes: none since there are no common roots in the numerator and denominator.

vertical asymptotes: $x=2, x=-2, x=-3, x=3$ since these are the roots of the denominator.

horizontal asymptote: $y=0$ since the degree of the numerator is less than the degree of the denominator.

x-intercepts: $0, 1, -5$

y-intercept: 0

Three windows are needed.

$-5 \le x \le 4.4, -8 \le y \le 4$

$-2 \le x \le 2, -.5 \le y \le .5$

$-15 \le x \le -3, -.07 \le y \le .02$

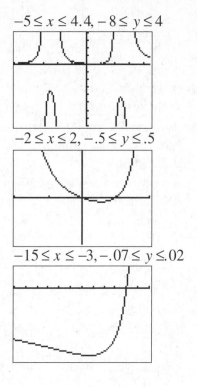

39. $h(x) = \dfrac{(2x+3)(x-2)}{x(x-2)(x+3)} = \dfrac{2x+3}{x(x+3)}, \ x \ne 2$

holes: $(2, 7/10)$ since there is a common factor

of $x-2$ in the numerator and the denominator.

vertical asymptotes: $x=0, x=-3$ since they are the roots of the denominator that are not roots of the numerator.

horizontal asymptote: $y=0$ since the degree of the numerator is less than the degree of the denominator.

x-intercept: $-3/2$

y-intercept: none

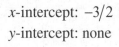

$-9.4 \le x \le 9.4; \ -4 \le y \le 4$

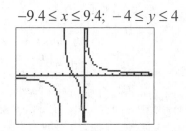

41. $g(x) = \dfrac{x-2}{(x-1)(x+1)(x-11)}$

holes: none since there are no common factors in the numerator and denominator.

vertical asymptotes: $x=1$, $x=-1$, $x=11$ since they are the roots of the denominator that are not roots of the numerator.

horizontal asymptote: $y=0$ since the degree of the numerator is less than the degree of the denominator.

x-intercept: 2

y-intercept: $-2/11$

$-5 < x < 15$, $-2 < y < 2$

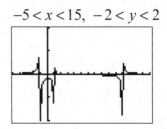

43. $\dfrac{f(x+h)-f(x)}{h} = \dfrac{\dfrac{1}{x+h} - \dfrac{1}{x}}{h}$

$= \dfrac{x-(x+h)}{hx(x+h)} = \dfrac{-h}{hx(x+h)} = \dfrac{-1}{x(x+h)}$

45. $\dfrac{f(x+h)-f(x)}{h} = \dfrac{\dfrac{3}{x+h-2} - \dfrac{3}{x-2}}{h}$

$= \dfrac{3(x-2) - 3(x+h-2)}{h(x-2)(x+h-2)}$

$= \dfrac{-3h}{h(x-2)(x+h-2)}$

$= \dfrac{-3}{(x-2)(x+h-2)}$

47. $\dfrac{g(x+h)-g(x)}{h} = \dfrac{\dfrac{3}{(x+h)^2} - \dfrac{3}{x^2}}{h}$

$= \dfrac{3x^2 - 3(x+h)^2}{hx^2(x+h)^2}$

$= \dfrac{-6x-3h}{x^2(x+h)^2} = -3\left(\dfrac{2x+h}{x^2(x+h)^2}\right)$

49. a. From 2 to 2.1: $x = 2, h = .1$ $\dfrac{-1}{x(x+h)} = \dfrac{-1}{2(2+.1)} = -\dfrac{1}{4.2} = -.2381$

From 2 to 2.01: $x = 2, h = .01$ $\dfrac{-1}{x(x+h)} = \dfrac{-1}{2(2+.01)} = -\dfrac{1}{4.02} = -.2488$

From 2 to 2.001: $x = 2, h = .001$ $\dfrac{-1}{x(x+h)} = \dfrac{-1}{2(2+.001)} = -\dfrac{1}{4.002} = -.2499$

Instantaneous rate of change: $-1/4$

b. From 3 to 3.1: $x = 3, h = .1$ $\dfrac{-1}{x(x+h)} = \dfrac{-1}{3(3+.1)} = -\dfrac{1}{9.3} = -.1075$

From 3 to 3.01: $x = 3, h = .01$ $\dfrac{-1}{x(x+h)} = \dfrac{-1}{3(3+.01)} = -\dfrac{1}{9.03} = -.1107$

From 3 to 3.001: $x = 3, h = .001$ $\dfrac{-1}{x(x+h)} = \dfrac{-1}{3(3+.001)} = -\dfrac{1}{9.003} = -.1111$

Instantaneous rate of change: $-1/9$

c. They are identical.

51. a. $g(x) = \dfrac{x-1}{x-2}$ **b.** $g(x) = \dfrac{x-1}{-x-2}$

c. $x = 2$ and $x = -2$ are vertical asymptotes for the two "halves "of the graph. Graph using the window $-4.7 \le x \le 4.7, -6 \le y \le 6$.

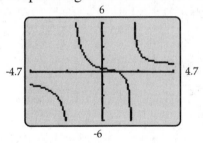

53. For the top, area $= x^2$, cost $= 2.5x^2$

For the bottom, area $= x^2$, cost $= 2.5x^2$

For the sides, area $= 4 \cdot \dfrac{1000}{x}$, cost $= 1.5 \cdot 4 \cdot \dfrac{1000}{x} = \dfrac{6000}{x}$

Cost $= c(x) = 2.5x^2 + 2.5x^2 + \dfrac{6000}{x}$

$c(x) = 5x^2 + \dfrac{6000}{x}$

Graph in the window $0 \le x \le \sqrt{500}, \ 0 \le y \le 5000$ to obtain minimum

$x = 8.4343$ inches. Then $h = \dfrac{1000}{8.4343^2} = 14.057$ inches.

8.4343in. by 8.4343in. by 14.057in.

55. a. In 1980, Joseph has $20 to spend and each figure cost $2 He can buy $20/2 = 10$ action figures.

   b. In 2010, Joseph has $\$20 + \$5(30) = \$170$ to spend and the cost of each action figure is $\$2 + \$0.25(30) = \$9.50$. He can buy $170/9.5 \approx 17$.

   c. Number of figures $= \dfrac{20 + 5t}{2 + .25t}$

   d.

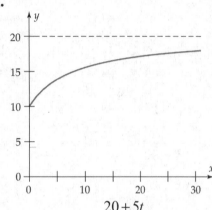

   e. $18 = \dfrac{20 + 5t}{2 + .25t}$ Yes, in the year 2012.

   $$18(2 + .25t) = 20 + 5t$$
   $$36 + 4.5t = 20 + 5t$$
   $$36 - .5t = 20$$
   $$-.5t = -16$$
   $$t = 32$$

   He will never be able to buy 21 since the horizontal asymptote is at 20.

57. a. Let $y =$ the length of the non-parallel sides. Then $xy = 200$, $y = \dfrac{200}{x}$.

$$\text{Total length} = P(x) = x + 2y = x + 2\left(\frac{200}{x}\right) = x + \frac{400}{x}$$

b. Graph $P(x)$ and $y = 60$ in the window $0 \le x \le 60$, $0 \le y \le 100$. Then use the intersection routine in the calculator.

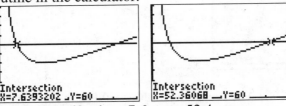

Intersection
X=7.6393202 Y=60

Intersection
X=52.36068 Y=60

The function is less than 60 when $7.6 < x < 52.4$.

Find the exact value of the intersection points by solving the equation

$$x + \frac{400}{x} = 60 \qquad\qquad x = \frac{-(-60) \pm \sqrt{(-60)^2 - 4(1)(400)}}{2(1)}$$

$$x^2 + 400 = 60x \qquad\qquad x = \frac{60 \pm \sqrt{2000}}{2}$$

$$x^2 - 60x + 400 = 0 \qquad\qquad x = \frac{60 \pm 20\sqrt{5}}{2}$$

$$x = 30 \pm 10\sqrt{5}$$

c. Graph $P(x)$ in the same window, then use the minimum routine to find $x \approx 20$. The dimensions are then 20 meters by 10 meters.

59. a. $h_1 = h - 2$

b. Since volume $= 150 = \pi r^2 h$, $h = \dfrac{150}{\pi r^2}$. Hence $h_1 = \dfrac{150}{\pi r^2} - 2$.

c. Set $r_1 = r - 1$. Then $V = \pi r_1^2 h_1$, hence

$$V = \pi (r-1)^2 \left(\frac{150}{\pi r^2} - 2\right)$$

d. Since the walls are 1 foot thick, r must be greater than 1.

e. Graph V in the window $0 \le r \le 5$, $0 \le V \le 150$, then use the maximum routine to find $r = 2.88$ feet. Hence

$$h = \frac{150}{\pi r^2} = 5.76 \text{ feet.}$$

61. a. $g(0) = \dfrac{3.987 \times 10^{14}}{\left(6.378 \times 10^{6}\right)^{2}} = 9.801$ meters per second^{2}.

b. Use the window $0 \le r \le 9{,}000{,}000$, $0 \le g(r) \le 10$.

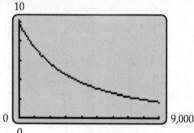

c. Since the r-axis is a horizontal asymptote, and the graph never crosses the axis, as $r \to \infty$, g approaches 0 but never reaches 0. You can never escape the pull of gravity.

4.5.A Other Rational Functions

1.   Use long division or synthetic division.

$$
\begin{array}{r}
x \\
x^2-4\,\overline{\smash{\big)}\,x^3+0x^2+0x-1} \\
\underline{x^3-4x} \\
4x-1
\end{array}
$$

$f(x) = x + \dfrac{4x-1}{x^{2}-4}$. Then the nonvertical asymptote is $y = x$.

Use viewing window $-14 \le x \le 14$, $-14 \le y \le 14$.

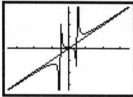

(vertical segments are erroneous; the calculator incorrectly connects points on opposite sides of the vertical asymptotes.)

3.   Use synthetic division or long division

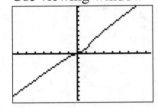

$$f(x) = 2x + 3 + \frac{-22}{x^2 - 3x + 8}.$$

Then the nonvertical asymptote is $y = 2x + 3$.

Use viewing window $-15 \le x \le 17, -30 \le y \le 30$.

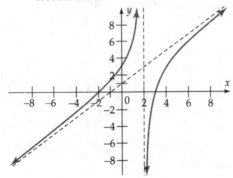

(vertical segments are erroneous; the calculator incorrectly connects points on opposite sides of the vertical asymptotes.)

5. $f(x) = \dfrac{x^2 - x - 6}{x - 2} = \dfrac{(x-3)(x+2)}{x-2}$

$\qquad = x + 1 - 4/(x - 2)$

vertical asymptote: $x = 2$

nonvertical asymptote: $y = x + 1$

holes: none

7. $Q(x) = \dfrac{4x^2 - 8x - 21}{2x - 5} = \dfrac{(2x-7)(2x+3)}{2x-5}$

$= 2x + 1 + \dfrac{-16}{2x - 5}$

vertical asymptote: $x = \frac{5}{2}$

nonvertical asymptote: $y = 2x + 1$

holes: none

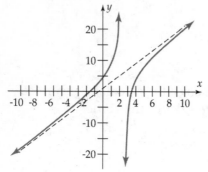

9. $f(x) = \dfrac{x^3 - 2}{x - 1} = x^2 + x + 1 + \dfrac{-1}{x - 1}$.

vertical asymptote: $x = 1$

nonvertical asymptote: $y = x^2 + x + 1$

holes: none

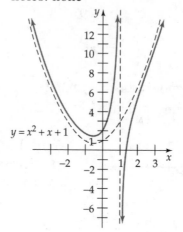

11. $q(x) = \dfrac{x^5 + 3x^4 - 11x^3 - 3x^2 + 10x + 1}{(x-1)(x+1)(x-2)}$

$= x^2 + 5x + \dfrac{10x + 1}{x^3 - 2x^2 - x + 2}$.

vertical asymptote: $x = 2, x = 1, x = -1$

nonvertical asymptote: $y = x^2 + 5x$

holes: none

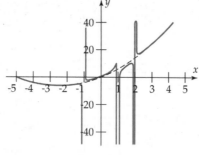

13. $f(x) = \dfrac{2x^2 + 5x + 2}{2x + 7} = x - 1 + \dfrac{9}{2x + 7}$

vertical asymptote: $x = -\frac{7}{2}$

nonvertical asymptote: $y = x - 1$

$-15.5 \le x \le 8.5, -16 \le y \le 8$

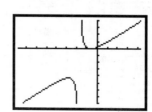

15. $h(x) = \dfrac{x^3 - 32x^2 + 341x - 1212}{(x-9)(x-11)}$

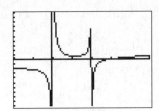

$= x - 12 + \dfrac{2x - 24}{x^2 - 20x + 99}$

vertical asymptotes: $x = 9, x = 11$

nonvertical asymptote: $y = x - 12$

$7 \le x \le 14, -30 \le y \le 30$

17. $g(x) = \dfrac{2x^4 + 7x^3 + 7x^2 + 2x}{x^3 - x + 50} = 2x + 7 + \dfrac{9x^2 - 91x - 350}{x^3 - x + 50}$

To find the vertical asymptote, solve $x^3 - x + 50 = 0$ to obtain $x = -3.774494$.

nonvertical asymptote: $y = 2x + 7$

Two windows are needed:

$-13 \le x \le 7, -20 \le y \le 20$ 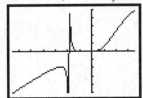 $-2.5 \le x \le 1, -.02 \le y \le .02$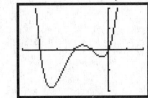

19. a.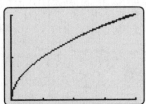

b. Shown is the graph of $r(x) - s(x)$ in the window $0 \le x \le 4, -.01 \le y \le .02$ together with the horizontal line $y = .01$. The graph is below the line, and thus $r(x)$ differs from $s(x)$ by less than $.01$, in the interval $[.06, 2.78]$.

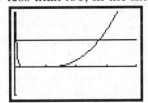

21. $f(x) = \dfrac{rx^3 + (b-2r)x^2 - (2b+10r)x + 1}{x^2 - 2x - 10}$

$$
\begin{array}{r}
rx + b \\
x^2 - 2x - 10 \overline{\smash{\big)}\, rx^3 + (b-2r)x^2 - (2b+10r)x + 1} \\
\underline{rx^3 \qquad -2rx^2 \qquad -10rx} \\
bx^2 \qquad -2bx + 1 \\
\underline{bx^2 \qquad -2bx - 10b} \\
10b + 1
\end{array}
$$

The nonvertical asymptote is $y = rx + b$.

4.6 Polynomial and Rational Inequalities

1. $2x + 4 \le 7$

$2x \le 3$

$x \le \dfrac{3}{2}$

$(-\infty, 3/2]$

3. $3 - 5x < 13$

$-5x < 10$

$x > -2$

$(-2, \infty)$

5. $6x + 3 \le x - 5$

$5x + 3 \le -5$

$5x \le -8$

$x \le -\dfrac{8}{5}$

$\left(-\infty, -\dfrac{8}{5}\right]$

7. $5 - 7x < 2x - 4$

$5 - 9x < -4$

$-9x < -9$

$x > 1$

$(1, \infty)$

9. $2 < 3x - 4 < 8$

$6 < 3x < 12$

$2 < x < 4$

$(2, 4)$

11. $0 < 5 - 2x \le 11$

$-5 < -2x \le 6$

$\dfrac{5}{2} > x \ge -3$

$\left[-3, \dfrac{5}{2}\right)$

13. $5x + 6(-8x - 1) < 2(x - 1)$

$5x - 48x - 6 < 2x - 2$

$-45x - 6 < -2$

$-45x < 4$

$x > -4/45$

$(-4/45, \infty)$

15. $\dfrac{x+1}{2} - 3x \le \dfrac{x+5}{3}$

$3(x+1) - 18x \le 2(x+5)$

$3x + 3 - 18x \le 2x + 10$

$-15x + 3 \le 2x + 10$

$-17x \le 7$

$x \ge -7/17$

$[-7/17, \infty)$

17. Solve $2x + 3 \le 5x + 6$ and $5x + 6 < -3x + 7$

$$-3x \le 3 \qquad\qquad 8x < 1$$
$$x \ge -1 \qquad\qquad x < 1/8$$

$[-1, 1/8)$

19. Solve $3 - x < 2x + 1$ and $2x + 1 \le 3x - 4$

$$-3x < -2 \qquad\qquad -x \le -5$$
$$x > 2/3 \qquad\qquad x \ge 5$$

Combining these yields $x \ge 5$: $[5, \infty)$

21. $ax - b < c$

$$ax < b + c$$

$$x < \frac{b + c}{a}$$

23. $0 < x - c < a$

$$c < x < a + c$$

25. The solutions of $x^2 - 4x + 3 \le 0$ are the numbers x at which the graph of
$f(x) = x^2 - 4x + 3 = (x - 1)(x - 3)$ lies on or below the x-axis. This condition holds
when x is between the roots 1 and 3. Solution: $[1, 3]$ or $1 \le x \le 3$.

27. $8 + x - x^2 \le 0$

$$x^2 - x - 8 \ge 0$$

$x^2 - x - 8$ has roots $\dfrac{1 \pm \sqrt{33}}{2}$. The solutions of $x^2 - x - 8 \ge 0$ is

$$x < \frac{1 - \sqrt{33}}{2} \text{ or } x > \frac{1 + \sqrt{33}}{2}.$$

29. $x^3 - x \ge 0$

$x(x - 1)(x + 1) \ge 0$

Interval	$x < -1$	$-1 < x < 0$	$0 < x < 1$	$1 < x$
Test number	-2	$-\frac{1}{2}$	$\frac{1}{2}$	2
Sign	$-$	$+$	$-$	$+$
Graph	Below x-axis	Above x-axis	Below x-axis	Above x-axis

$[-1, 0] \cup [1, \infty)$; $-1 \le x \le 0$ or $x \ge 1$.

31. $x^3 + 3x^2 - x - 3 < 0$

$x^2(x+3) - (x+3) < 0$

$(x+3)(x^2-1) < 0$

$(x+3)(x+1)(x-1) < 0$

Interval	$x<-3$	$-3<x<-1$	$-1<x<1$	$1<x$
Test number	-4	-2	0	2
Sign	$-$	$+$	$-$	$+$
Graph	Below x-axis	Above x-axis	Below x-axis	Above x-axis

$(-\infty,-3)\cup(-1,1)$; $x<-3$ or $-1<x<1$.

33. $x^4 - 5x^2 + 4 < 0$

$(x-1)(x-2)(x+1)(x+2) < 0$

Interval	$x<-2$	$-2<x<-1$	$-1<x<1$	$1<x<2$	$2<x$
Test number	-3	-1.5	0	1.5	3
Sign	$+$	$-$	$+$	$-$	$+$
Graph	Above x-axis	Below x-axis	Above x-axis	Below x-axis	Above x-axis

$(-2,-1)\cup(1,2)$; $-2<x<-1$ or $1<x<2$.

35. $2x^4 + 3x^3 < 2x^2 + 4x - 2$

$2x^4 + 3x^3 - 2x^2 - 4x + 2 < 0$

Graph this equation in the window $-5 \le x \le 5, -1 \le y \le 10$ and apply the zero (root)

routine to find that the graph is below the x-axis in the interval $(.5,.84)$; $.5 < x < .84$.

37. $\dfrac{3x+1}{2x-4} > 0$

Interval	$x<-\frac{1}{3}$	$-\frac{1}{3}<x<2$	$2<x$
Test number	-1	0	3
Sign	$+$	$-$	$+$
Graph	Above x-axis	Below x-axis	Above x-axis

$\left(-\infty,-\frac{1}{3}\right)\cup(2,\infty)$; $x<-\dfrac{1}{3}$ or $x>2$.

39. $\dfrac{x-2}{x-1} < 1 \Rightarrow \dfrac{x-2}{x-1} - 1 < 0$ or $\dfrac{-1}{x-1} < 0$

Interval	$x<1$	$1<x$
Test number	0	2
Sign	$+$	$-$
Graph	Above x-axis	Below x-axis

$(1,\infty)$; $x>1$

41. $\dfrac{2}{x+3} \geq \dfrac{1}{x-1} \Rightarrow \dfrac{2}{x+3} - \dfrac{1}{x-1} \geq 0 \Rightarrow \dfrac{2(x-1)-(x+3)}{(x+3)(x-1)} \geq 0$ or $\dfrac{x-5}{(x+3)(x-1)} \geq 0$

Interval	$x < -3$	$-3 < x < 1$	$1 < x < 5$	$5 < x$
Test number	-4	0	2	6
Sign	$-$	$+$	$-$	$+$
Graph	Below x-axis	Above x-axis	Below x-axis	Above x-axis

$(-3,1) \cup [5,\infty)$; $\ -3 < x < 1$ or $x \geq 5$.

43. Here a complete solution is necessarily approximate. Graph the function

$$f(x) = \dfrac{x^3 - 3x^2 + 5x - 29}{x^2 - 7}$$

and the horizontal line $y = 3$ in the window $-10 \leq x \leq 10,\ -20 \leq y \leq 20$.

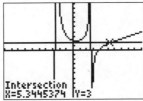

Observe that the graph of the function is above the horizontal line between the vertical asymptotes $x = \pm\sqrt{7}$ as well as for $x > 5.34$.

$\left(-\sqrt{7},\sqrt{7}\right) \cup (5.34,\infty)$; $\ -\sqrt{7} < x < \sqrt{7}$ or x>5.34.

45. $\dfrac{2x^2 + 6x - 8}{2x^2 + 5x - 3} < 1 \Rightarrow \dfrac{2x^2 + 5x - 3 + x - 5}{2x^2 + 5x - 3} < 1$

$1 + \dfrac{x-5}{2x^2 + 5x - 3} < 1 \Rightarrow \dfrac{x-5}{(2x-1)(x+3)} < 0$

Interval	$x < -3$	$-3 < x < \frac{1}{2}$	$\frac{1}{2} < x < 5$	$5 < x$
Test number	-4	0	1	6
Sign	$-$	$+$	$-$	$+$
Graph	Below x-axis	Above x-axis	Below x-axis	Above x-axis

$(-\infty,-3) \cup \left(\frac{1}{2},5\right)$; $\ x < -3$ or $\dfrac{1}{2} < x < 5$.

47. $x(x-1)^3(x-2)^4 \geq 0$

Interval	$x < 0$	$0 < x < 1$	$1 < x < 2$	$2 < x$
Test number	-1	$1/2$	$3/2$	3
Sign	$+$	$-$	$+$	$+$
Graph	Above x-axis	Below x-axis	Above x-axis	Above x-axis

$(-\infty,0] \cup [1,\infty)$; $\ x \leq 0$ or $x \geq 1$

49. $x > -1.43$ **51.** $x \le -3.79$ or $x \ge .79$

53. Let x = rate per kwh. Then

Cost of first freezer = $723.95 + 144(90x)$ over 12 years (144 months)

Cost of second freezer = $600 + 144(100x)$

Solve $723.95 + 144(90x) < 600 + 144(100x)$

$$-1440x < -123.95$$

$$x > \$.08608 \approx 8.608 \text{ cents per kwh.}$$

55. Let x = total sales per month.

1st compensation = $.1x$

2nd compensation = $3000 + .02x$

$.1x > 3000 + .02x$

$.08x > 3000$

$x > 37500$

More than $37,500.

57. a) Let x = number of months.

Cost to remain with AT&T bill = $60x$

Cost to switch to Virgin Mobile = $40 + 40 + 175 + .18(100)x = 255 + 18x$

$255 + 18x < 60x$

$255 < 42x$

$6.07 < x$

They would have to talk more than six months.

b) If they talk zero minutes they would not have any fee, saving then 60 dollars a month. If they talk 100 minutes, the fee would be 18 dollars saving then 42 dollars. The range of savings would be 42 to 60 dollars per month, not counting the up-front fees to switch.

59. Let x = one number, $y = 20 - x$ = other number. Then

$$x^2 + (20 - x)^2 < 362 \Rightarrow 2x^2 - 40x + 400 < 362$$

$$2x^2 - 40x + 38 < 0 \Rightarrow x^2 - 20x + 19 < 0.$$

The solutions are the numbers x for which the graph of $f(x) = x^2 - 20x + 19 = (x - 1)(x - 19)$ lies below the x-axis. This condition holds when x is between the roots 1 and 19. Thus any two numbers between 1 and 19 whose sum is 20, e.g. 2 and 18, 3 and 17, and $20 - \pi$, etc., satisfy the conditions. $\Rightarrow 1 < x < 19$ and $y = 20 - x$

61. Cost $= 350 + 5x$ and Revenue $= x(50 - x)$

Profit=Revenue − Cost

For the profit to be positive, then:

$$x(50-x)-(350+5x)>0 \Rightarrow 50x-x^2-350-5x>0$$

$$x^2-45x+350<0.$$

The solutions are the numbers x for which the graph of
$f(x) = x^2 - 45x + 350 = (x-10)(x-35)$ lies below the x-axis. This condition holds
when x is between the roots 10 and 35. More than 10 but fewer than 35 medallions
$\Rightarrow 10 < x < 35$

63. $h = -16t^2 + 80t$. Solve
$$-16t^2 + 80t \geq 64$$

$$-16t^2 + 80t - 64 \geq 0$$

$$t^2 - 5t + 4 \leq 0$$

The solutions are the numbers x for which the graph of
$f(t) = t^2 - 5t + 4 = (t-1)(t-4)$ lies below the t-axis. This condition holds when
t is between the roots 1 and 4.
$1 \leq t \leq 4$

65. $h = -16t^2 + 120$. Solve $39 < -16t^2 + 120 < 56$

Graph the equation $h = -16t^2 + 120$ in the window $0 \leq x \leq 5, 0 \leq y \leq 120$ together
with the horizontal lines $y = 39$ and $y = 56$ and apply the intersection routine twice
to obtain $2 < t < 2.25$. So t must be between 2 and 2.25 seconds.

67. a. $x^2 < x$ $x^2 > x$

$x^2 - x < 0$ $x^2 - x > 0$

$x(x-1) < 0$ $x(x-1) > 0$

Interval	$x < 0$	$0 < x < 1$	$1 < x$
Test number	-1	$\frac{1}{2}$	2
Sign	$+$	$-$	$+$
Graph	Above x-axis	Below x-axis	Above x-axis

Thus $x^2 < x$: if $0 < x < 1$

$x^2 > x$: if $x < 0$ or $x > 1$

b. If $0 < c < 1$, then $c^2 < c$. Since $|c| = c$, $c^2 < |c|$.
If $-1 < c < 0$, then $-c > c^2$. Since $|c| = -c$, $c^2 < |c|$.
Thus if $-1 < c < 1$, $c \neq 0$, $c^2 < |c|$.

c. If $c > 1$, then $c^2 > c$.
If $c > -1$, then $c^2 > -c$. Since $c < 0$, $-c > c$, hence $c^2 > c$.
Thus if $|c| > 1$, $c^2 > c$.

69. $4x - 5 \geq 4x + 2$

$-5 \geq 2$

False; no solutions.

71. $3x - 4 \geq 3x - 4$

$0 \geq 0$

True; all real numbers are solution; $(-\infty, \infty)$

73. $(x+2)^2 (x-3)^2 < 0$

Interval	$x < -2$	$-2 < x < 3$	$3 < x$
Test number	-3	0	4
Sign	$+$	$+$	$+$
Graph	Above x-axis	Above x-axis	Above x-axis

No solutions

75. $(x+1)^2 < 0$

Interval	$x < -1$	$-1 < x$
Test number	-2	0
Sign	$+$	$+$
Graph	Above x-axis	Above x-axis

No solutions

77. $8 \leq 4x - 2 \leq 8$

$10 \leq 4x \leq 10$

$5/2 \leq x \leq 5/2$

$x = \dfrac{5}{2}$

4.6.A Absolute Value Inequalities

1. $|3x+2| \le 2$

$-2 \le 3x+2 \le 2$

$-4 \le 3x \le 0$

$-4/3 \le x \le 0$

3. $|3-2x| < 2/3$

$-2/3 < 3-2x < 2/3$

$-11 < -6x < -7$

$11/6 > x > 7/6$

$7/6 < x < 11/6$

5. $|5x+2| \ge 3/4$

$5x+2 \ge 3/4$ or $5x+2 \le -3/4$

$5x \ge -5/4$ $5x \le -11/4$

$x \ge -1/4$ $x \le -11/20$

$x \le -11/20$ or $x \ge -1/4$

7. $|12/5+2x| > 1/4$

$12/5+2x > 1/4$ or $12/5+2x < -1/4$

$48+40x > 5$ $48+40x < -5$

$40x > -43$ $40x < -53$

$x > -43/40$ $x < -53/40$

$x < -53/40$ or $x > -43/40$

9. $\left|\dfrac{x-1}{x+2}\right| \le 3$. Graph $\left|\dfrac{x-1}{x+2}\right| - 3$ using the window $-10 \le x \le 5, -5 \le y \le 10$.

The graph is at or below the x-axis on the intervals $(-\infty, -7/2]$ and $[-5/4, \infty)$;

$x \le -7/2$ or $x \ge -5/4$

11. $\left|\dfrac{1-4x}{2+3x}\right| < 1$. Graph $\left|\dfrac{1-4x}{2+3x}\right| - 1$ using the window $-3 \le x \le 8, -2 \le y \le 2$.

The graph is below the x-axis on the interval $(-1/7, 3)$;

$-1/7 < x < 3$

13. $|x^2-2| < 1$. Graph $|x^2-2| - 1$ using the window $-5 \le x \le 5, -2 \le y \le 5$. Since the roots of the function are the solutions of $x^2-2=1$ and $x^2-2=-1$, that is, $\pm\sqrt{3}$ and $\pm\sqrt{1}$, the graph is below the x-axis on the intervals $(-\sqrt{3}, -1) \cup (1, \sqrt{3})$;

$-\sqrt{3} < x < -1$ or $1 < x < \sqrt{3}$.

15. $|x^2-2| > 4$. Graph $|x^2-2| - 4$ using the window $-5 \le x \le 5, -2 \le y \le 10$. Since the roots of the function are the solutions of $|x^2-2| = 4$, that is, $\pm\sqrt{6}$, the graph is above the x-axis on the intervals $(-\infty, -\sqrt{6}) \cup (\sqrt{6}, \infty)$; $x < -\sqrt{6}$ or $x > \sqrt{6}$.

17. $\left|x^2+x-1\right|\geq 1$. Graph $\left|x^2+x-1\right|-1$. using the window $-4\leq x\leq 3,\ -2\leq y\leq 5$. Since the roots of the function are the solutions of $\left|x^2+x-1\right|=1$, that is, $-2,-1,0$ and 1, the graph is at or above the x-axis on the intervals $(-\infty,-2]\cup[-1,0]\cup[1,\infty)$; $x\leq 2$ or $-1\leq x\leq 0$ or $x\geq 1$.

19. $\left|x^5-x^3+1\right|<2$. Graph $\left|x^5-x^3+1\right|-2$ using the window $-2\leq x\leq 2,\ -3\leq y\leq 3$. Since the roots of the function are approximately -1.43 and 1.24, the graph is below the x-axis on the interval $(-1.43,1.24)$; $-1.43<x<1.24$.

21. $\left|x^4-x^3+x^2-x+1\right|>4$. Graph $\left|x^4-x^3+x^2-x+1\right|-4$ using the window $-2\leq x\leq 2,\ -5\leq y\leq 5$. Since the roots of the function are approximately -0.89 and 1.56, the graph is above the x-axis on the intervals $(-\infty,-0.89)\cup(1.56,\infty)$; $x<-0.89$ or $x>1.56$.

23. $\dfrac{x+2}{|x-3|}\leq 4$. Graph $\dfrac{x+2}{|x-3|}-4$ using the window $-5\leq x\leq 10,\ -5\leq y\leq 5$. Since the roots of the function are 2 and $14/3$, the graph is at or below the x-axis on the intervals $(-\infty,2]\cup[14/3,\infty)$; $x\leq 2$ or $x\geq 14/3$.

25. Graph $\left|\dfrac{2x^2+2x-12}{x^3-x^2+x-2}\right|-2$ using the window $-3\leq x\leq 3,\ -3\leq y\leq 5$.

Since the roots of the function are approximately -1.13 and 1.67, the graph is undefined at the real root of $x^3-x^2+x-2=0$, approximately 1.35. Thus the graph is above the x-axis on the intervals $(-1.13,1.35)\cup(1.35,1.67)$; $-1.13<x<1.35$ or $1.35<x<1.67$.

27. Since the absolute value of a real number is always non negative, the solution to the absolute value inequality are $(-\infty,\infty)$, all real numbers.

29. Since the absolute value of a real number is always non negative, we need only find when x^2-3x+2 equals zero.
$$x^2-3x+2=0\Rightarrow(x-2)(x-1)=0$$
$$x-2=0\ \text{ or }\ x-1=0$$
$$x=2\quad\text{ or }\ x=1$$
The solution to the absolute value inequality is $x=0$ or $x=2$.

31. If $|x-3|<\dfrac{E}{5}\Rightarrow-\dfrac{E}{5}<x-3<\dfrac{E}{5}\Rightarrow-E<5x-15<E\Rightarrow-E<(5x-11)-4<E\Rightarrow\left|(5x-11)-4\right|<E$

4.7 Complex Numbers

1. $(2+3i)+(6-i)=2+3i+6-i=8+2i$

3. $(2-8i)-(4+2i)=2-8i-4-2i=-2-10i$

5. $\dfrac{5}{4}-\left(\dfrac{7}{4}+2i\right)=\dfrac{5}{4}-\dfrac{7}{4}-2i=-\dfrac{1}{2}-2i$

7. $\left(\dfrac{\sqrt{2}}{2}+i\right)-\left(\dfrac{\sqrt{3}}{2}-i\right)=\dfrac{\sqrt{2}}{2}+i-\dfrac{\sqrt{3}}{2}+i=\dfrac{\sqrt{2}-\sqrt{3}}{2}+2i$

9. $(2+i)(3+5i)=6+10i+3i+5i^2=6+13i+5(-1)=1+13i$

11. $(0-6i)(5+0i)=(-6i)(5)=-30i$

13. $(2-5i)^2=2^2-2\cdot 2(5i)+(5i)^2=4-20i+25i^2=4-20i+25(-1)=-21-20i$

15. $\left(\sqrt{3}+i\right)\left(\sqrt{3}-i\right)=\left(\sqrt{3}\right)^2-i^2=3-(-1)=4$

17. $i^{19}=\left(i^4\right)^4 i^3=1^4(-i)=-i$

19. $i^{33}=\left(i^4\right)^8 i=1^8 i=i$

21. $(-i)^{107}=\left[(-i)^4\right]^{26}(-i)^3=1^{26}\left(-i^3\right)=-(-i)=i$

23. $\dfrac{1}{3+2i}=\dfrac{1}{(3+2i)}\dfrac{(3-2i)}{(3-2i)}=\dfrac{3-2i}{9-4i^2}=\dfrac{3-2i}{9+4}=\dfrac{3-2i}{13}=\dfrac{3}{13}-\dfrac{2}{13}i$

25. $\dfrac{4}{3i}=\dfrac{4}{3i}\dfrac{(-i)}{(-i)}=\dfrac{-4i}{3\left(-i^2\right)}=\dfrac{-4i}{3\cdot 1}=-\dfrac{4}{3}i$

27. $\dfrac{3}{4+5i}=\dfrac{3}{(4+5i)}\dfrac{(4-5i)}{(4-5i)}=\dfrac{12-15i}{16-25i^2}=\dfrac{12-15i}{16-(-25)}=\dfrac{12-15i}{41}=\dfrac{12}{41}-\dfrac{15}{41}i$

29. $\dfrac{1}{i(4+5i)}=\dfrac{1}{4i+5i^2}=\dfrac{1}{-5+4i}=\dfrac{1}{(-5+4i)}\dfrac{(-5-4i)}{(-5-4i)}=\dfrac{-5-4i}{25-16i^2}=\dfrac{-5-4i}{41}=-\dfrac{5}{41}-\dfrac{4}{41}i$

31. $\dfrac{2+3i}{i(4+i)} = \dfrac{2+3i}{4i+i^2} = \dfrac{2+3i}{-1+4i} = \dfrac{(2+3i)}{(-1+4i)}\dfrac{(-1-4i)}{(-1-4i)} = \dfrac{-2-8i-3i-12i^2}{1-16i^2}$

$\qquad = \dfrac{-2-11i+12}{1+16} = \dfrac{10-11i}{17} = \dfrac{10}{17} - \dfrac{11}{17}i$

33. $\dfrac{2+i}{1-i} + \dfrac{1}{1+2i} = \dfrac{(2+i)(1+2i)}{(1-i)(1+2i)} + \dfrac{1-i}{(1-i)(1+2i)} = \dfrac{2+4i+i+2i^2+1-i}{1+2i-i-2i^2}$

$\qquad = \dfrac{3+4i+2i^2}{1+i-2i^2} = \dfrac{1+4i}{3+i} = \dfrac{(1+4i)}{(3+i)}\dfrac{(3-i)}{(3-i)} = \dfrac{3+12i-i-4i^2}{9-i^2}$

$\qquad = \dfrac{7+11i}{10} = \dfrac{7}{10} + \dfrac{11}{10}i$

35. $\dfrac{i}{3+i} - \dfrac{3+i}{4+i} = \dfrac{i(4+i)-(3+i)(3+i)}{(3+i)(4+i)} = \dfrac{4i+i^2-9-6i-i^2}{12+3i+4i+i^2} = \dfrac{-9-2i}{11+7i}$

$\qquad = \dfrac{(-9-2i)}{(11+7i)}\dfrac{(11-7i)}{(11-7i)} = \dfrac{-99+63i-22i+14i^2}{121-49i^2} = \dfrac{-113+41i}{170} = -\dfrac{113}{170} + \dfrac{41}{170}i$

37. $6i$

39. $\sqrt{14}i$

41. $-4i$

43. $\sqrt{-16} + \sqrt{-49} = 4i + 7i = 11i$

45. $\sqrt{-15} - \sqrt{-18} = \sqrt{15}i - \sqrt{18}i = \left(\sqrt{15} - 3\sqrt{2}\right)i$

47. $\sqrt{-16}/\sqrt{-36} = \sqrt{16}i/\sqrt{36}i = 4i/6i = 2/3$

49. $\left(\sqrt{-25}+2\right)\left(\sqrt{-49}-3\right) = (5i+2)(7i-3) = 35i^2 - 15i + 14i - 6 = -41 - i$

51. $\left(2+\sqrt{-5}\right)\left(1-\sqrt{-10}\right) = \left(2+\sqrt{5}i\right)\left(1-\sqrt{10}i\right) = 2 - 2\sqrt{10}i + \sqrt{5}i - \sqrt{50}i^2$

$\qquad = \left(2+\sqrt{50}\right) + \left(\sqrt{5}-2\sqrt{10}\right)i = \left(2+5\sqrt{2}\right) + \left(\sqrt{5}-2\sqrt{10}\right)i$

53. $\dfrac{1}{1+\sqrt{-5}} = \dfrac{1}{1+i\sqrt{5}} = \dfrac{1}{\left(1+i\sqrt{5}\right)}\dfrac{\left(1-i\sqrt{5}\right)}{\left(1-i\sqrt{5}\right)} = \dfrac{1-i\sqrt{5}}{1-5i^2} = \dfrac{1-i\sqrt{5}}{6} = \dfrac{1}{6} - \dfrac{\sqrt{5}}{6}i$

55. $3x - 4i = 6 + 2yi;$
$3x = 6$ and $-4 = 2y$
$x = 2 \qquad y = -2$

57. $3 + 4xi = 2y - 3i;$
$3 = 2y$ and $4x = -3$
$y = 3/2 \qquad x = -3/4$

59. $3x^2 - 2x + 5 = 0$
$x = \dfrac{-(-2) \pm \sqrt{(-2)^2 - 4(3)(5)}}{2(3)}$
$x = \dfrac{2 \pm \sqrt{-56}}{6}$
$x = \dfrac{1 \pm i\sqrt{14}}{3}$
$\dfrac{1}{3} + \dfrac{\sqrt{14}}{3}i, \dfrac{1}{3} - \dfrac{\sqrt{14}}{3}i$

61. $x^2 + 5x + 6 = 0$
$x = \dfrac{-5 \pm \sqrt{5^2 - 4(1)(6)}}{2(1)}$
$x = \dfrac{-5 \pm \sqrt{1}}{2} = \dfrac{-5 \pm 1}{2}$
$-2 + 0i, -3 + 0i$

63. $2x^2 - x = -4$
$2x^2 - x + 4 = 0$
$x = \dfrac{-(-1) \pm \sqrt{(-1)^2 - 4(2)(4)}}{2(2)}$
$x = \dfrac{1 \pm \sqrt{-31}}{4} = \dfrac{1 \pm i\sqrt{31}}{4}$
$\dfrac{1}{4} + \dfrac{\sqrt{31}}{4}i, \dfrac{1}{4} - \dfrac{\sqrt{31}}{4}i$

65. $x^2 + 1770.25 = -84x$
$x^2 + 84x + 1770.25 = 0$
$x = \dfrac{-84 \pm \sqrt{84^2 - 4(1)(1770.25)}}{2(1)}$
$x = \dfrac{-84 \pm \sqrt{-25}}{2} = \dfrac{-84 \pm 5i}{2}$
$= -42 \pm 2.5i$
$-42 + 2.5i, -42 - 2.5i$

67. $x^3 - 8 = 0$
$(x - 2)(x^2 + 2x + 4) = 0$
$x - 2 = 0$ or $x^2 + 2x + 4 = 0$
$x = 2 \qquad x = \dfrac{-2 \pm \sqrt{2^2 - 4(1)(4)}}{2(1)}$
$x = \dfrac{-2 \pm \sqrt{12}i}{2}$
$x = -1 \pm \sqrt{3}i$
$2 + 0i, -1 + \sqrt{3}i, -1 - \sqrt{3}i$

69. $x^4 - 1 = 0$
$(x^2 - 1)(x^2 + 1) = 0$
$x^2 - 1 = 0$ or $x^2 + 1 = 0$
$x = \pm 1 \qquad x = \pm\sqrt{-1} = \pm i$
$1 + 0i, -1 + 0i, 0 + i, 0 + -i$

71. $i + i^2 + i^3 + \cdots + i^{15} = i + (-1) + (-i) + 1$

$\qquad\qquad\qquad + i + (-1) + (-i) + 1$

$\qquad\qquad\qquad + i + (-1) + (-i) + 1$

$\qquad\qquad\qquad + i + (-1) + (-i)$

$\qquad\qquad\qquad = 4i - 4 - 4i + 3$

$\qquad\qquad\qquad = -1$

73. a. i. $\operatorname{mod}(3 - 4i) = \sqrt{3^2 + (-4)^2} = \sqrt{9 + 16} = \sqrt{25} = 5$

$\quad$ **ii.** $\operatorname{mod}(24 + 7i) = \sqrt{24^2 + 7^2} = \sqrt{576 + 49} = \sqrt{625} = 25$

$\quad$ **iii.** $\operatorname{mod}(8 + 0i) = \sqrt{8^2 + 0^2} = \sqrt{64} = 8$

$\quad$ **iv.** $\operatorname{mod}(-8 + 0i) = \sqrt{(-8)^2 + 0^2} = \sqrt{64} = 8$

$\quad$ **v.** $\operatorname{mod}(0 + 8i) = \sqrt{0^2 + 8^2} = \sqrt{64} = 8$

$\quad$ **b.** $\operatorname{mod}(5 + 12i) = \sqrt{5^2 + 12^2} = \sqrt{25 + 144} = \sqrt{169} = 13$

$\qquad \operatorname{mod}(11 + 6i) = \sqrt{11^2 + 6^2} = \sqrt{121 + 36} = \sqrt{157} \approx 12.53$

$\qquad \operatorname{mod}(5 + 12i) > \operatorname{mod}(11 + 6i)$

75. $\overline{zw} = \overline{(a + bi)(c + di)}$

$\qquad = \overline{(ac - bd) + (bc + ad)i}$

$\qquad = (ac - bd) - (bc + ad)i$

$\bar{z} \cdot \bar{w} = \overline{(a + bi)} \cdot \overline{(c + di)}$

$\qquad = (a - bi)(c - di)$

$\qquad = (ac - bd) - (bc + ad)i$

Thus $\overline{zw} = \bar{z} \cdot \bar{w}$

77. $\bar{\bar{z}} = \overline{\overline{a + bi}} = \overline{a - bi} = a + bi = z$

79. a. $\dfrac{z + \bar{z}}{2} = \dfrac{(a + bi) + (a - bi)}{2} = \dfrac{2a}{2} = a$ $\qquad$ **b.** $\dfrac{z - \bar{z}}{2i} = \dfrac{(a + bi) - (a - bi)}{2i} = \dfrac{2bi}{2i} = b$

81. a. (i) $(a, b) + (c, d) = (a + c, b + d) = (c + a, d + b) = (c, d) + (a, b)$

$\qquad$ (ii) $\big[(a, b) + (c, d)\big] + (e, f) = (a + c, b + d) + (e, f) = \big((a + c) + e, (b + d) + f\big)$

$\qquad\qquad = \big(a + (c + e), b + (d + f)\big) = (a, b) + \big[(c, d) + (e, f)\big]$

$\qquad$ (iii) $(a, b) + (0, 0) = (a + 0, b + 0) = (a, b)$

$\qquad$ (iv) $(a, b) + (-a, -b) = \big(a + (-a), b + (-b)\big) = (0, 0)$

Continued on next page

81. continued

   b. (i) $(a,b)(c,d) = (ac-bd, bc+ad) = (ca-db, cb+da) = (c,d)(a,b)$

      (ii) $\left[(a,b)(c,d)\right](e,f) = (ac-bd, bc+ad)(e,f)$

$$= (ace-adf-bcf-bde, acf+ade+bce-bdf)$$

$$= (a,b)(ce-df, cf+de) = (a,b)\left[(c,d)(e,f)\right]$$

      (iii) $(a,b)(1,0) = (a\cdot1-b\cdot0, a\cdot0+b\cdot1) = (a,b)$

      (iv) $(a,b)(0,0) = (a\cdot0-b\cdot0, a\cdot0+b\cdot0) = (0,0)$

   c. (i) $(a,0)+(c,0) = (a+c, 0+0) = (a+c, 0)$

      (ii) $(a,0)(c,0) = (a\cdot c-0\cdot0, a\cdot0+c\cdot0) = (ac,0)$

   d. (i) $(0,1)(0,1) = (0\cdot0-1\cdot1, 0\cdot1+0\cdot1) = (-1,0)$

      (ii) $(b,0)(0,1) = (b\cdot0-0\cdot1, b\cdot1+0\cdot0) = (0,b)$

      (iii) $(a,0)+(b,0)(0,1) = (a,0)+(0,b) = (a+0, 0+b) = (a,b)$

4.8 Theory of Equations

1. The remainder is $f(1) = 1^{15} + 1^{10} + 1^2 = 3$.

3. The remainder is $f(-1) = (-1)^5 + (-1)^4 + (-1)^3 + (-1)^2 = -1+1-1+1 = 0$.

5. The remainder is $f(2) = 2^5 - 3(2)^3 - 2(2)^2 = 32-24-8 = 0$. Since the remainder is zero, $x-2$ is a factor.

7. The remainder is $f(x) = (3+i)i^3 + (1-2i)i^2 + (2+i)i + (1-i)$

$$= (3+i)(-i) + (1-2i)(-1) + 2i + i^2 + 1 - i$$

$$= -3i - i^2 - 1 + 2i + 2i - 1 + 1 - i$$

$$= 0$$

Since the remainder is zero, $x-i$ is a factor.

9. 0 — multiplicity 54; $-\frac{4}{5}$ — multiplicity 1.

11. 0 — multiplicity 15; — multiplicity 14; $+1$ — multiplicity 13.

13. $x^2 - 2x + 5 = 0$

$\qquad (x-1)^2 = -4$

$\qquad\qquad x = 1 \pm 2i$

$[x-(1+2i)][x-(1-2i)]$

15. $3x^2 + 18x + 27 = 0$

$3(x^2 + 6x + 9) = 0$

$3(x+3)^2 = 0$

$x = -3$

-3 with multiplicity 2

$3(x+3)^2$ or $3(x+3)(x+3)$

17. $x^3 - 27 = 0$

$(x-3)(x^2 + 3x + 9) = 0$

$x = 3$ or $x = \dfrac{-3 \pm \sqrt{-27}}{2}$

$x = \dfrac{-3 \pm 3\sqrt{3}i}{2}$

$(x-3)\left(x - \dfrac{-3 + 3\sqrt{3}i}{2}\right)\left(x - \dfrac{-3 - 3\sqrt{3}i}{2}\right)$

19. $x^3 + 8 = 0$

$(x+2)(x^2 - 2x + 4) = 0$

$x = -2$ or $x = \dfrac{-(-2) \pm \sqrt{(-2)^2 - 4(1)(4)}}{2(1)}$

$x = \dfrac{2 \pm \sqrt{-12}}{2}$

$x = 1 \pm \sqrt{3}i$

$(x+2)\left[x - \left(1 + \sqrt{3}i\right)\right]\left[x - \left(1 - \sqrt{3}i\right)\right]$

21. $x^4 - 1 = 0$

$(x^2 - 1)(x^2 + 1) = 0$

$x^2 - 1 = 0$ or $x^2 + 1 = 0$

$x = \pm 1 \qquad x = \pm i$

$(x-1)(x+1)(x-i)(x+i)$

23. $0 = x^4 + 2x^3 + x^2 - 2x - 2$

$0 = x^2\left(x^2 + 2x + 1\right) - 2(x+1)$

$0 = x^2(x+1)(x+1) - 2(x+1)$

$0 = (x+1)\left(x^2(x+1) - 2\right)$

$0 = (x+1)\left(x^3 + x^2 - 2\right)$

$0 = (x+1)(x-1)\left(x^2 + 2x + 2\right)$

$x+1 = 0$ or $x-1 = 0$ or $x^2 + 2x + 2 = 0$

$x = -1$ or $x = 1$ or $x = \dfrac{-2 \pm \sqrt{2^2 - 4(1)(2)}}{2(1)}$

$\qquad\qquad\qquad = \dfrac{-2 \pm \sqrt{-4}}{2}$

$\qquad\qquad\qquad = \dfrac{-2 \pm 2i}{2}$

$\qquad\qquad\qquad = -1 \pm i$

$(x+1)(x-1)\big(x-(-1+i)\big)\big(x-(-1-i)\big)$

25. $(x-1)(x-7)(x+4)$

27. $(x-1)^2(x-2)^2(x-\pi)^2$

29. Let $f(x) = a(x+3)x(x-4)$
Since $f(5) = 80$,

$\qquad 80 = a(5+3)5(5-4)$

$\qquad\ \ a = 2$

$f(x) = 2x(x+3)(x-4)$

31. $\big[x-(2+i)\big]\big[x-(2-i)\big] = (x-2)^2 - i^2 = x^2 - 4x + 5$

33. $1+i$ and $1-2i$ must also be roots:
$$(x+3)\left[x-(1-i)\right]\left[x-(1+i)\right]\left[x-(1-2i)\right]\left[x-(1+2i)\right]$$
$$= (x+3)(x^2-2x+2)(x^2-2x+5)$$
$$= (x+3)(x^4-4x^3+11x^2-14x+10)$$
$$= x^5-x^4-x^3+19x^2-32x+30$$

35. $\left[x-(1+2i)\right]\left[x-(1-2i)\right]=(x-1)^2-4i^2=x^2-2x+5$

37. $(x-4)^2\left[x-(3+i)\right]\left[x-(3-i)\right]=(x^2-8x+16)(x^2-6x+10)$
$$= x^4-14x^3+74x^2-176x+160$$

39. $x^3(x-3)\left[x-(1+i)\right]\left[x-(1-i)\right]=(x^4-3x^3)(x^2-2x+2)= x^6-5x^5+8x^4-6x^3$

41. $1-i$ must also be a root:
$f(x)=a\left[x-(1+i)\right]\left[x-(1-i)\right]$
Since $f(0)=6$,
$6= a\left[-(1+i)\right]\left[-(1-i)\right]$
$6= 2a$
$a=3$
$f(x)=3\left[x-(1+i)\right]\left[x-(1-i)\right]$
$$= 3(x^2-2x+2)=3x^2-6x+6$$

43. $(x-i)\left[x-(1-2i)\right]$
$$= x^2+(-1+i)x+(2+i)$$

45. $(x-3)(x-i)\left[x-(2-i)\right]$
$$= x^3-5x^2+(7+2i)x+(-3-6i)$$

47. Since 3 is a root, $x-3$ is a factor. Synthetic division by $x-3$ yields x^2+x+1, which has roots $\dfrac{-1\pm\sqrt{3}i}{2}$.

49. Since i is a root, $-i$ is also a root. Division by $(x-i)(x+i)$, or x^2+1, yields x^2+3x+2, which has roots -1 and -2.

51. Since $2-i$ is a root, $2+i$ is also a root. Division by $\left[x-(2-i)\right]\left[x-(2+i)\right]$, or x^2-4x+5, yields x^2+1, which has roots $\pm i$.

53. a. $\overline{z+w} = \overline{(a+bi)+(c+di)}$

$= \overline{(a+c)+(b+d)i}$

$= (a+c)-(b+d)i$

$= a-bi+c-di$

$= \overline{z}+\overline{w}$

b. $\overline{z \cdot w} = \overline{(a+bi) \cdot (c+di)}$

$= \overline{(ac-bd)+(bc+ad)i}$

$= (ac-bd)-(bc+ad)i$

$= (a-bi)(c-di)$

$= \overline{z} \cdot \overline{w}$

55. a. If z is a root of $f(x)$, $f(z)=0$. Then $\overline{f(z)} = \overline{0} = 0$. Also

$\overline{f(z)} = \overline{az^3 + bz^2 + cz + d}$

$= \overline{az^3} + \overline{bz^2} + \overline{cz} + \overline{d}$

$= \overline{a}\,\overline{z^3} + \overline{b}\,\overline{z^2} + \overline{c}\,\overline{z} + \overline{d}$

$= a\overline{z}^3 + b\overline{z}^2 + c\overline{z} + d = f(\overline{z})$

b. Since $\overline{f(z)} = 0$, $a\overline{z}^3 + b\overline{z}^2 + c\overline{z} + d = 0$, therefore $\overline{z}$ is a root of $f(x)$.

57. For each non-real complex root z, there must be two factors of the polynomial: $(x-z)$ and $(x-\overline{z})$. This yields an even number of factors. There will remain at least one factor, hence at least one root, which must be real.

Chapter 4 Review Exercises

1. The graph of $f(x)$ is the graph of $g(x) = x^2$ shifted horizontally 2 units to the right and shifted 3 units upward. The vertex of the graph is $(2,3)$.

3. We rewrite the rule of f as follows:
$$f(x) = x^2 - 8x + 12$$
$$= \left(x^2 - 8x + 16 - 16\right) + 12$$
$$= \left(x^2 - 8x + 16\right) - 16 + 12$$
$$= (x-4)^2 - 4$$
The graph of $f(x)$ is the graph of $g(x) = x^2$ shifted horizontally 4 units to the right and shifted 4 units downward. The vertex of the graph is $(4,-4)$.

5. The graph is an upward-opening parabola (because f is a quadratic function and the coefficient of x^2 is positive). The x-coordinate of its vertex is
$$-\frac{b}{2a} = -\frac{-9}{2(3)} = \frac{9}{6} = 1.5$$
The y-coordinate of its vertex is $f(1.5) = 3(1.5)^2 - 9(1.5) + 1 = -5.75$
The vertex of the graph is $(1.5, -5.75)$.

7. **a.** The fence consists of four pieces, of lengths x, $x - 70$, y, and $y - 50$, respectively. Therefore, the total amount of fencing is given by: $x + (x - 70) + y + (y - 50) = 400$
Solving this equation for y in terms of x, we obtain:
$$2x + 2y - 120 = 400$$
$$2y = 520 - 2x$$
$$y = 260 - x$$
b. Since the area of a rectangle is the product of the length and the width, the area of the playground is the area of the rectangle with dimensions x and y, minus the area of the school itself. Thus,
$$A(x) = xy - (70)(50) = x(260 - x) - 3500 = -x^2 + 260x - 3500$$
c. Since the function in part **b** is quadratic, and the coefficient of x^2 is negative, the function attains its maximum value at its vertex. The x-coordinate of its vertex is
$$-\frac{b}{2a} = -\frac{260}{2(-1)} = 130$$
Therefore, the dimensions of the playground with the largest possible area are $x = 130$ feet and $y = 130$ feet.

9. Sketch a figure.

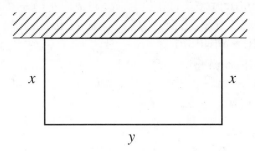

Let x be the width and y be the length of the garden. From the figure it is clear that:
$2x + y = 120$; $y = 120 - 2x$

$$A = xy = x(120 - 2x)$$

Rewriting this as $A(x) = 120x - 2x^2 = -2x^2 + 120x$, we have a quadratic function where the coefficient of x^2 is negative. The function attains its maximum value at the vertex of the graph of $A(x)$, when

$$x = -\frac{b}{2a} = -\frac{120}{2(-2)} = 30$$

Thus, the lengths of the other three sides are given by $x = 30$ feet, $y = 60$ feet, and $x = 30$ feet.

11. **a**, **c**, **d**, and **f** are polynomials.

13. To find the remainder when $f(x) = x^{112} - 2x^8 + 9x^5 - 4x^4 + x - 5$ is divided by $x - 1$, we apply the Remainder Theorem with $c = 1$. The remainder is
$f(1) = 1^{112} - 2 \cdot 1^8 + 9 \cdot 1^5 - 4 \cdot 1^4 + 1 - 5 = 1 - 2 + 9 - 4 + 1 - 5 = 0$.

15. The synthetic division is performed as follows:

$$
\begin{array}{r|rrrrrr}
2 & 1 & -5 & 8 & 1 & -17 & 16 & -4 \\
  &   & 2 & -6 & 4 & 10 & -14 & 4 \\
\hline
  & 1 & -3 & 2 & 5 & -7 & 2 & \underline{0}
\end{array}
$$

The last row shows that the quotient is $\square x^5 - 3x^4 + 2x^3 + 5x^2 - 7x + 2$. This is the other factor; since the remainder is zero, $x - 2$ is a factor.

17. A general rule for a polynomial function of degree 3 is $f(x) = ax^3 + bx^2 + cx + d$. In this case, since $f(0) = 5$, substitution gives $5 = d$; thus $f(x) = ax^3 + bx^2 + cx + 5$. Furthermore, since $f(1) = 0$ and $f(-1) = 0$, substitution gives $0 = a + b + c + 5$ and $0 = -a + b - c + 5$. Therefore $2b + 10 = 0$ and $b = -5$, and $c = -a$. Thus any polynomial of the form $f(x) = ax^3 - 5x^2 - ax + 5$ meets the requirements of the problem, for example, choosing $a = 1$, $f(x) = x^3 - 5x^2 - x + 5$.

19. The roots of the polynomial $3x^2 - 2x - 5$ are the solutions of the quadratic equation $3x^2 - 2x - 5 = 0$. Solving this equation yields:
$$3x^2 - 2x - 5 = 0$$
$$(3x - 5)(x + 1) = 0$$
$$3x - 5 = 0 \quad \text{or} \quad x + 1 = 0$$
$$x = \tfrac{5}{3} \qquad\qquad x = -1$$

21. The roots of $x^6 - 4x^3 + 4$ are the solutions of $x^6 - 4x^3 + 4 = 0$. Solving this equation yields:
$$x^6 - 4x^3 + 4 = 0$$
$$\left(x^3 - 2\right)\left(x^3 - 2\right) = 0$$
$$x^3 - 2 = 0$$
The only real solution of this equation is $x = \sqrt[3]{2}$, and this is the only real root of the polynomial.

23. Clearly, 0 is a root of multiplicity 3 for $3y^3\left(y^4 - y^2 - 5\right)$. To find the other real roots, solve the equation in quadratic form $y^4 - y^2 - 5 = 0$, first for y^2. The quadratic formula applied to this equation yields:
$$y^2 = \frac{1 \pm \sqrt{21}}{2}$$
Since only the choice of the positive sign yields a positive value for y^2, the only real roots are the solutions of
$$y^2 = \frac{1 + \sqrt{21}}{2},$$
that is,
$$y = \pm\sqrt{\frac{1 + \sqrt{21}}{2}} \quad \text{or} \quad y = \pm\frac{\sqrt{2 + 2\sqrt{21}}}{2}$$

25. a. If the polynomial has a rational root r / s, then by the Rational Root Test, r must be a factor of the constant term 3. Therefore r must be one of ± 1 or ± 3. Similarly s must be a factor of the leading coefficient 2, so s must be one of ± 1 or ± 2. The only possible rational roots are therefore

$$1, -1, 3, -3, \tfrac{1}{2}, -\tfrac{1}{2}, \tfrac{3}{2}, -\tfrac{3}{2}$$

b. Graph the polynomial in a viewing window that includes all of these numbers on the x-axis, say $-4 \le x \le 4$ and $-5 \le y \le 5$.

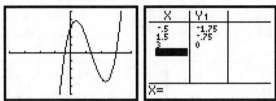

From the graph, the only numbers on the list that could possibly be roots are $-\tfrac{1}{2}$, $\tfrac{3}{2}$, and 3. The table shows that only 3 is a root. The other roots must be irrational numbers.

c. Synthetic division by $x - 3$ yields

$$
\begin{array}{r|rrrr}
3 & 2 & -8 & 5 & 3 \\
 &   & 6 & -6 & -3 \\
\hline
 & 2 & -2 & -1 & \underline{\,0}
\end{array}
$$

The other roots are therefore the solutions of the equation $2x^2 - 2x - 1 = 0$.

Solving this equation by the quadratic formula yields $x = \dfrac{1 \pm \sqrt{3}}{2}$ in addition to the root $x = 3$.

27. Since $x^3 + 4x = x\left(x^2 + 4\right)$, and $x^2 + 4$ has no real factors, the only real root of the polynomial is 0.

29. The equation $x^4 - 11x^2 + 18 = 0$ can be solved by factoring as follows:

$$x^4 - 11x^2 + 18 = 0$$

$$\left(x^2 - 9\right)\left(x^2 - 2\right) = 0$$

$$(x - 3)(x + 3)\left(x^2 - 2\right) = 0$$

$$x - 3 = 0 \quad \text{or} \quad x + 3 = 0 \quad \text{or} \quad x^2 - 2 = 0$$

$$x = 3 \qquad\qquad x = -3 \qquad\qquad x = \pm\sqrt{2}$$

Thus, the roots of the polynomial are 3, –3, $\sqrt{2}$, and $-\sqrt{2}$.

31. Synthetic division by $x - 5$ yields

$$
\begin{array}{r|rrrr}
5 & 1 & -4 & 0 & 16 & -16 \\
  &   & 5 & 5 & 25 & 205 \\
\hline
  & 1 & 1 & 5 & 41 & \underline{|189}
\end{array}
$$

Since every number in the last row in the synthetic division is nonnegative, 5 is an upper bound for the real roots of the polynomial.

33. If the polynomial has a rational root, by the Rational Root Test, it can only be ± 1. Graph the polynomial in a viewing window that includes all of these numbers on the x-axis, say $-2 \le x \le 2$ and $-5 \le y \le 5$. The graph shows, and a table confirms, that -1 and 1 are both roots.

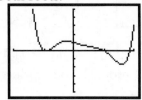 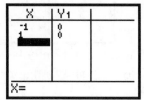

   Synthetic division by $x - 1$ yields a quotient of $x^5 - x^4 - 2x^3 + x^2 - 1$. Synthetic division of this reduced polynomial by $x + 1$ yields a quotient of $x^4 - 2x^3 + x - 1$. This polynomial has no further rational roots; use the zero (root) routine to find two further real roots of -0.867 and 1.867.

35. The difference quotient is computed as follows:

$$
\frac{f(x+h) - f(x)}{h} = \frac{\left[(x+h)^2 + (x+h)\right] - \left(x^2 + x\right)}{h} = \frac{x^2 + 2xh + h^2 + x + h - x^2 - x}{h}
$$

$$
= \frac{2xh + h^2 + h}{h} = \frac{h(2x + h + 1)}{h} = 2x + h + 1
$$

37. The graph below could not possibly be the graph of a polynomial function, because it has a sharp corner.

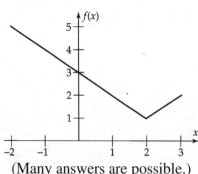

(Many answers are possible.)

39. Statement **c** is false. $(f \circ f)(0) = f(f(0)) = f(2)$ is negative.

41. Use the window $-1 \leq x \leq 2, -10 \leq y \leq 60$.

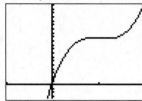

43. The window $-2 \leq x \leq 18, -500 \leq y \leq 1200$ provides a complete graph of the function.

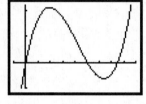

45. a. Entering the data into the TI-83 yields the following:

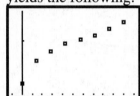

b. Applying the cubic regression routine yields:

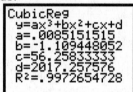

c. The cost of making the seventy-first statue is $C(71) - C(70)$. Using the calculator values for $C(70)$ and $C(71)$:
$$C(71) - C(70) = 3466.53 - 3439.74 = \$26.79.$$

d. Average Cost $= C(x)/x$

$$\frac{C(35)}{35} = \frac{2992.3125}{35} = \$85.49$$

$$\frac{C(75)}{75} = \frac{3588.3168}{75} = \$47.84$$

47. x-intercepts: $-3, 0, 3$
local maximum: $(-1.732, 10.392)$
local minimum: $(1.732, -10.392)$

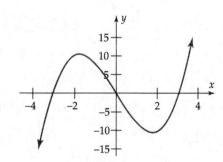

49. x-intercepts: $(-4,0),(-3,0),(-2,0)$

 y-intercept: $(0,48)$

 local maximum: $(-.71,55.33)$

 local minima: $(-3.54,-7.624)$,

 $(2,0)$

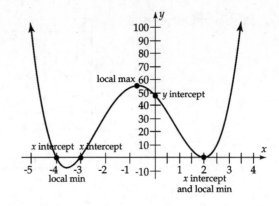

51. x-intercepts: none
vertical asymptote: $x=-4$
horizontal asymptote: $y=0$
holes: none

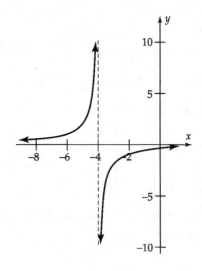

53. x-intercepts: -2.5
vertical asymptote: $x=3$
horizontal asymptote: $y=\frac{4}{3}$
holes: none

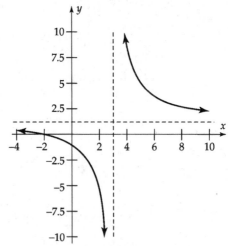

55. $f(x) = \dfrac{(x+1)(x-1)}{(x-1)(x+2)(x-3)} = \dfrac{x+1}{(x+2)(x-3)}$

vertical asymptotes: $x = -2$, $x = 3$

horizontal asymptote: $y = 0$

57. Use windows $-4.7 \le x \le 4.7$, $-5 \le y \le 5$ and $2 \le x \le 20$, $-.2 \le y \le .1$

59. Use window $-18.8 \le x \le 18.8$, $-8 \le y \le 8$

61. Use $C(x) = .5x + 500$, $R(x) = xp = x\left(1.95 - \dfrac{x}{2000}\right)$.

Then $P(x) = R(x) - C(x) = 1.45x - 500 - \dfrac{x^2}{2000}$

Graph using the window $0 \le x \le 3000$, $0 \le y \le 600$ and use the zero (root) routine to find that $P(x)$ is positive if x is between 400 and 2500 bags of Munchies. If $x = 400$, $p = \$1.75$; if $x = 2500$, $p = \$.70$.

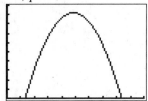

63. a. Graph the function and the horizontal line $s = .21$ using the window $0 \le x \le 10$, $0 \le y \le 1$, and use the intersection routine to find that the survival rate will be above .21 between $x = .4547$ and $x = 5.4976$.

b. Use the maximum routine to find the maximum survival rate of .395 when $x = 1.5811$ meters.

65. $\dfrac{g(x+h) - g(x)}{h} = \dfrac{\dfrac{1}{(x+h)^2 + 1} - \dfrac{1}{x^2 + 1}}{h}$

$= \dfrac{x^2 + 1 - \left[(x+h)^2 + 1\right]}{h(x^2 + 1)\left[(x+h)^2 + 1\right]}$

$= \dfrac{-2xh - h^2}{h(x^2 + 1)\left[(x+h)^2 + 1\right]}$

$= \dfrac{-2x - h}{(x^2 + 1)\left[(x+h)^2 + 1\right]}$

67. $-3(x-4) \le 5 + x$

$-3x + 12 \le 5 + x$

$-4x \le -7$

$x \ge 7/4$

69. $-4 < 2x + 5 < 9$

$-9 < 2x < 4$

$-9/2 < x < 2$

71. To solve $\dfrac{2}{x+1} < x$, graph $\dfrac{2}{x+1} - x$ using the window $-6 \le x \le 4, -5 \le y \le 5$. The graph is below the x-axis on the interval $(-2,-1) \cup (1,\infty)$.

73. $x^2 + x > 12$

$x^2 + x - 12 > 0$

$(x+4)(x-3) > 0$

Interval	$x < -4$	$-4 < x < 3$	$3 < x$
Test number	-5	0	5
Sign	$+$	$-$	$+$
Graph	Above x-axis	Below x-axis	Above x-axis

Solution: $(-\infty, -4) \cup (3, \infty)$; $x < -4$ or $x > 3$.

75. **(c)** is false.

77. $|3x + 2| \ge 2$

$3x + 2 \ge 2$ or $3x + 2 \le -2$

$\qquad 3x \ge 0 \qquad\qquad\qquad 3x \le -4$

$\qquad\quad x \ge 0 \qquad\qquad\qquad\;\; x \le -4/3$

79. $\dfrac{x-2}{x+4} \le 3$ is equivalent to $\dfrac{x-2-3(x+4)}{x+4} \le 0$, that is, to $\dfrac{-2x-14}{x+4} \le 0$.

Interval	$x < -7$	$-7 < x < -4$	$-4 < x$
Test number	-8	-5	0
Sign	$-$	$+$	$-$
Graph	Below x-axis	Above x-axis	Below x-axis

Solution: $(-\infty, -7] \cup (-4, \infty)$; $x \le -7$ or $x > -4$.

81.

$\dfrac{x^2 + x - 9}{x+3} < 1$

$\dfrac{x^2 + x - 9}{x+3} - 1 < 0$

$\dfrac{x^2 + x - 9}{x+3} - \dfrac{x+3}{x+3} < 0$

$\dfrac{x^2 - 12}{x+3} < 0$

Graph function whose rule is given by the fractional expression using the window $-5 \le x \le 5, -10 \le y \le 10$. The graph is below the x-axis on the interval $\left(-\infty, -2\sqrt{3}\right) \cup \left(-3, 2\sqrt{3}\right)$; $x < -2\sqrt{3}$ or $-3 < x < 2\sqrt{3}$.

83.
$$\frac{x^2 - x - 5}{x^2 + 2} > -2$$

$$\frac{x^2 - x - 5}{x^2 + 2} + 2 > 0$$

$$\frac{x^2 - x - 5}{x^2 + 2} + \frac{2x^2 + 4}{x^2 + 2} > 0$$

$$\frac{3x^2 - x - 1}{x^2 + 2} > 0$$

Graph the function whose rule is given by the fractional expression using the window $-5 \le x \le 5, \ -1 \le y \le 4$. The graph is above the x-axis on the interval $\left(-\infty, \dfrac{1 - \sqrt{13}}{6}\right) \cup \left(\dfrac{1 + \sqrt{13}}{6}, \infty\right)$;

$x < \dfrac{1 + \sqrt{13}}{6}$ or $x > \dfrac{1 + \sqrt{13}}{6}$. (the radical expressions being the zeros of the numerator).

85. $x^2 + 3x + 10 = 0$

$$x = \frac{-3 \pm \sqrt{3^2 - 4 \cdot 1 \cdot 10}}{2 \cdot 1}$$

$$x = \frac{-3 \pm \sqrt{-31}}{2 \cdot 1}$$

$$x = -\frac{3}{2} \pm i\frac{\sqrt{31}}{2}$$

87. $5x^2 + 2 = 3x$

$5x^2 - 3x + 2 = 0$

$$x = \frac{-(-3) \pm \sqrt{(-3)^2 - 4 \cdot 5 \cdot 2}}{2 \cdot 5}$$

$$x = \frac{3 \pm \sqrt{-31}}{10}$$

$$x = \frac{3}{10} \pm i\frac{\sqrt{31}}{10}$$

89. $3x^4 + x^2 - 2 = 0$

$(3x^2 - 2)(x^2 + 1) = 0$

$3x^2 - 2 = 0$ or $x^2 + 1 = 0$

$$x = \pm\sqrt{\frac{2}{3}} \qquad x = \pm\sqrt{-1}$$

$$x = \pm\frac{\sqrt{6}}{3} \qquad x = \pm i$$

91. $x^3 + 8 = 0$

$(x + 2)(x^2 - 2x + 4) = 0$

$x + 2 = 0 \quad x^2 - 2x + 4 = 0$

$x = -2$
$$x = \frac{-(-2) \pm \sqrt{(-2)^2 - 4 \cdot 1 \cdot 4}}{2 \cdot 1}$$

$$x = \frac{2 \pm \sqrt{-12}}{2}$$

$$x = 1 \pm i\sqrt{3}$$

93. Since the polynomial has real coefficients, $-i$ must also be a root. Divide by $(x - i)(x + i) = x^2 + 1$ to obtain $x^2 - x - 2 = (x - 2)(x + 1)$. Thus the roots are $i, -i, 2$ and -1

95. $x^2 \left[x - (1+i) \right] \left[x - (1-i) \right] = x^2 \left(x^2 - 2x + 2 \right) = x^4 - 2x^3 + 2x^2$ is one possibility.

Chapter 4 Test

1. The object is at the start when $f(t) = 0$.

$$15t^4 - 5t^2 + 6t^3 - 2t = 0$$

$$5t^2 \left(3t^2 - 1 \right) + 2t \left(3t^2 - 1 \right) = 0$$

$$\left(3t^2 - 1 \right) \left(5t^2 + 2t \right) = 0$$

$$t \left(3t^2 - 1 \right) \left(5t + 2 \right) = 0$$

$t = 0$ or $3t^2 - 1 = 0$ or $5t + 2 = 0$

$t = 0$ or $3t^2 = 1$ or $5t = -2$

$t = 0$ or $t = \pm \sqrt{\dfrac{1}{3}}$ or $t = -\dfrac{2}{5}$

The only positive solution is $t = \sqrt{1/3} = .57735$ minutes

2. a.

$$\begin{array}{r} x^2+x+1 \\ x+4{\overline{\smash{\big)}\,x^3+5x^2+5x+7}} \end{array}$$

$$\underline{x^3+4x^2}$$
$$x^2+5x$$
$$\underline{x^2+4x}$$
$$x+7$$
$$\underline{x+4}$$
$$3$$

Quotient: x^2+x+1
Remainder: 3

b.

$$\begin{array}{r|rrrrr} 2 & 2 & -5 & 2 & -1 & 2 \\ & & 4 & -2 & 0 & -2 \\ \hline & 2 & -1 & 0 & -1 & \underline{|0} \end{array}$$

Quotient: $2x^3-x^2-1$
Remainder: 0

c.

$$\begin{array}{r} x^3-2x^2+x-1 \\ x^2-3{\overline{\smash{\big)}\,x^5-2x^4-2x^3+5x^2-7x+3}} \end{array}$$

$$\underline{x^5\qquad\quad -3x^3}$$
$$-2x^4+x^3+5x^2$$
$$\underline{-2x^4\qquad +6x^2}$$
$$x^3-x^2-7x$$
$$\underline{x^3\qquad -3x}$$
$$-x^2-4x+3$$
$$\underline{-x^2\qquad +3}$$
$$-4x$$

Quotient: x^3-2x^2+x-1
Remainder: $-4x$

3. a. $f(x)=(x-1)^2+2$

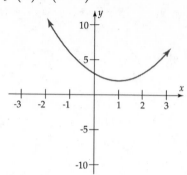

b. $f(x)=(x+1)^2+2$

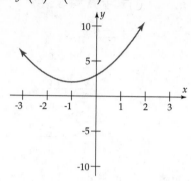

c. $f(x)=(x-1)^2-2$

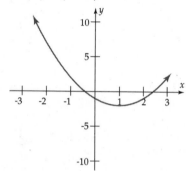

d. $f(x)=(x+1)^2-2$

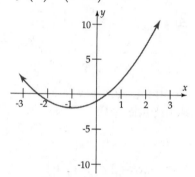

e. $f(x)=-(x-1)^2+2$

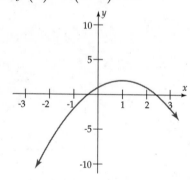

f. $f(x)=-(x+1)^2+2$

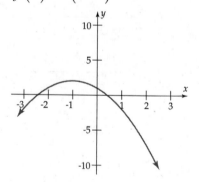

g. $f(x)=-(x-1)^2-2$

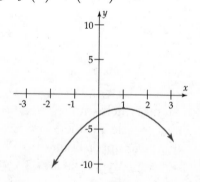

h. $f(x)=-(x+1)^2-2$

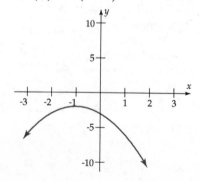

4. a. Let $x =$ the price of the ticket and $y =$ the number of 25 cent decreases. If $x = 10 - .25y$, then number of tickets sold is $75 + 5y$. The income,

$$I = (\text{price}) \times (\text{number of tickets sold})$$

$$= x(75 + 5y) = x(75 + 5(40 - 4x)) = x(75 + 20(10 - x))$$

$$= x(75 + 200 - 20x) = -20x^2 + 275x$$

b. Since the function that models the income is a down facing parabola, the maximum income will occur at the vertex of the parabola.

$$-\frac{b}{2a} = -\frac{275}{2(-20)} \approx \$6.88$$

5. a. $3x^3 - x^2 - 3x + 1 = 0 \Rightarrow x^2(3x - 1) - (3x - 1) = 0$

$(3x - 1)(x^2 - 1) = 0 \Rightarrow (3x - 1)(x - 1)(x + 1) = 0$

$x = \frac{1}{3}; \ x = 1; \ x = -1$

b.
$$x^4 - 11x^2 + 18 = 0$$

$$(x^2 - 9)(x^2 - 2) = 0$$

$$(x - 3)(x + 3)(x - \sqrt{2})(x - \sqrt{2}) = 0$$

$$x = 3; \ x = -3; \ x = \sqrt{2}; \ x = -\sqrt{2}$$

c. Using the Rational Root Test, the possible rational roots are

$$\frac{\pm 1}{\pm 1}, \frac{\pm 2}{\pm 1}, \frac{\pm 3}{\pm 1}, \frac{\pm 6}{\pm 1}.$$

Eliminating duplicates we have possible rational roots of $1, -1, 2, -2, 3, -3, 6, -6$.
Graph g in the window $-5 \le x \le 5, \ -10 \le y \le 10$.

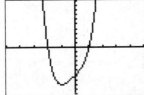 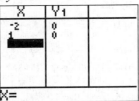

Using the table we can see that -2 and 1 are the only rational roots.
Since the polynomial is of 4th degree, there are possibly two more roots. We use the bounds test with -3 as a lower bound and 2 as an upper bound to be sure that we have all of the roots.

-3	1	1	1	3	-6		2	1	1	1	3	-6
		-3	6	-21	54				2	6	14	34
	1	-2	7	-18	48			1	3	7	17	28

Synthetic division by $x + 3$ yields alternating signs which implies that -3 is a lower bound for the real roots of the polynomial, while synthetic division by $x - 2$ yields all positive signs which implies that 2 is an upper bound for the real roots of the polynomial. Thus the only real roots of the polynomial are -2 and 1.

6. $x + y = 30$ and $xy > 200$

$$xy > 200 \Rightarrow x(30 - x) > 200 \Rightarrow -x^2 + 30x - 200 > 0$$

$$x^2 - 30x + 200 < 0 \Rightarrow (x - 10)(x - 20) < 0$$

Interval	$x < 10$	$10 < x < 20$	$20 < x$
Test number	5	15	25
Sign	+	−	+
Graph	Above x-axis	Below x-axis	Above x-axis

$10 < x < 20$

7. a.
$$2x + 3 - 4i = 3x + 6 + 2i$$
$$2x - 3x + 3 - 4i = 3x - 3x + 6 + 2i$$
$$-x + 3 - 3 - 4i + 4i = 6 - 3 + 2i + 4i$$
$$-x = 3 + 6i$$
$$x = -3 - 6i$$

b. $x^2 - 4x + 13 = 0$

$$x = \frac{-(-4) \pm \sqrt{(-4)^2 - 4(1)(13)}}{2(1)} = \frac{4 \pm \sqrt{16 - 52}}{2} = \frac{4 \pm \sqrt{-36}}{2} = \frac{4 \pm 6i}{2} = 2 \pm 3i$$

$2 + 3i, \ 2 - 3i$

c. $4 + x^2 = 5x^2 + 5 \Rightarrow -1 = 4x^2 \Rightarrow \dfrac{-1}{4} = \dfrac{4x^2}{4} \Rightarrow \pm\sqrt{-\dfrac{1}{4}} = \sqrt{x^2} \Rightarrow \pm\dfrac{1}{2}i = x$

$\dfrac{1}{2}i, \ -\dfrac{1}{2}i$

8. a. $f(0) = \dfrac{0 + 1}{0^2 + 2(0) - 8} = \dfrac{1}{-8} = -\dfrac{1}{8}$

b. $x^2 + 2x - 8 = 0 \Rightarrow (x + 4)(x - 2) = 0 \Rightarrow x = -4, \ x = 2$

The vertical asymptotes are at $x = -4$ and $x = 2$. Since the degree of the numerator is less that the degree of the denominator, the horizontal asymptote is at $y = 0$.

c. $x + 1 = 0$ The zero of the function is at $(-1, 0)$

$x = -1$

d.

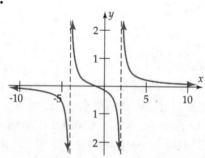

9. **a.** $3x - 6 \geq 4 - 2x \Rightarrow 5x \geq 10 \Rightarrow \dfrac{5x}{5} \geq \dfrac{10}{5} \Rightarrow x \geq 2$

b. $x^2 - 3x + 2 < 0$

$(x - 1)(x - 2) < 0$

Interval	$x < 1$	$1 < x < 2$	$2 < x$
Test number	0	1.5	3
Sign	+	−	+
Graph	Above x-axis	Below x-axis	Above x-axis

$1 < x < 2$

c. $\dfrac{3x - 1}{x + 2} > \dfrac{1}{x + 2} \Rightarrow \dfrac{3x - 1}{x + 2} - \dfrac{1}{x + 2} > 0 \Rightarrow \dfrac{3x - 2}{x + 2} > 0$

Interval	$x < -2$	$-2 < x < \frac{2}{3}$	$\frac{2}{3} < x$
Test number	−3	0	1
Sign	+	−	+
Graph	Above x-axis	Below x-axis	Above x-axis

$x > \frac{2}{3}$ or $x < -2$

d. $|2x - 1| < 5 \Rightarrow -5 < 2x - 1 < 5 \Rightarrow -4 < 2x < 6 \Rightarrow -2 < x < 3$

e. $|4x - 6| \geq 12$

$4x - 6 \geq 12$ or $4x - 6 \leq -12$

$4x \geq 18$ or $4x \leq -6$

$x \geq \dfrac{9}{2}$ or $x \leq -\dfrac{3}{2}$

10. $x - 1 \overline{) 3x^2 - 3x + 1}$ with quotient $3x$ The non vertical asymptote is at $y = 3x$.

$\underline{3x^2 - 3x}$

1

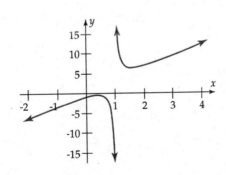

11. $f(x) = x\big(x - (1 + 2i)\big)\big(x - (1 - 2i)\big)\big(x - (5 + i)\big)\big(x - (5 - i)\big)$

$= x^5 - 12x^4 + 51x^3 - 102x^2 + 130x$

Chapter 5
Exponential and Logarithmic Functions

5.1 Radical and Rational Exponents

1. $\left(25k^2\right)^{\frac{3}{2}}\left(16k^{\frac{1}{3}}\right)^{\frac{3}{4}} = \left(25^{\frac{3}{2}}k^3\right)\left(16^{\frac{3}{4}}k^{\frac{1}{4}}\right) = 125k^3 \cdot 8 \cdot k^{\frac{1}{4}} = 1000k^{\frac{13}{4}}$

3. $\left(c^{\frac{2}{5}}d^{-\frac{2}{3}}\right)\left(c^6d^3\right)^{\frac{4}{3}} = c^{\frac{2}{5}}d^{-\frac{2}{3}}c^8d^4 = c^{\frac{42}{5}}d^{\frac{10}{3}}$

5. $\dfrac{\left(x^2\right)^{\frac{1}{3}}\left(y^2\right)^{\frac{2}{3}}}{3x^{\frac{2}{3}}y^2} = \dfrac{x^{\frac{2}{3}}y^{\frac{4}{3}}}{3x^{\frac{2}{3}}y^2} = \dfrac{1}{3y^{\frac{2}{3}}}$

7. $\dfrac{(7a)^2(5b)^{\frac{3}{2}}}{(5a)^{\frac{3}{2}}(7b)^4} = \dfrac{7^2a^2 5^{\frac{3}{2}}b^{\frac{3}{2}}}{5^{\frac{3}{2}}a^{\frac{3}{2}}7^4b^4} = \dfrac{a^{\frac{1}{2}}}{49b^{\frac{5}{2}}}$

9. $\left(a^{x^2}\right)^{\frac{1}{x}} = a^x$

11. $x^{\frac{1}{2}}\left(x^{\frac{2}{3}} - x^{\frac{4}{3}}\right) = x^{\frac{1}{2}}x^{\frac{2}{3}} - x^{\frac{1}{2}}x^{\frac{4}{3}} = x^{\frac{7}{6}} - x^{\frac{11}{6}}$

13. $\left(x^{\frac{1}{2}} + y^{\frac{1}{2}}\right)\left(x^{\frac{1}{2}} - y^{\frac{1}{2}}\right) = x^{\frac{1}{2}}x^{\frac{1}{2}} - y^{\frac{1}{2}}y^{\frac{1}{2}} = x - y$

15. $(x+y)^{\frac{1}{2}}\left[(x+y)^{\frac{1}{2}} - (x+y)\right] = (x+y)^{\frac{1}{2}}(x+y)^{\frac{1}{2}} - (x+y)^{\frac{1}{2}}(x+y) = (x+y) - (x+y)^{\frac{3}{2}}$

17. $\left(x^{\frac{1}{3}} + 3\right)\left(x^{\frac{1}{3}} - 2\right)$

19. $\left(x^{\frac{1}{2}} + 1\right)\left(x^{\frac{1}{2}} + 3\right)$

21. $x^{\frac{4}{5}} - 81 = \left(x^{\frac{2}{5}} - 9\right)\left(x^{\frac{2}{5}} + 9\right) = \left(x^{\frac{1}{5}} - 3\right)\left(x^{\frac{1}{5}} + 3\right)\left(x^{\frac{2}{5}} + 9\right)$

23. $x^{-\frac{1}{2}}$

25. $a(a+b)^{\frac{1}{2}}$

27. $t^{\frac{1}{5}}16^{\frac{1}{2}}t^{\frac{1}{2}} = 4t^{\frac{27}{10}}$

29. $\sqrt{80} = \sqrt{16 \cdot 5} = \sqrt{16}\sqrt{5} = 4\sqrt{5}$

31. $\sqrt{6}\sqrt{12} = \sqrt{6}\sqrt{6}\sqrt{2} = 6\sqrt{2}$

33. $\dfrac{-6+\sqrt{99}}{15} = \dfrac{-6+3\sqrt{11}}{15} = \dfrac{3\left(-2+\sqrt{11}\right)}{15} = \dfrac{-2+\sqrt{11}}{5}$

35. $\sqrt{50} - \sqrt{72} = \sqrt{25 \cdot 2} - \sqrt{36 \cdot 2} = 5\sqrt{2} - 6\sqrt{2} = -\sqrt{2}$

37. $5\sqrt{20} - \sqrt{45} + 2\sqrt{80} = 5\sqrt{4 \cdot 5} - \sqrt{9 \cdot 5} + 2\sqrt{16 \cdot 5} = 10\sqrt{5} - 3\sqrt{5} + 8\sqrt{5} = 15\sqrt{5}$

39. $\sqrt{16a^8b^{-2}} = 4a^4b^{-1} = \dfrac{4a^4}{b}$

41. $\dfrac{\sqrt{c^2d^6}}{\sqrt{4c^3d^{-4}}} = \sqrt{\dfrac{c^2d^6}{4c^3d^{-4}}} = \sqrt{\dfrac{cd^{10}}{4c^2}} = \dfrac{\sqrt{c}\sqrt{d^{10}}}{\sqrt{4c^2}} = \dfrac{d^5\sqrt{c}}{2c}$

43. $\dfrac{\sqrt[3]{a^5b^4c^3}}{\sqrt[3]{a^{-1}b^2c^6}} = \sqrt[3]{\dfrac{a^5b^4c^3}{a^{-1}b^2c^6}} = \sqrt[3]{\dfrac{a^6b^2}{c^3}} = \dfrac{a^2b^{2/3}}{c} = \dfrac{a^2\sqrt[3]{b^2}}{c}$

45. $\dfrac{3}{\sqrt{8}} = \dfrac{3}{\sqrt{8}}\dfrac{\sqrt{2}}{\sqrt{2}} = \dfrac{3\sqrt{2}}{\sqrt{16}} = \dfrac{3\sqrt{2}}{4}$

47. $\dfrac{3}{2+\sqrt{12}} = \dfrac{3}{\left(2+\sqrt{12}\right)}\dfrac{\left(2-\sqrt{12}\right)}{\left(2-\sqrt{12}\right)} = \dfrac{6-3\sqrt{12}}{4-12} = \dfrac{6-6\sqrt{3}}{-8} = \dfrac{3\sqrt{3}-3}{4}$

49. $\dfrac{2}{\sqrt{x}+2} = \dfrac{2}{\left(\sqrt{x}+2\right)}\dfrac{\left(\sqrt{x}-2\right)}{\left(\sqrt{x}-2\right)} = \dfrac{2\left(\sqrt{x}-2\right)}{x-4}$

51. $\dfrac{10}{\sqrt[3]{2}}\cdot\dfrac{\sqrt[3]{2^2}}{\sqrt[3]{2^2}} = \dfrac{10\sqrt[3]{2^2}}{2} = 5\sqrt[3]{4}$

53. $\dfrac{1}{\sqrt[3]{3}+1}\cdot\dfrac{\left(\sqrt[3]{3^2}-\sqrt[3]{3}+1\right)}{\left(\sqrt[3]{3^2}-\sqrt[3]{3}+1\right)} = \dfrac{\sqrt[3]{9}-\sqrt[3]{3}+1}{3+1} = \dfrac{\sqrt[3]{9}-\sqrt[3]{3}+1}{4}$

55. $\dfrac{1}{\sqrt[3]{4}-\sqrt[3]{2}+1}\cdot\dfrac{\left(\sqrt[3]{2}+1\right)}{\left(\sqrt[3]{2}+1\right)} = \dfrac{\sqrt[3]{2}+1}{2+1} = \dfrac{\sqrt[3]{2}+1}{3}$

57. $\dfrac{f(x+h)-f(x)}{h} = \dfrac{\sqrt{x+h+1}-\sqrt{x+1}}{h}$

$= \dfrac{\sqrt{x+h+1}-\sqrt{x+1}}{h}\dfrac{\sqrt{x+h+1}+\sqrt{x+1}}{\sqrt{x+h+1}+\sqrt{x+1}}$

$= \dfrac{(x+h+1)-(x+1)}{h\left(\sqrt{x+h+1}+\sqrt{x+1}\right)} = \dfrac{1}{\sqrt{x+h+1}+\sqrt{x+1}}$

59.
$$\frac{f(x+h)-f(x)}{h}=\frac{\sqrt{(x+h)^2+1}-\sqrt{x^2+1}}{h}$$

$$=\frac{\sqrt{(x+h)^2+1}-\sqrt{x^2+1}}{h}\cdot\frac{\sqrt{(x+h)^2+1}+\sqrt{x^2+1}}{\sqrt{(x+h)^2+1}+\sqrt{x^2+1}}$$

$$=\frac{(x+h)^2+1-(x^2+1)}{h\left(\sqrt{(x+h)^2+1}+\sqrt{x^2+1}\right)}$$

$$=\frac{2xh+h^2}{h\left(\sqrt{(x+h)^2+1}+\sqrt{x^2+1}\right)}$$

$$=\frac{2x+h}{\sqrt{(x+h)^2+1}+\sqrt{x^2+1}}$$

61. $92.8935(.24)^{.6669}=35,863,131$ mi

63. $92.8935(29.46)^{.6669}=886,781,537$ mi

65. In 1800, $x=10$. $3.9572(1.0299)^{10}=5,312,985$

67. In 1845, $x=55$. $3.9572(1.0299)^{55}=20,003,970$

69. a. Since even powers of non-zero numbers, positive or negative, are positive, x^{2n} is positive and cannot equal -4.
b. $(-8)^{1/3}=-2$ but $(-8)^{1/6}=\sqrt[6]{(-8)^2}=\sqrt[6]{64}=2$.

71. a.

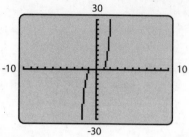

The function is one-to-one, as indicated by the fact
that the graph passes the horizontal line test.

b. Let $f(x) = x^5$ and $g(x) = x^{1/5}$. Then $(g \circ f)(x) = g(x^5) = (x^5)^{1/5} = x$

and $(f \circ g)(x) = f(x^{1/5}) = (x^{1/5})^5 = x$. Thus $g(x) = x^{1/5}$ is the inverse of $f(x) = x^5$.

c. No, $f(x) = x^6$ is not one-to-one so therefore does not have an inverse.

73. The graph of $g(x) = \sqrt{x+3}$ is the graph of $f(x) = \sqrt{x}$ shifted horizontally 3 units to
the left.

75. The graph of $k(x) = \sqrt{x+4} - 4$ is the graph of $f(x) = \sqrt{x}$ shifted horizontally
4 units to the left, then vertically 4 units down.

77. a.

L	C	$Q = L^{1/4}C^{3/4}$
10	7	7.65
20	14	15.31
30	21	22.96
40	28	30.61
60	42	45.92

b. If both labor and capital are doubled, output is doubled. If both are tripled, output
is tripled.

79. a.

L	C	$Q = L^{1/2}C^{3/4}$
10	7	13.61
20	14	32.37
30	21	53.73
40	28	76.98
60	42	127.79

b. If both labor and capital are doubled, output is multiplied by $2^{5/4}$. If both are
tripled, output is multiplied by $3^{5/4}$.

5.1.A Radical Equations

1.
$$\sqrt{x+2} = 3$$
$$\left(\sqrt{x+2}\right)^2 = 3^2$$
$$x+2 = 9$$
$$x = 7$$
The solution 7 checks.

3.
$$\sqrt{4x+9} = 0$$
$$\left(\sqrt{4x+9}\right)^2 = 0$$
$$4x+9 = 0$$
$$x = -\frac{9}{4}$$
The solution $-\frac{9}{4}$ checks.

5.
$$\sqrt[3]{5-11x} = 3$$
$$\left(\sqrt[3]{5-11x}\right)^3 = 3^3$$
$$5-11x = 27$$
$$-11x = 22$$
$$x = -2$$
The solution -2 checks.

7.
$$\sqrt[3]{x^2-1} = 2$$
$$\left(\sqrt[3]{x^2-1}\right)^3 = 2^3$$
$$x^2-1 = 8$$
$$x^2 = 9$$
$$x = \pm 3$$
The solutions 3 and -3 check.

9.
$$\sqrt{x^2-x-1} = 1$$
$$\left(\sqrt{x^2-x-1}\right)^2 = 1^2$$
$$x^2-x-1 = 1$$
$$x^2-x-2 = 0$$
$$x = 2 \ \text{ or } \ x = -1$$
The solutions 2 and -1 check.

11.
$$\sqrt{x+7} = x-5$$
$$\left(\sqrt{x+7}\right)^2 = (x-5)^2$$
$$x+7 = x^2-10x+25$$
$$0 = x^2-11x+18$$
$$x = 2 \ \text{ or } \ x = 9$$
The solution 9 checks. (2 is extraneous.)

13.
$$\sqrt{3x^2+7x-2} = x+1$$
$$\left(\sqrt{3x^2+7x-2}\right)^2 = (x+1)^2$$
$$3x^2+7x-2 = x^2+2x+1$$
$$2x^2+5x-3 = 0$$
$$(2x-1)(x+3) = 0$$
$$x = \frac{1}{2} \ \text{ or } \ x = -3$$
The solution $\frac{1}{2}$ checks. (−3 is extraneous.)

15.
$$\sqrt[3]{x^3+x^2-4x+5} = x+1$$
$$\left(\sqrt[3]{x^3+x^2-4x+5}\right)^3 = (x+1)^3$$
$$x^3+x^2-4x+5 = x^3+3x^2+3x+1$$
$$0 = 2x^2+7x-4$$
$$0 = (2x-1)(x+4)$$
$$x = \frac{1}{2} \ \text{ or } \ x = -4$$
The solutions $\frac{1}{2}$ and −4 check.

17. Graph $y = \sqrt[5]{9-x^2}$ and $y = x^2+1$ using the window $-4 \le x \le 4, -2 \le y \le 5$ and apply the intersection routine to find $x = .730$ and, by symmetry, $x = -.730$.

19. Graph $y = \sqrt[3]{x^5-x^3-x}$ and $y = x+2$ using the window $-2 \le x \le 4, -2 \le y \le 6$ and apply the intersection routine to find $x = -1.17, -1$, and 2.59.

21. $\sqrt{x^2+x-1}=\sqrt{14-x}$

$\left(\sqrt{x^2+x-1}\right)^2=\left(\sqrt{14-x}\right)^2$

$x^2+x-1=14-x$

$x^2+2x-15=0$

$(x+5)(x-3)=0$

$x=-5$ or $x=3$

The solution -5 and 3 check.

23. $\sqrt{5x+6}=3+\sqrt{x+3}$

$5x+6=9+6\sqrt{x+3}+x+3$

$4x-6=6\sqrt{x+3}$

$2x-3=3\sqrt{x+3}$

$4x^2-12x+9=9x+27$

$4x^2-21x-18=0$

$(4x+3)(x-6)=0$

$x=-\dfrac{3}{4}$ or $x=6$

The solution 6 checks.

$\left(-\frac{3}{4}\text{ is extraneous.}\right)$

25. $\sqrt{2x-5}=1+\sqrt{x-3}$

$2x-5=1+2\sqrt{x-3}+x-3$

$x-3=2\sqrt{x-3}$

$x^2-6x+9=4x-12$

$x^2-10x+21=0$

$(x-3)(x-7)=0$

$x=3$ or $x=7$

The solutions 3 and 7 check.

27. Use the given formula with $h=5$, $S=100$

$100=\pi r\sqrt{r^2+25}$

A calculator graph in the window $0\le x\le10,\,0\le y\le150$:

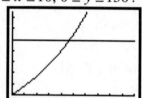

shows the root at 4.658 inches.

29. Use the given formula with $h=\frac{1}{3}r$, volume =180.

$180=\pi r^2\left(\dfrac{1}{3}r\right)/3$

$180=\dfrac{\pi r^3}{9}$

$r^3=\dfrac{1620}{\pi}=8.019$ inches

31. $A=\sqrt{1+\dfrac{a^2}{b^2}}$

$A^2=1+a^2/b^2$

$A^2-1=a^2/b^2$

$b^2\left(A^2-1\right)=a^2$

$b^2=a^2/\left(A^2-1\right)$

$b=\sqrt{\dfrac{a^2}{A^2-1}}=\dfrac{a}{\sqrt{A^2-1}}$

33.
$$y = \frac{1}{\sqrt{1-x^2}}$$

$$y^2 = \frac{1}{1-x^2}$$

$$y^2\left(1-x^2\right) = 1$$

$$1-x^2 = \frac{1}{y^2}$$

$$x = \sqrt{1 - \frac{1}{y^2}}$$

35. $x - 4x^{\frac{1}{2}} + 4 = 0$

Let $u = x^{\frac{1}{2}}$, $u^2 = x$

$u^2 - 4u + 4 = 0$

$(u-2)^2 = 0$

$u = 2$

$x^{\frac{1}{2}} = 2$

$x = 4$

The solution 4 checks.

37. $2x - 8\sqrt{x} - 24 = 0$

Let $u = \sqrt{x}$, $u^2 = x$

$2u^2 - 8u - 24 = 0$

$2\left(u^2 - 4u - 12\right) = 0$

$u^2 - 4u - 12 = 0$

$(u-6)(u+2) = 0$

$u = 6$ or $u = -2$

$\sqrt{x} = 6 \qquad \sqrt{x} = -2$

$x = 36 \qquad$ Impossible

The solution 36 checks.

39. $x^{\frac{2}{3}} + 3x^{\frac{1}{3}} + 2 = 0$

Let $u = x^{\frac{1}{3}}$, $u^2 = x^{\frac{2}{3}}$

$u^2 + 3u + 2 = 0$

$(u+1)(u+2) = 0$

$u = -1 \quad$ or $\quad u = -2$

$x^{\frac{1}{3}} = -1 \qquad x^{\frac{1}{3}} = -2$

$x = -1 \qquad x = -8$

The solutions -1 and -8 check.

41. $x^{\frac{1}{2}} - x^{\frac{1}{4}} - 2 = 0$

Let $u = x^{\frac{1}{4}}$, $u^2 = x^{\frac{1}{2}}$

$u^2 - u - 2 = 0$

$(u+1)(u-2) = 0$

$u = -1 \quad$ or $\quad u = 2$

$x^{\frac{1}{4}} = -1 \qquad x^{\frac{1}{4}} = 2$

impossible $\qquad x = 16$

The solution 16 checks.

43. Let $u = x^{\frac{1}{5}}$. Then the equation becomes $u^3 - 2u^2 + u - 6 = 0$. Graph using the standard window and apply the zero (root) routine to find $u = 2.537656$. Then $x^{\frac{1}{5}} = 2.537656$, thus $x = (2.537656)^5 = 105.236$.

45. Since x cannot equal zero, multiply both sides by x^3 to obtain $1 + 2x - 4x^2 + 5x^3 = 0$. Graph using the window $-2 \le x \le 2$, $-10 \le y \le 10$ and apply the zero (root) routine to find $x = -.283$.

47. Label the text figure:

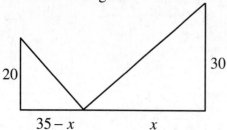

From the Pythagorean Theorem, the length of the rope is given by
$$L = \sqrt{20^2 + (35-x)^2} + \sqrt{30^2 + x^2}$$

a. Solve $63 = \sqrt{20^2 + (35-x)^2} + \sqrt{30^2 + x^2}$. Graph the left and right sides of the equation in the window $0 \le x \le 35, 50 \le y \le 70$ and apply the intersection routine to obtain $x = 11.47$ feet and $x = 29.91$ feet.

b. Graph the equation for L in the same window and apply the minimum routine to obtain $x = 21$ feet.

49. Redraw the text figure:

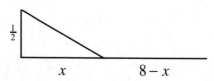

Cost (in thousands) = Underwater Cost (in thousands) + Land Cost (in thousands)
$$72 = 12\sqrt{x^2 + \left(\tfrac{1}{2}\right)^2} + 8(8-x)$$

Solve by graphing the left and right sides of the equation in the window $0 \le x \le 8, 0 \le y \le 100$ and apply the intersection routine to obtain $x = 1.795$ miles.

51. a. $10 = \dfrac{1}{4}\sqrt{h_0}$

$40 = \sqrt{h_0}$

$1600 \text{ ft} = h_0$

b. $t = \dfrac{1}{4}\sqrt{1450_0}$

$t = 9.52$ seconds

5.2 Exponential Functions

1.

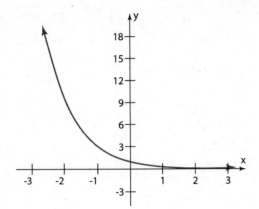

3.

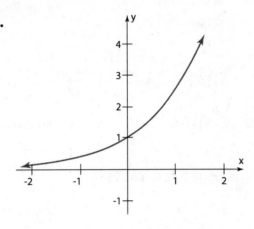

5.

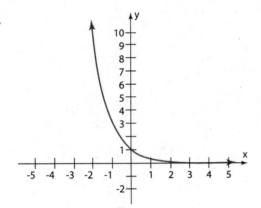

7.

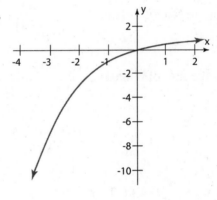

9.

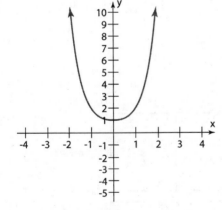

11. Shift vertically 5 units down.

13. Stretch away from the x-axis by a factor of 3.

15. Shift horizontally 2 units to the left, then 5 units down.

17. $A - j(x); B - g(x); C - h(x); D - f(x)$

19. $f(-x)=10^{-x}.$ $f(x)\neq f(-x)$, so the function is not even.
$-f(x)=-10^{x}.$ $-f(x)\neq f(-x)$, so the function is not odd.
Neither

21. $f(-x)=\dfrac{e^{-x}+e^{-(-x)}}{2}=\dfrac{e^{-x}+e^{x}}{2}=f(x)$, so the function is even.

23. $f(-x)=e^{-(-x)^{2}}=e^{-x^{2}}=f(x)$, so the function is even.

25. average rate of change $=\dfrac{f(3)-f(1)}{3-1}=\dfrac{3\cdot 4^{3}-3\cdot 4^{1}}{2}=\dfrac{192-12}{2}=90$

27. average rate of change $=\dfrac{g(1)-g(-1)}{1-(-1)}=\dfrac{3^{1^{2}-1-3}-3^{(-1)^{2}-(-1)-3}}{2}=-\dfrac{4}{27}$

29. average rate of change $=\dfrac{h(1.001)-h(1)}{1.001-1}=\dfrac{2^{1.001}-2}{.001}=1.387$

31. average rate of change$=\dfrac{h(1.001)-h(1)}{1.001-1}=\dfrac{e^{1.001}-e}{.001}=2.720$

33. $\dfrac{f(x+h)-f(x)}{h}=\dfrac{10^{x+h}-10^{x}}{h}$

35. $\dfrac{f(x+h)-f(x)}{h}=\dfrac{\left(2^{x+h}+2^{-(x+h)}\right)-\left(2^{x}+2^{-x}\right)}{h}$

37. $-4\leq x\leq 4,-1\leq y\leq 10$

39. $-4\leq x\leq 4,-1\leq y\leq 10$

41. $-10\leq x\leq 10,-10\leq y\leq 21$

43. $-5\leq x\leq 10,-1\leq y\leq 6$

45. The negative x-axis is an asymptote. There is no vertical asymptote.
minimum: $(-1.443,-.531)$

47. There are no asymptotes. minimum: (0,1)

49. The *x*-axis is an asymptote. There is no vertical asymptote. maximum: (0,1)

51. a. today: $p(0) = 100 \cdot 12^{0/10} = 100$ flies

    b. next week: $p(7) = 100 \cdot 12^{7/10} = 569$ flies

      two weeks: $p(14) = 100 \cdot 12^{14/10} = 3242$ flies

    c. Solve $250 = 100 \cdot 12^{t/10}$ to obtain $t = 13$ weeks

    d. No, at some point there will not be enough space for the fruit flies and the rate the fly population grows will decrease.

53. a. Since $p(0) = 15$, $15 = ke^{-.0000425(0)}$, $15 = k$.

    b. Use $p(x) = 15e^{-.0000425x}$ to obtain:

      $p(5000) = 15e^{-.0000425(5000)} = 12.13$ pounds per square inch

    c. $p(160,000) = 15e^{-.0000425(160,000)} = .0167$ pounds per square inch

55. a. $D(1980) = \dfrac{79.257}{1 + 9.7135 \times 10^{24} e^{-.0304(1980)}} = 74.1$ years

      $D(2000) = \dfrac{79.257}{1 + 9.7135 \times 10^{24} e^{-.0304(2000)}} = 76.3$ years

    b. Solve: $60 = \dfrac{79.257}{1 + 9.7135 \times 10^{24} e^{-.0304t}}$ to obtain $t = 1930$

57. a. At $t = 0$, $p(0) = \dfrac{2000}{1 + 199e^{-.5544(0)}} = \dfrac{2000}{1 + 199} = 10$ beavers

      At $t = 5$, $p(5) = \dfrac{2000}{1 + 199e^{-.5544(5)}} \approx 149$ beavers

    b. Solve $1000 = \dfrac{2000}{1 + 199e^{-.5544t}}$ to obtain $t \approx 9.5$ years

59. a.

Time	Number of Cells
0	1
.25	2
.5	4
.75	8
1	16

    b. $C(t) = 2^{4t}$ or $C(t) = 16^{t}$

61. a. Since $f(1) = 18$, $Pa = 18$. Since $f(2) = 54$, $Pa^2 = 54$.
   Therefore, $a = 3$, $P = 6$. $f(x) = 6 \cdot 3^x$
   b. 3
   c. The park can accomodate $6.2 \times 5280^2 \div 10$, or 17,284,608 frogs. Since
   $f(12) = 3,188,646$, and $f(14) = 28,697,814$, there will be enough space after
   12 weeks, but not after 14 weeks.

63. a.

Fold	Thickness
0	.002
1	.004
2	.008
3	.016
4	.032

   b. $f(x) = (.002)2^x$
   c. $f(20) = (.002)2^{20} = 2097.15$ inches or 174.76 feet
   d. Solve $243,000 \times 5280 \times 12 = (.002)2^x$ to obtain $x = 43$ folds.

65. a. $f(x) = 1200(1.04)^x$
   b. After 3 years, $f(3) = 1200(1.04)^3 = \$1349.84$
   After 5 years 9 months, $f(5.75) = 1200(1.04)^{5.75} = \$1503.57$
   c. Solve $1850 = 1200(1.04)^x$ to obtain approximately $x = 11$ years and one month.

67. a. $f(x) = 100.4(1.014)^x$
   b. In 2010, $x = 10$. $f(10) = 100.4(1.014)^{10} = 115.38$ million
   c. Solve $125 = 100.4(1.014)^x$ to obtain $x = 15.8$. This corresponds to 2016.

69. a. Set $f(x) = 32.44a^x$. Since $f(50) = 98.23$, solve $98.23 = 32.44a^{50}$ to obtain
   $a = 1.0224$. $f(x) = 32.44(1.0224)^x = 32.44e^{.02216x}$.
   b. In 2010, $x = 10$. $f(10) = 32.44(1.0224)^{10} = 40.49$ million
   In 2025, $x = 25$. $f(25) = 32.44(1.0224)^{25} = 56.45$ million
   c. Solve $55 = 32.44(1.0224)^x$ to obtain $x = 23.8$. This corresponds to 2024.

71. Since the population grows by a factor of $205/200 = 1.025$ every hour, we can write
   $P(x) = 200(1.025)^x$.
   After 10 hours $P(10) = 256$. After 2 days, or 48 hours, $P(48) = 200(1.025)^{48} = 654$.

73. a. $f(x) = (1 - .25)^x = .75^x$ **b.** Solve $1 - .90 = .75^x$ to obtain $x = 8$ feet.

75. a. Use the formula $M(x) = c\left(.5^{x/h}\right)$ with $c = 5$ and $h = 5730$. $M(x) = 5\left(.5^{x/5730}\right)$

   b. After 4000 years, $M(4000) = 5\left(.5^{4000/5730}\right) = 3.08$ grams.

   After 8000 years, $M(8000) = 5\left(.5^{8000/5730}\right) = 1.90$ grams.

   c. Solve $1 = 5\left(.5^{x/5730}\right)$ to obtain $x = 13,305$ years.

77. a. $f(x) = a^x$, for any nonnegative constant a.

79. a. 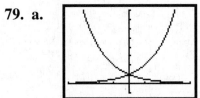 The graphs are reflections of each other in the y-axis.

   b. As above, the graphs are reflections of each other in the y-axis.

81. a. The graphs appear to coincide near $x = 0$, but diverge further away.

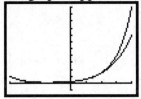

   b. $f_8(x)$ is a good fit in this window on the TI-83.

   c. This polynomial will not be a good fit, but, eventually, $f_{12}(x)$ will be.

5.2.A Compound Interest and the Number e

1. **a.** Compounded annually for 5 years represents 5 periods.

    $A = 1000(1 + .08)^5 = \$1469.33$

   b. Compounded quarterly for 5 years represents $4 \times 5 = 20$ periods.
    $A = 1000(1 + .08/4)^{20} = \$1485.95$

   c. Compounded monthly for 5 years represents $12 \times 5 = 60$ periods.
    $A = 1000(1 + .08/12)^{60} = \$1489.85$

   d. Compounded weekly for 5 years represents $\dfrac{365}{7} \times 5 = \dfrac{1825}{7}$ periods.

    $A = 1000\left(1 + .08 \div \tfrac{365}{7}\right)^{1825/7} = \$1491.37$

3. $A = 500(1 + .03)^8 = \$633.39$

5. $A = 500(1 + .03/4)^{10 \times 4} = \$674.17$

7. $A = 500(1 + .02477/4)^{20 \times 4} = \$819.32$

9. $A = 500e^{.03(127/12)} = \$686.84$

11. Use $A = P(1 + r)^t$ with
    $A = 5000, r = .06, t = 7$.
    $5000 = P(1 + .06)^7$

    $P = \dfrac{5000}{1.06^7}$

    $P = \$3325.29$

13. Use $A = P(1 + r)^t$ with $A = 4800$,
    $r = .072 \div 4$ or $.018, t = 5 \times 4$ or 20.
    $4800 = P(1 + .018)^{20}$

    $P = \dfrac{4800}{1.018^{20}}$

    $P = \$3359.59$

15. **a.** Use $A = Pe^{rt}$ with $A = 8000, r = .04, t = 3$.

    Solve $8000 = Pe^{.04 \times 3}$ to obtain $P = \$7095.36$.

    b. $A = 7050e^{.04 \times 3} = \$7948.85$; No it is not a better deal

17. Fund A: $10,000(1 + .132)^2 = 10,000(1.132)^2 = \$12,814.24$
    Fund B: $10,000(1 + .127/4)^{4 \times 2} = 10,000(1.03175)^8 = \$12,840.91$
    Fund C: $10,000(1 + .126/12)^{12 \times 2} = 10,000(1.0105)^{24} = \$12,849.07$
    Fund C will return the most money.

19. First calculate the amount of money paid: $A = 1200(1 + .14/12)^{12 \times 2} = \$1585.18$
    Since $1200 of this is principal, $1585.18 − $1200, or $385.18, is interest.

21. Use $A = P(1+r)^t$ with $A = 1.5$ mil,
$r = .064 \div 12$, $t = 12 \times 4$.
$1,500,000 = P(1+.064/12)^{12\times4}$

$$P = \frac{1,500,000}{(1+.064/12)^{48}}$$

$P = \$1,162,003.14$

23. Use $A = P(1+r)^t$ with
$A = 1407.1$, $P = 1000$, $t = 7$.
$1407.1 = 1000(1+r)^7$

$(1+r)^7 = 1.4071$

$r = 1.4071^{1/7} - 1$

$r = .05$ or 5%

25. Use $A = P(1+r)^t$ with $A = 4000$, $P = 3000$, $t = 5$.
$4000 = 3000(1+r)^5$

$(1+r)^5 = 4/3$

$$r = \left(\frac{4}{3}\right)^{1/5} - 1$$

$r = .0592$ or 5.92%

27. Use $A = P(1+r)^t$ with $r = .08$.

a. $A = 200$, $P = 100$: Solve $200 = 100(1.08)^t$. t=9 years.

b. $A = 1000$, $P = 500$: Solve $1000 = 500(1.08)^t$. t=9 years.

c. $A = 2400$, $P = 1200$: Solve $2400 = 1200(1.08)^t$. t=9 years.

d. The doubling time is independent of the investment.

29. Use $A = Pe^{rt}$ with $A = 1000$, $P = 500$, $r = .07$.
Solve $1000 = 500e^{.07t}$ to obtain $t = 9.9$ years.

31. a. Use $I = PRT$ with $P = 1000$, $R = .0363$, $T = .5$. To obtain an semiannual interest payment of $90.75. Total interest will be $90.75(10) = \$907.50$.

b. Use $A = P\left(1 + \dfrac{r}{n}\right)^{nt}$ with $P = 5000$, $r = .034$, $n = 2$, $t = 5$. To obtain a final amount of $5918.06, or $918.06 in interest on the CD which is better than the bond.

c. The first interest payment of $90.75 will be paid on October 1, 2007, and you will add this same amount to the account every six months until October 1, 2012.

$$A = 90.75(1 + .03)^{.5} = 92.10; \ \$92.10 + \$90.75 = 182.85$$

$$A = 182.85(1 + .03)^{.5} = 185.57; \ \$185.87 + \$90.75 = 276.32$$

$$A = 276.32(1 + .03)^{.5} = 280.43; \ \$280.43 + \$90.75 = 371.18$$

$$A = 371.18(1 + .03)^{.5} = 376.71; \ \$376.71 + \$90.75 = 467.46$$

$$A = 467.46(1 + .03)^{.5} = 474.42; \ \$474.42 + \$90.75 = 565.17$$

$$A = 565.17(1 + .03)^{.5} = 573.58; \ \$573.58 + \$90.75 = 664.33$$

$$A = 664.33(1 + .03)^{.5} = 674.22; \ \$674.22 + \$90.75 = 764.97$$

$$A = 764.97(1 + .03)^{.5} = 776.36; \ \$776.36 + \$90.75 = 867.11$$

$$A = 867.11(1 + .03)^{.5} = 880.02; \ \$880.02 + \$90.75 = 970.77$$

Thus, with this option you end up with $970.77, which is a better option.

5.3 Common and Natural Logarithmic Functions

1. To what power must 10 be raised to obtain 10,000? 4. log 10,000 = 4

3. To what power must 10 be raised to obtain $\sqrt{10}/1000$? $-5/2$, since
$$\sqrt{10}/1000 = \sqrt{10}/\left(\sqrt{10}\right)^6 = 1/\left(\sqrt{10}\right)^5 = 10^{-5/2}. \quad \log\sqrt{10}/1000 = -5/2$$

5. $10^3 = 1000$ 7. $10^{2.88} = 750$ 9. $e^{1.0986} = 3$ 11. $e^{-4.6052} = .01$

13. $e^{z+w} = x^2 + 2y$ 15. log .01 = -2 17. log 3 = .4771   19. ln 25.79 = 3.25

21. ln 5.5527 = 12/7   23. ln w = 2/r 25. $\sqrt{43}$ 27. 15

29. $\ln\sqrt{e} = \ln e^{\frac{1}{2}}$ 31. 931   33. $x + y$ 35. x^2
$\quad = 1/2$
$\quad = .5$

37. $4(25)^x$ 39. $-16(30.5)^x$ 41. $\left(\dfrac{1}{e^3}\right)^x = .0498^x$
$\quad = 4\left(e^{\ln 25}\right)^x$ $\quad = -16\left(e^{\ln 30.5}\right)^x$
$\quad = 4e^{(\ln 25)x}$ $\quad = -16e^{(\ln 30.5)x}$
$\quad = 4e^{3.2189x}$ $\quad = -16e^{3.4177x}$

43. $x + 1 > 0$ hence the domain is $x > -1$.   45. $-x > 0$ hence the domain is $x < 0$.

47. **a.** The graphs are identical when $x > 0$, but not everywhere.

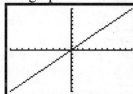   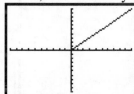

   b. The functions $f(x) = e^x$ and $g(x) = \ln x$ are inverse functions of one another. This
   means that $(f \circ g)(x) = e^{\ln x} = x$ everywhere *on the domain of g*. The domain of
   $g(x) = \ln x$ is precisely $x > 0$.

49. The graphs are identical when $x > 0$, but not everywhere.

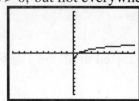

The domain of $f(x) = \log(x^2)$ is all real numbers except 0. The domain of $g(x) = 2\log x$ is only $x > 0$.

51. Stretch by a factor of 2 away from the x-axis.

53. Shift horizontally 4 units to the right.

55. Shift horizontally 3 units to the left, then 4 units vertically down.

57.

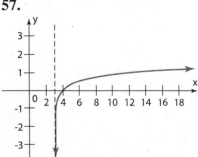

59.

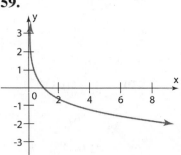

61. $0 \le x \le 20, -10 \le y \le 10$

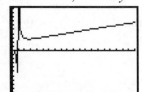

Then, $0 \le x \le 2, -20 \le y \le 20$

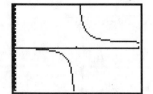

63. $-10 \le x \le 10, -3 \le y \le 3$

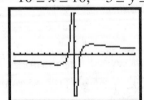

65. $0 \le x \le 20, -6 \le y \le 3$

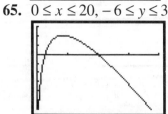

67. $-10 \le x \le 10$ and $-10 \le y \le 10$

69. average rate of change

$$= \frac{f(5) - f(3)}{5 - 3} = \frac{\ln(5-2) - \ln(3-2)}{2} = \frac{\ln 3 - \ln 1}{2} = \frac{\ln 3}{2} \approx .5493$$

71. average rate of change

$$= \frac{f(-3) - f(-5)}{(-3) - (-5)} = \frac{\ln\left((-3)^2 + (-3) + 1\right) - \ln\left((-5)^2 + (-5) + 1\right)}{2} = \frac{\ln 7 - \ln 21}{2} \approx -.5493$$

73. a. $\dfrac{\ln(2+h) - \ln 2}{h}$ **b.** $0.49875415; 0.499875042; 0.49998750; 0.49999875$

75. If $f(x) = \dfrac{1}{1 + e^{-x}}$ and $g(x) = \ln\left(\dfrac{x}{1-x}\right)$, then

$$(f \circ g)(x) = f\left(\ln\left(\frac{x}{1-x}\right)\right) = \frac{1}{1 + e^{-\ln\left(\frac{x}{1-x}\right)}} = \frac{1}{1 + \frac{1-x}{x}} = \frac{x}{x + 1 - x} = x, \text{ and}$$

$$(g \circ f)(x) = g\left(\frac{1}{1 + e^{-x}}\right) = \ln\left(\frac{\frac{1}{1 + e^{-x}}}{1 - \frac{1}{1 + e^{-x}}}\right) = \ln\left(\frac{1}{1 + e^{-x} - 1}\right) = \ln\left(\frac{1}{e^{-x}}\right) = \ln\left(e^x\right) = x.$$

Hence, $g(x)$ is the inverse of $f(x)$.

77. Let $f(x) = A \ln x + B$. Since $f(1) = 10$, $10 = A \ln 1 + B$, $10 = B$.

Thus $f(x) = A \ln x + 10$. Since $f(e) = 1$, $1 = A \ln e + 10$, $1 = A + 10$, $A = -9$.

Since $f(e) = 1$, $1 = A \ln e + 10$, $1 = A + 10$, $A = -9$

Thus $f(x) = -9 \ln x + 10$.

79. Use the given formula with $t = 7$, $c = 75.126$, $p = 44$ to obtain
$$h = (30 \cdot 7 + 8000) \ln(75.126 / 44) = 4392 \text{ meters}$$

81. a. 1999: $x = 99$
       $f(99) = -154.41 + 39.38 \ln 99$

       $f(99) = 26.55$ billion pounds

     2002: $x = 102$
       $f(102) = -154.41 + 39.38 \ln 102$

       $f(102) = 27.72$ billion pounds

b. Solve $35 = -154.41 + 39.38 \ln x$

       $189.41 = 39.38 \ln x$

       $\dfrac{189.41}{39.38} = \ln x$

         $x = e^{189.41/39.38}$

         $x = 122.7$

Approximately, 2022.

83. a. Use the given formula with $x = 6000$. $T = -.93 \ln \left[\dfrac{7000 - 6000}{6999(6000)} \right] = 9.9$

b. Solve $14 = -.93 \ln \left[\dfrac{7000 - x}{6999x} \right]$ to obtain $x = 6986$

85. a. $d = 0$:
$N = 51 + 100 \ln(0/100 + 2)$

$N = 120$ bikes
$d = 1000$:
$N = 51 + 100 \ln(1000/100 + 2)$

$N = 299$ bikes
$d = 10,000$:
$N = 51 + 100 \ln(10,000/100 + 2)$

$N = 51$ bikes

b. Spending $1000 on advertising, selling 299 bikes at $25 each, generates a profit of $7475. This is worthwhile. Spending $9000 more to generate a profit of $25 \times (513 - 299)$, or $5350 more, might not be worthwhile.

c. Spending $1000 on advertising selling 299 bikes at $35 each, generates a profit of $10,465. This is worthwhile. Spending $9000 more to generate a profit of $35 \times (513 - 299)$, or $7490 more, might not be worthwhile.

87. $n = 30$ gives an approximation with a maximum error of .00001 when $-.7 \le x \le .7$

89. a. Answers will vary

b. lower bound: $\ln(10 \cdot 60 \cdot 60 \cdot 24 \cdot 356 \cdot 5007) \approx 28.087$

upper bound: $\ln(10 \cdot 60 \cdot 60 \cdot 24 \cdot 356 \cdot 5007) = 1 \approx 29.087$

c. $28.187 < \ln n + 1$

$n \approx 2529$

5.4 Properties of Logarithms

1. $\ln x^2 + 3\ln y = \ln x^2 + \ln y^3 = \ln\left(x^2 y^3\right)$

3. $\log\left(x^2 - 9\right) - \log(x+3) = \log\left(\dfrac{x^2-9}{x+3}\right) = \log(x-3)$

5. $\begin{aligned}[t] 2(\ln x) - 3\left(\ln x^2 + \ln x\right) &= 2\ln x - 3\ln x^2 - 3\ln x \\ &= -\ln x - 3\ln x^2 \\ &= -\ln x - \ln x^6 \\ &= \ln\frac{1}{x \cdot x^6} \\ &= \ln\frac{1}{x^7} \\ &= \ln\left(x^{-7}\right) \end{aligned}$

7. $\begin{aligned}[t] 3\ln\left(e^2 - e\right) - 3 &= 3\ln\left(e^2 - e\right) - 3\ln e \\ &= 3\ln\frac{e^2 - e}{e} \\ &= 3\ln(e-1) \\ &= \ln(e-1)^3 \end{aligned}$

9. $\begin{aligned}[t] \log(10x) + \log(20y) - 1 &= \log(200xy) - \log 10 \\ &= \log\frac{200xy}{10} \\ &= \log 20xy \end{aligned}$

11. $\ln\left(x^2 y^5\right) = \ln x^2 + \ln y^5 = 2\ln x + 5\ln y = 2u + 5v$

13. $\ln\left(\sqrt{x} \cdot y^2\right) = \ln x^{1/2} + \ln y^2 = \frac{1}{2}\ln x + 2\ln y = \frac{1}{2}u + 2v$

15. $\ln\sqrt[3]{x^2\sqrt{y}} = \ln\left(x^2 y^{1/2}\right)^{1/3} = \ln\left(x^{2/3}y^{1/6}\right) = \frac{2}{3}\ln x + \frac{1}{6}\ln y = \frac{2}{3}u + \frac{1}{6}v$

17. Since $\ln\left|\frac{1}{e}\right| = -1$ and $\left|\ln\frac{1}{e}\right| = 1$, the statement is false.

19. The statement is true by the Power Law for Logarithms.

21. Since $\ln e^3 = 3$ and $(\ln e)^3 = 1$, the statement is false.

23. False

25. Many possible answers. For example, $a = 10$ and $b = 10$.

27. Since $v = e^{\ln v}$ and $w = e^{\ln w}$, we have $\dfrac{v}{w} = \dfrac{e^{\ln v}}{e^{\ln v}} = e^{\ln v} - e^{\ln v}$ by the division property of exponents. So raising e to the exponent $(\ln v - \ln w)$ produces $\dfrac{v}{w}$. But the exponent to which e must be raised to produce $\dfrac{v}{w}$ is, by definition, $\ln\left(\dfrac{v}{w}\right)$. Therefore, $\ln\left(\dfrac{v}{w}\right) = \ln v - \ln w$.

29. $R\left(10^{4.7} i_0\right) = \log\left(\dfrac{10^{4.7} i_0}{i_0}\right) = 4.7$

31. $R\left(1500 i_0\right) = \log\left(\dfrac{1500 i_0}{i_0}\right) \approx 3.176$

33. $L\left(10,000 i_0\right) = 10 \cdot \log\left(\dfrac{10,000 i_0}{i_0}\right) = 40$ decibels

35. $L\left(1 \times 10^9 i_0\right) = 10 \cdot \log\left(\dfrac{1 \times 10^9 i_0}{i_0}\right) = 100$ decibels

37. I must be squared

39. Following the hint: $10^{\log c} = c$ because $\log c$ is the exponent to which 10 must be raised to produce c. taking the natural logarithm of both sides produces $\ln\left(10^{\log c}\right) = \ln c$, but $\ln\left(10^{\log c}\right) = \log c \cdot \ln 10$ by the power law for logarithms. Thus, $\log c \cdot \ln 10 = \ln c$, and dividing both of the sides of this equation by $\ln 10$ produces the desired result.

41. If c is a positive number, then c can be written in scientific notation as $c = b \times 10^k$ where $1 \le b \le 10$ and k is an integer.
Then $\log c = \log\left(b \times 10^k\right) = \log b + \log)\left(10^k\right) = \log b + k$.

43. a

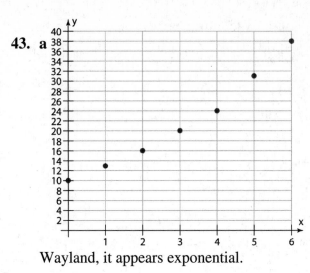

Wayland, it appears exponential.

b.

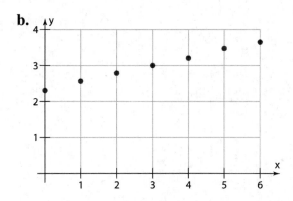

c. Wayland. It now appears linear in part b.

5.4.A Logarithmic Functions to Other Bases

1.

x	0	1	2	4
$f(x) = \log_4 x$	Not Defined	0	.5	1

3

x	1/36	1/6	1	216
$h(x) = \log_6 x$	-2	-1	0	3

5.

x	0	1/7	$\sqrt{7}$	49
$f(x) = 2\log_7 x$	Not Defined	-2	1	4

7.

x	-2.75	-1	1	29
$h(x) = 3\log_2 (x+3)$	-6	3	6	15

9. $\log .01 = -2$

11. $\log \sqrt[3]{10} = \frac{1}{3}$

13. $\log r = 7k$

15. $\log_7 5,764,801 = 8$

17. $\log_3 \left(\frac{1}{9}\right) = -2$

19. $10^4 = 10,000$

21. $10^{2.88} \approx 750$

23. $5^3 = 125$

25. $2^{-2} = 1/4$

27. $10^{z+w} = x^2 + 2y$

29. $\sqrt{97}$

31. $x^2 + y^2$

33. $\log_{16} 4 = \log_{16} 16^{\frac{1}{2}} = 1/2$

35. $\log_{\sqrt{3}} 27 = \log_{\sqrt{3}} \left(\sqrt{3}\right)^6 = 6$

37. Since $\log_b 3 = 1$, $b = 3$.

39. Since $\log_b .05 = -1$, $b^{-1} = .05$, $b = 20$.

41. $\log_3 243 = \log_3 3^5 = 5$.

43. $\log_{27} x = 1/3$
$x = 27^{\frac{1}{3}}$
$x = 3$

45. $\log_x 64 = 3$
$x^3 = 64$
$x = 4$

47. $2\log x + 3\log y - 6\log z = \log x^2 + \log y^3 - \log z^6 = \log \dfrac{x^2 y^3}{z^6}$

49. $\log x - \log(x+3) + \log(x^2-9) = \log\dfrac{x(x^2-9)}{x+3} = \log x(x-3) = \log(x^2-3x)$

51. $\dfrac{1}{2}\log_2(25c^2) = \log_2(25c^2)^{1/2} = \log_2(5c)$

53. $-2\log_4(7c) = \log_4(7c)^{-2} = \log_4\dfrac{1}{49c^2}$

55. $2\ln(x+1) - \ln(x+2) = \ln(x+1)^2 - \ln(x+2) = \ln\left[\dfrac{(x+1)^2}{x+2}\right]$

57. $\log_2(2x) - 1 = \log_2(2x) - \log_2 2 = \log_2\dfrac{2x}{2} = \log_2 x$

59. $2\ln(e^2-e) - 2 = 2\ln(e^2-e) - 2\ln e = 2\ln\dfrac{e^2-e}{e} = 2\ln(e-1) = \ln(e-1)^2$

61. $\log_2 10 = \dfrac{\ln 10}{\ln 2} = 3.3219$ **63.** $\log_7 5 = \dfrac{\ln 5}{\ln 7} = .8271$

65. $\log_{500} 1000 = \dfrac{\ln 1000}{\ln 500} = 1.1115$ **67.** $\log_{12} 56 = \dfrac{\ln 56}{\ln 12} = 1.6199$

69. True by the quotient law for logarithms.

71. True by the power law for logarithms.

73. False. For example, let $x = 1$, $\log_5 5 \cdot 1 = 1$ but $5\log_5 1 = 0$.

75. Since $398\log 397 \approx 1034.3$ and $397\log 398 \approx 1032.1$,
$$398\log 397 > 397\log 398$$
$$10^{398\log 397} > 10^{397\log 398} \quad \left(\text{since } 10^x \text{ is an increasing function}\right)$$
$$397^{398} > 398^{397}$$

77. $\log_b u = \dfrac{\log_a u}{\log_a b}$ by the change-of-base formula.

79. $\log_{100} u = \dfrac{\log_{10} u}{\log_{10} 100} = \dfrac{\log_{10} u}{2}$. So, $\log_{10} u = 2\log_{100} u$.

81. $\log_b x = \dfrac{1}{2}\log_b v + 3 = \log_b v^{\frac{1}{2}} + \log_b b^3 = \log_b b^3\sqrt{v}$

$x = b^3\sqrt{v}$ (since $\log_b x$ is a one-to-one function)

83. The graphs are not identical, and the statement is false. The window $0 \le x \le 10$, $-3 \le y \le 3$ is shown.

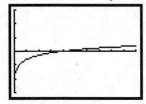 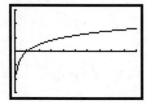

85. Graph using the window $-5 \le x \le 5, -5 \le y \le 5$.

Local maximum: $x = (-.3679, .3195)$

Local minimum: $x = (.3679, -.3195)$

Hole: $(0, 0)$

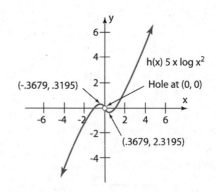

87. a. 5^{150} is greater because $5^{150} > 5^{143.0667} = 10^{100}$

b.

$2^{3322} > \left(10^{100}\right)^{10} = 10^{1,000}$ this is because $\log 2^{3322} = 3322\log 2 \approx 1000.02$ and $\log 10^{1000} = 1000$

5.5 Algebraic Solutions of Exponential and Logarithmic Equations

1. $3^x = 81$
$3^x = 3^4$
$x = 4$

3. $3^{x+1} = 9^{5x}$
$3^{x+1} = 3^{2(5x)}$
$x + 1 = 2(5x)$
$x = 1/9$

5. $3^{5x}9^{x^2} = 27$
$3^{5x} \cdot 3^{2x^2} = 3^3$
$5x + 2x^2 = 3$
$5x + 2x^2 - 3 = 0$
$(2x - 1)(x + 3) = 0$
$x = 1/2 \text{ or } x = -3$

7. $9^{x^2} = 3^{-5x-2}$
$3^{2x^2} = 3^{-5x-2}$
$2x^2 = -5x - 2$
$2x^2 + 5x + 2 = 0$
$(2x + 1)(x + 2) = 0$
$x = -1/2 \text{ or } x = -2$

9. $3^x = 5$
$\ln 3^x = \ln 5$
$x \ln 3 = \ln 5$
$x = \dfrac{\ln 5}{\ln 3}$
$\quad = 1.465$

11. $2^x = 3^{x-1}$
$\ln 2^x = \ln 3^{x-1}$
$x \ln 2 = (x - 1) \ln 3$
$x \ln 2 = x \ln 3 - \ln 3$
$-\ln 3 = x(\ln 2 - \ln 3)$
$x = \dfrac{-\ln 3}{\ln 2 - \ln 3} = \dfrac{\ln 3}{\ln 3 - \ln 2}$
$\quad = \dfrac{\ln 3}{\ln \frac{3}{2}} = \dfrac{\ln 3}{\ln 1.5} = 2.7095$

13. $3^{1-2x} = 5^{x+5}$
$\ln 3^{1-2x} = \ln 5^{x+5}$
$(1 - 2x) \ln 3 = (x + 5) \ln 5$
$\ln 3 - 2x \ln 3 = x \ln 5 + 5 \ln 5$
$\ln 3 - 5 \ln 5 = x(2 \ln 3 + \ln 5)$
$x = \dfrac{\ln 3 - 5 \ln 5}{2 \ln 3 + \ln 5}$
$x = -1.825$

15. $2^{1-3x} = 3^{x+1}$
$\ln 2^{1-3x} = \ln 3^{x+1}$
$(1 - 3x) \ln 2 = (x + 1) \ln 3$
$\ln 2 - 3x \ln 2 = x \ln 3 + \ln 3$
$\ln 2 - \ln 3 = x(3 \ln 2 + \ln 3)$
$x = \dfrac{\ln 2 - \ln 3}{3 \ln 2 + \ln 3} = -.1276$

17. $e^{2x} = 5$
$2x = \ln 5$
$x = \dfrac{\ln 5}{2}$
$x = .805$

19. $6e^{-1.4x} = 21$
$e^{-1.4x} = 3.5$
$-1.4x = \ln 3.5$
$x = \dfrac{\ln 3.5}{-1.4} = -.895$

21. $2.1e^{(x/2)\ln 3} = 5$

$$e^{(x/2)\ln 3} = \frac{5}{2.1}$$

$$\frac{x}{2}\ln 3 = \ln\frac{5}{2.1}$$

$$x = \frac{2\ln\frac{5}{2.1}}{\ln 3}$$

$$x = 1.579$$

25. $e^{2x} - 5e^x + 6 = 0$
Let $e^x = u$, $e^{2x} = u^2$
$u^2 - 5u + 6 = 0$
$(u-3)(u-2) = 0$
$u = 3$ or $u = 2$
$e^x = 3$ $e^x = 2$
$x = \ln 3$ $x = \ln 2$
$x = 1.099$ $x = .693$

29. $4^x + 6 \cdot 4^{-x} = 5$
$(4^x)^2 + 6 = 5 \cdot 4^x$
Let $4^x = u$, $(4^x)^2 = u^2$
$u^2 + 6 = 5u$
$u^2 - 5u + 6 = 0$
$(u-3)(u-2) = 0$
$u = 3$ or $u = 2$
$4^x = 3$ $4^x = 2$
$x = \ln 3/\ln 4$ $x = .5$
$x = .792$

23. $9^x - 4 \cdot 3^x + 3 = 0$
Let $u = 3^x$, $u^2 = 9^x$
$u^2 - 4u + 3 = 0$
$(u-1)(u-3) = 0$
$u = 1$ or $u = 3$
$3^x = 1$ $3^x = 3$
$x = 0$ $x = 1$

27. $6e^{2x} - 16e^x = 6$
Let $e^x = u$, $e^{2x} = u^2$
$6u^2 - 16u = 6$
$3u^2 - 8u - 3 = 0$
$(u-3)(3u+1) = 0$
$u = 3$ or $u = -1/3$
$e^x = 3$ $e^x = -1/3$
$x = \ln 3$ impossible
≈ 1.099

31. $\dfrac{e^x - e^{-x}}{2} = t$. Let $e^x = u$

$$\frac{u - u^{-1}}{2} = t$$

$$\frac{u^2 - 1}{2u} = t$$

$$u^2 - 2ut - 1 = 0$$

$$u = \frac{2t \pm \sqrt{4t^2 + 4}}{2}$$

$u = t \pm \sqrt{t^2 + 1}$ Since $t - \sqrt{t^2 + 1}$ is negative, the only solution is
$e^x = t + \sqrt{t^2 + 1}$
$x = \ln\left(t + \sqrt{t^2 + 1}\right)$

33. a. If $\ln u = \ln v$, then $u = e^{\ln u} = e^{\ln v} = v$.

  b. No, if u and v are negative then $\ln u$ and $\ln v$ are undefined.

35. $\ln(3x - 5) = \ln 11 + \ln 2$
$\ln(3x - 5) = \ln 22$
$3x - 5 = 22$
$x = 9$

37. $\log(3x-1)+\log 2 = \log 4 + \log(x+2)$
$$\log 2(3x-1) = \log 4(x+2)$$
$$6x-2 = 4x+8$$
$$x = 5$$

39. $2\ln x = \ln 36$
$$\ln x^2 = \ln 36$$
$$x^2 = 36$$
$$x = \pm 6$$
Since x must be positive, the only solution is 6.

41. $\ln x + \ln(x+1) = \ln 3 + \ln 4$
$$\ln x(x+1) = \ln 12$$
$$x(x+1) = 12$$
$$x^2 + x - 12 = 0$$
$$(x+4)(x-3) = 0$$
$x = -4$ or $x = 3$. Since x must be positive, the only solution is 3.

43. $\ln x = \ln 3 - \ln(x+5)$
$$\ln x = \ln \frac{3}{x+5}$$
$$x = \frac{3}{x+5}$$
$$x^2 + 5x = 3$$
$$x^2 + 5x - 3 = 0$$
$$x = \frac{-5 \pm \sqrt{37}}{2} \quad \text{Since } x \text{ must be}$$
positive, the only solution
is $\dfrac{-5 + \sqrt{37}}{2}$.

45. $\ln(x+9) - \ln x = 1$
$$\ln(x+9)/x = 1$$
$$(x+9)/x = e$$
$$x + 9 = ex$$
$$x = \frac{9}{e-1}$$

47. $\log x + \log(x-3) = 1$
$$\log x(x-3) = 1$$
$$x(x-3) = 10$$
$$x^2 - 3x - 10 = 0$$
$$(x-5)(x+2) = 0$$
$x = 5$ or $x = -2$ Since x must be greater than 3, the only solution is 5.

49. $\log \sqrt{x^2-1} = 2$
$$\sqrt{x^2-1} = 10^2$$
$$x^2 - 1 = 10,000$$
$$x^2 = 10,001$$
$$x = \pm\sqrt{10,001}$$

51. $\ln(x^2+1)-\ln(x-1)=1+\ln(x+1)$

$\ln\dfrac{x^2+1}{x-1}=\ln e+\ln(x+1)$

$\ln\dfrac{x^2+1}{x-1}=\ln e(x+1)$

$\dfrac{x^2+1}{x-1}=e(x+1)$

$x^2+1=e(x^2-1)$

$x^2+1=ex^2-e$

$e+1=x^2(e-1)$

$x^2=\dfrac{e+1}{e-1}$

$x=\pm\sqrt{\dfrac{e+1}{e-1}}$ Since x must be

greater than 1, the only solution is

$\sqrt{\dfrac{e+1}{e-1}}$.

53. Use $M(x)=c(.5)^{x/h}$ with $c=300$, $x=.26$, $M(.26)=200$.

$200=300(.5)^{.26/h}$

$2/3=(.5)^{.26/h}$

$\ln 2/3=(.26/h)\ln .5$

$h=\dfrac{.26\ln .5}{\ln 2/3}$

$=.444$ bil.

or 444,000,000 yrs

55. Use $M(x)=c(.5)^{x/h}$ with $c=3$, $x=23.7$ and $M(23.7)=1$.

$1=3(.5)^{23.7/h}$

$\dfrac{1}{3}=(.5)^{23.7/h}$

$\ln\dfrac{1}{3}=\dfrac{23.7}{h}\ln .5$

$h=\dfrac{23.7\ln .5}{\ln 1/3}$

$h=14.95$ days

57. Use $M(x)=c(.5)^{x/h}$ with $x=6$, and $M(6)=.336c$.

$.336c=c(.5)^{6/h}$

$.336=(.5)^{6/h}$

$\ln .336=(6/h)\ln .5$

$h=\dfrac{6\ln .5}{\ln .336}=3.8132$ days

59. Use $M(x)=c(.5)^{x/h}$ with $h=5730$ (half-life of carbon-14) and $M(c)=c-.36c=.64c$.

$.64c=c(.5)^{x/5730}$

$.64=(.5)^{x/5730}$

$\ln .64=\dfrac{x}{5730}\ln .5$

$x=\dfrac{5730\ln .64}{\ln .5}=3689$ years

61. Use $M(x)=c(.5)^{x/h}$ with $h=5730$ (half-life of carbon-14) and $M(c)=c-.264c=.736c$.

$.736c=c(.5)^{x/5730}$

$.736=(.5)^{x/5730}$

$\ln .736=\dfrac{x}{5730}\ln .5$

$x=\dfrac{5730\ln .736}{\ln .5}=2534$ years

63. Use $A = P(1+r)^t$ with $P = 1000$,
$A = 2000, t = 10 \times 4$, and $r = i/4$.

$$2000 = 1000(1 + i/4)^{40}$$

$$2 = (1 + i/4)^{40}$$

$$1 + i/4 = 2^{1/40}$$

$$i = 4(2^{1/40} - 1)$$

$$i = .0699 \text{ or } 6.99\%$$

65. a. Use $A = P(1+r)^t$ with $P = 500$,
$A = 1500$, and $r = .05$.

$$1500 = 500(1 + .05)^t$$

$$3 = 1.05^t$$

$$\ln 3 = t \ln 1.05$$

$$t = \frac{\ln 3}{\ln 1.05} = 22.5 \text{ years}$$

b. Use $A = P(1+r)^t$ with $P = 500$,
$A = 1500$, and $r = .05/4, t = 4n$.

$$1500 = 500(1 + .05/4)^{4n}$$

$$3 = 1.0125^{4n}$$

$$\ln 3 = 4n \ln 1.0125$$

$$n = \frac{\ln 3}{4 \ln 1.0125} = 22.1 \text{ years}$$

67. Use $A = P(1+r)^t$ with $A = 5000$,
$t = 4 \times 9 = 36$, and $r = .05/4$.

$$5000 = P(1 + .05/4)^{36}$$

$$P = \frac{5000}{1.0125^{36}}$$

$$P = \$3197.05$$

69. Use $f(x) = Pe^{kx}$ with $P = 364$, $k = .004$, $f(x) = 500$.

$$500 = 364e^{.004x}$$

$$\frac{500}{364} = e^{.004x}$$

$$\ln \frac{500}{364} = .004x$$

$$x = \frac{1}{.004} \ln \frac{500}{364}$$

$$x = 79.36 \text{ years}$$

71. a. Use $f(x) = Pe^{kx}$ with $P = 151$, $x = 12$, $f(12) = 180$.

$$180 = 151e^{k(12)}$$
$$180/151 = e^{12k}$$
$$\ln(180/151) = 12k$$
$$k = \frac{\ln\left(180/151\right)}{12}$$
$$k = .0146 \text{ or } 1.46\% \text{ per year}$$

b. Use $f(x) = Pe^{kx}$ with $P = 151$, $f(x) = 250$ and this value of k.

$$250 = 151e^{.0146x}$$
$$250/151 = e^{.0146x}$$
$$\ln(250/151) = .0146x$$
$$x = \frac{\ln\left(250/151\right)}{.0146}$$
$$x = 34 \text{ years or } 2024$$

73. a. Use $P = e^{kt}$ with $P = 25$, $t = .15$

$$25 = e^{k(.15)}$$
$$\ln 25 = .15k$$
$$k = \frac{\ln 25}{.15} = 21.46$$

b. Use $P = e^{kt}$ with $P = 50$ and this value of k.

$$50 = e^{21.46t}$$
$$\ln 50 = 21.46t$$
$$t = \frac{\ln 50}{21.46} = .182$$

75. a. When $x = 1$, $f(1) = 100$. When $x = 2$, $f(2) = 500$. Since the number of bacteria is multiplied by 5 every hour, $f(0) = 100/5 = 20$.
When $x = 3$,
$$f(3) = 5 \times 500 = 2500 \text{ bacteria}.$$

b. Use $f(x) = 20 \cdot 5^x$ with $f(x) = 40$.

$$40 = 20 \cdot 5^x$$
$$2 = 5^x$$
$$\ln 2 = x \ln 5$$
$$x = \frac{\ln 2}{\ln 5} = .43 \text{ hours}$$

77. a. At $t = 0$,
$$f(0) = \frac{45,000}{1 + 224e^{-.899(0)}} = 200$$
At $t = 3$,
$$f(3) = \frac{45,000}{1 + 224e^{-.899(3)}} = 2795$$

b. Use $f(t) = 22,500$.

$$22,500 = \frac{45,000}{1 + 224e^{-.899t}}$$
$$1 + 224e^{-.899t} = 2$$
$$e^{-.899t} = 1/224$$
$$t = \frac{1}{-.899}\ln\left(\frac{1}{224}\right) = 6 \text{ wks}$$

79. a. $1000(.75)^x = 10$

$$.75^x = .01$$
$$x \ln .75 = \ln .01$$
$$x = \frac{\ln .75}{\ln .01} \approx 16 \text{ years}$$

b. $1000(.75)^x = 1$

$$.75^x = .001$$
$$x \ln .75 = \ln .001$$
$$x = \frac{\ln .75}{\ln .001} \approx 24 \text{ years}$$

81. a. $50 = c\left(1 - e^{-4k}\right)$ and $70 = c\left(1 - e^{-8k}\right)$

Thus, $\dfrac{70}{50} = \dfrac{c\left(1 - e^{-8k}\right)}{c\left(1 - e^{-4k}\right)} = \dfrac{c\left(1 - e^{-4k}\right)\left(1 + e^{-4k}\right)}{c\left(1 - e^{-4k}\right)}$

$\dfrac{7}{5} = 1 + e^{-4k}$

$\dfrac{2}{5} = e^{-4k}$

$k = -\dfrac{1}{4}\ln\left(\dfrac{2}{5}\right)$

Then, $50 = c\left(1 - e^{-4\left(-\frac{1}{4}\ln\left(\frac{2}{5}\right)\right)}\right)$

$50 = c\left(\dfrac{3}{5}\right)$

$\dfrac{250}{3} = c$

b. $50 = c\left(1 - e^{-4k}\right)$ and $90 = c\left(1 - e^{-8k}\right)$

Thus, $\dfrac{90}{50} = \dfrac{c\left(1 - e^{-8k}\right)}{c\left(1 - e^{-4k}\right)} = \dfrac{c\left(1 - e^{-4k}\right)\left(1 + e^{-4k}\right)}{c\left(1 - e^{-4k}\right)}$

$\dfrac{9}{5} = 1 + e^{-4k}$

$\dfrac{4}{5} = e^{-4k}$

$k = -\dfrac{1}{4}\ln\left(\dfrac{4}{5}\right)$

Then, $50 = c\left(1 - e^{-4\left(-\frac{1}{4}\ln\left(\frac{4}{5}\right)\right)}\right)$; $250 = c$. Thus, $N = 250\left(1 - e^{-\left(-\frac{1}{4}\ln\left(\frac{4}{5}\right)\right)t}\right)$.

Continued on next page

81. continued

When $N=125$, $125 = 250\left(1 - e^{-\left(-\frac{1}{4}\ln\left(\frac{4}{5}\right)\right)t}\right)$

$$.5 = 1 - e^{\left(\frac{1}{4}\ln\left(\frac{4}{5}\right)\right)t}$$

$$.5 = e^{\left(\frac{1}{4}\ln\left(\frac{4}{5}\right)\right)t}$$

$$.5 = \left(\frac{4}{5}\right)^{t/4}$$

$$\ln(.5) = \ln\left(\frac{4}{5}\right)^{t/4}$$

$$\ln(.5) = \frac{t}{4}\ln\left(\frac{4}{5}\right)$$

$$\frac{\ln(.5)}{\ln(.8)} = \frac{t}{4}$$

$$\frac{4\ln(.5)}{\ln(.8)} = t$$

$t = 12.43$ weeks.

5.6 Exponential, Logarithmic, and Other Models

1. A linear model is probably best.

3. A logarithmic model is probably best.

5. An exponential, quadratic, cubic, or power model will work.

7. A logistic or cubic model will work.

9. A quadratic or cubic model will work.

11. Successive ratios: $\dfrac{15.2}{3} = 5.07$, $\dfrac{76.9}{15.2} = 5.06$, $\dfrac{389.2}{76.9} = 5.06$, $\dfrac{1975.5}{389.2} = 5.08$, and
$\dfrac{9975.8}{1975.5} = 5.05$. Since these ratios are very nearly the same, an exponential model
would be appropriate.

13. a. The logistic model in Example 2 is $y = \dfrac{442.1}{1 + 56.33e^{-.0216x}}$. As x increases, $e^{-.0216x}$
decreases from a maximum of 1 toward 0. The denominator therefore decreases
toward 1, and thus y increases toward 442.1 (in millions) but can never reach or
exceed this value.
b. $0 \le x \le 550, 0 \le y \le 500$

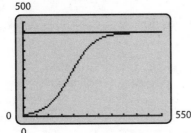

15. a. $0 \le y \le 5$ **b.** $0 \le y \le 10$ **c.** $0 \le y \le 40$

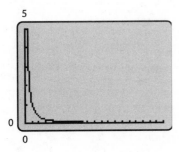   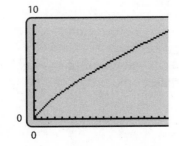   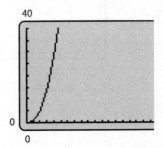

17. Shown is the graph of x vs. $\ln y$ using the window $0 \le x \le 12, 0 \le y \le 8$.

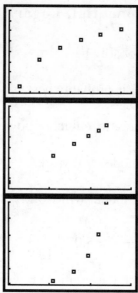

Shown next is the graph of $\ln x$ vs. $\ln y$ using the window $0 \le x \le 3, 0 \le y \le 8$.

Shown next is the graph of $\ln x$ vs. y using the window $0 \le x \le 3, 0 \le y \le 500$

Of these, the most linear is the graph of $\ln x$ vs. $\ln y$ indicating that a power model would be most appropriate.

19. Shown is the graph of x vs. $\ln y$ using the window $0 \le x \le 40, 0 \le y \le 4$.

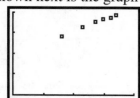

Shown next is the graph of $\ln x$ vs. $\ln y$ using the window $0 \le x \le 4, 0 \le y \le 4$.

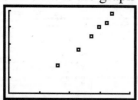

Shown next is the graph of $\ln x$ vs. y using the window $0 \le x \le 4, 0 \le y \le 50$.

Of these, the most linear is the graph of $\ln x$ vs. $\ln y$ indicating that a power model would be most appropriate.

21. a. Use the window $0 \le x \le 8, 90,000 \le y \le 105,000$.

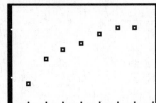

b. The logarithmic model is shown on the left, the logistic model on the right.

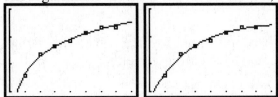

c. In the table below, the logarithmic model is Y_1 and the logistic model is Y_2

X	Y₁	Y₂
11	104103	102375
12	104498	102426
13	104862	102458
14	105199	102480
15	105513	102494
16	105806	102503
17	106082	102509

X=16

Thus for 2000 ($x = 12$) the logarithmic model predicts 104,498 while the logistic model predicts 102,426. 2005 is shown as $x = 17$.

d. As a general rule, logistic models are better for ordinary population growth data, but in this case the prediction of slow steadily decreasing growth from the logarithmic model might be better than the prediction of levelling off at 102,519 from the logistic model.

23. a. Use the window $15 \le x \le 110, 0 \le y \le 80$.

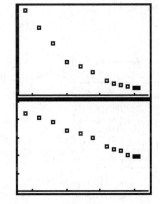

b. Use the window $15 \le x \le 110, 0 \le y \le 5$.

c. An exponential model seems best. The TI-83 result is shown, as well as the result of graphing it and the original data in the original window.

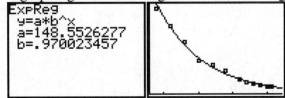

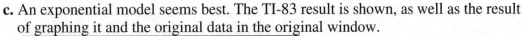

25. a. Use the window
$0 \le x \le 25, 0 \le y \le 36$

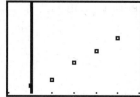

b. Entering the data into the TI-83 and applying the ExpReg feature yields the following:

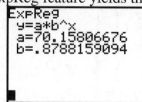

ExpReg
y=a*b^x
a=70.15806676
b=.8788159094

c. In the year 2003, the model yields 0.7 students/computer.
d. 2012
e. Any projection of future trends becomes tenuous with passing time. The availability of new technology could have unforeseen effects.

27. a. Use the window $0 \le x \le 5, 130,000 \le y \le 150,000$.

b. Successive ratios: $\dfrac{1330007}{1316333} = 1.01$, $\dfrac{1367547}{1330007} = 1.03$, $\dfrac{1392796}{1367547} = 1.02$,

and $\dfrac{1421911}{1392796} = 1.02$. Since these ratios are very nearly the same, an exponential model would be appropriate.

c.
ExpReg
y=a*b^x
a=1311520.253
b=1.020245499

d. With $x = 0$ corresponding to 2000:

X	Y1
0	1.31E6
1	1.34E6
2	1.37E6
3	1.39E6
4	1.42E6
5	1.45E6
6	1.48E6

X=0

e. The table shown in part **d** predicts 1,602,588.

29. a. Entering the data into the TI-83 and applying the LnReg feature yields:

LnReg
y=a+blnx
a=21.34235303
b=12.63831192
r²=.9888406528
r=.9944046725

b. The model predicts 79.03 years ($x = 96$).
c. The model predicts 2031 the life expectancy of a woman born in that year will be 83.

Chapter 5 Review Exercises

1. $\sqrt{\sqrt[5]{a^{20}}} = \left(\left(a^{20}\right)^{\frac{1}{5}}\right)^{\frac{1}{2}} = a^{2}$

3. $\left(x^{\frac{2}{5}}y^{-\frac{2}{3}}\right)\left(x^{\frac{2}{3}}y^{4}\right)^{\frac{1}{2}} = x^{\frac{2}{5}}y^{-\frac{2}{3}}x^{\frac{1}{3}}y^{2} = x^{\frac{11}{15}}y^{\frac{4}{3}}$

5. $\left(x^{\frac{1}{3}}+y^{\frac{1}{3}}\right)\left(x^{\frac{1}{3}}-y^{\frac{1}{3}}\right) = x^{\frac{2}{3}}-y^{\frac{2}{3}}$

7. $\dfrac{\sqrt[3]{7x^{-4}y^{11}}}{\sqrt[3]{56x^{2}y^{2}}} = \sqrt[3]{\dfrac{7x^{-4}y^{11}}{56x^{2}y^{2}}} = \sqrt[3]{\dfrac{y^{9}}{8x^{6}}} = \dfrac{y^{3}}{2x^{2}}$

9. $\dfrac{\sqrt{3x+3h-5}-\sqrt{3x-5}}{h} \cdot \dfrac{\sqrt{3x+3h-5}+\sqrt{3x-5}}{\sqrt{3x+3h-5}+\sqrt{3x-5}} = \dfrac{(3x+3h-5)-(3x-5)}{h\left(\sqrt{3x+3h-5}+\sqrt{3x-5}\right)}$

$$= \dfrac{3}{\sqrt{3x+3h-5}+\sqrt{3x-5}}$$

11. $\sqrt{2x-1} = 3x-4$

$2x-1 = 9x^{2}-24x+16$

$9x^{2}-26x+17 = 0$

$(x-1)(9x-17) = 0$

$x = 1 \text{ (impossible)}$

$x = \dfrac{17}{9}$

13. $3\sqrt{x+3}+\sqrt{5x-1} = 2$

$3\sqrt{x+3} = 2-\sqrt{5x-1}$

$9(x+3) = 4-4\sqrt{5x-1}+5x-1$

$4x+24 = -4\sqrt{5x-1}$

$16x^{2}+192x+576 = 16(5x-1)$

$16x^{2}+112x+592 = 0$

This quadratic has no real solutions so therefore there is no solution to this problem.

15. Graph $y = \sqrt[3]{x^{4}-2x^{3}+6x-7}$ and $y = x+3$ using the window $-10 \le x \le 10$, $-5 \le y \le 15$ and apply the intersection routine to find $x = -1.733$ and $x = 5.521$.

17. $-3 \le x \le 3, 0 \le y \le 2$

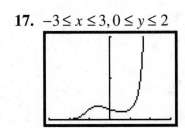

19. Horizontal asymptote: $y = -1$

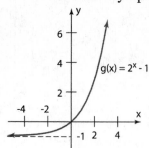

21. Vertical asymptote: $x = -4$

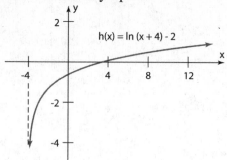

23. Graph the function using the window $0 \le x \le 20, 0 \le y \le 100$.

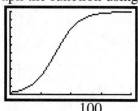

a. $P(0) = \dfrac{100}{1 + 48.2e^{-.52(0)}} = 2.03\%$

b. $P(6) = \dfrac{100}{1 + 48.2e^{-.52(6)}} = 31.97\%$

c. The graph seems steepest between $t = 6$ and $t = 10$.

d. Since $P = 100\%$ is a horizontal asymptote for the graph, the program will never be completely mastered.

25. $\ln 23 = 3.14$

27. $\ln(4a - b) = x - 7y$

29. $\log 177 = 2.248$

31. $e^{6.0014} = 404$

33. $e^{15t} = 4xy$

35. $6^y = 3x - 4$

37. 2/3

39. $x/2$

41. $4\ln \sqrt[5]{x} + \left(\dfrac{1}{5}\right)\ln x$

$= \ln x^{4/5} + \ln x^{1/5}$

$= \ln x$

43. $\ln 6a - 4\ln b - \ln 2a = \ln 6a - \ln b^4 - \ln 2a = \ln\left(\dfrac{6a}{2ab^4}\right) = \ln\left(\dfrac{3}{b^4}\right)$

45. $3\ln x - 4\left(\ln x^3 - 5\ln x\right) = \ln x^3 - 4\ln x^{-2} = \ln x^3 + \ln x^8 = \ln x^{11}$

47. $\log_{30} 900 = \log_{30} 30^2 = 2$

49. (c)

51. (c)

53. $\log_4 16^{5x^2} = 160$

$\quad\quad 4^{160} = 16^{5x^2}$

$\quad\quad 4^{160} = 4^{10x^2}$

$\quad\quad 160 = 10x^2$

$\quad\quad x = \pm 4$

55. $9^{9-x} = 3^{x^2-5x}$

$\quad\quad 3^{18-2x} = 3^{x^2-5x}$

$\quad\quad 18 - 2x = x^2 - 5x$

$\quad\quad x^2 - 3x - 18 = 0$

$\quad\quad (x-6)(x+3) = 0$

$\quad\quad x = 6 \ \text{ or } \ x = -3$

57. $2 \cdot 27^x - 7 = -\frac{19}{3}$

$\quad\quad 27^x = \frac{1}{3}$

$\quad\quad 3^{3x} = 3^{-1}$

$\quad\quad x = -\frac{1}{3}$

59. $2a = 3b - c\ln x$

$\quad\quad c\ln x = 3b - 2a$

$\quad\quad \ln x = \dfrac{3b-2a}{c}$

$\quad\quad x = e^{\frac{3b-2a}{c}}$

61. $\ln 5x + \ln(2x-4) = \ln 30$

$\quad\quad \ln(5x(2x-4)) = \ln 30$

$\quad\quad 10x^2 - 20x = 30$

$\quad\quad 10(x^2 - 2x - 3) = 0$

$\quad\quad (x-3)(x+1) = 0$

$\quad\quad x = 3 \ \text{ or } \ x = -1$

Since ln must be positive, 3 is the only solution.

63. $\log(4x^2-9) = 2 + \log(2x-3)$

$\quad\quad \log(4x^2-9) - \log(2x-3) = 2$

$\quad\quad \log\left(\dfrac{4x^2-9}{2x-3}\right) = 2$

$\quad\quad \log(2x+3) = 2$

$\quad\quad 2x + 3 = 10^2$

$\quad\quad x = \frac{97}{2} = 48.5$

65. Use $M(x) = c(.5)^{x/h}$ with $c = 10, h = 140, x = 365$.

$\quad\quad M(365) = 10(.5)^{365/140} = 1.64$ milligrams

67. If P is the initial amount, then after 1 year $P - .0559P = .9441P$ remains. Use $M(x) = P(.5)^{x/h}$ with $x = 1, M(1) = .9441P$.

$\quad\quad .9441P = P(.5)^{1/h}$

$\quad\quad .9441 = (.5)^{1/h}$

$\quad\quad \dfrac{1}{h} = \dfrac{\ln .9441}{\ln .5}$

$\quad\quad h = \dfrac{\ln .5}{\ln .9441}$

$\quad\quad h = 12.05$ years

69. Use $A = P(1 + r)^t$ with $A = 5000$,

$$5000 = P\left(1 + \frac{.045}{4}\right)^{24}$$

$P = \$3822.66$

71. Increasing the intensity by a factor of 10^3 increases the Richter magnitude by 3 units; the second earthquake measures 7.6.

73. a. The table shows that a 20-mph wind makes 25°F feel like 11°F.   **b.**

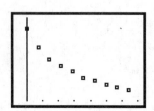

c. The plotted data strongly suggest **d.** exponential decay. One might argue that the ratios of successive values are very roughly equal.

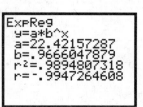

e. Using the Table feature of the TI-83, the model yields 10.3°F.

Chapter 5 Test

1. a. $\dfrac{1}{\sqrt{2+h}}\cdot\dfrac{\sqrt{2+h}}{\sqrt{2+h}}=\dfrac{\sqrt{2+h}}{2+h}$

 b. $\dfrac{a-b}{\sqrt{a}+\sqrt{b}}\cdot\dfrac{\sqrt{a}-\sqrt{b}}{\sqrt{a}-\sqrt{b}}=\dfrac{(a-b)(\sqrt{a}-\sqrt{b})}{a-b}$

$$=\sqrt{a}-\sqrt{b}$$

c. $\dfrac{3}{\sqrt{x}-4}\cdot\dfrac{\sqrt{x}+4}{\sqrt{x}+4}=\dfrac{3(\sqrt{x}+4)}{x-16}$

2. a. $\sqrt{x^3+x^2-4x+5}=x+1$

$x^3+x^2-4x+5=(x+1)^2$

$x^3+x^2-4x+5=x^2+2x+1$

$x^3-6x+4=0$

by synthetic division

$(x-2)(x^2+2x-2)=0$

$x-2=0$ or $x^2+2x-2=0$

$x=2$ and by the quadratic

formula $x=-1\pm\sqrt{3}$

The two solutions that work are

$x=2,-1+\sqrt{3}$

 b. $(x+1)^{2/3}=4$

$x+1=\pm4^{3/2}$

$x+1=\pm8$

$x=-9,7$

3. a. Use $P=P_0 e^{.01t}$

$1000=P_0 e^{.01(10)}$

$\dfrac{1000}{e^{.01(10)}}=P_0$

$P_0\approx905$

 b. The initial population is 905.

 c. $P=905e^{.01(20)}$

$P\approx1105$

4. a. k is positive, because the function approaches zero as it goes to the left, and gets larger as it goes to the right.

 b. Substitute the point $(0,-3)$ into $y=Ae^{kt}$ to obtain: $-3=Ae^{0\cdot k}$

$$-3=A$$

5. **a.** $\dfrac{10+\sqrt{75}}{5} = \dfrac{10+5\sqrt{3}}{5}$

$$= \dfrac{5\left(2+\sqrt{3}\right)}{5}$$

$$= 2+\sqrt{3}$$

b. $\dfrac{3ab^2 - \sqrt[3]{a^4b^7}}{a} = \dfrac{3ab^2 - a^{4/3}b^{7/3}}{a}$

$$= \dfrac{a\left(3b^2 - a^{1/3}b^{7/3}\right)}{a}$$

$$= 3b^2 - a^{1/3}b^{7/3}$$

$$= b^2\left(3 - a^{1/3}b^{1/3}\right)$$

$$= b^2\left(3 - \sqrt[3]{ab}\right)$$

c. $\dfrac{a^{1/2}\left(a^{3/2}b^{5/2}\right)}{b^{-1/2}} = \dfrac{a^{4/2}b^{5/2}}{b^{-1/2}}$

$$= a^2b^{5/2}b^{1/2}$$

$$= a^2b^{6/2}$$

$$= a^2b^3$$

6. **a.** The equation $y = 1,000,000 \cdot 2^{x/2}$, where x is the number of years since 1990. So, $y = 1,000,000 \cdot 2^{10/2} = 32,000,000$.

   b. $y = 1,000,000 \cdot 2^{-5/2} = 176,777$.

   c. $y = 1,000,000 \cdot 2^{30/2} = 32,768,000,000$.

7. $3000\left(1 + \dfrac{.08}{4}\right)^4 \approx \$3247.30$

8. **a.** $\log_5 625 = x$

$$5^x = 625$$

$$x = 4$$

   b. $11^{\log_{11}(x+h)} = x + h$

   c. $\log_3\left(9^{1138}\right) = x$

$$3^x = 9^{1138}$$

$$3^x = \left(3^2\right)^{1138}$$

$$3^x = 3^{2276}$$

$$x = 2276$$

9. Use $f(x) = Pe^{kt}$, to obtain: $.5P = Pe^{5730k}; .5 = e^{5730k}; \ln.5 = 5730k; \dfrac{\ln.5}{5730} = k$, $k \approx -.000121$. Substitute the value found for k to obtain: $.9P = Pe^{-.000121t}; .9 = e^{-.000121t}; \ln.9 = -.000121t; \dfrac{\ln.9}{-.000121} = t \approx 871$ years.

10. **a.** $\ln\left(e^{x^2+y^2}\right) = x^2 + y^2$ **b.** $e^{\ln a - 2\ln b} = e^{\ln\left(a/b^2\right)} = \dfrac{a}{b^2}$

     c. $\ln\left(\dfrac{1}{\sqrt[3]{e}}\right) = \ln\left(\dfrac{1}{e^{1/3}}\right) = \ln\left(e^{-1/3}\right) = -\dfrac{1}{3}$

11. Use the point $(1,2)$ to obtain: $2 = A\log(1) + B$

$$2 = 0 + B$$
$$B = 2$$

     Then use the point $(10,5)$ to obtain: $5 = A\log(10) + 2$

$$5 = A + 2$$
$$A = 3$$

12. **a.** $\dfrac{\ln(x) + \ln(a)}{\ln(ax)} = \dfrac{\ln(xa)}{\ln(ax)} = 1$ **b.** $\ln\left(\dfrac{2}{b^4}\right) + 4\ln b = \ln\left(2b^{-4}\right) + \ln\left(b^4\right)$

$$= \ln\left(2b^{-4}b^4\right)$$
$$= \ln 2$$

     c. $\ln\left(3 + x^2\right)$ cannot be simplified

13. **a.** If $t = 0$, then $P = 230e^{.6(0)} = 230$, mayflies initially.

     b. $10,000 = 230e^{.6t}$

$$\dfrac{1000}{23} = e^{.6t}$$

$$\ln\left(\tfrac{1000}{23}\right) = .6t$$

$$\dfrac{\ln\left(\tfrac{1000}{23}\right)}{.6} = t \approx 6.29 \text{ days.}$$

14. **a.** Use the TI , where x is in months, and $x = 1$ corresponds to January
to obtain: $y = 120 + 15\ln x$.

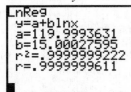

     b. According to the model, her score in February will be 130.40, and in December
her score will be 157.27.

Chapter 6
Trigonometric Functions
6.1 Angles and Their Measurement

1. $\dfrac{1}{9}(2\pi) = \dfrac{2\pi}{9}$ **3.** $\dfrac{1}{18}(2\pi) = \dfrac{\pi}{9}$ **5.** $\dfrac{1}{36}(2\pi) = \dfrac{\pi}{18}$

7. $\dfrac{\pi}{3}$ **9.** $-\dfrac{3\pi}{4}$

11. Four of many possibilities would be $\dfrac{\pi}{4} + 2\pi = \dfrac{9\pi}{4}$, $\dfrac{\pi}{4} + 4\pi = \dfrac{17\pi}{4}$, $\dfrac{\pi}{4} - 2\pi = -\dfrac{7\pi}{4}$, $\dfrac{\pi}{4} - 4\pi = -\dfrac{15\pi}{4}$.

13. Four of many possibilities would be $-\dfrac{\pi}{6} + 2\pi = \dfrac{11\pi}{6}$, $-\dfrac{\pi}{6} + 4\pi = \dfrac{23\pi}{6}$, $-\dfrac{\pi}{6} - 2\pi = -\dfrac{13\pi}{6}$, $-\dfrac{\pi}{6} - 4\pi = -\dfrac{25\pi}{6}$.

15. No **17.** Yes

19. Add 2π: $-\dfrac{\pi}{3} + 2\pi = \dfrac{5\pi}{3}$. **21.** $\dfrac{19\pi}{4} - 4\pi = \dfrac{3\pi}{4}$

23. $-\dfrac{7\pi}{5} + 2\pi = \dfrac{3\pi}{5}$ **25.** $7 - 2\pi$

27. $6° \cdot \dfrac{\pi}{180°} = \dfrac{\pi}{30}$ **29.** $-12° \cdot \dfrac{\pi}{180°} = -\dfrac{\pi}{15}$

31. $75° \cdot \dfrac{\pi}{180°} = \dfrac{5\pi}{12}$ **33.** $135° \cdot \dfrac{\pi}{180°} = \dfrac{3\pi}{4}$

35. $-225° \cdot \dfrac{\pi}{180°} = -\dfrac{5\pi}{4}$ **37.** $930° \cdot \dfrac{\pi}{180°} = \dfrac{31\pi}{6}$

39. $\dfrac{\pi}{5} \cdot \dfrac{180°}{\pi} = 36°$ **41.** $-\dfrac{\pi}{10} \cdot \dfrac{180°}{\pi} = -18°$

43. $\dfrac{3\pi}{4} \cdot \dfrac{180°}{\pi} = 135°$ **45.** $\dfrac{\pi}{45} \cdot \dfrac{180°}{\pi} = 4°$

47. $-\dfrac{5\pi}{12} \cdot \dfrac{180°}{\pi} = -75°$ **49.** $\dfrac{27\pi}{5} \cdot \dfrac{180°}{\pi} = 972°$

51. $\dfrac{40}{60} \cdot 2\pi = \dfrac{4\pi}{3}$ **53.** $\dfrac{35}{60} \cdot 2\pi = \dfrac{7\pi}{6}$

55. $\left(3 + \dfrac{25}{60}\right) \cdot 2\pi = \dfrac{41\pi}{6}$

57. 3.5 minutes at 1 rpm equals 3.5 revolutions, or $3.5(2\pi) = 7\pi$ radians.

59. 1 minute at 2 rpm equals 2 revolutions, or $2(2\pi) = 4\pi$ radians.

61. 4.25 minutes at 5 rpm equals $4.25 \times 5 = 21.25$ revolutions, or $21.25(2\pi) = 42.5\pi$ rad.

63. 1 minute at k rpm equals k revolutions, or $2\pi k$ radians.

6.1.A Arc Length and Angular Speed

1. $s = r\theta$

$s = 5 \cdot 1$

$s = 5$

3. $s = r\theta$

$s = 5 \cdot 1.75$

$s = 8.75$

5. $s = r\theta$

$s = 6 \cdot \dfrac{4\pi}{3}$

$s = 25.13$

7. $s = r\theta$

$175 = 70\theta$

$\theta = 2.5\text{rad}$

9. $s = r\theta$

$30 = 15\theta$

$\theta = 2\text{rad}$

11. $s = r\theta$

$9 = 1.5r$

$r = 6$

13. $s = r\theta$

$25\pi = \left(2\pi - \dfrac{\pi}{3}\right)r$

$25\pi = \dfrac{5\pi}{3}r$

$r = 15$

15. $s = r\theta$

$200 = 36\theta$

$\theta = 5.56\text{rad}$

17. $s = r\theta$

$72,000 = 36\theta$

$\theta = 2000\text{rad}$

19. $s = r\theta$

$s = 3960\left(90° - 40°\right) \cdot \dfrac{\pi}{180°}$

$s = 3455.8 \text{ mi.}$

21. $s = r\theta$

$s = 3960\left(41.5° - 28°\right) \cdot \dfrac{\pi}{180°}$

$s = 933.1 \text{ mi.}$

23. $A = \dfrac{1}{2}r^2\theta$

$A = \dfrac{1}{2}(5)^2(1)$

$A = 12.5\text{ft}^2$

25. $A = \dfrac{1}{2}r^2\theta$

$A = \dfrac{1}{2}(8)^2(2.4)$

$A = 76.8\text{m}^2$

27. $A = \dfrac{1}{2}r^2\theta$

$4 = \dfrac{1}{2}(2)^2\theta$

$\theta = 2\text{rad}$

29. $A = \dfrac{1}{2}r^2\theta$

$4.32 = \dfrac{1}{2}(1.5)r^2$

$r = 2.4\text{ft}$

31. First convert $7.2°$ into $\theta = .04\pi$ rad. Then use $s = \theta r$ with $s = .04\pi r$ and $s = 500$, thus $r = 3978.8736$ miles. Then $C = 2\pi r = 2\pi(3978.8736) = 25,000.001$ miles.

33. Use $s = r\theta$ with $s = 6$ and $r = 2$. $6 = 2\theta$, thus $\theta = 3$ radians.

35. a. 200 revolutions per minute equals $200(2\pi) = 400\pi$ radians per minute.
  b. linear speed $= r(\text{angular speed}) = 2(400\pi) = 800\pi$
$$= 2513.3 \text{ inches per minute}$$
$$= 209.4 \text{ feet per minute}$$

37. a. 2.5 revolutions per second equals $2.5(2\pi) = 5\pi$ radians per second.
  b. linear speed $= r(\text{angular speed})$
$$= 7.5(5\pi)$$
$$= 37.5\pi \text{ inches per second}$$
$$= 37.5\pi \frac{\text{inches}}{\text{second}} \times \frac{1}{12} \frac{\text{foot}}{\text{inches}} \times \frac{1}{5280} \frac{\text{miles}}{\text{foot}} \times 3600 \frac{\text{seconds}}{\text{hour}}$$
$$= 6.69 \text{ miles per hour}$$

39. 6 revolutions per minute equals $6(2\pi) \div 60$ or $\pi/5$ radians per second.
Since linear speed $= r(\text{angular speed})$
$$10 = r(\pi/5)$$
$$r = 50/\pi$$
$$r = 15.9 \text{ feet}$$

6.2 The Sine, Cosine and Tangent Functions

1. $\sin\left(\dfrac{3\pi}{2}\right) = -1$ **3.** $\cos\left(\dfrac{3\pi}{2}\right) = 0$ **5.** $\tan 4\pi = \dfrac{\sin 4\pi}{\cos 4\pi} = \dfrac{0}{1} = 0$

7. $\cos\left(-\dfrac{3\pi}{2}\right) = 0$ **9.** $\cos\left(-\dfrac{11\pi}{2}\right) = 0$

11. $\sin t = y = \dfrac{1}{\sqrt{5}}$, $\cos t = x = -\dfrac{2}{\sqrt{5}}$, $\tan t = \dfrac{y}{x} = \dfrac{1}{\sqrt{5}} \div -\dfrac{2}{\sqrt{5}} = -\dfrac{1}{2}$

13. $\sin t = y = -\dfrac{4}{5}$, $\cos t = x = -\dfrac{3}{5}$, $\tan t = \dfrac{y}{x} = -\dfrac{4}{5} \div -\dfrac{3}{5} = \dfrac{4}{3}$

15. $\sin\dfrac{\pi}{3} = \dfrac{\sqrt{3}}{2}$, $\cos\dfrac{\pi}{3} = \dfrac{1}{2}$, $\tan\dfrac{\pi}{3} = \sin\dfrac{\pi}{3} \div \cos\dfrac{\pi}{3} = \dfrac{\sqrt{3}}{2} \div \dfrac{1}{2} = \sqrt{3}$

17. $\sin\dfrac{7\pi}{4} = -\dfrac{\sqrt{2}}{2}$, $\cos\dfrac{7\pi}{4} = \dfrac{\sqrt{2}}{2}$, $\tan\dfrac{7\pi}{4} = \sin\dfrac{7\pi}{4} \div \cos\dfrac{7\pi}{4} = -\dfrac{\sqrt{2}}{2} \div \dfrac{\sqrt{2}}{2} = -1$

19. $\sin\dfrac{3\pi}{4} = \dfrac{\sqrt{2}}{2}$, $\cos\dfrac{3\pi}{4} = -\dfrac{\sqrt{2}}{2}$, $\tan\dfrac{3\pi}{4} = -1$

21. $\sin\dfrac{5\pi}{6} = \dfrac{1}{2}$, $\cos\dfrac{5\pi}{6} = -\dfrac{\sqrt{3}}{2}$, $\tan\dfrac{5\pi}{6} = \sin\dfrac{5\pi}{6} \div \cos\dfrac{5\pi}{6} = \dfrac{1}{2} \div -\dfrac{\sqrt{3}}{2} = -\dfrac{\sqrt{3}}{3}$

23. $\theta = -\dfrac{23\pi}{6}$ is coterminal with $-\dfrac{23\pi}{6} + 4\pi = \dfrac{\pi}{6}$.

$\sin\left(-\dfrac{23\pi}{6}\right) = \sin\dfrac{\pi}{6} = \dfrac{1}{2}$, $\cos\left(-\dfrac{23\pi}{6}\right) = \cos\dfrac{\pi}{6} = \dfrac{\sqrt{3}}{2}$, $\tan\left(-\dfrac{23\pi}{6}\right) = \tan\dfrac{\pi}{6} = \dfrac{\sqrt{3}}{3}$

25. $\theta = -\dfrac{19\pi}{3}$ is coterminal with $-\dfrac{19\pi}{3} + 8\pi = \dfrac{5\pi}{3}$. This angle has terminal side in

quadrant IV and the angle formed by the terminal side and the x-axis is $\dfrac{\pi}{3}$.

Therefore,

$\sin\left(-\dfrac{19\pi}{3}\right) = -\sin\dfrac{\pi}{3} = -\dfrac{\sqrt{3}}{2}$, $\cos\left(-\dfrac{19\pi}{3}\right) = \cos\dfrac{\pi}{3} = \dfrac{1}{2}$, $\tan\left(-\dfrac{19\pi}{3}\right) = -\tan\dfrac{\pi}{3} = -\sqrt{3}$

27. $\theta = -\dfrac{15\pi}{4}$ is coterminal with $-\dfrac{15\pi}{4} + 4\pi = \dfrac{\pi}{4}$.

$\sin\left(-\dfrac{15\pi}{4}\right) = \sin\dfrac{\pi}{4} = \dfrac{\sqrt{2}}{2}$, $\cos\left(-\dfrac{15\pi}{4}\right) = \cos\dfrac{\pi}{4} = \dfrac{\sqrt{2}}{2}$, $\tan\left(-\dfrac{15\pi}{4}\right) = \tan\dfrac{\pi}{4} = 1$

29. The terminal side of $\theta = -17\pi/2$ in standard position is along the negative y-axis.
Using the point $(0,-1)$ on this terminal side yields

$$\sin\left(-\frac{17\pi}{2}\right) = -1, \quad \cos\left(-\frac{17\pi}{2}\right) = 0, \quad \tan\left(-\frac{17\pi}{2}\right) \text{ is undefined}.$$

31. $\sin\dfrac{\pi}{3}\cos\pi + \sin\pi\cos\dfrac{\pi}{3} = \dfrac{\sqrt{3}}{2}(-1) + 0\left(\dfrac{1}{2}\right) = -\dfrac{\sqrt{3}}{2}$

33. $\cos\dfrac{\pi}{2}\cos\dfrac{\pi}{4} - \sin\dfrac{\pi}{2}\sin\dfrac{\pi}{4} = 0\left(\dfrac{\sqrt{2}}{2}\right) - 1\left(\dfrac{\sqrt{2}}{2}\right) = -\dfrac{\sqrt{2}}{2}$

35. $\sin\dfrac{3\pi}{4}\cos\dfrac{5\pi}{6} - \cos\dfrac{3\pi}{4}\sin\dfrac{5\pi}{6} = \dfrac{\sqrt{2}}{2}\left(-\dfrac{\sqrt{3}}{2}\right) - \left(-\dfrac{\sqrt{2}}{2}\right)\left(\dfrac{1}{2}\right) = \dfrac{-\sqrt{6}+\sqrt{2}}{4}$

37. $(x,y) = (3,5); \ r = \sqrt{x^2+y^2} = \sqrt{3^2+5^2} = \sqrt{34}$

$\sin t = \dfrac{y}{r} = \dfrac{5}{\sqrt{34}}, \quad \cos t = \dfrac{x}{r} = \dfrac{3}{\sqrt{34}}, \quad \tan t = \dfrac{y}{x} = \dfrac{5}{3}$

39. $(x,y) = (-4,-5); \ r = \sqrt{x^2+y^2} = \sqrt{(-4)^2+(-5)^2} = \sqrt{41}$

$\sin t = \dfrac{y}{r} = \dfrac{-5}{\sqrt{41}}, \quad \cos t = \dfrac{x}{r} = \dfrac{-4}{\sqrt{41}}, \quad \tan t = \dfrac{y}{x} = \dfrac{5}{4}$

41. $(x,y) = \left(\sqrt{3},-8\right); \ r = \sqrt{x^2+y^2} = \sqrt{\left(\sqrt{3}\right)^2+(-8)^2} = \sqrt{67}$

$\sin t = \dfrac{y}{r} = \dfrac{-8}{\sqrt{67}}, \quad \cos t = \dfrac{x}{r} = \sqrt{\dfrac{3}{67}}, \quad \tan t = \dfrac{y}{x} = \dfrac{-8}{\sqrt{3}}$

43. a. $\sin\left(\dfrac{.8}{2}\right) = \dfrac{w}{1200}$ **b.** $\sin\left(\dfrac{.7}{2}\right) = \dfrac{500}{p}$

$w = 467$ mph $p = 1458$ mph

45. a.

Date	Jan 1	Mar 1	May 1	July 1	Sept 1	Nov 1
Average Temperature	31.7	43.8	68.3	80.9	69.0	44.5

b.

Date	June 1	June 4	June 7	June 10	June 13
Average Temperature	77.4	78.1	78.6	79.1	79.6

Date	June 16	June 19	June 22	June 25	June 28
Average Temperature	80.0	80.3	80.5	80.7	80.8

47. $\dfrac{\cos(\pi)-\cos\left(\pi/2\right)}{\pi-\pi/2}=-\dfrac{2}{\pi}$

49. $\dfrac{\sin\left(11\pi/3\right)-\sin\left(\pi/6\right)}{11\pi/3-\pi/6}=\dfrac{-\sqrt{3}-1}{7\pi}$

51. $\dfrac{\cos\left(\pi/4\right)-\cos\left(-5\pi/4\right)}{\pi/4-\left(-5\pi/4\right)}=\dfrac{2\sqrt{2}}{3\pi}$

53. $\dfrac{\tan\left(\pi/3\right)-\tan\left(\pi/6\right)}{\pi/3-\pi/6}=\dfrac{4\sqrt{3}}{\pi}$

55. a. -.420686; -.416601; -.416192; -.416151

b. $\cos(2)\approx-.416147$, thus the instantaneous rate of change of $\sin t$ at $t=2$ is $\cos(2)$.

57. The line has equation $y=-2x$.

Choose $(1,-2)$, $r=\sqrt{x^2+y^2}=\sqrt{1+(-2)^2}=\sqrt{5}$

$\sin t=\dfrac{y}{r}=\dfrac{-2}{\sqrt{5}}$, $\cos t=\dfrac{x}{r}=\dfrac{1}{\sqrt{5}}$, $\tan t=\dfrac{y}{x}=-2$

59. $m=\dfrac{15-5}{-9-(-3)}=\dfrac{10}{-6}=-\dfrac{5}{3}$; so $y-5=-\dfrac{5}{3}(x-(-3))$, solving for y give $y=-\dfrac{5}{3}x$. A

point on this line in Quadrant 4 is $(5,-3)$. Now, $r=\sqrt{5^2+(-3)^2}=\sqrt{34}$.

$\sin t=\dfrac{y}{r}=\dfrac{-3}{\sqrt{34}}$, $\cos t=\dfrac{x}{r}=\dfrac{5}{\sqrt{34}}$, $\tan t=\dfrac{y}{x}=-\dfrac{3}{5}$

61. The terminal side of the angle is on the line $2y+x=0$, with x negative.

Let $(x,y)=(-2,1)$; $r=\sqrt{x^2+y^2}=\sqrt{(-2)^2+1^2}=\sqrt{5}$.

$\sin t=\dfrac{y}{r}=\dfrac{1}{\sqrt{5}}$, $\cos t=\dfrac{x}{r}=\dfrac{-2}{\sqrt{5}}$, $\tan t=\dfrac{y}{x}=\dfrac{1}{-2}=-\dfrac{1}{2}$

63. Quadrant I: $\sin t$, $\cos t$, $\tan t$ all $+$ Quadrant II: $\sin t\,+$, $\cos t\,-$, $\tan t\,-$
Quadrant III: $\sin t\,-$, $\cos t\,-$, $\tan t\,+$ Quadrant IV: $\sin t\,-$, $\cos t\,+$, $\tan t\,-$

65. Since $0 < 1 < \pi/2$, the terminal side of an angle of 1 radian lies in the first quadrant, so $y = \sin 1$ is positive.

67. Since $\pi/2 < 3 < \pi$, the terminal side of an angle of 3 radians lies in the second quadrant, so $y/x = \tan 3$ is negative.

69. Since $0 < 1/5 < \pi/2$, the terminal side of an angle of 1.5 radians lies in the first quadrant, so $y/x = \tan 1.5$ is positive.

71. The terminal side of t must pass through $(0,1)$. Hence $t = \pi/2 + 2\pi n$, n any integer.

73. The terminal side of t must pass through $(1,0)$ or $(-1,0)$. Hence $t = \pi n$, n any integer.

75. The terminal side of t must pass through $(0,-1)$ or $(0,1)$. Hence $t = \pi/2 + \pi n$, n any integer.

77. $\sin(\cos 0) = \sin 1$ $\cos(\sin 0) = \cos 0 = 1$. Since $\sin 1 < 1$, $\sin(\cos 0) < \cos(\sin 0)$.

79. **a.** Horse A will reach the position occupied by horse B at a time t after rotation of $\pi/4$ radians, which is an eighth of a complete circle. This takes 1/8 of a minute. Thus $B(t) = A(t + 1/8)$.

   b. Horse A will reach the position occupied by horse C at a time t after rotation of 1/3 of a complete circle. This takes 1/3 of a minute. Thus $C(t) = A(t + 1/3)$.

   c. $E(t) = D(t + 1/8)$; $F(t) = D(t + 1/3)$

   d. $D(t) / A(t) = 5/1$; $D(t) = 5A(t)$; $D(t) = 5A(t)$

   e. $E(t) = D(t + 1/8) = 5A(t + 1/8)$; $F(t) = D(t + 1/3) = 5A(t + 1/3)$

   f. In t minutes, the merry-go-round rotates $2\pi t$ radians. Therefore, $A(t) = \sin(2\pi t)$.

   g. $B(t) = \sin 2\pi(t + 1/8) = \sin(2\pi t + \pi/4)$; $C(t) = \sin 2\pi(t + 1/3) = \sin(2\pi t + 2\pi/3)$

   h. $D(t) = 5\sin 2\pi t$; $E(t) = 5\sin(2\pi t + \pi/4)$; $F(t)\,5\sin(2\pi t + 2\pi/3)$

6.3 Algebra and Identities

1. $(fg)(t) = f(t)g(t) = 3\sin t(\sin t + 2\cos t) = 3\sin^2 t + 6\sin t \cos t$

3. $(fg)(t) = f(t)g(t) = 3\sin^2 t(\sin t + \tan t) = 3\sin^3 t + 3\sin^2 t \tan t$

5. $(\cos t + 2)(\cos t - 2)$ **7.** $(\sin t + \cos t)(\sin t - \cos t)$

9. $(\tan t + 3)(\tan t + 3) = (\tan t + 3)^2$ **11.** $(3\sin t + 1)(2\sin t - 1)$

13. $\cos^4 t + 4\cos^2 t - 5 = (\cos^2 t + 5)(\cos^2 t - 1) = (\cos^2 t + 5)(\cos t + 1)(\cos t - 1)$

15. $(f \circ g)(t) = f(g(t)) = \cos(g(t)) = \cos(2t + 4)$
$(g \circ f)(t) = g(f(t)) = 2f(t) + 4 = 2\cos t + 4$

17. $(f \circ g)(t) = f(g(t)) = \tan(g(t) + 3) = \tan(t^2 - 1 + 3) = \tan(t^2 + 2)$
$(g \circ f)(t) = g(f(t)) = (f(t))^2 - 1 = \tan^2(t + 3) - 1$

19. By the Pythagorean identity, $\sin^2 t + \cos^2 t = 1$. Since $(5/13)^2 + (12/13)^2 = 1$, these values are possible.

21. By the Pythagorean identity, $\sin^2 t + \cos^2 t = 1$. Since $(-1)^2 + 1^2 = 2$, these values are not possible.

23. By the Pythagorean identity, $\sin^2 t + \cos^2 t = 1$. Also, $\tan t = \dfrac{\sin t}{\cos t}$, so $1 = \dfrac{1}{\cos t}$ and $\cos t = 1$ in this case. But since $1^2 + 1^2 = 2$, these values are not possible.

25. By the Pythagorean identity,
$\sin^2 t = 1 - \cos^2 t = 1 - (-.5)^2 = 3/4$. Hence $\sin t = -\sqrt{3/4}$ or $\sin t = \sqrt{3/4}$.
Since $\pi < t < 3\pi/2$, $\sin t$ is negative, hence $\sin t = -\sqrt{3/4} = -\sqrt{3}/2$.

27. By the Pythagorean identity, $\sin^2 t = 1 - \cos^2 t = 1 - (1/2)^2 = 3/4$.
Hence $\sin t = -\sqrt{3/4}$ or $\sin t = \sqrt{3/4}$.
Since $0 < t < \pi/2$, $\sin t$ is positive, hence $\sin t = \sqrt{3/4} = \sqrt{3}/2$.

29. $\sin(-t) = -\sin t = -3/5$ **31.** $\sin(2\pi - t) = -\sin t = -3/5$

33. By the Pythagorean identity, $\cos^2 t = 1 - \sin^2 t = 1 - (3/5)^2 = 16/25$.

Since $0 < t < \pi/2$, $\cos t$ is positive, hence $\cos t = \sqrt{16/25} = 4/5$.

$\tan t = \sin t \div \cos t = 3/5 \div 4/5 = 3/4$

35. $\tan(2\pi - t) = \tan(-t) = -\tan t = -3/4$ (using Exercise **33**)

37. By the Pythagorean identity, $\sin^2 t = 1 - \cos^2 t = 1 - (-2/5)^2 = 21/25$.

Since $\pi < t < 3\pi/2$, $\sin t$ is negative, hence $\sin t = -\sqrt{21/25} = -\sqrt{21}/5$.

39. $\cos(2\pi - t) = \cos(-t) = \cos t = -2/5$

41. $\sin(4\pi + t) = \sin t = -\sqrt{21}/5$ (using Exercise **37**)

43. By the Pythagorean identity,

$$\cos^2 \frac{\pi}{8} = 1 - \sin^2 \frac{\pi}{8} = 1 - \left(\frac{\sqrt{2-\sqrt{2}}}{2}\right)^2 = \frac{4 - 2 + \sqrt{2}}{4} = \frac{2 + \sqrt{2}}{4}.$$

Since $0 < \frac{\pi}{8} < \frac{\pi}{2}$, $\cos \frac{\pi}{8}$ is positive, hence $\cos \frac{\pi}{8} = \frac{\sqrt{2+\sqrt{2}}}{2}$.

45. $\sin \frac{17\pi}{8} = \sin \frac{\pi}{8} = \frac{\sqrt{2-\sqrt{2}}}{2}$

47. $\sin^2 t - \cos^2 t$

49. $\tan t \cos t = \frac{\sin t}{\cos t} \cos t = \sin t$

51. $\sqrt{\sin^3 t \cos t}\sqrt{\cos t} = \sqrt{\sin^3 t \cos^2 t} = |\sin t \cos t|\sqrt{\sin t}$

53. $\left(\frac{4\cos^2 t}{\sin^2 t}\right)\left(\frac{\sin t}{4\cos t}\right)^2 = \frac{4\cos^2 t}{\sin^2 t} \cdot \frac{\sin^2 t}{16\cos^2 t} = \frac{1}{4}$

55. $\dfrac{\cos^2 t + 4\cos t + 4}{\cos t + 2} = \dfrac{(\cos t + 2)^2}{\cos t + 2} = \cos t + 2$

57. $\dfrac{1}{\cos t} - \sin t \tan t = \dfrac{1}{\cos t} - \sin t \dfrac{\sin t}{\cos t} = \dfrac{1 - \sin^2 t}{\cos t} = \dfrac{\cos^2 t}{\cos t} = \cos t$

59. a. Tables for $f(t) = 22.7\sin(.52x - 2.18) + 49.6$ and

$$f(t + 12.083) = 22.7\sin(.52(x + 12.083) - 2.18) + 49.6$$

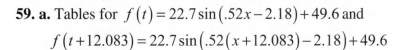

X	Y₂
1	26.99
2	28.974
3	36.411
4	47.334
5	58.856
6	67.932
7	72.161

Y₂∎22.7sin(.52X...

X	Y₂
8	70.426
9	63.185
10	52.353
11	40.794
12	31.562
13	27.099
14	28.584

Y₂∎22.7sin(.52X...

X	Y₃
1	26.99
2	28.974
3	36.411
4	47.334
5	58.856
6	67.932
7	72.161

Y₃∎Y₂(X+12.083)

X	Y₃
8	70.426
9	63.185
10	52.353
11	40.794
12	31.562
13	27.099
14	28.584

Y₃∎Y₂(X+12.083)

b. Yes, the period is 12.083, which is a reasonable approximation of the expected period of 12 months.

61. a. 0%; about 76.5%; about 51.3%
   b. Tables for

$$Y_1 = g(t) = .5\left(1 - \cos\frac{2\pi t}{29.5}\right) \text{ and } Y_2 = g(t + 29.5) = .5\left(1 - \cos\frac{2\pi(t + 29.5)}{29.5}\right)$$

X	Y₄	Y₅
0	0	0
1	.0113	.0113
2	.04468	.04468
3	.09864	.09864
4	.17074	.17074
5	.25772	.25772
6	.35565	.35565

Y₄∎.5(1-cos(2πX...

X	Y₄	Y₅
7	.46011	.46011
8	.56636	.56636
9	.66962	.66962
10	.76521	.76521
11	.84882	.84882
12	.91666	.91666
13	.96567	.96567

Y₅∎Y₄(X+29.5)

X	Y₄	Y₅
14	.99363	.99363
15	.99929	.99929
16	.98238	.98238
17	.94368	.94368
18	.88492	.88492
19	.80876	.80876
20	.71865	.71865

Y₄∎.5(1-cos(2πX...

X	Y₄	Y₅
21	.61866	.61866
22	.51331	.51331
23	.40736	.40736
24	.30559	.30559
25	.21261	.21261
26	.13261	.13261
27	.06922	.06922

Y₅∎Y₄(X+29.5)

X	Y₄	Y₅
28	.0253	.0253
29	.00283	.00283
30	.00283	.00283
31	.0253	.0253
32	.06922	.06922
33	.13261	.13261
34	.21261	.21261

Y₄∎.5(1-cos(2πX...

   c. Late on day 14 ($t = 14.75$)

63. Use $k = 2\pi/3$. $f(t + k) = \cos 3(t + 2\pi/3) = \cos(3t + 2\pi) = \cos 3t = f(t)$ for all t.

65. Use $k = 2$. $f(t + k) = \sin(\pi(t + 2)) = \sin(\pi t + 2\pi) = \sin \pi t = f(t)$ for all t.

67. Use $k = \pi/2$. $f(t + k) = \tan 2(t + \pi/2) = \tan(2t + \pi) = \tan 2t = f(t)$ for all t in the domain.

69. a. If $0 < k < 2\pi$, there is no number such that $\cos k = 1$.

  b. If $\cos(t + k) = \cos t$ for every number t, then use $t = 0$.

  $$\cos(0 + k) = \cos 0$$

  $$\cos k = 1$$

  c. Therefore there is no number k with $0 < k < 2\pi$ such that $\cos(t + k) = \cos t$ for every number t.

6.4 Basic Graphs

1. $t = \pi k$ where n is any integer.

3. $t = \dfrac{\pi}{2} + 2\pi k$ where n is any integer.

5. $t = \pi + 2\pi k$ where n is any integer.

7. Since $\dfrac{y}{x} = 11$, $\tan t = 11$.

9. Since $\dfrac{y}{x} = 1.4$, $\tan t = 1.4$.

11. Shift 3 units vertically up.

13. Reflect the graph in the horizontal axis.

15. Shift 5 units vertically up.

17. Stretch by a factor of 3 away from the horizontal axis.

19. Stretch by a factor of 3 away from the horizontal axis, then shift 2 units vertically up.

21. Shift 2 units horizontally to the right.

23. D   **25.** B   **27.** F   **29.** G

31. Graph $y = \sin x$ and the horizontal line $y = \frac{3}{5}$ using the window $0 \le x \le 2\pi$, $-1 \le y \le 1$. Since there are two points of intersection, conclude that the equation has two solutions in the interval.

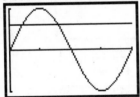

33. Graph $y = \tan x$ and the horizontal line $y = 4$ using the window $0 \le x \le 2\pi$, $-5 \le y \le 5$. Ignore the erroneous vertical lines near the asymptotes of $y = \tan x$. Since there are two points of intersection, conclude that the equation has two solutions in the interval.

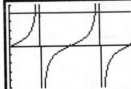

35. Graph $y = \sin x$ and the horizontal line $y = -\frac{1}{2}$ using the window $0 \le x \le 2\pi$, $-1 \le y \le 1$. Since there are two points of intersection, conclude that the equation has two solutions in the interval.

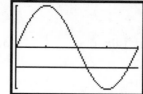

37. Graph $y = \cos x$ and any horizontal line, say $y = \frac{1}{2}$, using the window $0 \le x \le 2\pi$, $-1 \le y \le 1$. Since there are two points of intersection, conclude that any such equation has two solutions in the interval.

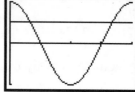

39. Graph $y = \sin(-x)$ and $y = -\sin x$ using the window $0 \le x \le 2\pi$, $-1 \le y \le 1$.
The graphs appear to coincide, hence the equation could be an identity.

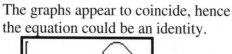
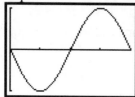

41. Graph $y = \sin^2 x + \cos^2 x$ and $y = 1$ using the window $0 \le x \le 2\pi$, $-1 \le y \le 1$. The graphs appear to coincide, hence the equation could be an identity (and is a well-known one).

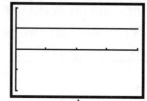

43. Graph $y = \sin x$ and $y = \cos\left(x - \frac{\pi}{2}\right)$ using the window $0 \le x \le 2\pi$, $-1 \le y \le 1$. The graphs appear to coincide, hence the equation could be an identity.

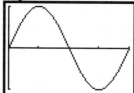

45. Graph $y = \dfrac{\sin x}{1 + \cos x}$ and $y = \tan x$ using the window $0 \le x \le 2\pi$, $-5 \le y \le 5$. The graphs do not coincide, hence the equation could not be an identity.

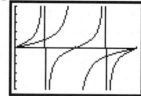

47. Graph $y = \cos\left(\frac{\pi}{2} + x\right)$ and $y = -\sin x$ using the window $0 \le x \le 2\pi$, $-1 \le y \le 1$. The graphs appear to coincide, hence the equation could be an identity.

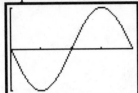

49. Graph $y = (1 + \tan x)^2$ and $y = 1/\cos x$ using the window $0 \le x \le 2\pi$, $-5 \le y \le 5$. The graphs do not coincide, hence the equation could not be an identity.

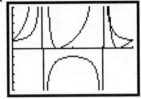

51. No

53. Yes, period 2π

55. Yes, period 2π

57. No

59. No

61. a. The graphs will appear identical, but the functions are not the same.
  b. The calculator plots 95 points, each of which has an x-coordinate that is a multiple of 2π. Since all y-coordinates are 1, the graph looks like the horizontal line $y = 1$.

63. a. There should be 80 full waves in an interval of length $80 \cdot 2\pi$.
  b. The calculator graph can't show 80 waves, since it plots only 95 points. As a signal that something is wrong, note that the waves shown are different.

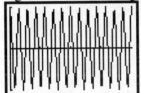

65. a. Shown is the window $-2\pi < x \le 2\pi$, $-2 \le y \le 2$. Agreement is good between $x = -\pi$ and $x = \pi$.

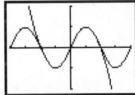

  b. Agreement is good over the entire interval for f_{15}. In fact: $\sin(2) = .9092974268$. $f_{15}(2) = .9092974265$

67. Since $\tan t = \dfrac{\sin t}{\cos t}$, a function of the form $\dfrac{f_m(t)}{g_n(t)}$, where the numerator and denominator functions are as defined in Exercises 65 and 66, is needed. The graph of $y = \dfrac{f_{15}(t)}{g_{16}(t)}$ is indistinguishable from the graph of $y = \tan t$ in the specified window.

69. The *y*-coordinate of the point on the graph of the cosine function is the same as the *x*-coordinate of the point on the unit circle, because of the definition of the cosine function as the *x*-coordinate of a point on the unit circle.

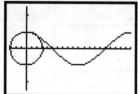

6.5 Periodic Graphs and Simple Harmonic Motion

1. Amplitude $= |3| = 3$
Period $= 2\pi/2 = \pi$
Phase shift $= -\dfrac{-\pi}{2} = \dfrac{\pi}{2}$

3. Amplitude $= |-5| = 5$
Period $= 2\pi/5$
Phase shift $= -\dfrac{1}{5} \div 5 = -\dfrac{1}{25}$

5. Amplitude $= 1$
Period $= \dfrac{2\pi}{2\pi} = 1$
Phase shift $= 0$

7. Amplitude $= |6| = 6$
Period $= \dfrac{2\pi}{3\pi} = \dfrac{2}{3}$
Phase shift $= -\dfrac{1}{3\pi}$

9. Period $= \dfrac{2\pi}{b} = \dfrac{\pi}{4}$. Thus $b = 8$. Phase shift $= -\dfrac{c}{b} = \dfrac{\pi}{5}$. Thus $c = -\dfrac{\pi}{5}b = -\dfrac{8\pi}{5}$.
$f(t) = 3\sin(8t - 8\pi/5)$ or $f(t) = 3\cos(8t - 8\pi/5)$

11. Period $= \dfrac{2\pi}{b} = 2$. Thus $b = \pi$. Phase shift $= -\dfrac{c}{b} = 0$. Thus $c = 0$.
$f(t) = (3/4)\sin \pi t$ or $f(t) = (3/4)\cos \pi t$

13. Period $= \dfrac{2\pi}{b} = \dfrac{5}{3}$. Thus $b = \dfrac{6\pi}{5}$. Phase shift $= -\dfrac{c}{b} = -\dfrac{\pi}{2}$. Thus $c = \dfrac{\pi}{2}b = \dfrac{3\pi^2}{5}$.
$f(t) = 7\sin\left(\dfrac{6\pi}{5}t + \dfrac{3\pi^2}{5}\right)$ or $f(t) = 7\cos\left(\dfrac{6\pi}{5}t + \dfrac{3\pi^2}{5}\right)$

15. Amplitude $= 2$. Period $= \dfrac{2\pi}{b} = \dfrac{\pi}{2}$. Thus $b = 4$. The shape is that of a sine function:
$f(t) = 2\sin 4t$

17. Amplitude $= 1.5$. Period $= \dfrac{2\pi}{b} = 4\pi$. Thus $b = \dfrac{1}{2}$. The shape is that of a cosine function: $f(t) = 1.5\cos(\tfrac{1}{2}t)$

19. a. Period $= 2\pi/200 = \pi/100$.
b. The graph should have 200 complete waves between 0 and 2π.
c. Shown is the window $0 \le x \le \pi/25, -2 \le y \le 2$.

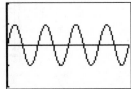

21. a. Period $= 2\pi / 900 = \pi / 450$.

   b. The graph should have 900 complete waves between 0 and 2π.

   c. Shown is the window $0 \le x \le 2\pi / 225, -2 \le y \le 2$.

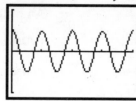

23. a. The amplitude is $A = 12$ and the period is $\pi / 5$, hence $2\pi / b = \pi / 5$; $b = 10$.

   The phase shift is 1/4 of a period to the right, hence

$$-\frac{c}{b} = \frac{\pi / 5}{4}$$

$$c = -\frac{\pi}{20} b = -\frac{\pi}{20} \cdot 10 = -\frac{\pi}{2}.$$

   Thus, $f(t) = 12 \sin\left(10t - \frac{\pi}{2}\right)$ or $f(t) = -12 \sin\left(10t + \frac{\pi}{2}\right)$

   b. This is a standard cosine curve reflected about the x-axis, with amplitude and period as above and no phase shift. Thus, $g(t) = -12 \cos(10t)$.

25. a. This is a standard sine curve reflected about the x-axis. The amplitude is $A = 1$ and the period is π, hence

$$\frac{2\pi}{b} = \pi$$

$$b = 2$$

   Thus, $f(t) = -\sin(2t)$

   b. As in **a**, the amplitude is $A = 1$ and the period is π, but there is a phase shift of 1/4 of a period to the left, hence

$$-\frac{c}{b} = -\frac{\pi}{4}$$

$$c = \frac{\pi}{4} b = \frac{\pi}{4} \cdot 2$$

$$c = \frac{\pi}{2}$$

   Thus, $g(t) = \cos\left(2t + \frac{\pi}{2}\right)$ or $g(t) = -\cos\left(2t - \frac{\pi}{2}\right)$.

27. Use the window $0 \le x \le 2\pi, -3 \le y \le 3$.   **29.** Use the window $0 \le x \le 2\pi, -1 \le y \le 1$.

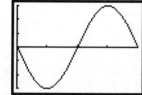
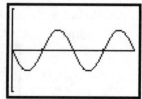

31. Use the window $0 \le x \le 2\pi$, $-3 \le y \le 3$.

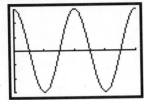

33. Use the window $0 \le x \le 2\pi$, $-1 \le y \le 1$.

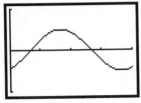

Since there is a phase shift of $\pi/3$, the local maximum and the local minimum must be at

$$x = \frac{\pi}{2} + \frac{\pi}{3} = \frac{5\pi}{6} \approx 2.6180 \text{ and at}$$

$$x = \frac{3\pi}{2} + \frac{\pi}{3} = \frac{11\pi}{6} \approx 5.7596 \text{ respectively.}$$

35. Use the window $0 \le x \le 2\pi$, $-3 \le y \le 3$.

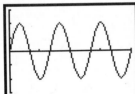

The maxima and minima must occur when

$$3x - \pi = -\frac{\pi}{2}, \frac{3\pi}{2}, \frac{7\pi}{2}$$

$$x = \frac{\pi}{6} \approx .5236, \frac{5\pi}{6} \approx 2.6180, \frac{3\pi}{2} \approx 4.7124$$

and

$$3x - \pi = \frac{\pi}{2}, \frac{5\pi}{2}, \frac{9\pi}{2}$$

$$x = \frac{\pi}{2} \approx 1.5708, \frac{7\pi}{6} \approx 3.6652, \frac{11\pi}{6} \approx 5.7596$$

respectively.

37. Graph the function using the window $-2\pi \le x \le 2\pi$, $-6 \le y \le 6$

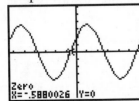 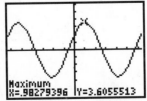

Note that the function has a maximum value of $A = 3.606$. One complete period occurs between $x = -.5880$ and $x = -.5880 + 2\pi$, hence $b = 1$ and $-c/b = -.5880$, thus $c = .5880$. $f(t) = 3.606 \sin(t + .5880)$

39. Graph the function using the window $-2\pi \le x \le 2\pi, -6 \le y \le 6$

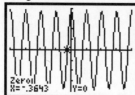 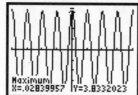

Note that the function has a maximum value of $A = 3.8332$. Four complete periods occur between $x = 0$ and $x = 2\pi$, hence $b = 4$. One complete period occurs between $x = -.3643$ and $x = -.3643 + \pi / 2$, hence $-c / b = -c / 4 = -.3643$, thus $c = 1.4572$. $f(t) = 3.8332 \sin(4t + 1.4572)$

41. Graph the function using the window $0 \le x \le 2\pi, -4 \le y \le 4$.

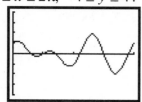

The waves do not have the same amplitude.

43. a. The person's blood pressure is 134/92
   b. The pulse rate is 75.

45. a. Maximum is 79.5 cubic inches; minimum is 30.5 cubic inches.
   b. Every 4 seconds.
   c. 15

47. The period of the function $f(t) = A\sin(1{,}960{,}000\pi t)$ is
$$\frac{2\pi}{1{,}960{,}000\pi} = \frac{1}{980{,}000} \text{ seconds.}$$
The frequency is the reciprocal of the period, that is, 980,000 cycles per second (known as 980 KiloHertz).

49. The amplitude is $A = 125$. If time is measured in minutes, the period is 10, hence $b = 2\pi / 10 = \pi / 5$. At time $t = 0$ the car is at $y = 0$, hence the phase shift is 0.
$$f(t) = 125 \sin\left(\frac{\pi}{5}t\right)$$

51. Since the wheel is rotating 10 times per second, the period is 1/10 second, we can
write $2\pi / b = 1/10$, $b = 20\pi$, and the coordinates of W are given by
$(\cos(20\pi t), \sin(20\pi t))$. The coordinates of the shadow S of W on the x-axis are given
by $(\cos(20\pi t), 0)$. From the Pythagorean Theorem the distance from S to P is given
by $\sqrt{16 - \sin^2(20\pi t)}$ and hence the x-coordinate of P is
$\cos(20\pi t) + \sqrt{16 - \sin^2(20\pi t)}$.

53. Since the weight starts at equilibrium, use the sine function. $A = 6$. $2\pi / b = 4$,
$b = \pi / 2$. $h(t) = 6\sin(\pi t / 2)$.

55. Since the weight starts above equilibrium, use the cosine function with positive coefficient. $h(t) = 6\cos(\pi t / 2)$

57. $A = \frac{1}{2}(20) = 10$. $2\pi / b = 4$, $b = \pi / 2$. $d(t) = 10\sin(\pi t / 2)$.

59. a. For a very rough picture, plot the maximum, the minimum, and the two end
points, four in all.
  b. There are 100 waves in this interval. After plotting the first end point, 3 are
needed for each succeeding wave, hence 301 would be needed.
  c. Every calculator is different; the TI-83 plots 95 points, others plot as many as 239.
  d. Obviously, 236 points (or fewer) is not enough when the absolute minimum is
301.

61. a.

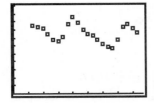

  b. Yes, the data appears to be approximately periodic. An appropriate model is
$f(t) = 1.1371\sin(.6352t - .6366) + 7.4960$.

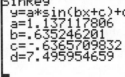

  c. No, the model is only a fair approximation of the data
and, in any case, unemployment is hard to predict.

63. a. The SinReg feature of the TI-83 yields:

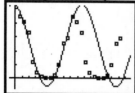

b. Using this value of b yields a period of $2\pi/b = 2\pi/.4 = 15.7$ months. This seems very unlikely for precipitation figures.

c. Plotting another year of the same data versus the model yields:

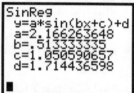

Clearly the 15.7 month period produces a very poor fit.

d. Now we obtain the following model and the following graph:

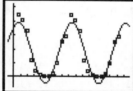

e. Using this value of b yields a period of $2\pi/b = 2\pi/.513333335 = 12.2$ months. This seems more reasonable, and the model fits the data better.

65. a. If the period is 2, then $\dfrac{2\pi}{\omega} = 2$, $\omega = \pi$. Solve $\pi = \sqrt{\dfrac{9.8}{k}}$ to obtain $k = .9929$ meters.

b. If k is increased by .01% to $1.0001k$, then the period becomes approximately 2.000099998 seconds, hence the clock loses about .000099998 seconds every 2 seconds. Multiply this by 7,948,800/2 to obtain a loss of about 397.43 seconds (6.62 minutes) during the three months.

6.5.A Other Trigonometric Graphs

1. Graph the function using the window $-2\pi \le x \le 2\pi, -3 \le y \le 3$

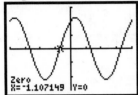 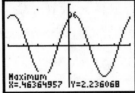

Note that the function has a maximum value of $A = 2.2361$. One complete period occurs between $x = 0$ and $x = 2\pi$, hence $b = 1$. The start of one period is at $x = -1.1071$, hence $-c/b = -c/1 = -1.1071$, thus $c = 1.1071$.
$$f(t) = 2.2361 \sin(t + 1.1071)$$

3. Graph the function using the window $-2\pi \le x \le 2\pi, -6 \le y \le 6$

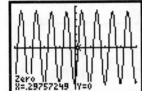 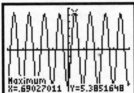

Note that the function has a maximum value of $A = 5.3852$. Four complete periods occur between $x = 0$ and $x = 2\pi$, hence $b = 4$. The start of one period is at $x = .29757$, hence $-c/b = -c/4 = .29757$, thus $c = -1.1903$.
$$f(t) = 5.3852 \sin(4t - 1.1903)$$

5. Graph the function using the window $-2\pi \le x \le 2\pi, -6 \le y \le 6$

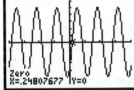 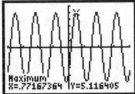

Note that the function has a maximum value of $A = 5.1164$. Three complete periods occur between $x = 0$ and $x = 2\pi$, hence $b = 3$. The start of one period is at $x = .24808$, hence $-c/b = -c/3 = .24808$, thus $c = -.7442$.
$$f(t) = 5.1164 \sin(3t - .7442)$$

7. $0 \le x \le 2\pi, -5 \le y \le 5$

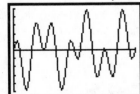

9. $-10 \le x \le 10, -10 \le y \le 10$

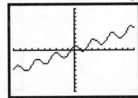

11. $0 \le x \le \pi/25, -2 \le y \le 2$

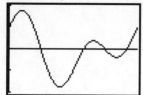

13. $0 \le x \le .04, -10 \le y \le 10$

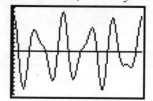

15. $0 \le x \le 20, -11 \le y \le 11$

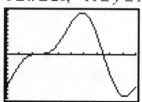

17. The graph has amplitude 1 and has one complete wave between $x = \ln(2\pi)$ and $x = \ln(4\pi)$, and then between $x = \ln(2n\pi)$ and $x = \ln((2n+2)\pi)$, where n is any positive integer. The negative x-axis is a horizontal asymptote. The window $-3 \le x \le \ln(8\pi), -2 \le y \le 2$ is shown.

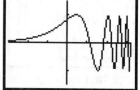

19. The function has one complete wave every 2π units. The graph lies between the graph of $y = \sqrt{|t|}$ and the graph of $y = -\sqrt{|t|}$ as shown in the right-hand graph. Note that this is an even function, hence the graph has y-axis symmetry. The window $-30 \le x \le 30, -6 \le y \le 6$ is shown.

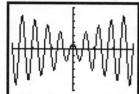
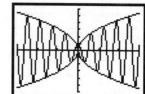

21. The function has one complete wave every 2π units. The graph lies between the graph of $y = 1/t$ and the graph of $y = -1/t$ as shown in the right-hand graph. There is a hole at the point $(0,1)$. The x-axis is a horizontal asymptote (even though the graph crosses this horizontal asymptote an infinite number of times). The window $-30 \le x \le 30, -.3 \le y \le 1$ is shown.

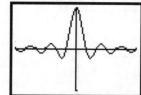
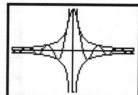

23. The function has one complete period every π units. There is a vertical asymptote at each point where cos $t = 0$, that is, where $t = \frac{\pi}{2} + nt$, where n is an any integer. window $-2\pi \le x \le 2\pi, -3 \le y \le 1$ is shown.
The window $-2\pi \le x \le 2\pi, -3 \le y \le 1$ is shown.

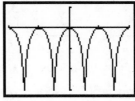

25. a.

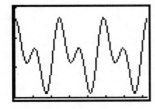

   b. Maximum a $t = .2$, $t = 24.4$, and $t = 48.6$, that is shortly after midnight. Minimum at $t = 17.1$, $t = 41.3$, $t = 65.5$, that is around 5:30p.

   c. 24.2 hours

6.6 Other Trigonometric Functions

1. IV   **3.** II   **5.** IV

7. $r = \sqrt{x^2 + y^2} = \sqrt{3^2 + 4^2} = 5$

$\sin t = y/r = 4/5 \quad \cos t = x/r = 3/5 \quad \tan t = y/x = 4/3$

$\cot t = x/y = 3/4 \quad \sec t = r/x = 5/3 \quad \csc t = r/y = 5/4$

9. $r = \sqrt{x^2 + y^2} = \sqrt{(-5)^2 + 12^2} = 13$

$\sin t = y/r = 12/13 \quad \cos t = x/r = -5/13 \quad \tan t = y/x = -12/5$

$\cot t = x/y = -5/12 \quad \sec t = r/x = -13/5 \quad \csc t = r/y = 13/12$

11. $r = \sqrt{x^2 + y^2} = \sqrt{(-1/5)^2 + 1^2} = \sqrt{26/25} = \sqrt{26}/5$

$\sin t = \dfrac{y}{r} = 1 \div \dfrac{\sqrt{26}}{5} = \dfrac{5}{\sqrt{26}}$ $\qquad \cot t = \dfrac{x}{y} = \left(-\dfrac{1}{5}\right) \div 1 = -\dfrac{1}{5}$

$\cos t = \dfrac{x}{r} = \left(-\dfrac{1}{5}\right) \div \dfrac{\sqrt{26}}{5} = -\dfrac{1}{\sqrt{26}}$ $\quad \sec t = \dfrac{r}{x} = \dfrac{\sqrt{26}}{5} \div \left(-\dfrac{1}{5}\right) = -\sqrt{26}$

$\tan t = \dfrac{y}{x} = 1 \div \left(-\dfrac{1}{5}\right) = -5$ $\qquad \csc t = \dfrac{r}{y} = \dfrac{\sqrt{26}}{5} \div 1 = \dfrac{\sqrt{26}}{5}$

13. $r = \sqrt{x^2 + y^2} = \sqrt{\left(\sqrt{2}\right)^2 + \left(\sqrt{3}\right)^2} = \sqrt{5}$

$\sin t = \dfrac{y}{r} = \dfrac{\sqrt{3}}{\sqrt{5}} = \dfrac{\sqrt{15}}{5}$ $\qquad \cot t = \dfrac{x}{y} = \dfrac{\sqrt{2}}{\sqrt{3}} = \dfrac{\sqrt{6}}{3}$

$\cos t = \dfrac{x}{r} = \dfrac{\sqrt{2}}{\sqrt{5}} = \dfrac{\sqrt{10}}{5}$ $\qquad \sec t = \dfrac{r}{x} = \dfrac{\sqrt{5}}{\sqrt{2}} = \dfrac{\sqrt{10}}{2}$

$\tan t = \dfrac{y}{x} = \dfrac{\sqrt{3}}{\sqrt{2}} = \dfrac{\sqrt{6}}{2}$ $\qquad \csc t = \dfrac{r}{y} = \dfrac{\sqrt{5}}{\sqrt{3}} = \dfrac{\sqrt{15}}{3}$

15. $r = \sqrt{x^2 + y^2} = \sqrt{\left(1 + \sqrt{2}\right)^2 + 3^2} = \sqrt{12 + 2\sqrt{2}}$

$\sin t = \dfrac{y}{r} = \dfrac{3}{\sqrt{12 + 2\sqrt{2}}}$ $\quad \cos t = \dfrac{x}{r} = \dfrac{1 + \sqrt{2}}{\sqrt{12 + 2\sqrt{2}}}$ $\quad \tan t = \dfrac{y}{x} = \dfrac{3}{1 + \sqrt{2}}$

$\cot t = \dfrac{x}{y} = \dfrac{1 + \sqrt{2}}{3}$ $\quad \sec t = \dfrac{r}{x} = \dfrac{\sqrt{12 + 2\sqrt{2}}}{1 + \sqrt{2}}$ $\quad \csc t = \dfrac{r}{y} = \dfrac{\sqrt{12 + 2\sqrt{2}}}{3}$

17. $\dfrac{156.25 \tan^2(1.4)}{\sec^2(1.4)} + 3 \approx 154.74$ ft

19. a. $0 \le x \le 150, -5 \le y \le 20$ **b.** about 15.16 ft   **c.** about 143 ft

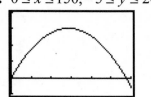

21. The terminal side of an angle of $\dfrac{4\pi}{3}$ in standard position lies in quadrant III.

$$\sin\frac{4\pi}{3} = -\sin\frac{\pi}{3} = -\frac{\sqrt{3}}{2} \qquad \cot\frac{4\pi}{3} = \cot\frac{\pi}{3} = \frac{1}{\sqrt{3}}$$

$$\cos\frac{4\pi}{3} = -\cos\frac{\pi}{3} = -\frac{1}{2} \qquad \sec\frac{4\pi}{3} = -\sec\frac{\pi}{3} = -2$$

$$\tan\frac{4\pi}{3} = \tan\frac{\pi}{3} = \sqrt{3} \qquad \csc\frac{4\pi}{3} = -\csc\frac{\pi}{3} = -\frac{2}{\sqrt{3}}$$

23. The terminal side of an angle of $\dfrac{7\pi}{4}$ in standard position lies in quadrant IV.

$$\sin\frac{7\pi}{4} = -\sin\frac{\pi}{4} = -\frac{\sqrt{2}}{2} \qquad \cot\frac{7\pi}{4} = -\cot\frac{\pi}{4} = -1$$

$$\cos\frac{7\pi}{4} = \cos\frac{\pi}{4} = \frac{\sqrt{2}}{2} \qquad \sec\frac{7\pi}{4} = \sec\frac{\pi}{4} = \sqrt{2}$$

$$\tan\frac{7\pi}{4} = -\tan\frac{\pi}{4} = -1 \qquad \csc\frac{7\pi}{4} = -\csc\frac{\pi}{4} = -\sqrt{2}$$

25. The terminal side of an angle of $\dfrac{-11\pi}{4}$ in standard position lies in quadrant IV.

$$\sin\frac{-11\pi}{4} = \sin\frac{5\pi}{4} = -\frac{\sqrt{2}}{2} \qquad \cot\frac{-11\pi}{4} = \cot\frac{5\pi}{4} = 1$$

$$\cos\frac{-11\pi}{4} = \cos\frac{5\pi}{4} = -\frac{\sqrt{2}}{2} \qquad \sec\frac{-11\pi}{4} = \sec\frac{5\pi}{4} = -\sqrt{2}$$

$$\tan\frac{-11\pi}{4} = \tan\frac{5\pi}{4} = 1 \qquad \csc\frac{-11\pi}{4} = \csc\frac{5\pi}{4} = -\sqrt{2}$$

27. average rate of change $= \dfrac{\cot 3 - \cot 1}{3 - 1} = \dfrac{1}{2}\left(\dfrac{1}{\tan 3} - \dfrac{1}{\tan 1}\right) = -3.8287$

29. a. average rate of change $= \dfrac{\tan 2.01 - \tan 2}{h}$

$h = .01 \qquad \text{rate} = \dfrac{\tan 2.01 - \tan 2}{.01} = 5.6511$

$h = .001 \qquad \text{rate} = \dfrac{\tan 2.001 - \tan 2}{.001} = 5.7618$

$h = .0001 \qquad \text{rate} = \dfrac{\tan 2.0001 - \tan 2}{.0001} = 5.7731$

$h = .00001 \quad \text{rate} = \dfrac{\tan 2.00001 - \tan 2}{.00001} = 5.7743$

b. $(\sec 2)^2 = 5.7744$. This seems to be the limit of the process in part **a.**

31. $\cos t \sin t(\csc t + \sec t) = \cos t \sin t \csc t + \cos t \sin t \sec t$

$$= \cos t \sin t \frac{1}{\sin t} + \cos t \sin t \frac{1}{\cos t} = \cos t + \sin t$$

33. $(1 - \sec t)^2 = 1 - 2\sec t + \sec^2 t$

35. $(\cot t - \tan t)(\cot^2 t + 1 + \tan^2 t) = \cot^3 t - \tan^3 t$

37. $\csc t(\sec t - \csc t)$

39. $\tan^4 t - \sec^4 t = (\tan^2 t - \sec^2 t)(\tan^2 t + \sec^2 t) = (-1)(\tan^2 t + \sec^2 t) = -\tan^2 t - \sec^2 t$

41. $\cos^3 t - \sec^3 t = (\cos t - \sec t)(\cos^2 t + \cos t \sec t + \sec^2 t)$

$$= (\cos t - \sec t)(\cos^2 t + 1 + \sec^2 t)$$

43. $\dfrac{\cos^2 t \sin t}{\sin^2 t \cos t} = \dfrac{\cos t}{\sin t} = \cot t$

45. $\dfrac{4\tan t \sec t + 2\sec t}{6\sin t \sec t + 2\sec t} = \dfrac{2\sec t(2\tan t + 1)}{2\sec t(3\sin t + 1)} = \dfrac{2\tan t + 1}{3\sin t + 1}$

47. $4 - \tan t$

49. $\tan t = \dfrac{\sin t}{\cos t} = \dfrac{1}{\cos t/\sin t} = \dfrac{1}{\cot t}$

51. $1 + \cot^2 t = 1 + \dfrac{\cos^2 t}{\sin^2 t}$

$$= \dfrac{\sin^2 t + \cos^2 t}{\sin^2 t}$$

$$= \dfrac{1}{\sin^2 t} = \csc^2 t$$

53. $\sec(-t) = \dfrac{1}{\cos(-t)} = \dfrac{1}{\cos t} = \sec t$

55. $\sin t = +\sqrt{1 - \cos^2 t} = \sqrt{1 - \left(-\dfrac{1}{2}\right)^2} = \dfrac{\sqrt{3}}{2}$

$\cos t = -1/2$

$\tan t = \dfrac{\sin t}{\cos t} = \dfrac{\sqrt{3}/2}{-1/2} = -\sqrt{3}$

$\cot t = \dfrac{1}{\tan t} = \dfrac{1}{-\sqrt{3}} = -\dfrac{1}{\sqrt{3}}$

$\sec t = \dfrac{1}{\cos t} = \dfrac{1}{-1/2} = -2$

$\csc t = \dfrac{1}{\sin t} = \dfrac{1}{\sqrt{3}/2} = \dfrac{2}{\sqrt{3}}$

57. $\sin t = 1$
$\cos t = 0$

$\tan t = \dfrac{\sin t}{\cos t} = \dfrac{1}{0}$ is undefined

$\cot t = \dfrac{1}{\tan t} = \dfrac{0}{1} = 0$

$\sec t = \dfrac{1}{\cos t} = \dfrac{1}{0}$ is undefined

$\csc t = \dfrac{1}{\sin t} = \dfrac{1}{1} = 1$

59. $\tan t = -\sqrt{\sec^2 t - 1}$

$= -\sqrt{(-13/5)^2 - 1} = -12/5$

$\cos t = \dfrac{1}{\sec t} = \dfrac{1}{-13/5} = -\dfrac{5}{13}$

$\sin t = \tan t \cos t = (-12/5)(-5/13) = 12/13$

$\cot t = \dfrac{1}{\tan t} = \dfrac{1}{-12/5} = -\dfrac{5}{12}$

$\sec t = -13/5$

$\csc t = \dfrac{1}{\sin t} = \dfrac{1}{12/13} = \dfrac{13}{12}$

61. Graph $y = \tan x$ and $y = \cot\left(\frac{\pi}{2} - x\right)$ using the window $-2\pi \le x \le 2\pi$, $-5 \le y \le 5$. The graphs appear to coincide, hence the equation could be an identity.

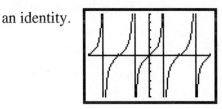

63. Graph $y = \dfrac{\sin x}{1 - \cos x}$ and $y = \cot x$ using the window $0 \le x \le 2\pi$, $-5 \le y \le 5$. The graphs do not coincide, hence the equation could not be an identity.

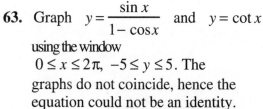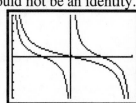

65. Graphing $y = \sec x$ and $y = x$ using the window $-2\pi \le x \le 2\pi$, $-5 \le y \le 5$ shows two, possibly three solutions, and indicates that there are an infinite number; however, none of them lie between $-\pi/2$ and $\pi/2$.

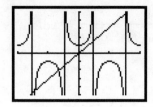

67. a. area $OCA = \frac{1}{2}AC \cdot OA = \frac{1}{2}\sin\theta\cos\theta$

    b. area $ODB = \frac{1}{2}DB \cdot OB = \frac{1}{2}\tan\theta \cdot 1 = \frac{1}{2}\tan\theta$

    c. area $OCB = \frac{1}{2}OB^2 \cdot \theta = \frac{1}{2}1^2 \cdot \theta = \frac{1}{2}\theta$

Chapter 6 Review Exercises

1. $-\dfrac{23\pi}{3}+8\pi=\dfrac{\pi}{3}$

3. $\dfrac{9\pi}{5}\cdot\dfrac{180°}{\pi}=324°$

5. $220°\cdot\dfrac{\pi}{180°}=\dfrac{11\pi}{9}$ radians

7. $-\dfrac{11\pi}{4}\cdot\dfrac{180°}{\pi}=-495°$

9. In the second quadrant, cosine is negative. Hence.

$$\cos v=-\sqrt{1-\sin^2 v}=-\sqrt{1-\left(\sqrt{\tfrac{8}{9}}\right)^2}=-\sqrt{\tfrac{1}{9}}=-\tfrac{1}{3}$$

11. $\sin(-13\pi)=\sin\pi=0$

12. $\dfrac{\tan(t+\pi)}{\sin(t+\pi)}=\dfrac{\tan t}{-\sin t}=\dfrac{\sin t/\cos t}{-\sin t}=-\dfrac{1}{\cos t}=-\sec t$

13. .809

15. $\sqrt{3}/3$

17. $\sqrt{3}/2$

19. 1.701

21. $-\sqrt{3}/3$

23. -2

25.

t	0	$\pi/6$	$\pi/4$	$\pi/3$	$\pi/2$
$\sin t$	0	$1/2$	$\sqrt{2}/2$	$\sqrt{3}/2$	1
$\cos t$	1	$\sqrt{3}/2$	$\sqrt{2}/2$	$1/2$	0

27. $\left(\sin\dfrac{\pi}{6}+1\right)^2=\left(\dfrac{1}{2}+1\right)^2=\left(\dfrac{3}{2}\right)^2=\dfrac{9}{4}$

29. $(f\circ g)(\pi)=f(g(\pi))=f(-\cos\pi)=f(1)=\log_{10}1=0$

31. Since the terminal side lies in the second quadrant, cosine is negative.

$$\cos t=-\sqrt{1-\sin^2 t}=-\sqrt{1-\left(\dfrac{1}{\sqrt{3}}\right)^2}=-\sqrt{\dfrac{2}{3}}\ \text{ or }\ -\dfrac{\sqrt{6}}{3}$$

33. Since the terminal side lies in the third quadrant, cosine is negative.

$$\cos t=-\sqrt{1-\sin^2 t}=-\sqrt{1-\left(-\dfrac{4}{5}\right)^2}=-\sqrt{\dfrac{9}{25}}=-\dfrac{3}{5}$$

35. Since the terminal side lies in the second quadrant, cosine is negative.

$$\cos t=-\sqrt{1-\sin^2 t}=-\sqrt{1-\left(\dfrac{5}{13}\right)^2}=-\sqrt{\dfrac{144}{169}}=-\dfrac{12}{13}$$

37. $\cos\dfrac{2\pi}{3} = -\cos\dfrac{\pi}{3} = -\dfrac{1}{2}$ **39.** $\dfrac{\sqrt{3}}{2}$ **41. (c)**

43. $r = \sqrt{x^2 + y^2} = \sqrt{\left(\dfrac{-3}{\sqrt{50}}\right)^2 + \left(\dfrac{7}{\sqrt{50}}\right)^2} = \sqrt{\dfrac{9}{50} + \dfrac{49}{50}} = \sqrt{\dfrac{58}{50}} = \dfrac{\sqrt{58}}{\sqrt{50}}$

$\sin t = \dfrac{y}{r} = \dfrac{7}{\sqrt{50}} \div \dfrac{\sqrt{58}}{\sqrt{50}} = \dfrac{7}{\sqrt{58}}$

45. $\tan t = \dfrac{y}{x} = \dfrac{7}{\sqrt{50}} \div -\dfrac{3}{\sqrt{50}} = -\dfrac{7}{3}$

47. $\cos t = \dfrac{r}{x} = \dfrac{\sqrt{58}}{\sqrt{50}} \div -\dfrac{3}{\sqrt{50}} = -\dfrac{\sqrt{58}}{3}$ (using Exercise **43**)

49. The slope of the line is $\tan\left(\frac{5\pi}{3}\right) = -\sqrt{3}$. The line passes through the origin. Hence $y = -\sqrt{3}x$ is the required equation.

51. $\tan w = \dfrac{\sin w}{\cos w} = -\dfrac{3}{\sqrt{13}} \div \dfrac{2}{\sqrt{13}} = -\dfrac{3}{2}$ **53.** $\cos(-w) = \cos w = \dfrac{2}{\sqrt{13}}$

55. $\csc(-w) = \dfrac{1}{\sin(-w)} = -\dfrac{1}{\sin(w)} = -1 \div -\dfrac{3}{\sqrt{13}} = \dfrac{\sqrt{13}}{3}$

57. The dashed line is the graph of $h(t) = \csc t$; the solid curve is the graph of $f(t) = \sin t$.

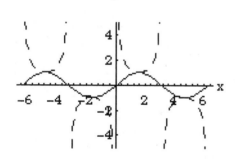

59. (d)   **61.** $\tan\theta = y/x = 3/(-2) = -3/2$ **63.** 0

65. (d)   **67.** $\cot\left(\dfrac{2\pi}{3}\right) = -\cot\left(\dfrac{\pi}{3}\right) = -\dfrac{1}{\sqrt{3}}$

69. a. 3/2   **b.** $f(t) = 0$ if $5t = n\pi$. The smallest possible value is therefore $t = \pi/5$.

71.

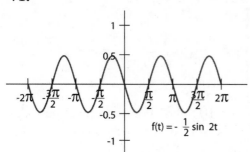

73. Graph $y = \cos x$ and $y = \sin\left(x - \frac{\pi}{2}\right)$ using the window $0 \le x \le 2\pi$, $-1 \le y \le 1$. The graphs do not coincide, hence the equation could not be an identity.

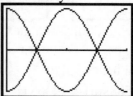

75. Graph $y = \dfrac{\sin x - \sin 3x}{\cos x + \cos 3x}$ and $y = -\tan x$ using the window $0 \le x \le 2\pi$, $-5 \le y \le 5$. The graphs appear to coincide, hence the equation could be an identity.

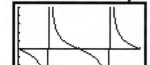

77. Graph $y = \sec^2 x + \csc^2 x$ and $y = \left(\sec^2 x\right)\left(\csc^2 x\right)$ using the window $0 \le x \le 2\pi$, $0 \le y \le 10$. The graphs appear to coincide, hence the equation could be an identity.

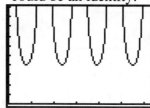

79. The period of $\sin bt$ is $\dfrac{2\pi}{b}$, hence the period of $g(t)$ is $\dfrac{2\pi}{4\pi} = \dfrac{1}{2}$.

81. a. On the days with the most daylight there are approximately 15.1 hours of daylight. On the days with the least daylight there are approximately 8.9 hours.

b. Between January 1 and March 2, and between October 9 and December 31.

83. The graph has the general appearance of a cosine graph.
Amplitude = 2
Period $= \dfrac{4\pi}{5} = \dfrac{2\pi}{b}$, hence $b = \dfrac{5}{2}$.
The rule can be written $g(t) = 2\cos\frac{5}{2}t$.

85. $A = 8$. $\dfrac{2\pi}{b} = 5$, hence $b = \dfrac{2\pi}{5}$. $-\dfrac{c}{b} = 14$, hence $c = -14b = -\dfrac{28\pi}{5}$.
$f(t) = 8\sin\left(\dfrac{2\pi}{5}t - \dfrac{28\pi}{5}\right)$ or $f(t) = 8\cos\left(\dfrac{2\pi}{5}t - \dfrac{28\pi}{5}\right)$

87. Graph the function using the window $-2\pi \le x \le 2\pi, -11 \le y \le 11$.

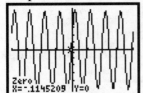 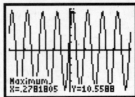

Note that the function has a maximum value of $A = 10.5588$. Four complete periods occurs between $x = 0$ and $x = 2\pi$, hence $b = 4$. The start of one period is at $x = -.11452$, hence $-c/b = -c/4 = -.11452$, thus $c = .4581$.

$$f(t) = 10.5588\sin(4t + .4581)$$

89. $0 \le x \le \pi / 50, -5 \le y \le 5$

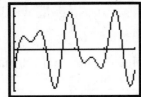

91. a.

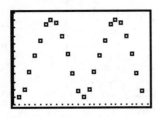

b.

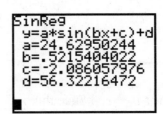

c. The period is 12.05 months, which is one year. This model is a good fit.

Chapter 6 Test

1. $\dfrac{1}{72}$ of a circle is $5°$. Then converting degrees to radians to get $5°\left(\dfrac{\pi}{360°}\right) = \dfrac{\pi}{36} \approx .0873$.

2.   **a.** $\sin\left(-\dfrac{13\pi}{3}\right) = \sin\left(\dfrac{5\pi}{3}\right) = -\dfrac{\sqrt{3}}{2}$ **b.** $\cos\left(-\dfrac{13\pi}{3}\right) = \cos\left(\dfrac{5\pi}{3}\right) = \dfrac{1}{2}$

   c. $\tan\left(-\dfrac{13\pi}{3}\right) = \tan\left(\dfrac{5\pi}{3}\right) = -\sqrt{3}$

3. $\sin^3 t - \sin t = \sin t\left(\sin^2 t - 1\right) = \sin t\left(\sin t + 1\right)\left(\sin t - 1\right)$

4.   First convert: $5km\left(\dfrac{1000\,m}{1\,km}\right)\left(\dfrac{100\,cm}{1\,m}\right) = 500,000\,cm.$

   Then, $\dfrac{500,000\,cm}{36\,cm/rev} \cdot \dfrac{2\pi\,rad}{1\,rev}; x = 13,888.9\,rev.$

5.   **a.** iii   **b.** i   **c.** iv

6.   Since $(-3, 2)$ is on the terminal side, by the Pythagorean Theorem,

   $r = \sqrt{(-3)^2 + 2^2} = \sqrt{9 + 4} = \sqrt{13}.$

   a. $\sin t = \dfrac{2}{\sqrt{13}} = \dfrac{2\sqrt{13}}{13}$ **b.** $\cos t = -\dfrac{3}{\sqrt{13}} = -\dfrac{3\sqrt{13}}{13}$ **c.** $\tan t = \dfrac{2/\sqrt{13}}{-3/\sqrt{13}} = -\dfrac{2}{3}$

7.   Since t is in quadrant IV, $\sin t$ is negative. Then,
   $\sin^2 t + \cos^2 t = 1$

   $\sin^2 t + \left(\tfrac{1}{\sqrt{17}}\right)^2 = 1$

   $\sin^2 t = 1 - \dfrac{1}{17}$

   $\sin t = -\sqrt{\dfrac{16}{17}} = -\dfrac{4}{\sqrt{17}} = -\dfrac{4\sqrt{17}}{17}$

8. $A = \dfrac{1}{2}r^2\theta = \dfrac{1}{2}(4)^2(2) = 16\ cm^2$

9.   **a.** $480°\left(\dfrac{\pi}{180°}\right) = \dfrac{8\pi}{3}$ **b.** $\dfrac{\pi}{40}\left(\dfrac{180°}{\pi}\right) = 4.5°$

10. First, find a point on the terminal side of t.

Since $7y - 2x = -4; 7y = 2x - 4; y = \dfrac{2}{7}x - \dfrac{4}{7}$ is parallel to the terminal side of the angle, the point $(7,2)$ is on the terminal side of the angle. By the Pythagorean Theorem $r = \sqrt{2^2 + 7^2} = \sqrt{4 + 49} = \sqrt{53}$. Then

a. $\sin t = \dfrac{2}{\sqrt{53}} = \dfrac{2\sqrt{53}}{53}$ **b.** $\cos t = \dfrac{7}{\sqrt{53}} = \dfrac{7\sqrt{53}}{53}$ **c.** $\tan t = \dfrac{2/\sqrt{53}}{7/\sqrt{53}} = \dfrac{2}{7}$

11. $\tan(5\pi - t) = \tan(-\pi - t) = \tan(-(\pi + t)) = \tan(-t) = -\tan(t)$

Since $\sin t = \dfrac{3}{5}$, and t lies in quadrant 1, $\cos^2 t = 1 - \sin^2 t$

$$\cos^2 t = 1 - \left(\tfrac{3}{5}\right)^2$$

$$\cos t = \sqrt{1 - \left(\tfrac{9}{25}\right)}$$

$$\cos t = \sqrt{\tfrac{16}{25}} = \dfrac{4}{5}$$

Then, $-\tan t = -\dfrac{\sin t}{\cos t} = -\dfrac{3/5}{4/5} = -\dfrac{3}{4}$

12. The second had makes a full revolution every 60 seconds, that is it moves an angle of 2π radians. So $\omega = \dfrac{2\pi\, rad}{1\,min} = 2\pi\, rad/min$. Thus, the linear speed is

$$\dfrac{\theta r}{t} = \dfrac{2\pi(5\,in)}{1\,min} = 10\pi\, in/min.$$

13. $\dfrac{8\cos t}{\sin^2 t} \cdot \dfrac{\sin^2 t - \sin t \cos t}{\sin^2 t - \cos^2 t} = \dfrac{8\cos t}{\sin^2 t} \cdot \dfrac{\sin t(\sin t - \cos t)}{(\sin t - \cos t)(\sin t + \cos t)}$

$$= \dfrac{8\cos t}{\sin t} \cdot \dfrac{1}{(\sin t + \cos t)}$$

$$= \dfrac{8\cos t}{\sin t(\sin t + \cos t)}$$

14. $\dfrac{f(t+h) - f(t)}{t+h} = \dfrac{f\left(\frac{\pi}{4}\right) - f\left(\frac{\pi}{6}\right)}{\frac{\pi}{4} - \frac{\pi}{6}} = \dfrac{\tan\left(\frac{\pi}{4}\right) - \tan\left(\frac{\pi}{6}\right)}{\frac{\pi}{12}} = \dfrac{1 - \frac{\sqrt{3}}{3}}{\frac{\pi}{12}} = \dfrac{12 - 4\sqrt{3}}{\pi}$

15. $\sin\left(\frac{-17\pi}{4}\right) = \sin\left(\frac{7\pi}{4}\right) = -\dfrac{\sqrt{2}}{2}$

16. $\dfrac{4\tan t \sin t - 2\sin t}{6\sin^2 t + 2\sin t} = \dfrac{2\sin t\,(2\tan t - 1)}{2\sin t\,(3\sin t + 1)} = \dfrac{2\tan t - 1}{3\sin t + 1}$

17. $6.2 = \dfrac{31}{5} = \dfrac{y}{x};\ \text{so}\ \tan t = \dfrac{31}{5} = 6.2$

18. $0 \le t \le 2\pi$, which implies $0 \le 2\pi t \le 4\pi^2\,(\approx 39.48)$. Now for $f(t) = \cos(2\pi t) = -1$;

$2\pi t = \pi, 3\pi, 5\pi, 7\pi, 9\pi, 11\pi\,(\approx 34.56)$

$t = \dfrac{1}{2}, \dfrac{3}{2}, \dfrac{5}{2}, \dfrac{7}{2}, \dfrac{9}{2},\ \text{and}\ \dfrac{11}{2}$

19. a. $\sin\left(-\dfrac{11\pi}{6}\right) = \sin\left(\dfrac{\pi}{6}\right) = \dfrac{1}{2}$ **b.** $\cos\left(-\dfrac{11\pi}{6}\right) = \cos\left(\dfrac{\pi}{6}\right) = \dfrac{\sqrt{3}}{2}$

c. $\tan\left(-\dfrac{11\pi}{6}\right) = \tan\left(\dfrac{\pi}{6}\right) = \dfrac{1}{\sqrt{3}} = \dfrac{\sqrt{3}}{3}$ **d.** $\cot\left(-\dfrac{11\pi}{6}\right) = \cot\left(\dfrac{\pi}{6}\right) = \sqrt{3}$

e. $\csc\left(-\dfrac{11\pi}{6}\right) = \csc\left(\dfrac{\pi}{6}\right) = 2$ **f.** $\sec\left(-\dfrac{11\pi}{6}\right) = \sec\left(\dfrac{\pi}{6}\right) = \dfrac{2}{\sqrt{3}} = \dfrac{2\sqrt{3}}{3}$

20.

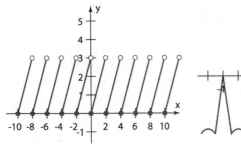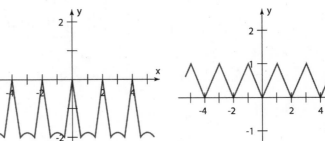

21. $0 \le x \le .034272, -1 \le y \le 1$

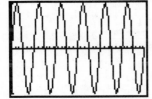

22. $(\sin t - \csc t)(\sin^2 t + \csc^2 t + 1) = \sin^3 t + \sin t \csc^2 t + \sin t - \csc t \sin^2 t - \csc^3 t - \csc t$

$= \sin^3 t + \csc t + \sin t - \sin t - \csc^3 t - \csc t$

$= \sin^3 t - \csc^3 t$

23. From the graph, there are four solutions.

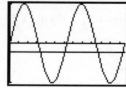

24. a. 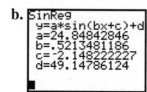 **b.**

    SinReg
    y=a*sin(bx+c)+d
    a=25.6295562
    b=.4914195876
    c=-1.945064679
    d=47.65157156

    SinReg
    y=a*sin(bx+c)+d
    a=24.84842846
    b=.5213481186
    c=-2.14822227
    d=49.14786124

c. The model in part (b) (its period is 12.05, whereas the model in part (a) has period 12.79).

25. $1+\tan^2 x = \sec^2 x$

$\sec^2 x = \sec^2 x$

26. From the graphs, it is not an identity.

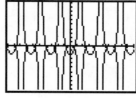

27. a. Amplitude: 3; Period: $\dfrac{\pi}{2}$; Phase Shift: $-\dfrac{7}{4}$

b. $f(x) = 6\cos(\pi x + \pi)$

28. a. $\cos t = \sqrt{1-\sin^2 t} = \sqrt{1-\left(-\frac{6}{7}\right)^2} = \dfrac{\sqrt{13}}{7}$ **b.** $\tan t = \dfrac{\sin t}{\cos t} = \dfrac{-6/7}{\sqrt{13}/7} = \dfrac{-6\sqrt{13}}{13}$

c. $\cot t = \dfrac{\cos t}{\sin t} = \dfrac{\sqrt{13}/7}{-6/7} = \dfrac{-\sqrt{13}}{6}$ **d.** $\csc t = \dfrac{1}{\sin t} = -\dfrac{7}{6}$

e. $\sec t = \dfrac{1}{\cos t} = \dfrac{7\sqrt{13}}{13}$

29. The highest occur when $\sin t = 1.$ So, solve

$1 = 2\sin(6t); \; 0 \le t \le \pi$

$\dfrac{1}{2} = \sin(6t); \; 0 \le 6t \le 6\pi$

$6t = \dfrac{\pi}{2}, \dfrac{5\pi}{2}, \dfrac{9\pi}{2}$

$t = \dfrac{\pi}{12}, \dfrac{5\pi}{12}, \dfrac{3\pi}{4}$

$\left(\dfrac{\pi}{12}, 2\right), \left(\dfrac{5\pi}{12}, 2\right), \left(\dfrac{3\pi}{4}, 2\right)$

30. The function $g(t) = 2\sin(3t-5) + 2\cos(3t+2)$ has a period of $\dfrac{2\pi}{3}$, thus the function

$f(t) = A\cos(bt+c)$ had a period of $\dfrac{2\pi}{3}$, giving $\dfrac{2\pi}{b} = \dfrac{2\pi\pi}{3}; b = 3.$ Graphing

$g(t)$ gives a maximum height (amplitude) of 1.6565 and that the since wave begins

at approximately $t=-.8090.$ Therefore the phase shift is

$-\dfrac{c}{b} = -\dfrac{c}{3} = -.8090; c = 2.4270.$

Then $f(t) = 1.6565\cos(3t + 2.4270).$

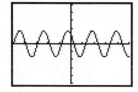

Chapter 7
Trigonometric Identities and Equations

7.1 Basic Identities and Proofs

1. If the functions $f(x) = \dfrac{\sec x - \cos x}{\sec x}$ and $g(x) = \sin^2 x$ are graphed in a window such as $0 \le x \le 2\pi$, $0 \le y \le 1$, they will appear identical, hence this equation could be an identity.

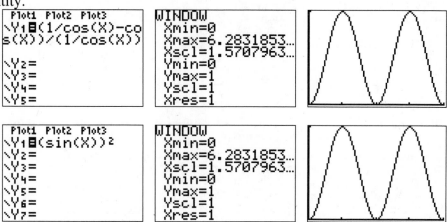

3. If the functions $f(x) = \dfrac{1 - \cos 2x}{2}$ and $g(x) = \sin^2 x$ are graphed in a window such as $0 \le x \le 2\pi$, $0 \le y \le 1$, they will appear identical, hence this equation could be an identity.

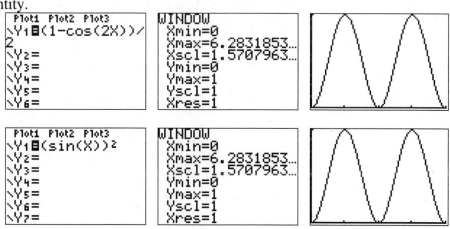

5.   B; If the functions $f(x) = \csc x \tan x$ and $g(x) = \sec x$ are graphed in a window such
     as $0 \le x \le 2\pi$, $-10 \le y \le 10$, they will appear identical.

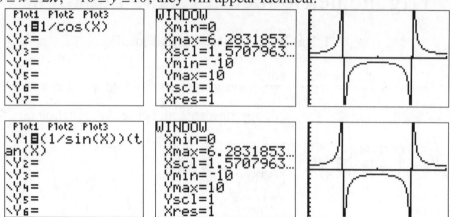

7.   E; If the functions $f(x) = \dfrac{\sin^4 x - \cos^4 x}{\sin x + \cos x}$ and $g(x) = \sin x - \cos x$ are graphed in a
     window such as $0 \le x \le 2\pi$, $-2 \le y \le 2$, they will appear identical.

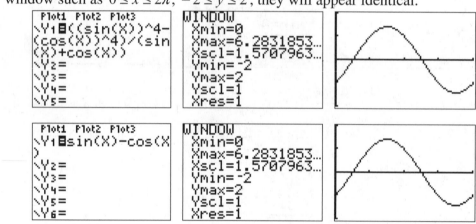

9. $\tan x \cos x = \dfrac{\sin x}{\cos x} \cdot \cos x$

$\qquad\quad = \sin x$

11. $\cos x \sec x = \cos x \cdot \dfrac{1}{\cos x}$

$\qquad\qquad\ = 1$

13. $\tan x \csc x = \dfrac{\sin x}{\cos x} \cdot \dfrac{1}{\sin x}$

$\qquad\qquad = \dfrac{1}{\cos x}$

$\qquad\qquad = \sec x$

15. $\dfrac{\tan x}{\sec x} = \dfrac{\sin x}{\cos x} \div \dfrac{1}{\cos x}$

$\qquad\quad = \dfrac{\sin x}{\cos x} \cdot \dfrac{\cos x}{1}$

$\qquad\quad = \sin x$

17. $(1 + \cos x)(1 - \cos x) = 1 - \cos^2 x = (\sin^2 x + \cos^2 x) - \cos^2 x = \sin^2 x$

19. $\cot x \sec x \sin x = \dfrac{\cos x}{\sin x} \cdot \dfrac{1}{\cos x} \cdot \sin x = 1$

21. $\tan x + \cot x = \dfrac{\sin x}{\cos x} + \dfrac{\cos x}{\sin x} = \dfrac{\sin^2 x}{\cos x \sin x} + \dfrac{\cos^2 x}{\sin x \cos x} = \dfrac{\sin^2 x + \cos^2 x}{\cos x \sin x}$

$\qquad = \dfrac{1}{\cos x \sin x} = \dfrac{1}{\cos x} \cdot \dfrac{1}{\sin x} = \sec x \csc x$

23. $\cos x - \cos x \sin^2 x = \cos x \left(1 - \sin^2 x\right) = \cos x \left[\left(\sin^2 x + \cos^2 x\right) - \sin^2 x \right]$

$\qquad = \cos x \left(\cos^2 x\right) = \cos^3 x$

25. This is not an identity. For example, let $x = \dfrac{3\pi}{2}$.

$\sin \dfrac{3\pi}{2} = -1$, but $\sqrt{1 - \cos^2\left(\dfrac{3\pi}{2}\right)} = 1$.

27. This is an identity.

$\dfrac{\sin(-x)}{\cos(-x)} = \dfrac{-\sin x}{\cos x} = -\tan x$

29. This is an identity.

$\cot(-x) = \dfrac{\cos(-x)}{\sin(-x)} = \dfrac{\cos x}{-\sin x} = -\dfrac{\cos x}{\sin x} = -\cot x$

31. This is not an identity. For example, let $x = 0$.

$1 + \sec^2 0 = 1 + 1^2 = 1 + 1 = 2$, but $\tan^2 0 = 0^2 = 0$.

33. This is an identity.

$\sec^2 x - \csc^2 x = \left(1 + \tan^2 x\right) - \left(1 + \cot^2 x\right) = 1 + \tan^2 x - 1 - \cot^2 x = \tan^2 x - \cot^2 x$

35. This is an identity.

$$\text{Since } \sin^2 x(\cot x+1)^2 = \sin^2 x\left(\frac{\cos x}{\sin x}+1\right)^2 = \sin^2 x\left(\frac{\cos x}{\sin x}+\frac{\sin x}{\sin x}\right)^2$$

$$= \sin^2 x\left(\frac{\cos x+\sin x}{\sin x}\right)^2 = \sin^2 x\cdot\frac{(\cos x+\sin x)^2}{\sin^2 x}$$

$$= (\cos x+\sin x)^2 = (\sin x+\cos x)^2$$

and

$$\cos^2 x(\tan x+1)^2 = \cos^2 x\left(\frac{\sin x}{\cos x}+1\right)^2 = \cos^2 x\left(\frac{\sin x}{\cos x}+\frac{\cos x}{\cos x}\right)^2$$

$$= \cos^2 x\left(\frac{\sin x+\cos x}{\cos x}\right)^2 = \cos^2 x\cdot\frac{(\sin x+\cos x)^2}{\cos^2 x} = (\sin x+\cos x)^2$$

This is an identity.

37. This is an identity.

$$\sin^2 x-\tan^2 x = \sin^2 x-\frac{\sin^2 x}{\cos^2 x} = \frac{\sin^2 x\cos^2 x}{\cos^2 x}-\frac{\sin^2 x}{\cos^2 x} = \frac{\sin^2 x\cos^2 x-\sin^2 x}{\cos^2 x}$$

$$= \frac{\sin^2 x(\cos^2 x-1)}{\cos^2 x} = \frac{\sin^2 x}{\cos^2 x}\cdot(\cos^2 x-1) = \tan^2 x(\cos^2 x-1)$$

$$= \tan^2 x\left[\cos^2 x-(\sin^2 x+\cos^2 x)\right] = \tan^2 x(\cos^2 x-\sin^2 x-\cos^2 x)$$

$$= \tan^2 x(-\sin^2 x) = -\sin^2 x\tan^2 x$$

39. This is an identity.

$$(\cos^2 x-1)(\tan^2 x+1) = (\cos^2 x-1)(\sec^2 x) = \cos^2 x\sec^2 x-\sec^2 x$$

$$= \cos^2 x\cdot\frac{1}{\cos^2 x}-\sec^2 x = 1-\sec^2 x = 1-(\tan^2 x+1)$$

$$= 1-\tan^2 x-1 = -\tan^2 x$$

41. This is an identity.
Starting with the right side we have:

$$\frac{\sec x}{\csc x} = \frac{1/\cos x}{1/\sin x} = \frac{1}{\cos x}\cdot\frac{\sin x}{1} = \frac{\sin x}{\cos x} = \tan x.$$

43. This is an identity.

$$\cos^4 x-\sin^4 x = (\cos^2 x)^2-(\sin^2 x)^2 = (\cos^2 x+\sin^2 x)(\cos^2 x-\sin^2 x)$$

$$= 1(\cos^2 x-\sin^2 x) = \cos^2 x-\sin^2 x$$

45. This is not an identity. For example, let $x = \dfrac{\pi}{4}$.

$$\left(\sin\frac{\pi}{4}+\cos\frac{\pi}{4}\right)^2 = \left(\frac{\sqrt{2}}{2}+\frac{\sqrt{2}}{2}\right)^2 = \left(\sqrt{2}\right)^2 = 2, \text{ but } \sin^2\frac{\pi}{4}+\cos^2\frac{\pi}{4}=1.$$

47. This is an identity.

Left side: $\quad \dfrac{1+\sin x}{\sin x} = \dfrac{1}{\sin x}+\dfrac{\sin x}{\sin x} = \csc x + 1;$

Right side: $\quad \dfrac{\cot^2 x}{\csc x-1} = \dfrac{\csc^2 x-1}{\csc x-1} = \dfrac{(\csc x+1)(\csc x-1)}{\csc x-1} = \csc x + 1$

Therefore, $\quad \dfrac{1+\sin x}{\sin x} = \dfrac{\cot^2 x}{\csc x-1}.$

49. This is an identity.
Starting with the right side we have:

$$\frac{\sec x+\tan x}{\sec x-\tan x} = \frac{\dfrac{1}{\cos x}+\dfrac{\sin x}{\cos x}}{\dfrac{1}{\cos x}-\dfrac{\sin x}{\cos x}} = \frac{\left(\dfrac{1}{\cos x}+\dfrac{\sin x}{\cos x}\right)\cdot\cos x}{\left(\dfrac{1}{\cos x}-\dfrac{\sin x}{\cos x}\right)\cdot\cos x} = \frac{1+\sin x}{1-\sin x}.$$

51. This is an identity.

$$\frac{\tan x+\sin x}{1+\cos x} = \frac{\dfrac{\sin x}{\cos x}+\sin x}{1+\cos x} = \frac{\left(\dfrac{\sin x}{\cos x}+\sin x\right)\cdot\cos x}{(1+\cos x)\cdot\cos x} = \frac{\sin x+\sin x\cos x}{\cos x+\cos^2 x}$$

$$= \frac{\sin x(1+\cos x)}{\cos x(1+\cos x)} = \frac{\sin x}{\cos x} = \tan x$$

53. This is an identity.
By Strategy 6, we need only prove $(1-\sin x)(1+\sin x) = \cos x\cos x.$

$$(1-\sin x)(1+\sin x) = 1+\sin x-\sin x-\sin^2 x = 1-\sin^2 x = \cos^2 x = \cos x\cos x$$

55. This is an identity.

$$\frac{\sec^2 x-1}{\sec^2 x} = \frac{1+\tan^2 x-1}{\sec^2 x} = \frac{\tan^2 x}{\sec^2 x} = \frac{\dfrac{\sin^2 x}{\cos^2 x}}{\dfrac{1}{\cos^2 x}} = \sin^2 x$$

57. This is an identity.

$$\frac{\sec x}{\csc x}+\frac{\sin x}{\cos x}=\frac{1/\cos x}{1/\sin x}+\tan x=\frac{1}{\cos x}\cdot\frac{\sin x}{1}+\tan x$$

$$=\frac{\sin x}{\cos x}+\tan x=\tan x+\tan x=2\tan x$$

59. This is an identity.

$$\frac{\sec x+\csc x}{1+\tan x}=\frac{\dfrac{1}{\cos x}+\dfrac{1}{\sin x}}{1+\dfrac{\sin x}{\cos x}}=\frac{\left(\dfrac{1}{\cos x}+\dfrac{1}{\sin x}\right)\cdot\sin x\cos x}{\left(1+\dfrac{\sin x}{\cos x}\right)\cdot\sin x\cos x}$$

$$=\frac{\sin x+\cos x}{(\cos x+\sin x)\sin x}=\frac{1}{\sin x}=\csc x$$

61. This is an identity.

$$\frac{1}{\csc x-\sin x}=\frac{1}{\dfrac{1}{\sin x}-\sin x}=\frac{1\cdot\sin x}{\left(\dfrac{1}{\sin x}-\sin x\right)\cdot\sin x}=\frac{\sin x}{1-\sin^2 x}$$

$$=\frac{\sin x}{\cos^2 x}=\frac{1}{\cos x}\cdot\frac{\sin x}{\cos x}=\sec x\tan x$$

63. This is not an identity. For example, let $x=\dfrac{\pi}{4}$.

$$\frac{\sin\dfrac{\pi}{4}-\cos\dfrac{\pi}{4}}{\tan\dfrac{\pi}{4}}=\frac{\dfrac{\sqrt{2}}{2}-\dfrac{\sqrt{2}}{2}}{1}=0,\quad\text{but}\quad\frac{\tan\dfrac{\pi}{4}}{\sin\dfrac{\pi}{4}+\cos\dfrac{\pi}{4}}=\frac{1}{\dfrac{\sqrt{2}}{2}+\dfrac{\sqrt{2}}{2}}=\frac{1}{\sqrt{2}}=\frac{\sqrt{2}}{2}.$$

65. The graph of the function is shown using the window $-2\pi\le x\le 2\pi,\,-1\le y\le 1$.

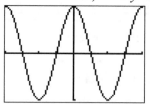

Conjecture: right side $=\cos x$

Proof: $1-\dfrac{\sin^2 x}{1+\cos x}=\dfrac{1+\cos x}{1+\cos x}-\dfrac{\sin^2 x}{1+\cos x}$

$$=\frac{1+\cos x-\sin^2 x}{1+\cos x}=\frac{\cos^2 x+\sin^2 x+\cos x-\sin^2 x}{1+\cos x}$$

$$=\frac{\cos^2 x+\cos x}{1+\cos x}=\frac{\cos x(\cos x+1)}{1+\cos x}=\cos x$$

67. The graph of the function is shown using the window $-2\pi \le x \le 2\pi, -5 \le y \le 5$.

Conjecture: right side $= \tan x$

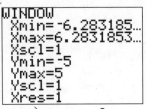 

Proof: $(\sin x + \cos x)(\sec x + \csc x) - \cot x - 2 =$

$$= (\sin x + \cos x)\left(\frac{1}{\cos x} + \frac{1}{\sin x}\right) - \cot x - 2$$

$$= \frac{\sin x}{\cos x} + 1 + 1 + \frac{\cos x}{\sin x} - \cot x - 2$$

$$= \tan x + 1 + 1 + \cot x - \cot x - 2$$

$$= \tan x$$

69. Starting with the right side we have:

$$\frac{\cos^3 x}{1+\sin x} = \frac{\cos^3 x (1-\sin x)}{(1+\sin x)(1-\sin x)} = \frac{\cos^3 x (1-\sin x)}{1-\sin^2 x} = \frac{\cos^3 x (1-\sin x)}{\cos^2 x}$$

$$= \cos x (1-\sin x) = \frac{1-\sin x}{\sec x}.$$

71. $\dfrac{\cos x}{1-\sin x} = \dfrac{\cos x (1+\sin x)}{(1-\sin x)(1+\sin x)} = \dfrac{\cos x (1+\sin x)}{1-\sin^2 x} = \dfrac{\cos x (1+\sin x)}{\cos^2 x}$

$$= \frac{1+\sin x}{\cos x} = \frac{1}{\cos x} + \frac{\sin x}{\cos x} = \sec x + \tan x$$

73. Use Strategy 6: Prove that $(\cos x \cot x)(\cos x \cot x) = (\cot x + \cos x)(\cot x - \cos x)$

$$(\cot x + \cos x)(\cot x - \cos x) = \cot^2 x - \cos^2 x = \frac{\cos^2 x}{\sin^2 x} - \cos^2 x = \frac{\cos^2 x}{\sin^2 x} - \frac{\cos^2 x \sin^2 x}{\sin^2 x}$$

$$= \frac{\cos^2 x - \cos^2 x \sin^2 x}{\sin^2 x} = \frac{\cos^2 x (1-\sin^2 x)}{\sin^2 x}$$

$$= \frac{\cos^2 x}{\sin^2 x}(1-\sin^2 x) = \cot^2 x \cos^2 x = (\cos x \cot x)(\cos x \cot x)$$

Therefore, $\dfrac{\cos x \cot x}{\cot x - \cos x} = \dfrac{\cot x + \cos x}{\cos x \cot x}.$

75. $\log_{10}(\cot x) = \log_{10}\dfrac{1}{\tan x} = \log_{10} 1 - \log_{10}\tan x = 0 - \log_{10}\tan x = -\log_{10}\tan x$

77. $\log_{10}(\csc x + \cot x) = \log_{10}\left(\dfrac{(\csc x + \cot x)(\csc x - \cot x)}{\csc x - \cot x}\right)$

$$= \log_{10}\frac{\csc^2 x - \cot^2 x}{\csc x - \cot x}$$

$$= \log_{10}\frac{1}{\csc x - \cot x}$$

$$= \log_{10} 1 - \log_{10}(\csc x - \cot x)$$

$$= 0 - \log_{10}(\csc x - \cot x)$$

$$= -\log_{10}(\csc x - \cot x)$$

79. $-\tan x \tan y(\cot x - \cot y) = -\tan x \cot x \tan y + \tan x \tan y \cot y$

$$= -1 \cdot \tan y + (\tan x)1 = -\tan y + \tan x = \tan x - \tan y$$

81. $\dfrac{\cos x - \sin y}{\cos y - \sin x} = \dfrac{\cos x - \sin y}{\cos y - \sin x} \cdot \dfrac{\cos x + \sin y}{\cos x + \sin y} = \dfrac{\cos^2 y - \sin^2 x}{(\cos y - \sin x)(\cos x + \sin y)}$

$$= \frac{(1 - \sin^2 x) - (1 - \cos^2 y)}{(\cos y - \sin x)(\cos x + \sin y)} = \frac{(\cos y - \sin x)(\cos y + \sin x)}{(\cos y - \sin x)(\cos x + \sin y)}$$

$$= \frac{\cos y + \sin x}{\cos x + \sin y}$$

7.2 Addition and Subtraction Identities

1.　$\cos\dfrac{\pi}{12}=\cos\left(\dfrac{\pi}{3}-\dfrac{\pi}{4}\right)=\cos\dfrac{\pi}{3}\cos\dfrac{\pi}{4}+\sin\dfrac{\pi}{3}\sin\dfrac{\pi}{4}=\dfrac{1}{2}\dfrac{\sqrt{2}}{2}+\dfrac{\sqrt{3}}{2}\dfrac{\sqrt{2}}{2}=\dfrac{\sqrt{2}+\sqrt{6}}{4}$

3.　$\sin\dfrac{5\pi}{12}=\sin\left(\dfrac{\pi}{6}+\dfrac{\pi}{4}\right)=\sin\dfrac{\pi}{6}\cos\dfrac{\pi}{4}+\cos\dfrac{\pi}{6}\sin\dfrac{\pi}{4}=\dfrac{1}{2}\dfrac{\sqrt{2}}{2}+\dfrac{\sqrt{3}}{2}\dfrac{\sqrt{2}}{2}=\dfrac{\sqrt{2}+\sqrt{6}}{4}$

5.　$\cot\dfrac{5\pi}{12}=\tan\left(\dfrac{\pi}{2}-\dfrac{5\pi}{12}\right)=\tan\dfrac{\pi}{12}=\tan\left(\dfrac{\pi}{3}-\dfrac{\pi}{4}\right)=\dfrac{\tan\dfrac{\pi}{3}-\tan\dfrac{\pi}{4}}{1+\tan\dfrac{\pi}{3}\tan\dfrac{\pi}{4}}$

$=\dfrac{\sqrt{3}-1}{1+\sqrt{3}\cdot 1}=\dfrac{\sqrt{3}-1}{\sqrt{3}+1}=\dfrac{\sqrt{3}-1}{\sqrt{3}+1}\cdot\left(\dfrac{\sqrt{3}-1}{\sqrt{3}-1}\right)=\dfrac{3-2\sqrt{3}+1}{3-1}=\dfrac{4-2\sqrt{3}}{2}=2-\sqrt{3}$

7.　$\tan\dfrac{7\pi}{12}=\tan\left(\dfrac{\pi}{3}+\dfrac{\pi}{4}\right)=\dfrac{\tan\dfrac{\pi}{3}+\tan\dfrac{\pi}{4}}{1-\tan\dfrac{\pi}{3}\tan\dfrac{\pi}{4}}=\dfrac{\sqrt{3}+1}{1-\sqrt{3}}=\dfrac{1+\sqrt{3}}{1-\sqrt{3}}\cdot\left(\dfrac{1+\sqrt{3}}{1+\sqrt{3}}\right)$

$=\dfrac{1+2\sqrt{3}+3}{1-3}=\dfrac{4+2\sqrt{3}}{-2}=-2-\sqrt{3}$

9.　$\cos\dfrac{11\pi}{12}=\cos\left(\dfrac{\pi}{4}+\dfrac{2\pi}{3}\right)=\cos\dfrac{\pi}{4}\cos\dfrac{2\pi}{3}-\sin\dfrac{\pi}{4}\sin\dfrac{2\pi}{3}$

$=\dfrac{\sqrt{2}}{2}\left(-\dfrac{1}{2}\right)-\dfrac{\sqrt{2}}{2}\dfrac{\sqrt{3}}{2}=\dfrac{-\sqrt{2}-\sqrt{6}}{4}$

11.　$\sin105°=\sin\left(60°+45°\right)=\sin60°\cos45°+\cos60°\sin45°=\dfrac{\sqrt{3}}{2}\dfrac{\sqrt{2}}{2}+\dfrac{1}{2}\dfrac{\sqrt{2}}{2}=\dfrac{\sqrt{6}+\sqrt{2}}{4}$

13.　$\sin\left(\dfrac{\pi}{2}+x\right)=\sin\dfrac{\pi}{2}\cos x+\cos\dfrac{\pi}{2}\sin x=1\cdot\cos x+0\cdot\sin x=\cos x$

15.　$\cos\left(x-\dfrac{3\pi}{2}\right)=\cos x\cos\dfrac{3\pi}{2}+\sin x\sin\dfrac{3\pi}{2}=\cos x\cdot 0+\sin x\cdot(-1)=-\sin x$

17. $\sec(x-\pi) = \dfrac{1}{\cos(x-\pi)} = \dfrac{1}{\cos x \cos \pi + \sin x \sin \pi} = \dfrac{1}{\cos x \cdot (-1) + \sin x \cdot 0}$

$= \dfrac{1}{-\cos x + 0} = \dfrac{1}{-\cos x} = -\dfrac{1}{\cos x}$

19. $\sin 3 \cos 5 - \cos 3 \sin 5 = \sin(3-5) = \sin(-2) = -\sin 2$

21. $\cos(x+y)\cos y + \sin(x+y)\sin y = \cos((x+y)-y) = \cos x$

23. $\cos(x+y) - \cos(x-y) = (\cos x \cos y - \sin x \sin y) - (\cos x \cos y + \sin x \sin y)$

$= \cos x \cos y - \sin x \sin y - \cos x \cos y - \sin x \sin y$

$= -2 \sin x \sin y$

25. First note that since $0 < x < \dfrac{\pi}{2}$, $\cos x > 0$ and $\cos x = \sqrt{1-\sin^2 x} = \sqrt{1 - \left(\dfrac{1}{3}\right)^2}$

$= \sqrt{1 - \dfrac{1}{9}} = \sqrt{\dfrac{8}{9}} = \dfrac{\sqrt{8}}{3} = \dfrac{2\sqrt{2}}{3}.$

$\sin\left(\dfrac{\pi}{4} + x\right) = \sin \dfrac{\pi}{4} \cos x + \cos \dfrac{\pi}{4} \sin x = \dfrac{\sqrt{2}}{2} \dfrac{2\sqrt{2}}{3} + \dfrac{\sqrt{2}}{2} \dfrac{1}{3} = \dfrac{4 + \sqrt{2}}{6}$

27. First note that since $\pi < x < \dfrac{3\pi}{2}$, $\sin x < 0$ and $\sin x = -\sqrt{1-\cos^2 x} = -\sqrt{1 - \left(-\dfrac{1}{5}\right)^2}$

$= -\sqrt{1 - \dfrac{1}{25}} = -\sqrt{\dfrac{24}{25}} = -\dfrac{\sqrt{24}}{5} = -\dfrac{2\sqrt{6}}{5}.$

$\sin\left(\dfrac{\pi}{3} - x\right) = \sin \dfrac{\pi}{3} \cos x - \cos \dfrac{\pi}{3} \sin x = \dfrac{\sqrt{3}}{2}\left(-\dfrac{1}{5}\right) - \dfrac{1}{2}\left(-\dfrac{2\sqrt{6}}{5}\right) = \dfrac{-\sqrt{3} + 2\sqrt{6}}{10}$

For Exercise 29 - 31, note that since $0 < x < \dfrac{\pi}{2}$, $\cos x > 0$. Moreover, $\cos x = \sqrt{1-\sin^2 x}$

$= \sqrt{1 - (.8)^2} = \sqrt{1 - .64} = \sqrt{.36} = .6$ and $\cos y = \sqrt{1 - \sin^2 x} = \sqrt{1 - \left(\sqrt{.75}\right)^2} = \sqrt{1 - .75}$

$= \sqrt{.25} = .5.$

29. $\sin(x+y) = \sin x \cos y + \cos x \sin y = .8(.5) + (.6)\sqrt{.75} = .4 + .6\sqrt{\dfrac{75}{100}}$

$= \dfrac{4}{10} + .6 \dfrac{\sqrt{75}}{10} = \dfrac{4}{10} + \dfrac{.6 \cdot 5\sqrt{3}}{10} = \dfrac{4 + 3\sqrt{3}}{10}$

31. $\sin(x-y)=\sin x\cos y-\cos x\sin y=.8(.5)-(.6)\sqrt{.75}=.4-.6\sqrt{\dfrac{75}{100}}$

$$=\dfrac{4}{10}-.6\dfrac{\sqrt{75}}{10}=\dfrac{4}{10}-\dfrac{.6\cdot5\sqrt3}{10}=\dfrac{4-3\sqrt3}{10}$$

33. From the figure, we have $r=\sqrt{3^2+4^2}=\sqrt{9+16}=\sqrt{25}=5,\ \sin t=\dfrac{y}{r}=\dfrac{4}{5}$, and

$\cos t=\dfrac{x}{r}=\dfrac{3}{5}$. Thus, we have $5\sin(x+t)=5(\sin x\cos t+\cos x\sin t)$

$=\sin x\cdot5\cos t+\cos x\cdot5\sin t\ =\sin x\cdot5\left(\dfrac{3}{5}\right)+\cos x\cdot5\left(\dfrac{4}{5}\right)=3\sin x+4\cos x.$

35. $\dfrac{f(x+h)-f(x)}{h}=\dfrac{\cos(x+h)-\cos x}{h}=\dfrac{\cos x\cos h-\sin x\sin h-\cos x}{h}$

$$=\dfrac{\cos x(\cos h-1)-\sin x\sin h}{h}=\cos x\left(\dfrac{\cos h-1}{h}\right)-\sin x\left(\dfrac{\sin h}{h}\right)$$

37. $\sin(x+y)=\cos\left[\dfrac{\pi}{2}-(x+y)\right]=\cos\left[\dfrac{\pi}{2}-x-y\right]=\cos\left[\left(\dfrac{\pi}{2}-x\right)-y\right]$

$$=\cos\left(\dfrac{\pi}{2}-x\right)\cos y+\sin\left(\dfrac{\pi}{2}-x\right)\sin y=\sin x\cos y+\cos x\sin y$$

Alternate Approach:
$\sin(x+y)=\sin(x-(-y))=\sin x\cos(-y)-\cos x\sin(-y)$

$$=\sin x\cos y-\cos x(-\sin y)=\sin x\cos y+\cos x\sin y$$

39. $\dfrac{\cos(x-y)}{\cos x\cos y}=\dfrac{\cos x\cos y+\sin x\sin y}{\cos x\cos y}$

$$=\dfrac{\cos x\cos y}{\cos x\cos y}+\dfrac{\sin x\sin y}{\cos x\cos y}=1+\dfrac{\sin x}{\cos x}\cdot\dfrac{\sin y}{\cos y}=1+\tan x\tan y$$

41. $\dfrac{\sin(x-y)}{\sin x\sin y}=\dfrac{\sin x\cos y-\cos x\sin y}{\sin x\sin y}$

$$=\dfrac{\sin x\cos y}{\sin x\sin y}-\dfrac{\cos x\sin y}{\sin x\sin y}=\dfrac{\cos y}{\sin y}-\dfrac{\cos x}{\sin x}=\cot y-\cot x$$

43. $\dfrac{\sin(x+y)}{\sin x\cos y}=\dfrac{\sin x\cos y+\cos x\sin y}{\sin x\cos y}$

$$=\dfrac{\sin x\cos y}{\sin x\cos y}+\dfrac{\cos x\sin y}{\sin x\cos y}=1+\dfrac{\cos x}{\sin x}\cdot\dfrac{\sin y}{\cos y}=1+\cot x\tan y$$

45. First note that x is in the first quadrant, $\cos x > 0$ and

$$\cos x = \sqrt{1 - \sin^2 x} = \sqrt{1 - \left(\frac{24}{25}\right)^2} = \sqrt{1 - \frac{576}{625}} = \sqrt{\frac{49}{625}} = \frac{7}{25}.$$ Also, since y is in the

second quadrant, and $\cos y < 0$ and $\cos y = -\sqrt{1 - \sin^2 y} = -\sqrt{1 - \left(\frac{4}{5}\right)^2} = -\sqrt{1 - \frac{16}{25}}$

$= -\sqrt{\frac{9}{25}} = -\frac{3}{5}.$ We now have $\tan x = \dfrac{\sin x}{\cos x} = \dfrac{24/25}{7/25} = \dfrac{24}{7}$ and $\tan y = \dfrac{\sin y}{\cos y}$

$= \dfrac{4/5}{-3/5} = -\dfrac{4}{3}.$ This gives us $\sin(x+y) = \sin x \cos y + \cos x \sin y = \dfrac{24}{25}\left(-\dfrac{3}{5}\right) + \dfrac{7}{25}\left(\dfrac{4}{5}\right)$

$= -\dfrac{72}{125} + \dfrac{28}{125} = \dfrac{-72+28}{125} = -\dfrac{44}{125}$ and $\tan(x+y) = \dfrac{\tan x + \tan y}{1 - \tan x \tan y} = \dfrac{\dfrac{24}{7} + \left(-\dfrac{4}{3}\right)}{1 - \dfrac{24}{7}\left(-\dfrac{4}{3}\right)}$

$= \dfrac{3 \cdot 24 - 7 \cdot 4}{21 + 24 \cdot 4} = \dfrac{72 - 28}{21 + 96} = \dfrac{44}{117}.$ Since $\sin(x+y)$ is negative and $\tan(x+y)$ is
positive, $x + y$ lies in quadrant III.

47. First note that x is in the first quadrant, $\cos x > 0$ and

$$\cos x = \sqrt{1 - \sin^2 x} = \sqrt{1 - \left(\frac{4}{5}\right)^2} = \sqrt{1 - \frac{16}{25}} = \sqrt{\frac{9}{25}} = \frac{3}{5}.$$ Also, since y is in the second

quadrant, and $\sin y > 0$ and $\sin y = \sqrt{1 - \cos^2 y} = \sqrt{1 - \left(-\frac{12}{13}\right)^2} = \sqrt{1 - \frac{144}{169}}$

$= \sqrt{\frac{25}{169}} = \frac{5}{13}.$ We now have $\tan x = \dfrac{\sin x}{\cos x} = \dfrac{4/5}{3/5} = \dfrac{4}{3}$ and $\tan y = \dfrac{\sin y}{\cos y} = \dfrac{5/13}{-12/13}$

$= -\dfrac{5}{12}.$ This gives us $\cos(x+y) = \cos x \cos y - \sin x \sin y = \dfrac{3}{5}\left(-\dfrac{12}{13}\right) - \dfrac{4}{5}\left(\dfrac{5}{13}\right)$

$= \dfrac{-36 - 20}{65} = -\dfrac{56}{65}$ and $\tan(x+y) = \dfrac{\tan x + \tan y}{1 - \tan x \tan y} = \dfrac{\dfrac{4}{3} + \left(-\dfrac{5}{12}\right)}{1 - \left(\dfrac{4}{3}\right)\left(-\dfrac{5}{12}\right)} = \dfrac{4 \cdot 12 - 5 \cdot 3}{36 + 4 \cdot 5}$

$= \dfrac{48 - 15}{36 + 20} = \dfrac{33}{56}.$ Since $\cos(x+y)$ is negative and $\tan(x+y)$ is positive, $x + y$ lies
in quadrant III.

49. $\sin(u+v+w) = \sin\left[(u+v)+w\right] = \sin(u+v)\cos w + \cos(u+v)\sin w$

$$= (\sin u\cos v + \cos u\sin v)\cos w + (\cos u\cos v - \sin u\sin v)\sin w$$

$$= \sin u\cos v\cos w + \cos u\sin v\cos w + \cos u\cos v\sin w - \sin u\sin v\sin w$$

51. If $x + y = \dfrac{\pi}{2}$, then $y = \dfrac{\pi}{2} - x$ and $\sin y = \sin\left(\dfrac{\pi}{2} - x\right) = \cos x$. Thus, we have

$$\sin^2 x + \sin^2 y = \sin^2 x + \cos^2 x = 1.$$

53. $\sin(x - \pi) = \sin x\cos\pi - \cos x\sin\pi = \sin x(-1) - \cos x(0) = -\sin x$

55. $\cos(\pi - x) = \cos\pi\cos x + \sin\pi\sin x = (-1)\cos x + (0)\sin x = -\cos x$

57. $\sin(x + \pi) = \sin x\cos\pi + \cos x\sin\pi = \sin x\cdot(-1) + \cos x\cdot(0) = -\sin x$

59. $\tan(x + \pi) = \dfrac{\tan x + \tan\pi}{1 - \tan x\tan\pi} = \dfrac{\tan x + 0}{1 - \tan x\cdot 0} = \tan x$

61. Starting with the right side we have the following.

$$\tfrac{1}{2}\left[\cos(x-y) - \cos(x+y)\right] = \tfrac{1}{2}\left[\cos x\cos y + \sin x\sin y - (\cos x\cos y - \sin x\sin y)\right]$$

$$= \tfrac{1}{2}(\cos x\cos y + \sin x\sin y - \cos x\cos y + \sin x\sin y)$$

$$= \tfrac{1}{2}(2\sin x\sin y) = \sin x\sin y$$

63. $\cos(x+y)\cos(x-y) = (\cos x\cos y - \sin x\sin y)(\cos x\cos y + \sin x\sin y)$

$$= \cos^2 x\cos^2 y - \sin^2 x\sin^2 y$$

65. Let $y = 0$, then graph both sides.

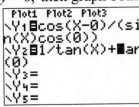

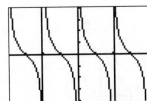

Since the two graphs appear to coincide, this could be an identity.

$\dfrac{\cos(x-y)}{\sin x\cos y} = \dfrac{\cos x\cos y + \sin x\sin y}{\sin x\cos y}$

$\qquad = \dfrac{\cos x\cos y}{\sin x\cos y} + \dfrac{\sin x\sin y}{\sin x\cos y}$

$\qquad = \dfrac{\cos x}{\sin x} + \dfrac{\sin y}{\cos y}$

$\qquad = \cot x + \tan y$

This is an identity.

67. Let $y = \dfrac{\pi}{2}$, then graph both sides.

 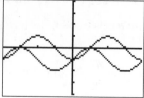

Since the graphs do not coincide, this is not an identity.

69. Let $y = 0$, then graph both sides.

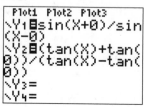

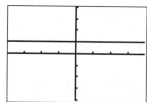

Since the two graphs appear to coincide, this could be an identity.

$$\frac{\sin(x+y)}{\sin(x-y)} = \frac{\sin x \cos y + \cos x \sin y}{\sin x \cos y - \cos x \sin y} = \frac{\dfrac{\sin x \cos y}{\cos x \cos y} + \dfrac{\cos x \sin y}{\cos x \cos y}}{\dfrac{\sin x \cos y}{\cos x \cos y} - \dfrac{\cos x \sin y}{\cos x \cos y}} = \frac{\tan x + \tan y}{\tan x - \tan y}$$

This is an identity.

71. Let $y = -\dfrac{\pi}{3}$, and graph both sides.

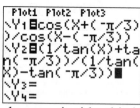 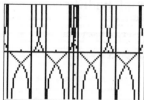

Since the graphs do not coincide, this is not an identity.

73. Let $y = \dfrac{\pi}{3}$, then graph both sides.

 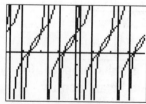

Since the two graphs do not coincide, this is not an identity.

7.2.A Lines and Angles

1. $\tan\theta = \text{slope} = \dfrac{y_2 - y_1}{x_2 - x_1} = \dfrac{5-2}{3-(-1)} = \dfrac{3}{3+1} = \dfrac{3}{4}$

3. $\tan\theta = \text{slope} = \dfrac{y_2 - y_1}{x_2 - x_1} = \dfrac{0-4}{6-1} = \dfrac{-4}{5} = -\dfrac{4}{5}$

5. $\tan\theta = \text{slope} = \dfrac{y_2 - y_1}{x_2 - x_1} = \dfrac{5-(-7)}{3-3} = \dfrac{5+7}{0} = \dfrac{12}{0}$ which is undefined.

7. $\tan\theta = \left|\dfrac{m-k}{1+mk}\right| = \left|\dfrac{\dfrac{3}{2}-(-1)}{1+\left(\dfrac{3}{2}\right)(-1)}\right| = \left|\dfrac{\dfrac{3}{2}+1}{1-\dfrac{3}{2}}\right| = \left|\dfrac{3+2}{2-3}\right| = \left|\dfrac{5}{-1}\right| = |-5| = 5$

Thus, $\theta = \tan^{-1} 5 \approx 1.37$ radians or 1.77 radians.

9. $\tan\theta = \left|\dfrac{m-k}{1+mk}\right| = \left|\dfrac{-1-0}{1+(-1)\cdot 0}\right| = \left|\dfrac{-1}{1+0}\right| = \left|\dfrac{-1}{1}\right| = |-1| = 1$

Thus, $\theta = \tan^{-1} 1 = \dfrac{\pi}{4}$ radians or $\dfrac{3\pi}{4}$ radians.

11. Line L: $m = \dfrac{6-2}{5-3} = \dfrac{4}{2} = 2$ and Line M: $k = \dfrac{0-3}{4-0} = \dfrac{-3}{4} = -\dfrac{3}{4}$.

$\tan\theta = \left|\dfrac{m-k}{1+mk}\right| = \left|\dfrac{2-\left(-\dfrac{3}{4}\right)}{1+2\left(-\dfrac{3}{4}\right)}\right| = \left|\dfrac{2+\dfrac{3}{4}}{1-\dfrac{6}{4}}\right| = \left|\dfrac{8+3}{4-6}\right| = \left|\dfrac{11}{-2}\right| = \dfrac{11}{2}$.

Thus, $\theta = \tan^{-1}\dfrac{11}{2} \approx 1.39$ radians or 1.75 radians.

13. Line L: $m = 3$ Line M: k satisfies $k(-.5) = -1$. Thus, $k = \dfrac{-1}{-.5} = 2$.

$\tan\theta = \left|\dfrac{m-k}{1+mk}\right| = \left|\dfrac{3-2}{1+3\cdot 2}\right| = \left|\dfrac{1}{1+6}\right| = \left|\dfrac{1}{7}\right| = \dfrac{1}{7}$.

Thus, $\theta = \tan^{-1}\dfrac{1}{7} \approx .142$ radians or 3.00 radians.

7.3 Other Identities

1. First note that since $0 < x < \dfrac{\pi}{2}$, $\cos x > 0$, and $\cos x = \sqrt{1-\sin^2 x} = \sqrt{1-\left(\dfrac{5}{13}\right)^2}$

$= \sqrt{1-\dfrac{25}{169}} = \sqrt{\dfrac{144}{169}} = \dfrac{12}{13}$. $\sin 2x = 2\sin x \cos x = 2\left(\dfrac{5}{13}\right)\left(\dfrac{12}{13}\right) = \dfrac{120}{169}$,

$\cos 2x = 1 - 2\sin^2 x = 1-2\left(\dfrac{5}{13}\right)^2 = 1-2\left(\dfrac{25}{169}\right) = 1-\dfrac{50}{169} = \dfrac{119}{169}$, and $\tan 2x = \dfrac{\sin 2x}{\cos 2x}$

$= \dfrac{120/169}{119/169} = \dfrac{120}{119}$.

3. First note that since $\pi < x < \dfrac{3\pi}{2}$, $\sin x < 0$, and $\sin x = -\sqrt{1-\cos^2 x}$

$= -\sqrt{1-\left(-\dfrac{3}{5}\right)^2} = -\sqrt{1-\dfrac{9}{25}} = -\sqrt{\dfrac{16}{25}} = -\dfrac{4}{5}$. $\sin 2x = 2\sin x \cos x = 2\left(-\dfrac{4}{5}\right)\left(-\dfrac{3}{5}\right) = \dfrac{24}{25}$,

$\cos 2x = 2\cos^2 x - 1 = 2\left(-\dfrac{3}{5}\right)^2 - 1 = 2\left(\dfrac{9}{25}\right) - 1 = \dfrac{18}{25} - 1 = -\dfrac{7}{25}$, and $\tan 2x = \dfrac{\sin 2x}{\cos 2x}$

$= \dfrac{24/25}{-7/25} = -\dfrac{24}{7}$.

5. First note that since $\pi < x < \dfrac{3\pi}{2}$, $\sec x < 0$, and $\sec x = -\sqrt{1+\tan^2 x}$

$= -\sqrt{1+\left(\dfrac{3}{4}\right)^2} = -\sqrt{1+\dfrac{9}{16}} = -\sqrt{\dfrac{25}{16}} = -\dfrac{5}{4}$. Thus, $\cos x = \dfrac{1}{\sec x} = \dfrac{1}{-5/4} = -\dfrac{4}{5}$. Since

$\pi < x < \dfrac{3\pi}{2}$, $\sin x < 0$, and $\sin x = -\sqrt{1-\cos^2 x} = -\sqrt{1-\left(-\dfrac{4}{5}\right)^2} = -\sqrt{1-\dfrac{16}{25}} = -\sqrt{\dfrac{9}{25}}$

$= -\dfrac{3}{5}$. $\sin 2x = 2\sin x \cos x = 2\left(-\dfrac{3}{5}\right)\left(-\dfrac{4}{5}\right) = \dfrac{24}{25}$, $\cos 2x = 2\cos^2 x - 1 = 2\left(-\dfrac{4}{5}\right)^2 - 1$

$= 2\left(\dfrac{16}{25}\right) - 1 = \dfrac{32}{25} - 1 = \dfrac{7}{25}$, and $\tan 2x = \dfrac{\sin 2x}{\cos 2x} = \dfrac{24/25}{7/25} = \dfrac{24}{7}$.

7. Since $\csc x = 4$, we have $\sin x = \dfrac{1}{\csc x} = \dfrac{1}{4}$. Since $0 < x < \dfrac{\pi}{2}$, $\cos x > 0$, and

$$\cos x = \sqrt{1-\left(\frac{1}{4}\right)^2} = \sqrt{1-\frac{1}{16}} = \sqrt{\frac{15}{16}} = \frac{\sqrt{15}}{4}. \quad \sin 2x = 2\sin x \cos x = 2\left(\frac{1}{4}\right)\left(\frac{\sqrt{15}}{4}\right) = \frac{\sqrt{15}}{8},$$

$$\cos 2x = 1 - 2\sin^2 x = 1 - 2\left(\frac{1}{4}\right)^2 = 1 - 2\left(\frac{1}{16}\right) = 1 - \frac{1}{8} = \frac{7}{8}, \text{ and } \tan 2x = \frac{\sin 2x}{\cos 2x}$$

$$= \frac{\sqrt{15}/8}{7/8} = \frac{\sqrt{15}}{7}.$$

9. a. The horizontal length of the rectangle is $2x$. The vertical height is y. The area is $2xy$.

b. Since $\cos t = \dfrac{x}{r} = \dfrac{x}{3}$, we have $x = 3\cos t$. Also, since $\sin t = \dfrac{y}{r} = \dfrac{y}{3}$, we have $y = 3\sin t$.

c. $A = 2xy = 2(3\cos t)(3\sin t) = 9(2\sin t \cos t) = 9\sin 2t$

11. $\cos\dfrac{\pi}{8} = \cos\left[\dfrac{1}{2}\left(\dfrac{\pi}{4}\right)\right] = \sqrt{\dfrac{1+\cos\left(\frac{\pi}{4}\right)}{2}} = \sqrt{\dfrac{1+\frac{\sqrt{2}}{2}}{2}} = \dfrac{\sqrt{2+\sqrt{2}}}{2}$

13. $\sin\dfrac{3\pi}{8} = \sin\left[\dfrac{1}{2}\left(\dfrac{3\pi}{4}\right)\right] = \sqrt{\dfrac{1-\cos\left(\frac{3\pi}{4}\right)}{2}} = \sqrt{\dfrac{1-\left(-\frac{\sqrt{2}}{2}\right)}{2}} = \dfrac{\sqrt{2+\sqrt{2}}}{2}$

15. $\tan\dfrac{\pi}{12} = \tan\left[\dfrac{1}{2}\left(\dfrac{\pi}{6}\right)\right] = \dfrac{1-\cos\left(\frac{\pi}{6}\right)}{\sin\left(\frac{\pi}{6}\right)} = \dfrac{1-\frac{\sqrt{3}}{2}}{\frac{1}{2}} = 2-\sqrt{3}$

17. $\cos\dfrac{\pi}{12} = \cos\left[\dfrac{1}{2}\left(\dfrac{\pi}{6}\right)\right] = \sqrt{\dfrac{1+\cos\left(\frac{\pi}{6}\right)}{2}} = \sqrt{\dfrac{1+\frac{\sqrt{3}}{2}}{2}} = \dfrac{\sqrt{2+\sqrt{3}}}{2}$

19. $\sin\dfrac{7\pi}{8} = \sin\left[\dfrac{1}{2}\left(\dfrac{7\pi}{4}\right)\right] = \sqrt{\dfrac{1-\cos\left(\frac{7\pi}{4}\right)}{2}} = \sqrt{\dfrac{1-\frac{\sqrt{2}}{2}}{2}} = \dfrac{\sqrt{2-\sqrt{2}}}{2}$

21. $\tan\dfrac{7\pi}{8} = \tan\left[\dfrac{1}{2}\left(\dfrac{7\pi}{4}\right)\right] = \dfrac{1-\cos\left(\dfrac{7\pi}{4}\right)}{\sin\left(\dfrac{7\pi}{4}\right)} = \dfrac{1-\dfrac{\sqrt{2}}{2}}{-\dfrac{\sqrt{2}}{2}} = \dfrac{2-\sqrt{2}}{-\sqrt{2}}$

$\qquad\quad = \dfrac{2-\sqrt{2}}{-\sqrt{2}}\cdot\dfrac{\sqrt{2}}{\sqrt{2}} = \dfrac{2\sqrt{2}-2}{-2} = 1-\sqrt{2}$

23. From Exercise 11, we have $\cos\dfrac{\pi}{8} = \dfrac{\sqrt{2+\sqrt{2}}}{2}$.

$\cos\dfrac{\pi}{16} = \cos\left[\dfrac{1}{2}\left(\dfrac{\pi}{8}\right)\right] = \sqrt{\dfrac{1+\cos\left(\dfrac{\pi}{8}\right)}{2}} = \sqrt{\dfrac{1+\dfrac{\sqrt{2+\sqrt{2}}}{2}}{2}} = \dfrac{\sqrt{2+\sqrt{2+\sqrt{2}}}}{2}$

25. From Exercise 17, we have $\cos\dfrac{\pi}{12} = \dfrac{\sqrt{2+\sqrt{3}}}{2}$.

$\sin\dfrac{\pi}{24} = \sin\left[\dfrac{1}{2}\left(\dfrac{\pi}{12}\right)\right] = \sqrt{\dfrac{1-\cos\left(\dfrac{\pi}{12}\right)}{2}} = \sqrt{\dfrac{1-\dfrac{\sqrt{2+\sqrt{3}}}{2}}{2}} = \dfrac{\sqrt{2-\sqrt{2+\sqrt{3}}}}{2}$

27. Since $0 < x < \dfrac{\pi}{2}$ we have $0 < \dfrac{x}{2} < \dfrac{\pi}{4}$. Hence, $\dfrac{x}{2}$ lies in the first quadrant. Thus,

$\sin\dfrac{x}{2} > 0$ and $\cos\dfrac{x}{2} > 0$.

$\sin\dfrac{x}{2} = \sqrt{\dfrac{1-\cos x}{2}} = \sqrt{\dfrac{1-.4}{2}} = \sqrt{\dfrac{.6}{2}} = \sqrt{.3} = \sqrt{\dfrac{30}{100}} = \dfrac{\sqrt{30}}{10}$

$\cos\dfrac{x}{2} = \sqrt{\dfrac{1+\cos x}{2}} = \sqrt{\dfrac{1+.4}{2}} = \sqrt{\dfrac{1.4}{2}} = \sqrt{.7} = \sqrt{\dfrac{70}{100}} = \dfrac{\sqrt{70}}{10}$

$\tan\dfrac{x}{2} = \dfrac{\sin\dfrac{x}{2}}{\cos\dfrac{x}{2}} = \dfrac{\dfrac{\sqrt{30}}{10}}{\dfrac{\sqrt{70}}{10}} = \dfrac{\sqrt{30}}{\sqrt{70}} = \sqrt{\dfrac{30}{70}} = \sqrt{\dfrac{3}{7}}$

29. First note that $\cos x = \sqrt{1 - \sin^2 x} = \sqrt{1 - \left(-\dfrac{3}{5}\right)^2} = \sqrt{1 - \dfrac{9}{25}} = \sqrt{\dfrac{16}{25}} = \dfrac{4}{5}$. Also, since

$\dfrac{3\pi}{2} < x < 2\pi$ we have $\dfrac{3\pi}{4} < \dfrac{x}{2} < \pi$. Hence, $\dfrac{x}{2}$ lies in the second quadrant. Thus,

$\sin\dfrac{x}{2} > 0$ and $\cos\dfrac{x}{2} < 0$.

$\sin\dfrac{x}{2} = \sqrt{\dfrac{1 - \cos x}{2}} = \sqrt{\dfrac{1 - 4/5}{2}} = \sqrt{\dfrac{5-4}{10}} = \sqrt{\dfrac{1}{10}} = \dfrac{1}{\sqrt{10}}$

$\cos\dfrac{x}{2} = -\sqrt{\dfrac{1 + \cos x}{2}} = -\sqrt{\dfrac{1 + \dfrac{4}{5}}{2}} = -\sqrt{\dfrac{5+4}{10}} = -\sqrt{\dfrac{9}{10}} = -\dfrac{3}{\sqrt{10}} =$

$\tan\dfrac{x}{2} = \dfrac{\sin\dfrac{x}{2}}{\cos\dfrac{x}{2}} = \dfrac{1/\sqrt{10}}{-3/\sqrt{10}} = \dfrac{1}{\sqrt{10}} \cdot \left(-\dfrac{\sqrt{10}}{3}\right) = -\dfrac{1}{3}$

31. Since $\pi < x < \dfrac{3\pi}{2}$ we have $\dfrac{\pi}{2} < \dfrac{x}{2} < \dfrac{3\pi}{4}$. Hence, $\dfrac{x}{2}$ lies in the second quadrant.

Thus, $\sec\dfrac{x}{2} < 0$ and $\sin\dfrac{x}{2} > 0$ and $\sec x < 0$.

First note that $\sec x = -\sqrt{1 + \tan^2 x} = -\sqrt{1 + (1/2)^2} = -\sqrt{5}/2$. Hence $\cos x = -2/\sqrt{5}$.

$\sin x = \cos x \tan x = \left(-\dfrac{2}{\sqrt{5}}\right)\left(\dfrac{1}{2}\right) = -\dfrac{1}{\sqrt{5}}$.

$\sin\dfrac{x}{2} = \sqrt{\dfrac{1 - \cos x}{2}} = \sqrt{\dfrac{1 - (-2/\sqrt{5})}{2}} = \sqrt{\dfrac{\sqrt{5}+2}{2\sqrt{5}}} = \dfrac{\sqrt{50 + 20\sqrt{5}}}{10}$

$\cos\dfrac{x}{2} = -\sqrt{\dfrac{1 + \cos x}{2}} = -\sqrt{\dfrac{1 + (-2/\sqrt{5})}{2}} = \sqrt{\dfrac{\sqrt{5}-2}{2\sqrt{5}}} = -\dfrac{\sqrt{50 - 20\sqrt{5}}}{10}$

$\tan\dfrac{x}{2} = \dfrac{1 - \cos x}{\sin x} = \dfrac{1 - (-2/\sqrt{5})}{-1/\sqrt{5}} = -(\sqrt{5} + 2)$

33. $\sin 4x \cos 6x = \dfrac{1}{2}\left[\sin(4x + 6x) + \sin(4x - 6x)\right] = \dfrac{1}{2}\left[\sin(10x) + \sin(-2x)\right]$
$\qquad = \dfrac{1}{2}\sin 10x - \dfrac{1}{2}\sin 2x$

35. $\cos 2x \cos 4x = \dfrac{1}{2}\left[\cos(2x + 4x) + \cos(2x - 4x)\right]$
$\qquad = \dfrac{1}{2}\left[\cos(6x) + \cos(-2x)\right] = \dfrac{1}{2}\cos 6x + \dfrac{1}{2}\cos 2x$

37. $\sin 17x \sin(-3x) = \dfrac{1}{2}\left[\cos(17x - (-3x)) - \cos(17x + (-3x))\right]$
$\qquad = \dfrac{1}{2}\left[\cos(20x) - \cos(14x)\right] = \dfrac{1}{2}\cos 20x - \dfrac{1}{2}\cos 14x$

39. $\sin 3x + \sin 5x = 2\sin\left(\dfrac{3x+5x}{2}\right)\cos\left(\dfrac{3x-5x}{2}\right) = 2\sin 4x\cos(-x) = 2\sin 4x\cos x$

41. $\sin 9x - \sin 5x = 2\cos\left(\dfrac{9x+5x}{2}\right)\sin\left(\dfrac{9x-5x}{2}\right) = 2\cos 7x\sin 2x$

43. $\cos 2x + \cos 5x = 2\cos\left(\dfrac{2x+5x}{2}\right)\cos\left(\dfrac{2x-5x}{2}\right) = 2\cos\left(\dfrac{7x}{2}\right)\cos\left(-\dfrac{3x}{2}\right)$

$$= 2\cos\left(\dfrac{7x}{2}\right)\cos\left(\dfrac{3x}{2}\right)$$

45. First note that $\cos x = \sqrt{1-\sin^2 x} = \sqrt{1-(.6)^2} = .8$.

$\sin 2x = 2\sin x\cos x = 2(.6)(.8) = .96$

47. $\cos 2x = 1 - 2\sin^2 x = 1 - 2(.6)^2 = .28$

49. $\sin\dfrac{x}{2} = \sqrt{\dfrac{1-\cos x}{2}} = \sqrt{\dfrac{1-.8}{2}} = \sqrt{.1} = .316$

51. $\cos 3x = \cos(2x + x)$

$= \cos 2x\cos x - \sin 2x\sin x$

$= \left(2\cos^2 x - 1\right)\cos x - (2\sin x\cos x)\sin x$

$= 2\cos^3 x - \cos x - 2\cos x\sin^2 x$

$= 2\cos^3 x - \cos x - 2\cos x\left(1 - \cos^2 x\right)$

$= 2\cos^3 x - \cos x - 2\cos x + 2\cos^3 x$

$= 4\cos^3 x - 3\cos x$

53. $\dfrac{\sin 2x}{2\sin x} = \dfrac{2\sin x\cos x}{2\sin x} = \cos x$

55. Since $2\sin x\cos x = \sin 2x$, $2\sin 2y\cos 2y = \sin 2(2y) = \sin 4y$

57. $(\sin x + \cos x)^2 - \sin 2x = \sin^2 x + 2\sin x\cos x + \cos^2 x - \sin 2x$

$$= \sin^2 x + \cos^2 x + 2\sin x\cos x - 2\sin x\cos = 1$$

59. Use the fourth product-to-sum identity with $\frac{1}{2}(x+y)$ in place of x and $\frac{1}{2}(x-y)$ in place of y.

$$2\cos\left[\frac{1}{2}(x+y)\right]\sin\left[\frac{1}{2}(x-y)\right]$$

$$=2\cdot\frac{1}{2}\left[\sin\left(\frac{1}{2}(x+y)+\frac{1}{2}(x-y)\right)-\sin\left(\frac{1}{2}(x+y)-\frac{1}{2}(x-y)\right)\right]$$

$$=\sin x-\sin y$$

61. Use the second product-to-sum identity with $\frac{1}{2}(x+y)$ in place of x and $\frac{1}{2}(x-y)$ in place of y.

$$-2\sin\left[\frac{1}{2}(x+y)\right]\sin\left[\frac{1}{2}(x-y)\right]$$

$$=-2\cdot\frac{1}{2}\left[\cos\left(\frac{1}{2}(x+y)-\frac{1}{2}(x-y)\right)-\cos\left(\frac{1}{2}(x+y)+\frac{1}{2}(x-y)\right)\right]$$

$$=-(\cos y-\cos x)$$

$$=\cos x-\cos y$$

63. Graph both sides.

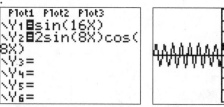 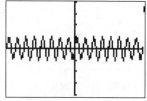

Since the graphs coincide, this could be an identity.

$\sin(16x)=\sin\left[2(8x)\right]=2\sin(8x)\cos(8x)$. This is an identity.

65. Graph both sides.

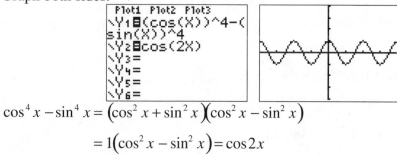

$$\cos^4 x-\sin^4 x=\left(\cos^2 x+\sin^2 x\right)\left(\cos^2 x-\sin^2 x\right)$$

$$=1\left(\cos^2 x-\sin^2 x\right)=\cos 2x$$

This is an identity.

67. Graph both sides.

 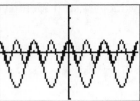

Since the graphs do not coincide, this is not an identity. For example, let $x = \pi/8$.

$$\cos 4x = \cos 4\left(\frac{\pi}{8}\right) = 0 \text{ but } 2\cos 2x - 1 = 2\cos 2\left(\frac{\pi}{8}\right) - 1 = \sqrt{2} - 1$$

69. Graph both sides.

 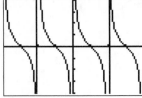

Since the graphs coincide, this could be an identity.

$$\frac{1 + \cos 2x}{\sin 2x} = \frac{1 + 2\cos^2 x - 1}{2\sin x \cos x}$$

$$= \frac{2\cos x \cos x}{2\sin x \cos x}$$

$$= \frac{\cos x}{\sin x} = \cot x$$

This is an identity.

71. Graph both sides.

 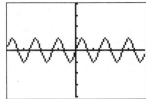

Since the graphs coincide, this could be an identity.

$$\sin 3x = \sin(2x + x)$$

$$= \sin 2x \cos x + \cos 2x \sin x$$

$$= 2\sin x \cos x \cos x + (1 - 2\sin^2 x)\sin x$$

$$= 2\sin x(1 - \sin^2 x) + (1 - 2\sin^2 x)\sin x$$

$$= \sin x(2 - 2\sin^2 x + 1 - 2\sin^2 x)$$

$$= \sin x(3 - 4\sin^2 x)$$

This is an identity.

73. Graph both sides.

 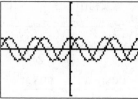

Since the graphs coincide, this is not an identity.
For example, let $x = 0$.

$$\cos 2x = \cos 2(0) = 1 \text{ but } \frac{2\tan x}{\sec^2 x} = \frac{2\tan 0}{\sec^2 0} = 0$$

75. Graph both sides.

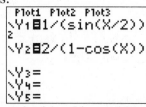 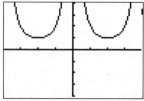

Since the graphs coincide, this could be an identity.

$$\csc^2\left(\frac{x}{2}\right) = \frac{1}{\sin^2\left(\frac{x}{2}\right)} = \frac{1}{\left(\pm\sqrt{\frac{1-\cos x}{2}}\right)^2} = \frac{1}{\frac{1-\cos x}{2}} = \frac{2}{1-\cos x}$$

This is an identity.

77. $\dfrac{\sin(5x) - \sin(3x)}{2\cos(4x)} = \dfrac{2\cos\left(\dfrac{8x}{2}\right)\sin\left(\dfrac{2x}{2}\right)}{2\cos(4x)} = \dfrac{2\cos(4x)\sin x}{2\cos(4x)} = \sin x.$

79. $\dfrac{\cos x + \cos(3x)}{2\cos(2x)} = \dfrac{2\cos 2x\cos(-x)}{2\cos(2x)} = \cos(-x) = \cos x$

81. $\dfrac{\sin(3x) - \sin x}{\cos x + \cos(3x)} = \dfrac{2\cos(2x)\sin x}{2\cos(2x)\cos(-x)} = \dfrac{\sin x}{\cos x} = \tan x$

83. $\dfrac{\sin x - \sin(3x)}{\cos x + \cos(3x)} = \dfrac{2\cos\left(\dfrac{x+3x}{2}\right)\sin\left(\dfrac{x-3x}{2}\right)}{2\cos\left(\dfrac{x+3x}{2}\right)\cos\left(\dfrac{x-3x}{2}\right)} = \dfrac{\sin(-x)}{\cos(-x)} = \dfrac{-\sin x}{\cos x} = -\tan x$

85. $\dfrac{\sin(4x)+\sin(6x)}{\cos(4x)-\cos(6x)} = \dfrac{2\sin\left(\dfrac{4x+6x}{2}\right)\cos\left(\dfrac{4x-6x}{2}\right)}{-2\sin\left(\dfrac{4x+6x}{2}\right)\sin\left(\dfrac{4x-6x}{2}\right)}$

$= \dfrac{\cos(-x)}{-\sin(-x)} = \dfrac{\cos x}{\sin x} = \cot x$

87. $\dfrac{\sin x+\sin y}{\cos x-\cos y} = \dfrac{2\sin\left(\dfrac{x+y}{2}\right)\cos\left(\dfrac{x-y}{2}\right)}{-2\sin\left(\dfrac{x+y}{2}\right)\sin\left(\dfrac{x-y}{2}\right)} = \dfrac{\cos\left(\dfrac{x-y}{2}\right)}{-\sin\left(\dfrac{x-y}{2}\right)} = -\cot\left(\dfrac{x-y}{2}\right)$

89. a. Use Strategy 6: Prove that $(1-\cos x)(1+\cos x) = \sin x\sin x$

$(1-\cos x)(1+\cos x) = 1-\cos^2 x = \sin^2 x = \sin x\sin x$

Therefore $\dfrac{1-\cos x}{\sin x} = \dfrac{\sin x}{1+\cos x}$

b. This follows precisely as stated.

91. $\cos\dfrac{\pi}{32} = \sqrt{\dfrac{1+\cos\dfrac{\pi}{16}}{2}} = \sqrt{\dfrac{1+\dfrac{\sqrt{2+\sqrt{2+\sqrt{2}}}}{2}}{2}}$

$= \sqrt{\dfrac{2}{2}\cdot\dfrac{1+\dfrac{\sqrt{2+\sqrt{2+\sqrt{2}}}}{2}}{2}} = \sqrt{\dfrac{2+\sqrt{2+\sqrt{2+\sqrt{2}}}}{4}}$

$= \dfrac{\sqrt{2+\sqrt{2+\sqrt{2+\sqrt{2}}}}}{2}$

7.4 Inverse Trigonometric Functions

1. Since $\sin\dfrac{\pi}{2} = 1$ and $-\dfrac{\pi}{2} \le \dfrac{\pi}{2} \le \dfrac{\pi}{2}$, $\sin^{-1}1 = \dfrac{\pi}{2}$.

3. Since $\tan\left(-\dfrac{\pi}{4}\right) = -1$ and $-\dfrac{\pi}{2} < -\dfrac{\pi}{4} < \dfrac{\pi}{2}$, $\tan^{-1}(-1) = -\dfrac{\pi}{4}$.

5. Since $\cos 0 = 1$ and $0 \le 0 \le \pi$, $\cos^{-1}1 = 0$.

7. Since $\tan\dfrac{\pi}{6} = \dfrac{\sqrt{3}}{3}$ and $-\dfrac{\pi}{2} < \dfrac{\pi}{6} < \dfrac{\pi}{2}$, $\tan^{-1}\left(\dfrac{\sqrt{3}}{3}\right) = \dfrac{\pi}{6}$.

9. Since $\sin\left(-\dfrac{\pi}{4}\right) = -\dfrac{\sqrt{2}}{2}$ and $-\dfrac{\pi}{2} \le -\dfrac{\pi}{4} \le \dfrac{\pi}{2}$, $\sin^{-1}\left(-\dfrac{\sqrt{2}}{2}\right) = -\dfrac{\pi}{4}$.

11. Since $\tan\left(-\dfrac{\pi}{3}\right) = -\sqrt{3}$ and $-\dfrac{\pi}{2} < -\dfrac{\pi}{3} < \dfrac{\pi}{2}$, $\tan^{-1}\left(-\sqrt{3}\right) = -\dfrac{\pi}{3}$.

13. Since $\cos\left(\dfrac{2\pi}{3}\right) = -\dfrac{1}{2}$ and $0 \le \dfrac{2\pi}{3} \le \pi$, $\cos^{-1}\left(-\dfrac{1}{2}\right) = \dfrac{2\pi}{3}$.

15. $.3576$ 17. -1.2728

19. $\sin^{-1}(\sin 7) = \sin^{-1}(.65698\ldots) = .7168$

21. $\tan^{-1}(\tan(-4)) = \tan^{-1}(-1.1578\ldots) = -.8584$

23. $\cos^{-1}(\cos(-8.5)) = \cos^{-1}(-.602011\ldots) = 2.2168$

25. Since $u = -\dfrac{\pi}{3}$, $\cos u = \cos\left(-\dfrac{\pi}{3}\right) = \dfrac{1}{2}$ and $\tan u = \tan\left(-\dfrac{\pi}{3}\right) = -\sqrt{3}$.

27. $\sin^{-1}(\cos 0) = \sin^{-1}1 = \dfrac{\pi}{2}$ 29. $\cos^{-1}\left(\sin\dfrac{4\pi}{3}\right) = \cos^{-1}\left(-\dfrac{\sqrt{3}}{2}\right) = \dfrac{5\pi}{6}$

31. $\sin^{-1}\left(\cos\dfrac{7\pi}{6}\right) = \sin^{-1}\left(-\dfrac{\sqrt{3}}{2}\right) = -\dfrac{\pi}{3}$ 33. $\sin^{-1}\left(\sin\dfrac{2\pi}{3}\right) = \sin^{-1}\left(\dfrac{\sqrt{3}}{2}\right) = \dfrac{\pi}{3}$

35. $\cos^{-1}\left(\cos\left(-\dfrac{\pi}{6}\right)\right) = \cos^{-1}\left(\dfrac{\sqrt{3}}{2}\right) = \dfrac{\pi}{6}$

37. Consider an angle of u radians in standard position whose x coordinate is 3 and whose distance from the origin is 5. By the point in the plane description, $\cos u = \dfrac{3}{5}$. Since $0 < u < \pi$, $u = \cos^{-1}\dfrac{3}{5}$. Thus,

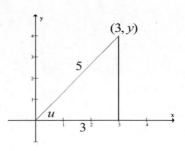

$\sin\left[\cos^{-1}\dfrac{3}{5}\right] = \sin u = \dfrac{y}{5}$. Using the picture to the right, the ordered pair forms a right triangle. We use Pythagorean theorem to find $y = 4$.

$$\sin\left[\cos^{-1}\dfrac{3}{5}\right] = \sin u = \dfrac{4}{5}.$$

39. Consider an angle of u radians in standard position whose y coordinate is -3 and whose x coordinate is 4.

By the point in the plane description, $\tan u = -\dfrac{3}{4}$.

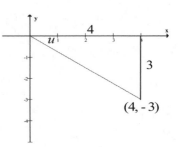

Since $-\dfrac{\pi}{2} < u < \dfrac{\pi}{2}$, $u = \tan^{-1}\left(-\dfrac{3}{4}\right)$.

Thus, $\cos\left[\tan^{-1}\left(-\dfrac{3}{4}\right)\right] = \cos u = \dfrac{4}{r}$. Using the picture to the right, the ordered pair forms a right triangle. We use Pythagorean theorem to find $r = 5$.

$$\cos\left[\tan^{-1}-\dfrac{3}{4}\right] = \cos u = \dfrac{4}{5}.$$

41. Consider an angle of u radians in standard position whose x coordinate is 5 and whose distance from the origin is 13. By the point in the plane description, $\cos u = \dfrac{5}{13}$. Since $0 < u < \pi$, $u = \cos^{-1}\dfrac{5}{13}$. Thus,

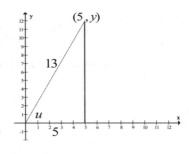

$\tan\left[\cos^{-1}\dfrac{5}{13}\right] = \tan u = \dfrac{y}{5}$. Using the picture to the right, the ordered pair forms a right triangle. We use Pythagorean theorem to find $y = 12$.

$$\tan\left[\cos^{-1}\dfrac{5}{13}\right] = \tan u = \dfrac{12}{5} = 2.4.$$

43. Consider an angle of u radians in standard position whose y coordinate is $\sqrt{3}$ and whose distance from the origin is 5. By the point in the plane description, $\sin u = \dfrac{\sqrt{3}}{5}$. Since $-\dfrac{\pi}{2} < u < \dfrac{\pi}{2}$, $u = \sin^{-1} \dfrac{\sqrt{3}}{5}$.

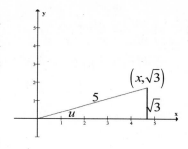

Thus, $\cos\left[\sin^{-1}\dfrac{\sqrt{3}}{5}\right] = \cos u = \dfrac{x}{5}$. Using the picture to the right, the ordered pair forms a right triangle. We use Pythagorean theorem to find $x = \sqrt{22}$.

$$\cos\left[\sin^{-1}\dfrac{\sqrt{3}}{5}\right] = \cos u = \dfrac{\sqrt{22}}{5}.$$

45. Consider an angle of u radians in standard position whose x coordinate is 3 and whose distance from the origin is $\sqrt{13}$. By the point in the plane description, $\cos u = \dfrac{3}{\sqrt{13}}$. Since $0 < u < \pi$, $u = \cos^{-1}\dfrac{3}{\sqrt{13}}$. Thus,

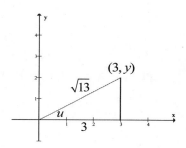

$\sin\left[\cos^{-1}\dfrac{3}{\sqrt{13}}\right] = \sin u = \dfrac{y}{\sqrt{13}}$. Using the picture to the right, the ordered pair forms a right triangle. We use Pythagorean theorem to find $y = 2$.

$$\sin\left[\cos^{-1}\dfrac{3}{\sqrt{13}}\right] = \sin u = \dfrac{2}{\sqrt{13}}.$$

47. Consider an angle of u radians in standard position whose y coordinate is $\sqrt{5}$ and whose x coordinate is 10. By the point in the plane description, $\tan u = \dfrac{\sqrt{5}}{10}$.

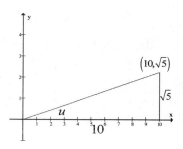

Since $-\dfrac{\pi}{2} < u < \dfrac{\pi}{2}$, $u = \tan^{-1}\dfrac{\sqrt{5}}{10}$.

Thus, $\sin\left[\tan^{-1}\dfrac{\sqrt{5}}{10}\right] = \sin u = \dfrac{\sqrt{5}}{r}$. Using the picture to the right, the ordered pair forms a right triangle. We use Pythagorean theorem to find $r = \sqrt{105}$.

$$\sin\left[\tan^{-1}\dfrac{\sqrt{5}}{10}\right] = \sin u = \dfrac{\sqrt{5}}{\sqrt{105}} = \sqrt{\dfrac{5}{105}} = \sqrt{\dfrac{1}{21}} = \dfrac{1}{\sqrt{21}}.$$

49. $\sin^{-1}v = u$, where $\sin u = v$ and $-\dfrac{\pi}{2} \le u \le \dfrac{\pi}{2}$. Hence,

$\cos\!\left(\sin^{-1}v\right) = \cos u = \sqrt{1-\sin^2 u} = \sqrt{1-v^2}\quad (-1 \le v \le 1).$

51. $\sin^{-1}v = u$, where $\sin u = v$ and $-\dfrac{\pi}{2} \le u \le \dfrac{\pi}{2}$. Hence,

$\tan\!\left(\sin^{-1}v\right) = \tan u = \dfrac{\sin u}{\cos u} = \dfrac{\sin u}{\sqrt{1-\sin^2 u}} = \dfrac{v}{\sqrt{1-v^2}}\quad (-1 < v < 1).$

53. $\tan^{-1}v = u$, where $\tan u = v$ and $-\dfrac{\pi}{2} \le u \le \dfrac{\pi}{2}$. Hence,

$\cos\!\left(\tan^{-1}v\right) = \cos u = \dfrac{1}{\sqrt{1+v^2}}.$

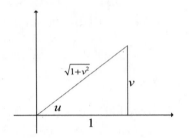

55. Let $\cos^{-1}v = u$,   Then $v = \cos u$, $\sqrt{1-v^2} = \sin u$, $\sin 2u = 2\sin u \cos u = 2v\sqrt{1-v^2}$

$(-1 \le v \le 1).$

57. By Exercise 50, we have $\tan\!\left(\cos^{-1}v\right) = \dfrac{\sqrt{1-v^2}}{v}.$

Starting from the right hand side, $\sin^{-1}v = u$, where $\sin u = v$ and $-\dfrac{\pi}{2} \le u \le \dfrac{\pi}{2}$.

Hence, $\cot\!\left(\sin^{-1}v\right) = \cot u = \dfrac{\cos u}{\sin u} = \dfrac{\sqrt{1-\sin^2 u}}{v} = \dfrac{\sqrt{1-v^2}}{v}$ when

$(-1 \le v < 0 \text{ or } 0 < v \le 1).$

Thus, $\tan\!\left(\cos^{-1}v\right) = \cot\!\left(\sin^{-1}v\right)$ when $(-1 \le v < 0 \text{ or } 0 < v \le 1)$

59.

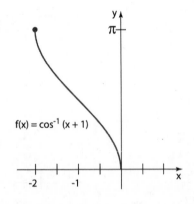

$f(x) = \cos^{-1}(x+1)$

61.

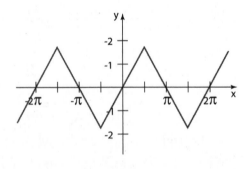

63. a. $V = V_{\max} \sin 2\pi ft$

$\sin 2\pi ft = \dfrac{V}{V_{\max}}$

$2\pi ft = \sin^{-1}\dfrac{V}{V_{\max}} + 2n\pi \quad (n \text{ any integer})$

$t = \dfrac{1}{2\pi f}\sin^{-1}\dfrac{V}{V_{\max}} + \dfrac{n}{f}$

b. The smallest positive value of t occurs when $n = 0$.

$\dfrac{1}{2\pi f}\sin^{-1}\dfrac{V}{V_{\max}} = \dfrac{1}{2\pi \cdot 120}\sin^{-1}\dfrac{8.5}{20} = 5.8219 \times 10^{-4}$ sec

65. a. $\theta = \sin^{-1}(40/x)$ **b.** $\theta = \sin^{-1}(40/250) = .16$ radians or $9.2°$

67. a. $t = \tan^{-1}\left(\dfrac{h}{4}\right)$

b. $h = .25$ mile; $t = \tan^{-1}\left(\dfrac{.25}{4}\right) = .0624$ radians.

$h = 1$ mile; $t = \tan^{-1}\left(\dfrac{1}{4}\right) = .2450$ radians.

$h = 2$ miles; $t = \tan^{-1}\left(\dfrac{2}{4}\right) = .4636$ radians.

c. $\dfrac{h}{4} = \tan(.4)$

$x = 4\tan(.4)$

$x \approx 1.6912$ miles

69. a.

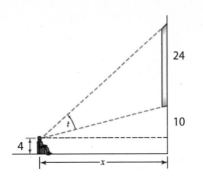
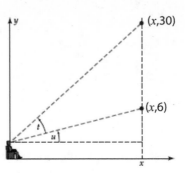

Imagine that the x-axis is at your eye level with your eye at the origin. (see the picture above) Since your eye is 4 feet above the ground, the bottom of the screen is 6 feet above your eye level. The point in the plane description shows that $\tan u = \dfrac{6}{x}$ and $\tan(u+t) = \dfrac{30}{x}$. Hence $u = \tan^{-1}\dfrac{6}{x}$ and $u + t = \tan^{-1}\dfrac{30}{x}$. So that

$$t = \tan^{-1}\frac{30}{x} - \tan^{-1}\frac{6}{x}$$

b. We graph the function and use the maximum feature to determine that t is largest when $x \approx 13.4$ ft.

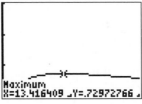

71. a. Let $\alpha =$ the angle formed by the road and the left edge of the light beam. Then $\tan\alpha = 10/x;\ \tan(\alpha+\theta) = (10+15)/x$.

Hence $\alpha = \tan^{-1}\dfrac{10}{x}$, $\alpha + \theta = \tan^{-1}\dfrac{25}{x}$, and $\theta = \tan^{-1}\dfrac{25}{x} - \tan^{-1}\dfrac{10}{x}$.

b. Graphing, we find that θ has maximum value at $x = 15.8$ ft.

73. The graph of $y = \csc(x)$ on this restricted domain is shown on the left. Since the graph passes the horizontal line test, we may conclude that the function is one-to-one and therefore has an inverse function. The graph of the inverse is shown on the right.

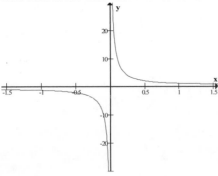
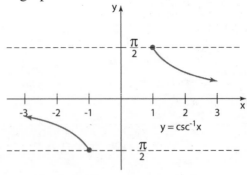

75. Let $\cos u = w$, $0 \le u \le \pi$. Then $u = \cos^{-1} w$, $\cos^{-1}(\cos u) = \cos^{-1} w = u$.
Let $u = \cos^{-1} v$, $\cos u = v$, $\cos(\cos^{-1} v) = \cos u = v$.

77. Let $u = \sin^{-1}(-x)$. Then

$$-x = \sin u \qquad -\pi/2 \le -u \le \pi/2$$

$$x = -\sin u$$

$$x = \sin(-u)$$

$$\sin^{-1} x = -u \qquad -\pi/2 \le -u \le \pi/2$$

$$\sin^{-1} x = -\sin^{-1}(-x)$$

$$\sin^{-1}(-x) = -\sin^{-1} x$$

79. Let $u = \cos^{-1}(-x)$. Then

$$-x = \cos u \qquad 0 \le u \le \pi, \text{ hence } -\pi \le -u \le 0, \ 0 \le \pi - u \le \pi$$

$$x = -\cos u$$

$$x = \cos(\pi - u) \quad \text{(from the stated identity)}$$

$$\pi - u = \cos^{-1} x$$

$$u = \pi - \cos^{-1} x$$

$$\cos^{-1}(-x) = \pi - \cos^{-1} x$$

81. Let $u = \cot x$. Then

$$\tan^{-1}(\pi/2 - x) = \cot x = u \qquad 0 < u < \pi, \text{ hence } -\pi/2 < \pi/2 - u < \pi/2$$

$$\pi/2 - x = \tan^{-1}(\cot x)$$

83. Let $u = \sin^{-1} x$. Then $\sin u = x \qquad -\pi/2 < u < \pi/2$

$$\tan u = \frac{\sin u}{\cos u} = \frac{\sin u}{\sqrt{1 - \sin^2 u}} = \frac{x}{\sqrt{1 - x^2}}$$

(positive square root, since u lies in quadrant I or IV)

$$u = \tan^{-1}\left(\frac{x}{\sqrt{1-x^2}}\right) \ ; \sin^{-1} x = \tan^{-1}\left(\frac{x}{\sqrt{1-x^2}}\right)$$

85. The statement is false. For example, let $x = 1/2$.

$$\tan^{-1}\frac{1}{2} \approx .4636 \text{ but } \frac{\sin^{-1}(1/2)}{\cos^{-1}(1/2)} = \frac{\pi/6}{\pi/3} = \frac{1}{2}$$

7.5 Trigonometric Equations

1. $\sin x = .465$. Since $\sin^{-1}(.465) = .4836$, and $\pi - (.4836) = 2.6580$, all solutions can be written as $.4836 + 2k\pi$ and $2.6580 + 2k\pi$.

3. $\cos x = -.564$. Since $\cos^{-1}(-.564) = 2.1700$, all solutions can be written as $2.1700 + 2k\pi$ and $-2.1700 + 2k\pi$.

5. $\tan x = -.354$. Since $\tan^{-1}(-.354) = -.3402$, all solutions can be written as $-.3402 + k\pi$.

7. $\cot x = 2.3$

$\dfrac{1}{\tan x} = 2.3$

$\tan x = \dfrac{1}{2.3}$

Since $\tan^{-1}\left(\frac{1}{2.3}\right) = .4101$, all solutions can be written as $.4101 + k\pi$.

9. $\sec x = -1.6$

$\dfrac{1}{\cos x} = -1.6$

$\cos x = -\dfrac{1}{1.6}$

Since $\cos^{-1}\left(-\frac{1}{1.6}\right) = 2.2459$, all solutions can be written as $2.2459 + 2k\pi$ and $-2.2459 + 2k\pi$.

11. $\sin x = .119$. Since $\sin^{-1}(.119) = .1193$, and $\pi - (.1193) = 3.0223$, these are all solutions in $[0, 2\pi)$.

13. $\tan x = 4$. Since $\tan^{-1} 4 = 1.3258$, and $1.3258 + \pi = 4.4674$, these are all solutions in $[0, 2\pi)$.

15. $\sin x = \dfrac{\sqrt{3}}{2}$. Since $\sin^{-1}\left(\dfrac{\sqrt{3}}{2}\right) = \dfrac{\pi}{3}$, and $\pi - \dfrac{\pi}{3} = \dfrac{2\pi}{3}$, all solutions can be written as $\dfrac{\pi}{3} + 2k\pi$ and $\dfrac{2\pi}{3} + 2k\pi$.

17. $\tan x = -\sqrt{3}$. Since $\tan^{-1}\left(-\sqrt{3}\right) = -\pi/3$, all solutions can be written as $-\pi/3 + k\pi$.

19. $2\cos x = -\sqrt{3}$

$\cos x = -\sqrt{3}/2$. Since $\cos^{-1}\left(-\sqrt{3}/2\right) = 5\pi/6$, all solutions can be written as $5\pi/6 + 2k\pi$ and $-5\pi/6 + 2k\pi$.

21. $2\sin x + 1 = 0$, $\sin x = -1/2$. Since $\sin^{-1}(-1/2) = -\pi/6$, and $\pi - (-\pi/6) = 7\pi/6$, all solutions can be written as $-\dfrac{\pi}{6} + 2k\pi$ and $\dfrac{7\pi}{6} + 2k\pi$.

23. $\csc x = 2$,

$\dfrac{1}{\sin x} = 2$

$\sin x = \dfrac{1}{2}$

$x = \dfrac{\pi}{6} + 2k\pi$ and $x = \pi - \left(\dfrac{\pi}{6}\right) + 2k\pi = \dfrac{5\pi}{6} + 2k\pi$

25. $\tan\theta = 7.95$. Since $\tan^{-1}(7.95) = 82.8°$, the solutions in the required interval are $82.8°$ and $82.8° + 180° = 262.8°$.

27. $\cos\theta = -.42$. Since $\cos^{-1}(-.42) = 114.8°$, the solutions in the required interval are $114.8°$ and $360° - 114.8° = 245.2°$.

29. $2\sin^2\theta + 3\sin\theta + 1 = 0$
$(2\sin\theta + 1)(\sin\theta + 1) = 0$

$\sin\theta = -1/2 \qquad \sin\theta = -1$
$\theta = 210°, 330° \qquad \theta = 270°$

31. $\tan^2\theta - 3 = 0$
$\tan\theta = \pm\sqrt{3}$
If $\tan\theta = \sqrt{3}$, $\theta = 60°$ or $240°$.
If $\tan\theta = -\sqrt{3}$, $\theta = 120°$ or $300°$.

33. $4\cos^2\theta + 4\cos\theta + 1 = 0$
$(2\cos\theta + 1)^2 = 0$
$2\cos\theta + 1 = 0$
$\cos\theta = -1/2, \quad \theta = 120°, 240°$

Calculator in degree mode for Exercises **35-41**

35. $m = 1.1$

$\sin\alpha = \dfrac{1}{1.1}$

$\alpha = \sin^{-1}\dfrac{1}{1.1}$

$\alpha = 65.4°$

37. $m = 2$

$\sin\alpha = \dfrac{1}{2}$

$\alpha = \sin^{-1}\dfrac{1}{2}$

$\alpha = 30°$

39. Use $\dfrac{\sin\theta_1}{\sin\theta_2}=\dfrac{v_1}{v_2}$

with $\dfrac{v_1}{v_2}=1.33,\ \theta_1=38°.$

$\dfrac{\sin 38°}{\sin\theta_2}=1.33$

$\sin\theta_2=\dfrac{\sin 38°}{1.33}$

$\theta_2=\sin^{-1}\!\left(\dfrac{\sin 38°}{1.33}\right)$

$\theta_2=27.6°$

41. Use $\dfrac{\sin\theta_1}{\sin\theta_2}=\dfrac{v_1}{v_2}$

with $\dfrac{v_1}{v_2}=1.66,\ \theta_1=24°.$

$\dfrac{\sin 24°}{\sin\theta_2}=1.66$

$\sin\theta_2=\dfrac{\sin 24°}{1.66}$

$\theta_2=\sin^{-1}\!\left(\dfrac{\sin 24°}{1.66}\right)$

$\theta_2=14.2°$

43. $\sin 2x=-\dfrac{\sqrt3}{2},\quad 2x=-\dfrac{\pi}{3}+2k\pi\ \text{and}\ 2x=\pi-\left(-\dfrac{\pi}{3}\right)+2k\pi=\dfrac{4\pi}{3}+2k\pi$

$x=-\dfrac{\pi}{6}+k\pi\qquad x=\dfrac{2\pi}{3}+k\pi$

45. $2\cos\dfrac{x}{2}=\sqrt2,\ \cos\dfrac{x}{2}=\dfrac{\sqrt2}{2}$

$\dfrac{x}{2}=\dfrac{\pi}{4}+2k\pi\quad\text{and}\quad\dfrac{x}{2}=-\dfrac{\pi}{4}+2k\pi$

$x=\dfrac{\pi}{2}+4k\pi\qquad x=-\dfrac{\pi}{2}+4k\pi$

47. $\tan 3x=-\sqrt3$

$3x=-\dfrac{\pi}{3}+k\pi$

$x=-\dfrac{\pi}{9}+\dfrac{k\pi}{3}$

49. $5\cos 3x=-3,\ \cos 3x=-\dfrac{3}{5}$

$3x=2.2143+2k\pi\ \text{and}\ 3x=-2.2143+2k\pi$

$x=.7381+2k\dfrac{\pi}{3}\qquad x=-.7381+\dfrac{2k\pi}{3}$

51. $4\tan\dfrac{x}{2}=8,$

$\tan\dfrac{x}{2}=2$

$\dfrac{x}{2}=1.10171+k\pi$

$x=2.2143+2k\pi$

53. $8=2(5)\sin\dfrac{t}{2}$

$\dfrac{4}{5}=\sin\dfrac{t}{2}$

$\sin^{-1}\dfrac{4}{5}=\dfrac{t}{2}$

$t=2\sin^{-1}\dfrac{4}{5}=1.8546$

$A=\dfrac{5^2}{2}(1.8546-\sin 1.8546)$

$=11.18$

55.
$$1.5 = 2(1)\sin\frac{t}{2}$$
$$\frac{1.5}{2} = \sin\frac{t}{2}$$
$$\sin^{-1}\frac{1.5}{2} = \frac{t}{2}$$
$$t = 2\sin^{-1}\frac{1.5}{2} = 1.6961$$
$$A = \frac{1^2}{2}(1.6961 - \sin 1.6961)$$
$$= .35$$

57. $50 = \dfrac{10^2}{2}(t - \sin t)$
$$50 = 50(t - \sin t)$$
$$1 = t - \sin t$$
$$0 = t - \sin t - 1$$
Enter $y = t - \sin t - 1$ in a graphing calculator and use the root finding feature to find that $t = 1.9346$.
$$L = 2(10)\sin\left(\frac{1.9346}{2}\right) = 16.47.$$

59. $20 = \dfrac{8^2}{2}(t - \sin t)$
$$20 = 32(t - \sin t)$$
$$\frac{20}{32} = t - \sin t$$
$$0 = t - \sin t - \frac{5}{8}$$
Enter $y = t - \sin t - \dfrac{5}{8}$ in a graphing calculator and use the root finding feature to find that $t = 1.6236$.
$$L = 2(8)\sin\left(\frac{1.6236}{2}\right) = 16.236.$$

61. a. $A = 2x(2\cos 2x) = 4x\cos 2x$.

b. $1 = 4x\cos 2x$
$$0 = 4x\cos 2x - 1$$
Enter $y = 4x\cos 2x - 1$ in a graphing calculator and use the root finding feature to find that $x = .3050$ and $x = .5490$ for $0 < x < 2$.

63. $3\sin^2 x - 8\sin x - 3 = 0$
$(3\sin x + 1)(\sin x - 3) = 0$
$\sin x = -\frac{1}{3}$ $\sin x = 3$
$\sin^{-1}(-\frac{1}{3}) = -.3398$ impossible
The two solutions in $[0, 2\pi)$ are
$-.3398 + 2\pi = 5.9433$ and
$\pi - (-.3398) = 3.4814$

65. $2\tan^2 x + 7\tan x + 5 = 0$
$(2\tan x + 5)(\tan x + 1) = 0$
$\tan x = -\frac{5}{2}$ $\tan x = -1$
$\tan^{-1}(-\frac{5}{2}) = 1.9513$ $\tan^{-1}(-1) = \frac{3\pi}{4}$
The solutions in $[0, 2\pi)$ are 1.9513 and
$1.9513 + \pi = 5.0929$
$\dfrac{3\pi}{4}$ and $\dfrac{3\pi}{4} + \pi = \dfrac{7\pi}{4}$

67. $\cot x \cos x = \cos x$

$\cot x \cos x - \cos x = 0$

$\cos x (\cot x - 1) = 0$

$\cos x = 0 \qquad \cot x = 1$

$x = \frac{\pi}{2}, \frac{3\pi}{2} \qquad x = \frac{\pi}{4}, \frac{5\pi}{4}$

69. $\cos x \csc x = 2 \cos x$

$\cos x (\csc x - 2) = 0$

$\cos x = 0 \qquad \csc x = 2$

$\qquad\qquad\qquad \sin x = 1/2$

$x = \frac{\pi}{2}, \frac{3\pi}{2} \qquad x = \frac{\pi}{6}, \frac{5\pi}{6}$

71. $4 \sin x \tan x - 3 \tan x + 20 \sin x - 15 = 0$

$\tan x (4 \sin x - 3) + 5 (4 \sin x - 3) = 0$

$(4 \sin x - 3)(\tan x + 5) = 0$

$\sin x = \frac{3}{4} \qquad\qquad \tan x = -5$

$\sin^{-1}\left(\frac{3}{4}\right) = .8481 \qquad \tan^{-1}(-5) = -1.3734$

The solutions in $[0, 2\pi)$ are $.8481$,

$\pi - .8481 = 2.2935$,

$-1.3734 + \pi = 1.7682$,

and $-1.3734 + 2\pi = 4.9098$.

73. $\sin^2 x + 2 \sin x - 2 = 0$

$\sin x = \dfrac{-2 \pm \sqrt{12}}{2}$

$\sin x = \dfrac{-2 - \sqrt{12}}{2} \quad \text{or} \quad \sin x = \dfrac{-2 + \sqrt{12}}{2}$

$\text{impossible} \qquad\qquad \sin^{-1}\left(\dfrac{-2 + \sqrt{12}}{2}\right)$

$\qquad\qquad\qquad\qquad = .8213$

The solutions in

$[0, 2\pi)$ are $.8213$, $\pi - .8213 = 2.3203$.

75. $\tan^2 x + 1 = 3 \tan x$

$\tan^2 x - 3 \tan x + 1 = 0$

$\tan x = \dfrac{3 \pm \sqrt{5}}{2}$

$\tan^{-1}\left(\dfrac{3 + \sqrt{5}}{2}\right) = 1.2059 \quad \tan^{-1}\left(\dfrac{3 - \sqrt{5}}{2}\right) = .3649$

The solutions in $[0, 2\pi)$ are 1.2059, $1.2059 + \pi = 4.3475$, $.3649$,

and $.3649 + \pi = 3.5065$.

77. $2\tan^2 x - 1 = 3\tan x$

$2\tan^2 x - 3\tan x - 1 = 0$

$\tan x = \dfrac{3 \pm \sqrt{17}}{4}$

$\tan^{-1}\left(\dfrac{3+\sqrt{17}}{4}\right) = 1.0591 \quad \tan^{-1}\left(\dfrac{3-\sqrt{17}}{4}\right) = -.2737$

The solutions in $[0, 2\pi)$ are $1.0591, 1.0591 + \pi = 4.2007, -.2737 + \pi = 2.8679$, and $-.2737 + 2\pi = 6.0095$.

79. $\sin^2 x + 3\cos^2 x = 0$ can have no solution, since the sum of two squares of non-zero numbers is never zero, and $\sin x$ and $\cos x$ are never both zero.

or

$\sin^2 x + 3\cos^2 x = 0 \Rightarrow \sin^2 x + \cos^2 x + 2\cos^2 x = 0 \Rightarrow 1 + 2\cos^2 x = 0$

$2\cos^2 x = -1 \Rightarrow \cos^2 x = -1/2 \Rightarrow \cos x = \pm\sqrt{-1/2}$

No real solution, thus there is no solution.

81. $\sin 2x + \cos x = 0$

$2\sin x \cos x + \cos x = 0$

$\cos x(2\sin x + 1) = 0$

$\cos x = 0 \qquad \sin x = -1/2$

$x = \dfrac{\pi}{2}, \dfrac{3\pi}{2} \qquad x = \dfrac{7\pi}{6}, \dfrac{11\pi}{6}$

83. $9 - 12\sin x = 4\cos^2 x$

$9 - 12\sin x = 4\left(1 - \sin^2 x\right)$

$4\sin^2 x - 12\sin x + 5 = 0$

$(2\sin x - 5)(2\sin x - 1) = 0$

$\sin x = 5/2 \qquad \sin x = 1/2$

impossible $\qquad x = \dfrac{\pi}{6}, \dfrac{5\pi}{6}$

85. $\cos^2 x - \sin^2 x + \sin x = 0$

$1 - \sin^2 x - \sin^2 x + \sin x = 0$

$2\sin^2 x - \sin x - 1 = 0$

$(2\sin x + 1)(\sin x - 1) = 0$

$\sin x = -1/2 \qquad \sin x = 1$

$x = \dfrac{7\pi}{6}, \dfrac{11\pi}{6} \qquad x = \dfrac{\pi}{2}$

87. $\sin\dfrac{x}{2} = 1 - \cos x$

$\sin\dfrac{x}{2} = 1 - \left(1 - 2\sin^2\dfrac{x}{2}\right)$

$2\sin^2\dfrac{x}{2} - \sin\dfrac{x}{2} = 0$

$\sin\dfrac{x}{2}\left(2\sin\dfrac{x}{2} - 1\right) = 0$

$\sin\dfrac{x}{2} = 0 \qquad \sin\dfrac{x}{2} = \dfrac{1}{2}$

$\dfrac{x}{2} = 0 \qquad \dfrac{x}{2} = \dfrac{\pi}{6}, \dfrac{5\pi}{6}$

$x = 0 \qquad x = \dfrac{\pi}{3}, \dfrac{5\pi}{3}$

89. Graph the function $f(x) = 4\sin 2x - 3\cos 2x - 2$ using the window $0 \le x \le 2\pi$, $-8 \le y \le 3$ and apply the zero (root) routine to obtain $x = .5275$ and $x = 1.6868$. Since the function is periodic with period π, all solutions are given by $x = .5275 + k\pi$ and $x = 1.6868 + k\pi$, where k is any integer.

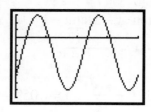

91. Graph the function $f(x) = 3\sin^3 2x - 2\cos x$ using the window $0 \le x \le 2\pi$, $-5 \le y \le 5$ and apply the zero (root) routine to obtain $x = .4959$, 1.2538, 1.5708, 1.8877, 2.6457, and 4.7124. Since the function is periodic with period 2π, all solutions are given by $x = .4959 + 2k\pi$, where k is any integer, and so on.

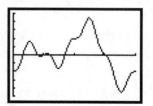

93. Graph the function $f(x) = \tan x + 5\sin x - 1$ using the window $0 \le x \le 2\pi$, $-5 \le y \le 5$, ignoring the erroneous vertical lines at the asymptotes, and apply the zero (root) routine to obtain $x = .1671$, 1.8256, 2.8867, and 4.5453. Since the function is periodic with period 2π, all solutions are given by $x = .1671 + 2k\pi$, where k is any integer, and so on.

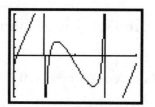

95. Graph the function $f(x) = \cos^3 x - 3\cos x + 1$ using the window $0 \le x \le 2\pi$, $-2 \le y \le 4$ and apply the zero (root) routine to obtain $x = 1.2161$ and $x = 5.0671$. Since the function is periodic with period 2π, all solutions are given by $x = 1.2161 + 2k\pi$ and $x = 5.0671 + 2k\pi$, where k is any integer.

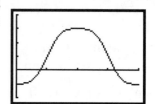

97. Graph the function $f(x) = \cos^4 x - 3\cos^3 x + \cos x - 1$ using the window $0 \le x \le 2\pi$, $-3 \le y \le 3$ and apply the zero (root) routine to obtain $x = 2.4620$ and $x = 3.8212$. Since the function is periodic with period 2π, all solutions are given by $x = 2.4620 + 2k\pi$ and $x = 3.8212 + 2k\pi$, where k is any integer.

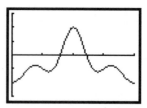

99. Graph the function $f(x) = \sin^3 x + 2\sin^2 x - 3\cos x + 2$ using the window $0 \le x \le 2\pi$, $-2 \le y \le 6$ and apply the zero (root) routine to obtain $x = .5166$ and $x = 5.6766$. Since the function is periodic with period 2π, all solutions are given by $x = .5166 + 2k\pi$ and $x = 5.6766 + 2k\pi$, where k is any integer.

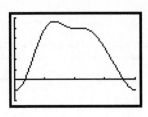

101.a. Graph $d(x) = 3\sin\left(\dfrac{2\pi}{365}(x - 80)\right) + 12$

and the horizontal line $y = 11$ and use the intersection routine twice to obtain days 60 and 282 (an algebraic solution is also possible, with the same result.)

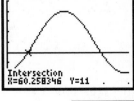 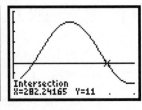

b. Graph $d(x) = 3\sin\left(\dfrac{2\pi}{365}(x - 80)\right) + 12$

and use the maximum routine to obtain day 171.

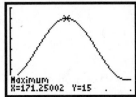

103. Use $d = \dfrac{v^2}{32}\sin 2\theta$ with

$d = 2500, v = 300.$

$$2500 = \frac{300^2}{32}\sin 2\theta$$

$$\sin 2\theta = \frac{2500 \cdot 32}{300^2} = \frac{8}{9}$$

$2\theta = 1.0949$ or $\pi - 1.0949$

$\theta = .5475$ or 1.0233

105. Use $d = \dfrac{v^2}{32}\sin 2\theta$ with

$d = 288, v = 98 \times \dfrac{5280}{3600}\left(\dfrac{\text{ft}}{\text{sec}} \div \dfrac{\text{mi}}{\text{hr}}\right).$

$$288 = \frac{1}{32}\left(98 \times \frac{5280}{3600}\right)^2 \sin 2\theta$$

$$\sin 2\theta = 288 \cdot 32 \div \left(98 \times \frac{5280}{3600}\right)^2$$

$$= .4461$$

$2\theta = 26.49°$ or $180° - 26.49°$

$\theta = 13.25°$ or $76.75°$

107. The basic equations $\sin x = c$ and $\cos x = c$ have no solutions if $c > 1$ or $c < -1$.

109. In dividing both sides by $\sin x$, the solutions such as 0 and π which correspond to $\sin x = 0$ are lost.

Chapter 7 Review Exercises

1. $$\frac{\sin^2 t + \left(\tan^2 t + 2\tan t - 4\right) + \cos^2 t}{3\tan^2 t - 3\tan t} = \frac{\tan^2 t + 2\tan t - 4 + 1}{3\tan^2 t - 3\tan t} = \frac{\tan^2 t + 2\tan t - 3}{3\tan^2 t - 3\tan t}$$

$$= \frac{(\tan t + 3)(\tan t - 1)}{3\tan t(\tan t - 1)} = \frac{\tan t + 3}{3\tan t} = \tfrac{1}{3} + \cot t$$

3. $$\frac{\tan^2 x - \sin^2 x}{\sec^2 x} = \left(\frac{\sin^2 x}{\cos^2 x} - \sin^2 x\right) \div \frac{1}{\cos^2 x}$$

$$= \frac{\sin^2 x - \sin^2 x \cos^2 x}{1}$$

$$= \sin^2 x\left(1 - \cos^2 x\right)$$

$$= \sin^2 x \sin^2 x = \sin^4 x$$

5. This is an identity.
$$\sin^4 t - \cos^4 t = \left(\sin^2 t + \cos^2 t\right)\left(\sin^2 t - \cos^2 t\right)$$
$$= 1\left(\sin^2 t - \left(1 - \sin^2 t\right)\right) = \sin^2 t - 1 + \sin^2 t = 2\sin^2 t - 1$$

7. This is an identity. Use Strategy 6. Prove
$\sin t \sin t = (1 - \cos t)(1 + \cos t)$ is an identity.
$\sin t \sin t = \sin^2 t = 1 - \cos^2 t$

$$= (1 - \cos t)(1 + \cos t)$$

Therefore $\dfrac{\sin t}{1 - \cos t} = \dfrac{1 + \cos t}{\sin t}$.

9. This is not an identity.
For example, let $t = \pi/6$.
$$\frac{\cos^2(\pi + \pi/6)}{\sin^2(\pi + \pi/6)} - 1 = 2$$

but $\dfrac{1}{\sin^2(\pi/6)} = 4.$

11. This is an identity.
$$(\sin x + \cos x)^2 - \sin 2x = \sin^2 x + 2\sin x \cos x + \cos^2 x - \sin 2x$$

$$= \sin^2 x + \cos^2 x + 2\sin x \cos x - 2\sin x \cos x$$

$$= 1 + 0 = 1$$

13. $$\frac{\tan x - \sin x}{2\tan x} = \frac{\sin x/\cos x - \sin x}{2(\sin x/\cos x)}$$

$$= \frac{\sin x - \sin x \cos x}{2\sin x}$$

$$= \frac{\sin x(1 - \cos x)}{2\sin x}$$

$$= \frac{1 - \cos x}{2}$$

$$= \sin^2(x/2)$$

15. $\cos(x+y)\cos(x-y) = (\cos x \cos y - \sin x \sin y)(\cos x \cos y + \sin x \sin y)$

$$= \cos x^2 \cos^2 y - \sin^2 x \sin^2 y$$

$$= \cos^2 x(1 - \sin^2 y) - (1 - \cos^2 x)\sin^2 y$$

$$= \cos^2 x - \cos^2 x \sin^2 y - \sin^2 y + \cos^2 x \sin^2 y$$

$$= \cos^2 x - \sin^2 y$$

17. Use Strategy 6. Prove that $(\sec x + 1)(\sec x - 1) = \tan x \tan x$ is an identity.

$(\sec x + 1)(\sec x - 1) = \sec^2 x - 1$

$$= \tan^2 x$$

$$= \tan x \tan x$$

Therefore $\dfrac{\sec x + 1}{\tan x} = \dfrac{\tan x}{\sec x - 1}$ is an identity.

19. $\dfrac{1 + \tan^2 x}{\tan^2 x} = \dfrac{1}{\tan^2 x} + \dfrac{\tan^2 x}{\tan^2 x}$

$$= \cot^2 x + 1$$

$$= \csc^2 x$$

21. $\tan^2 x - \sec^2 x = \sec^2 x - 1 - \sec^2 x$

$$= -1$$

$$= \csc^2 x - 1 - \csc^2 x$$

$$= \cot^2 x - \csc^2 x$$

23. $\tan x = \dfrac{5}{12}$; $\sec x = \sqrt{1 + \tan^2 x} = \sqrt{1 + \left(\dfrac{5}{12}\right)^2} = \dfrac{13}{12}$; $\cos x = \dfrac{12}{13}$, $\sin x = \dfrac{5}{13}$.

$\sin 2x = 2 \sin x \cos x = 2(5/13)(12/13) = 120/169$

25. First note that $\sin x, \cos x,$ and $\sin y$ are negative and $\cos y$ is positive.

$\sec x = -\sqrt{1 + \tan^2 x} = -\sqrt{1 + \left(\tfrac{4}{3}\right)^2} = -\tfrac{5}{3}$ $\quad \csc y = -\sqrt{1 + \cot^2 y} = -\sqrt{1 + \left(-\tfrac{5}{12}\right)^2} = -\tfrac{13}{12}$

$\cos x = \dfrac{1}{\sec x} = \dfrac{1}{-5/3} = -\dfrac{3}{5}$ $\qquad\qquad \sin y = \dfrac{1}{\csc y} = -\dfrac{12}{13}$

$\sin x = \tan x \cos x = \dfrac{4}{3}\left(-\dfrac{3}{5}\right) = -\dfrac{4}{5}$ $\qquad \cos y = \cot y \sin y = \left(-\dfrac{5}{12}\right)\left(-\dfrac{12}{13}\right) = \dfrac{5}{13}$

Then $\sin(x - y) = \sin x \cos y - \cos x \sin y$

$$= \left(-\dfrac{4}{5}\right)\left(\dfrac{5}{13}\right) - \left(-\dfrac{3}{5}\right)\left(-\dfrac{12}{13}\right)$$

$$= \dfrac{-20 - 36}{65}$$

$$= -\dfrac{56}{65}$$

27. First note that $\cos x = \sqrt{1 - \sin^2 x} = \sqrt{1 - \left(\dfrac{1}{4}\right)^2} = \dfrac{\sqrt{15}}{4}$

Then $\sin\left(\dfrac{\pi}{3} + x\right) = \sin\dfrac{\pi}{3}\cos x + \cos\dfrac{\pi}{3}\sin x$

$\qquad = \left(\dfrac{\sqrt{3}}{2}\right)\left(\dfrac{\sqrt{15}}{4}\right) + \left(\dfrac{1}{2}\right)\left(\dfrac{1}{4}\right)$

$\qquad = \dfrac{\sqrt{45} + 1}{8}$ or $\dfrac{3\sqrt{5} + 1}{8}$

29. Yes. If $\sin x = 0$, $\sin 2x = 2\sin x\cos x = 2\cdot 0\cos x = 0$.

31. Using the half-angle identity,

$\cos\dfrac{\pi}{12} = \cos\dfrac{1}{2}\left(\dfrac{\pi}{6}\right) = \sqrt{\dfrac{1 + \cos(\pi/6)}{2}} = \sqrt{\dfrac{1 + \sqrt{3}/2}{2}} = \dfrac{\sqrt{2 + \sqrt{3}}}{2}$

Using the subtraction identity for cosine,

$\cos\dfrac{\pi}{12} = \cos\left(\dfrac{\pi}{3} - \dfrac{\pi}{4}\right) = \cos\dfrac{\pi}{3}\cos\dfrac{\pi}{4} + \sin\dfrac{\pi}{3}\sin\dfrac{\pi}{4} = \dfrac{1}{2}\cdot\dfrac{\sqrt{2}}{2} + \dfrac{\sqrt{3}}{2}\cdot\dfrac{\sqrt{2}}{2} = \dfrac{\sqrt{2} + \sqrt{6}}{4}$

Hence $\dfrac{\sqrt{2 + \sqrt{3}}}{2} = \dfrac{\sqrt{2} + \sqrt{6}}{4}$ and $\sqrt{2 + \sqrt{3}} = \dfrac{\sqrt{2} + \sqrt{6}}{2}$.

33. $\sin\dfrac{5\pi}{12} = \sin\left(\dfrac{\pi}{4} + \dfrac{\pi}{6}\right) = \sin\dfrac{\pi}{4}\cos\dfrac{\pi}{6} + \cos\dfrac{\pi}{4}\sin\dfrac{\pi}{6}$

$\qquad = \dfrac{\sqrt{2}}{2}\cdot\dfrac{\sqrt{3}}{2} + \dfrac{\sqrt{2}}{2}\cdot\dfrac{1}{2}$

$\qquad = \dfrac{\sqrt{6} + \sqrt{2}}{4}$

35. (a)

37. First note that $\cos x = \sqrt{1 - \sin^2 x} = \sqrt{1 - (.6)^2} = .8$.
Then $\sin 2x = 2\sin x\cos x = 2(.6)(.8) = .96$.

39. $\dfrac{\cos(u/2+\pi/6)}{\cos(u/2)}=1.9$

$\dfrac{\cos(u/2)\cos(\pi/6)-\sin(u/2)\sin(\pi/6)}{\cos(u/2)}=1.9$

$\cos(\pi/6)-\tan(u/2)\sin(\pi/6)=1.9$

$\tan(u/2)=\dfrac{\cos(\pi/6)-1.9}{\sin(\pi/6)}$

$\dfrac{u}{2}=\tan^{-1}\left(\dfrac{\cos(\pi/6)-1.9}{\sin(\pi/6)}\right)$

$u=2\tan^{-1}\left(\dfrac{\cos(\pi/6)-1.9}{\sin(\pi/6)}\right)=-2.24076$

$u=-2.24076+k\pi \quad (k=0,\pm1,\pm2,...)$

41. The slope of the line is given by $\dfrac{y_2-y_1}{x_2-x_1}=\dfrac{2-6}{-2-2}=1$.

Therefore the inclination is the acute angle solution of $\tan\theta=1$, that is, $45°$ or $\pi/4$ radians.

43. $\dfrac{\pi}{4}$ **45.** $\dfrac{\pi}{6}$ **47.** $\cos^{-1}\left(\sin\dfrac{5\pi}{3}\right)=\cos^{-1}\left(-\dfrac{\sqrt{3}}{2}\right)=\dfrac{5\pi}{6}$

49. Since $-\dfrac{\pi}{2}\le.75\le\dfrac{\pi}{2}$, $\sin^{-1}(\sin.75)=.75$ **51.** $\sin^{-1}\left(\sin\dfrac{8\pi}{3}\right)=\sin^{-1}\dfrac{\sqrt{3}}{2}=\dfrac{\pi}{3}$

53.

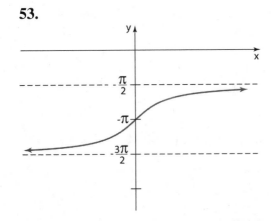

55. Let $u=\cos^{-1}(1/4)$. Then $\cos u=1/4$, $0\le u\le\pi$. Then

$\sin\left(\cos^{-1}(1/4)\right)=\sin u=\sqrt{1-\cos^2 u}=\sqrt{1-(1/4)^2}=\sqrt{15}/4$

57. $2\sin x = 1$

$\quad\;\; \sin x = 1/2$

Since $\sin^{-1}\left(\dfrac{1}{2}\right) = \dfrac{\pi}{6}$, and $\pi - \dfrac{\pi}{6} = \dfrac{5\pi}{6}$, all solutions can be written as

$x = \dfrac{\pi}{6} + 2k\pi$ and $x = \dfrac{5\pi}{6} + 2k\pi$.

59. $\cos x = \sqrt{2}/2$. Since $\cos^{-1}\left(\sqrt{2}/2\right) = \pi/4$, all solutions can be written as

$\quad\;\; x = \pi/4 + 2k\pi$ and $x = -\pi/4 + 2k\pi$.

61. $\tan x = -1$. Since $\tan^{-1}(-1) = -\pi/4$, all solutions can be written as $x = -\pi/4 + k\pi$.

63. $\sin 3x = -\sqrt{3}/2$

$\quad\; 3x = -\dfrac{\pi}{3} + 2k\pi$ and $3x = \pi - \left(-\dfrac{\pi}{3}\right) + 2k\pi = \dfrac{4\pi}{3} + 2k\pi$

$\quad\quad x = -\dfrac{\pi}{9} + \dfrac{2k\pi}{3} \qquad\quad x = \dfrac{4\pi}{9} + \dfrac{2k\pi}{3}$

65. $\sin x = .4$. Since $\sin^{-1}(.4) = .4115$, all solutions can be written as

$\quad\;\; x = .4115 + 2k\pi$ and $x = 2.7301 + 2k\pi$.

67. $\tan x = 15$. Since $\tan^{-1}(15) = 1.5042$, all solutions can be written as

$\quad\;\; x = 1.5042 + k\pi$.

69. $2\sin^2 x + 5\sin x = 3$

$\quad\;\; 2\sin^2 x + 5\sin x - 3 = 0$

$\quad\;\; (2\sin x - 1)(\sin x + 3) = 0$

$\quad\;\; \sin x = 1/2 \qquad\qquad\quad \sin x = -3$

$\quad\;\; \sin^{-1}(1/2) = \pi/6 \qquad\quad\;\; \text{impossible}$

$\quad\;\; x = \pi/6 + 2k\pi$ and $5\pi/6 + 2k\pi$

71. $2\sin^2 x - 3\sin x = 2$

$\quad\;\; 2\sin^2 x - 3\sin x - 2 = 0$

$\quad\;\; (2\sin x + 1)(\sin x - 2) = 0$

$\quad\;\; \sin x = -1/2 \qquad\qquad \sin x = 2$

$\quad\;\; \sin^{-1}(-1/2) = -\pi/6 \qquad\; \text{impossible}$

$\quad\;\; x = -\dfrac{\pi}{6} + 2k\pi$ and $\dfrac{7\pi}{6} + 2k\pi$

73. $\sec^2 x + 3\tan^2 x = 13$

$\quad\;\; \tan^2 x + 1 + 3\tan^2 x = 13$

$\quad\;\; 4\tan^2 x = 12$

$\quad\;\; \tan^2 x = 3$

$\quad\;\; \tan x = \sqrt{3}$ or $\tan x = -\sqrt{3}$

$\quad\;\; x = \pi/3 + k\pi \qquad x = -\pi/3 + k\pi$

75. $2\sin^2 x + \sin x - 2 = 0$

$\sin x = \dfrac{-1 \pm \sqrt{17}}{4}$

$\sin x = \dfrac{-1 + \sqrt{17}}{4}$ $\qquad\qquad$ $\sin x = \dfrac{-1 - \sqrt{17}}{4}$

$\sin x = .7808$ $\qquad\qquad\qquad$ impossible

$x = \sin^{-1} .7808 = .8959 + 2k\pi$

$x = \pi - \sin^{-1} .7808 = 2.2457 + 2k\pi$

77. Graph the function $f(x) = 5\tan x - 2\sin 2x$ using the window $0 \le x \le 2\pi$, $-5 \le y \le 5$, ignoring the erroneous vertical lines at the asymptotes. Solutions are observed precisely at $x = 0, \pi, 2\pi$ and since the function is periodic with period π, all solutions are given by $x = k\pi$, where k is any integer.

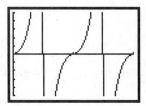

79. Graph the function $f(x) = \sin x + \sec^2 x - 3$ using the window $0 \le x \le 2\pi$, $-5 \le y \le 5$ and apply the zero (root) routine to obtain $x = .8419$, 2.2997, 4.1784, and 5.2463. Since the function is periodic with period 2π, all solutions are given by $x = .8419 + 2k\pi$, where k is any integer, and so on.

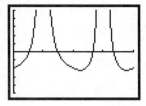

81. $\sin\theta = -.7133$. Since $\sin^{-1}(-.7133) = -45.5°$, all solutions in the required interval are given by $-45.5° + 360° = 314.5°$ and $180° - (-45.5°) = 225.5°$.

83. Use $d = \left(v^2/32\right)\sin 2\theta$ with $d = 3500$, $v = 600$.

$3500 = \left(600^2/32\right)\sin 2\theta$

$\sin 2\theta = \dfrac{3500 \cdot 32}{600^2}$

$\sin 2\theta = 14/45$

$2\theta = 18.13°$ or $180° - 18.13°$

$\theta = 9.06°$ or $80.94°$

Chapter 7 Test

1. $1 - \sin x \cos x \tan x = 1 - \sin x \cos x \dfrac{\sin x}{\cos x}$

$$= 1 - \sin^2 x$$

$$= \cos^2 x$$

2. $\cos\left(-\dfrac{5\pi}{12}\right) = \cos\left(\dfrac{\pi}{3} - \dfrac{3\pi}{4}\right) = \cos\left(\dfrac{\pi}{3}\right)\cos\left(\dfrac{3\pi}{4}\right) + \sin\left(\dfrac{\pi}{3}\right)\sin\left(\dfrac{3\pi}{4}\right)$

$$= \dfrac{1}{2} \cdot \dfrac{-\sqrt{2}}{2} + \dfrac{\sqrt{3}}{2} \cdot \dfrac{\sqrt{2}}{2} = \dfrac{\sqrt{6}}{4} - \dfrac{\sqrt{2}}{4} = \dfrac{\sqrt{6} - \sqrt{2}}{4}$$

3. $\tan\left(3\pi/8\right) = \tan\left[\dfrac{1}{2}\left(3\pi/4\right)\right] = \sqrt{\dfrac{1 - \cos\left(3\pi/4\right)}{1 + \cos\left(3\pi/4\right)}}$

$$= \sqrt{\dfrac{1 + \sqrt{2}/2}{1 - \sqrt{2}/2}} = \sqrt{\dfrac{\left(1 + \sqrt{2}/2\right)\sqrt{2}}{\left(1 - \sqrt{2}/2\right)\sqrt{2}}} = \sqrt{\dfrac{\sqrt{2} + 1}{\sqrt{2} - 1}}$$

4. $\cos^2 x - \cot^2 x = \cos^2 x - \dfrac{\cos^2 x}{\sin^2 x}$

$$= \cos^2 x\left(1 - \csc^2 x\right)$$

$$= -\cos^2 x\left(\csc^2 x - 1\right)$$

$$= -\cos^2 x \cot^2 x$$

Alternate approach:

$\cos^2 x - \cot^2 x = \cos^2 x - \dfrac{\cos^2 x}{\sin^2 x}$

$$= \dfrac{\cos^2 x \sin^2 x - \cos^2 x}{\sin^2 x}$$

$$= \dfrac{\cos^2 x\left(\sin^2 x - 1\right)}{\sin^2 x}$$

$$= \dfrac{\cos^2 x}{\sin^2 x}\left(\sin^2 x - 1\right)$$

$$= \cot^2 x\left(-\cos^2 x\right)$$

$$= -\cos^2 x \cot^2 x$$

5. $\sin(x-y)\sin x + \cos(x-y)\cos x$

$= (\sin x \cos y - \cos x \sin y)\sin x + (\cos x \cos y + \sin x \sin y)\cos x$

$= \sin^2 x \cos y - \sin x \cos x \sin y + \cos^2 x \cos y + \sin x \cos x \sin y$

$= \cos y (\sin^2 x + \cos^2 x)$

$= \cos y$

6. a. Since $\sin x = -\dfrac{40}{41}$ and x is in the third quadrant, $\cos x = -\dfrac{9}{41}$. Thus

$\cos 2x = \cos^2 x - \sin^2 x$

$= \left(-\dfrac{9}{41}\right)^2 - \left(-\dfrac{40}{41}\right)^2$

$= \dfrac{81}{1681} - \dfrac{1600}{1681} = -\dfrac{1519}{1681}$

b. $\sin 2x = 2\sin x \cos x$

$= 2\left(-\dfrac{40}{41}\right)\left(-\dfrac{9}{41}\right)$

$= \dfrac{720}{1681}$

c. $\tan x = \dfrac{\sin x}{\cos x} = \dfrac{-40/41}{-9/41} = \dfrac{40}{9}$

$\tan 2x = \dfrac{2\tan x}{1-\tan^2 x}$

$= \dfrac{2\left(\dfrac{40}{9}\right)}{1-\left(\dfrac{40}{9}\right)^2} = \dfrac{\dfrac{80}{9}}{\dfrac{-1519}{81}} = -\dfrac{720}{1519}$

7. $(\sin^2 x - 1)(\cot^2 x + 1) = (\sin^2 x - 1)\left(\dfrac{\cos^2 x}{\sin^2 x}+1\right) = \cos^2 x + \sin^2 x - \dfrac{\cos^2 x}{\sin^2 x} - 1$

$= 1 - \cot^2 x - 1 = -\cot^2 x$

Alternate Approach:

$(\sin^2 x - 1)(\cot^2 x + 1) = \sin^2 x \cot^2 x + \sin^2 x - \cot^2 x - 1$

$= \sin^2 x \dfrac{\cos^2 x}{\sin^2 x} + \sin^2 x - \cot^2 x - 1$

$= \cos^2 x + \sin^2 x - \cot^2 x - 1$

$= 1 - \cot^2 x - 1$

$= -\cot^2 x$

8. Since $\cos x = -\dfrac{1}{4}$ and x is in the third quadrant, $\sin x = -\dfrac{\sqrt{15}}{4}$. Thus

$$\cos\left(\frac{\pi}{6}-x\right)=\cos\frac{\pi}{6}\cos x+\sin\frac{\pi}{6}\sin x$$

$$=\frac{\sqrt3}{2}\cdot\frac{-1}{4}+\frac{1}{2}\cdot\frac{-\sqrt{15}}{4}$$

$$=\frac{-\sqrt3-\sqrt{15}}{8}$$

9. $2\cos^2\left(\dfrac{x}{2}\right)-1=2\left[\cos\left(\dfrac{x}{2}\right)\right]^2-1$

$$=2\left[\pm\sqrt{\frac{1+\cos x}{2}}\right]^2-1=2\left(\frac{1+\cos x}{2}\right)-1$$

$$=1+\cos x-1=\cos x$$

10. $\dfrac{1-\tan x}{\cot x-1}=\dfrac{1-\dfrac{\sin x}{\cos x}}{\dfrac{\cos x}{\sin x}-1}=\dfrac{\sin x\cos x-\sin^2 x}{\cos^2 x-\sin x\cos x}$

$$=\frac{\sin x(\cos x-\sin x)}{\cos x(\cos x-\sin x)}=\frac{\sin x}{\cos x}=\tan x$$

Alternate Approach:

$$\frac{1-\tan x}{\cot x-1}=\frac{1-\dfrac{1}{\cot x}}{\cot x-1}=\frac{\dfrac{\cot x-1}{\cot x}}{\cot x-1}=\frac{1}{\cot x}=\frac{1}{\dfrac{\cos x}{\sin x}}=\frac{\sin x}{\cos x}=\tan x$$

Alternate Approach:
Use strategy 6 of Section 7.1 to prove the equivalent identity.
$$(\cot x-1)\tan x=1-\tan x.$$

$$(\cot x-1)\tan x=\cot x\tan x-\tan x=\frac{\cos x}{\sin x}\cdot\frac{\sin x}{\cos x}-\tan x=1-\tan x$$

11. $\sin x\sin(\pi-x)=\sin x(\sin x)=\sin^2 x=1-\cos^2 x$

Alternative Approach:
$$\sin x\sin(\pi-x)=\sin x(\sin\pi\cos x-\cos\pi\sin x)$$

$$=\sin x(0\cos x-(-1)\sin x)$$

$$=\sin^2 x$$

$$=1-\cos^2 x$$

12. a. Since $\sin x = .4$ and x is in the second quadrant, $\cos x = \sqrt{.84}$. Thus

Since $\dfrac{\pi}{4} < \dfrac{x}{2} < \dfrac{\pi}{2}$, all the trigonometric functions of will be positive.

$$\sin\frac{x}{2} = \sqrt{\frac{1-\cos x}{2}} = \sqrt{\frac{1-\left(-\sqrt{.84}\right)}{2}}\sqrt{\frac{1+\sqrt{.84}}{2}}$$

b. $\cos\dfrac{x}{2} = \sqrt{\dfrac{1+\cos x}{2}} = \sqrt{\dfrac{1+\left(-\sqrt{.84}\right)}{2}}\sqrt{\dfrac{1-\sqrt{.84}}{2}}$

13. $\cos^{-1}\left(-\dfrac{\sqrt{3}}{2}\right) = \dfrac{5\pi}{6}$ since $\cos\dfrac{5\pi}{6} = -\dfrac{\sqrt{3}}{2}$ and $0 < \dfrac{5\pi}{6} < \pi$.

14. $7\cos x = 5$. Since $\cos^{-1}\left(\dfrac{5}{7}\right) = .7752$, all solutions can be written as

$x \approx .7752 + 2k\pi$ and $x \approx -.7752 + 2k\pi$.

15. $\cos^{-1}\left(\sin\dfrac{\pi}{4}\right) = \cos^{-1}\left(\dfrac{\sqrt{2}}{2}\right) = \dfrac{\pi}{4}$

16. Enter $y = \cos^2 x + 7\cos x - 7$ into a graphing calculator and use the root finder to find that $x = .4789$ and $x = 5.8042$.

17. By the point in the plane description $\tan t = \dfrac{4}{3}$. Thus, $t = \tan^{-1}\left(\dfrac{4}{3}\right) = .9273$.

18. $\sin\dfrac{x}{2} = \sqrt{2}/2$. Since $\dfrac{x}{2} = \sin^{-1}\left(\sqrt{2}/2\right) = \pi/4$, all solutions can be written as

$\dfrac{x}{2} = \pi/4 + 2k\pi$ and $\dfrac{x}{2} = 3\pi/4 + 2k\pi$.

$x = \pi/2 + 4k\pi$ and $x = 3\pi/2 + 4k\pi$.

19. By the cofunction identity, $\sin x = \cos\left(\dfrac{\pi}{2} - x\right)$. Since $-\dfrac{\pi}{2} < x < \dfrac{\pi}{2}$, it follows that

$0 < \dfrac{\pi}{2} - x < \pi$. Therefore, $\cos^{-1}(\sin x) = \cos^{-1}\left[\cos\left(\dfrac{\pi}{2} - x\right)\right] = \dfrac{\pi}{2} - x$ by the fact in the box on page 549.

20. $\dfrac{76 \text{ miles}}{\text{hour}} \times \dfrac{5280 \text{ feet}}{1 \text{ mile}} \times \dfrac{1 \text{ hour}}{3600 \text{ sec.}} = \dfrac{1672}{15}$ feet per second.

$$170 = \dfrac{(1672/15)^2}{32}\sin 2\theta$$

$$\sin 2\theta = \dfrac{170(32)}{(1672/15)^2}$$

$$2\theta = \sin^{-1}\left(\dfrac{170(32)}{(1672/15)^2}\right)$$

$$\theta = \dfrac{\sin^{-1}\left(\dfrac{170(32)}{(1672/15)^2}\right)}{2} \approx .2266 \text{ radians}$$

If 2θ is in the second quadrant $\theta = \dfrac{\pi - \sin^{-1}\left(\dfrac{170(32)}{(1672/15)^2}\right)}{2} \approx 1.3442$ radians. (An angle of approximately 1.3442 radians is theoretically possible, but seems unlikely when throwing a baseball.)

Chapter 8
Triangle Trigonometry

8.1 Trigonometric Functions of Angles

1. $r = \sqrt{x^2 + y^2} = \sqrt{2^2 + 3^2} = \sqrt{13}$

$$\sin\theta = \frac{y}{r} = \frac{3}{\sqrt{13}} \qquad \cot\theta = \frac{x}{y} = \frac{2}{3}$$

$$\cos\theta = \frac{x}{r} = \frac{2}{\sqrt{13}} \qquad \sec\theta = \frac{r}{x} = \frac{\sqrt{13}}{2}$$

$$\tan\theta = \frac{y}{x} = \frac{3}{2} \qquad \csc\theta = \frac{r}{y} = \frac{\sqrt{13}}{3}$$

3. $r = \sqrt{x^2 + y^2} = \sqrt{(-3)^2 + 7^2} = \sqrt{58}$

$$\sin\theta = \frac{y}{r} = \frac{7}{\sqrt{58}} \qquad \cot\theta = \frac{x}{y} = -\frac{3}{7}$$

$$\cos\theta = \frac{x}{r} = -\frac{3}{\sqrt{58}} \qquad \sec\theta = \frac{r}{x} = -\frac{\sqrt{58}}{3}$$

$$\tan\theta = \frac{y}{x} = -\frac{7}{3} \qquad \csc\theta = \frac{r}{y} = \frac{\sqrt{58}}{7}$$

5. $r = \sqrt{x^2 + y^2} = \sqrt{(-3)^2 + \left(-\sqrt{2}\right)^2} = \sqrt{11}$

$$\sin\theta = \frac{y}{r} = -\frac{\sqrt{2}}{\sqrt{11}} \qquad \cot\theta = \frac{x}{y} = \frac{3}{\sqrt{2}}$$

$$\cos\theta = \frac{x}{r} = -\frac{3}{\sqrt{11}} \qquad \sec\theta = \frac{r}{x} = -\frac{\sqrt{11}}{3}$$

$$\tan\theta = \frac{y}{x} = \frac{\sqrt{2}}{3} \qquad \csc\theta = \frac{r}{y} = -\frac{\sqrt{11}}{\sqrt{2}}$$

7.
$$\sin\theta = \frac{\text{opposite}}{\text{hypotenuse}} = \frac{\sqrt{2}}{\sqrt{11}}$$

$$\cos\theta = \frac{\text{adjacent}}{\text{hypotenuse}} = \frac{3}{\sqrt{11}}$$

$$\tan\theta = \frac{\text{opposite}}{\text{adjacent}} = \frac{\sqrt{2}}{3}$$

9.
$$\sin\theta = \frac{\sqrt{3}}{\sqrt{7}}$$
$$\cos\theta = \frac{2}{\sqrt{7}}$$
$$\tan\theta = \frac{\sqrt{3}}{2}$$

11.
$$\sin\theta = \frac{h}{m}$$
$$\cos\theta = \frac{d}{m}$$
$$\tan\theta = \frac{h}{d}$$

13. $\cos A = \dfrac{c}{b}$

$$\frac{12}{13} = \frac{c}{39}$$
$$c = 36$$

15. $\tan A = \dfrac{a}{c}$

$$\frac{5}{12} = \frac{15}{c}$$
$$c = 36$$

17. $\cot A = \dfrac{c}{a}$

$$6 = \frac{c}{1.4}$$
$$c = 8.4$$

19. $\cos 45° = \dfrac{h}{25}$

$$h = 25\cos 45°$$
$$h = \frac{25\sqrt{2}}{2}$$

21. $\sin 30° = \dfrac{150}{h}$

$$h = 150/\sin 30°$$
$$h = \frac{150}{1/2} = 300$$

23. $\cos 30° = \dfrac{h}{100}$

$$h = 100\cos 30°$$
$$h = 100\frac{\sqrt{3}}{2} = 50\sqrt{3}$$

25. $a/c = \tan 60°$

$c = \dfrac{a}{\tan 60°}$

$c = \dfrac{4}{\sqrt{3}} = \dfrac{4\sqrt{3}}{3}$

27. $a/c = \tan 30°$

$a = c\tan 30°$

$a = \dfrac{10}{\sqrt{3}} = \dfrac{10\sqrt{3}}{3}$

29. $\angle A = 90° - \angle C = 90° - 40° = 50°$

$\sin C = c/b$, $\sin 40° = c/10$, $c = 10\sin 40° = 6.4$

$\cos C = a/b$, $\cos 40° = a/10$, $a = 10\cos 40° = 7.7$

31. $\angle C = 90° - \angle A = 90° - 14° = 76°$

$\sin A = a/b$, $\sin 14° = 16/b$, $b = 16/\sin 14° = 66.1$

$\tan A = a/c$, $\tan 14° = 16/c$, $c = 16/\tan 14° \doteq 64.2$

33. $\angle C = 90° - \angle A = 90° - 65° = 25°$

$\cos A = \dfrac{c}{b}$, $\cos 65° = \dfrac{5}{b}$, $b = \dfrac{5}{\cos 65°} = 11.8$

$\tan A = \dfrac{a}{c}$, $\tan 65° = \dfrac{a}{5}$, $a = 5\tan 65° = 10.7$

35. $\angle C = 90° - \angle A = 90° - 72° = 18°$

$\cos A = \dfrac{c}{b}$, $\cos 72° = \dfrac{c}{3.5}$, $c = 3.5\cos 72° = 1.1$

$\sin A = \dfrac{a}{b}$, $\sin 72° = \dfrac{a}{3.5}$, $a = 3.5\sin 72° = 3.3$

37. $\sin \theta = 3/4$, $\theta = \sin^{-1}(3/4) = 48.6°$ **39.** $\cos \theta = 2/3$, $\theta = \cos^{-1}(2/3) = 48.2°$

41. $\tan A = a/c = 8/15$, $A = \tan^{-1}(8/15) = 28.1°$, $C = 90° - A = 61.9°$

43. $\sin A = a/b = 7/10$, $A = \sin^{-1}(7/10) = 44.4°$, $C = 90° - A = 45.6°$

45. $\cos A = c/b = 12/18$, $A = \cos^{-1}(12/18) = 48.2°$, $C = 90° - A = 41.8°$

47. $\tan A = a/c = 2.5/1.4$, $A = \tan^{-1}(2.5/1.4) = 60.8°$, $C = 90° - A = 29.2°$

49. a. $A = \dfrac{1}{2}ah$ **b.** $\sin \theta = \dfrac{h}{b}$ **c.** $h = b\sin \theta$ **d.** $A = \dfrac{1}{2}ab\sin \theta$

51. $A = \dfrac{1}{2}(14)(10)\sin 25° = 29.6$ **53.** $A = \dfrac{1}{2}(20)(44)\sin 30° = 220$

55. a. $\sin \theta = a/c = \cos \alpha$

b. Since $\theta + \alpha + 90° = 180°$, $\theta + \alpha = 90°$.

c. Since $\theta + \alpha = 90°$, $\alpha = 90° - \theta$ thus $\sin \theta = \cos(90° - \theta)$.

8.2 Applications of Right Triangle Trigonometry

1. $\angle A = 180° - 90° - 16° = 74°$

$\sin 16° = \dfrac{\text{opposite}}{\text{hypotenuse}} = \dfrac{8}{c}$

$c = \dfrac{8}{\sin 16°} = 29.0$

$\tan 16° = \dfrac{\text{opposite}}{\text{adjacent}} = \dfrac{8}{a}$

$a = \dfrac{8}{\tan 16°} = 27.9$

3. $\sin B = \dfrac{6}{7.5}$

$B = \sin^{-1}\left(\dfrac{6}{7.5}\right) = 53.1°$

$\angle A = 180° - 90° - 53.1° = 36.9°$

5. $\tan 40° = \dfrac{w}{120}$

$w = 120 \tan 40° = 100.7 \text{ ft}$

7. $\cos\theta = \dfrac{9}{24}$

$\theta = \cos^{-1}(9/24) = 68.0°$

9. $\alpha = 90° - 25.8° = 64.2°$

$\cos 64.2° = \dfrac{10}{w}$

$w = \dfrac{10}{\cos 64.2°} = 23.0 \text{ m}$

11. Let d = the distance required. Then $\sin 50° = \dfrac{d}{20}, d = 20\sin 50° = 15.3$ feet

13. Since 1 mile = 5280 feet, $\sin 3° = \dfrac{h}{5280},\ h = 5280\sin 3° = 276.3$ feet.

15. Let h = height. Then $\tan 80° = \dfrac{h}{600}, h = 600\tan 80° = 3402.8$ feet.

17. Let x = distance. Then $\tan 38° = \dfrac{x}{110}, x = 110\tan 38° = 85.9$ feet.

19. Let x = width. Then $\tan 54° = \dfrac{w}{20}, w = 20\tan 54° = 27.5$ feet. It is not safe for him to jump.

21. Let θ be the angle between the person and the mountain. Then $\theta = 90° - 62° = 28°$. Therefore, $\tan\theta = \dfrac{\text{arm length}}{\text{shoulder height}}$. Shoulder height = $5(12) = 60$ inches. Thus, $\tan 28° = \dfrac{x}{60}, x = 60\tan 28° = 31.9$ inches. Since the person's arm length is only 27 inches, (s)he cannot touch the mountain.

23. Let x = length. Then, $\sin 3° = \dfrac{450}{x}, x = \dfrac{450}{\sin 3°} = 8598.3$ feet (1.6 miles).

25. Let θ = angle required. Then, $\sin\theta = \dfrac{8}{16} = \dfrac{1}{2}$, hence $\theta = 30°$.

27. Let d = distance. The angle θ between the vertical tower to the base of the buoy is $\theta = 90° - 6.5° = 83.5°$. Then, $\tan 83.5° = \dfrac{x}{40}, x = 40\tan 83.5° = 351.1$ meters.

29. Let h = height. Then, $\dfrac{2}{h} = \cot 71° + \cot 46°, h = \dfrac{2}{\cot 71° + \cot 46°} = 1.53$ miles.

31. Let x = distance between the boats. Then, $\dfrac{x}{21} = \cot 27° - \cot 53°, x = 21(\cot 27° - \cot 53°) = 25.4$ meters.

33. Let h = height of the streetlight, and d = horizontal distance from the window to the streetlight. Then, $\tan 55° = \dfrac{35 - h}{d}$ and $\tan 57.8° = \dfrac{35}{d}$. Hence, $h = 35\left(1 - \dfrac{\tan 55°}{\tan 57.8°}\right) = 3.52$ miles.

35. Sketch the figure.

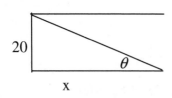

Note: θ is equal to the angle of depression, 7.4°; x is measured in feet. Then, $\tan\theta = \dfrac{20}{x}, x = \dfrac{20}{\tan 7.4°} = 153.99$ feet. So, the rate of the car is $\dfrac{x \text{ feet}}{2 \text{ sec}} \cdot \dfrac{60 \text{ miles}}{\text{hour}} \div 88\dfrac{\text{feet}}{\text{sec}} = \dfrac{153.99 \cdot 60}{2 \cdot 88} = 52.5$mph.

37. Let y = height of the streetlight. In similar triangles, shown, $\dfrac{5.\overline{3}}{4} = \dfrac{y}{10 + 4}$. Then, $y = 14\left(\dfrac{5.\overline{3}}{4}\right) = 18.67$ feet. So, $\tan\theta = \dfrac{5.\overline{3}}{4}$, $\theta = \tan^{-1}\left(\dfrac{5.\overline{3}}{4}\right) = 53.13°$.

39. Sketch the figure.

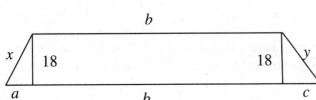

$18/a = \tan 15°$ $a = 18/\tan 15° = 67.177$
$18/c = \tan 21°$ $c = 18/\tan 21° = 46.892$
$18/x = \sin 15°$ $x = 18/\sin 15° = 69.547$
$18/y = \sin 21°$ $y = 18/\sin 21° = 50.228$

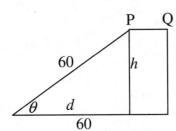

Since $a+b+c = 200, b = 200-a-c, b = 85.931$.
Then, the length of the walk is
$x+b+y = 205.7$ feet.

41. Since the angle between the path and West is $30°$, write d = distance,

then $\cos 30° = \dfrac{d}{200}, d = 200\cos 30° = 173.2$ miles.

43.   Sketch the figure.

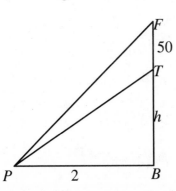

a. $\sin\theta = h/60, h = 60\sin 33° = 32.7$, so the height of P above the water is $24+h = 56.7$ feet.

b. $\cos\theta = d/60, d = 60\cos 33° = 50.3$, then $PQ = 60-d = 9.7$ feet.

45. **a.** Let x = half-width of the rectangle, y = height. Then $A = 2xy$. Since

$\cos t = \dfrac{x}{10}$, and $\sin t = \dfrac{y}{10}, x = 10\cos t$ and $y = 10\sin t$, hence

$A = 2(10\sin t)(10\cos t) = 200\sin t\cos t$.

b. Graph A using the window $0° \le t \le 90°, 0 \le y \le 200$ and apply the maximum capability of the graphing calculator to find $t = 45°$. Then, $2x = 2\cdot 10\cos 45° = 14.14$ feet, $y = 10\sin 45° = 7.07$ feet.

47. Sketch a figure.

In triangle PTB, $\dfrac{PB}{h} = \cot 40°$, $PB = h\cot 40°$

In triangle PFB,

$\dfrac{PB}{h+50} = \cot 43°$, $PB = (h+50)\cot 43°$

Then $h\cot 40° = (h+50)\cot 43°$

$h(\cot 40° - \cot 43°) = 50\cot 43°$

$h = \dfrac{50\cot 43°}{\cot 40° - \cot 43°} = \dfrac{50\tan 40°}{\tan 43° - \tan 40°} = 449.1$ feet

8.3 The Law of Cosines

As noted in the text, all decimals are printed in rounded-off form for reading convention, but no rounding is done in the actual computation until the final answer is stored.

1. $a^2 = b^2 + c^2 - 2bc\cos A;$ $a^2 = 10^2 + 7^2 - 2(10)(7)\cos 40°; a = 6.5$

$\cos B = \dfrac{a^2 + c^2 - b^2}{2ac};$ $B = 95.9°;$ $C = 180° - A - B = 44.1°$

3. $c^2 = a^2 + b^2 - 2ab\cos C;$ $c^2 = 6^2 + 12^2 - 2(6)(12)\cos 118°; c = 15.7$

$\cos B = \dfrac{a^2 + c^2 - b^2}{2ac};$ $B = 42.3°;$ $A = 180° - B - C = 19.7°$

5. $a^2 = b^2 + c^2 - 2bc\cos A;$ $a^2 = 12^2 + 14^2 - 2(12)(14)\cos 140°; a = 24.4$

$\cos B = \dfrac{a^2 + c^2 - b^2}{2ac};$ $B = 18.4°;$ $C = 180° - A - B = 21.6°$

7. $c^2 = a^2 + b^2 - 2ab\cos C;$ $c = 21.5$

$\cos B = \dfrac{a^2 + c^2 - b^2}{2ac};$ $B = 67.9°;$ $A = 180° - B - C = 33.5°$

9. $\cos B = \dfrac{a^2 + c^2 - b^2}{2ac};$ $B = 21.8°;$ $\cos C = \dfrac{a^2 + b^2 - c^2}{2ab};$ $C = 38.2°$

$A = 180° - B - C = 120°$

11. $\cos B = \dfrac{a^2 + c^2 - b^2}{2ac};$ $B = 68.2°;$ $\cos C = \dfrac{a^2 + b^2 - c^2}{2ab};$ $C = 82.1°$

$A = 180° - B - C = 29.7°$

13. $\cos B = \dfrac{a^2 + c^2 - b^2}{2ac};$ $B = 34.5°;$ $\cos C = \dfrac{a^2 + b^2 - c^2}{2ab};$ $C = 106.7°$

$A = 180° - B - C = 38.8°$

15. $\cos B = \dfrac{a^2 + c^2 - b^2}{2ac};$ $B = 53.2°;$ $\cos C = \dfrac{a^2 + b^2 - c^2}{2ab};$ $C = 91.2°$

$A = 180° - B - C = 35.6°$

17. The sides of the triangle are given by the distance formula as:

$$a = \sqrt{(5-0)^2 + (-2-0)^2} = \sqrt{29}; \quad b = \sqrt{(1-0)^2 + (-4-0)^2} = \sqrt{17}$$

$$c = \sqrt{(1-5)^2 + [-4-(-2)]^2} = \sqrt{20}$$

Then $\cos A = \dfrac{b^2 + c^2 - a^2}{2bc}$; $A = 77.5°$; $\cos B = \dfrac{a^2 + c^2 - b^2}{2ac}$; $B = 48.4°$

$C = 180° - A - B = 54.1°$

19. Using example 4 as a model, regard this situation as a triangle with $a = 90(2 + 45/60)$, $b = 55(2 + 45/60)$, and $C = 112°$. Then

$$c^2 = a^2 + b^2 - 2ab\cos C, \quad c = 334.9 \text{ km}.$$

21. Using example 6 as a model, regard this situation as a triangle with $a = 10.4$, $b = 9$,

and $c = 6$. $\cos A = \dfrac{b^2 + c^2 - a^2}{2bc}$; $A = 85.3°$. Then $\theta = 180° - 40° - 85.3° = 54.7°$

23. Apply the Law of Cosines to the triangle with $a = 60.5$, $b = 90$, and $C = 45°$.
$c^2 = a^2 + b^2 - 2ab\cos C$, $c = 63.7$ feet.

25. Apply the Alternate Law of Cosines to the triangle shown, with $a = b = 8$, $c = 10.8$.
$\cos C = \dfrac{a^2 + b^2 - c^2}{2ab}$, $C = 84.9°$

27. Using the diagram provided in the text, $EC = 3960$, $ES = 27960$, and $\angle CES = 60°$.
Then $CS^2 = EC^2 + ES^2 - 2ab\cos \angle CES$, $CS = 26,205$ miles.

29. Using the diagram provided in the text, $a = 65, b = 50, c = 100$. Using the Law of

Cosines we get $\cos A = \dfrac{b^2 + c^2 - a^2}{2bc}$; $A = 34.2°$, and

$\cos B = \dfrac{a^2 + c^2 - b^2}{2ac}$; $B = 25.6°$.

31. Regard this situation as a triangle with $a = 3$, $b = 6$, and $C = 180° - 45° = 135°$.
Then $c^2 = a^2 + b^2 - 2ab\cos C$, $c = 8.4$ km.

33. Apply the Law of Cosines to the triangle with $a = 120$, $c = 74$, and $B = 103°$.
$b^2 = a^2 + c^2 - 2ac\cos B$, $b = 154.5$ feet.

35. The length of the shorter cable can be obtained by solving a right triangle.
$400/a = \sin 70°$

$a = 400/\sin 70° = 425.7$ feet.

Then the length c of the longer cable can be obtained by applying the Law of
Cosines to the triangle with a as above, $b = 100$, and $C = 180° - 70° = 110°$.
$c^2 = a^2 + b^2 - 2ab\cos C$, $c = 469.4$ feet.

37. Apply the Law of Cosines to the triangle with $a = 350$, $b = 7 \cdot 18 = 126$ miles, and $C = 22°$. $c^2 = a^2 + b^2 - 2ab\cos C$, $c = 237.9$ miles. Now find the angle A of this triangle from the Alternate Law of Cosines. $\cos A = \left(b^2 + c^2 - a^2\right)/2bc$, $A = 146.6°$. The ship should turn through $180° - A = 33.44°$.

39. Construct a triangle as follows: Using the Law of Cosines, with $a = 3960$,

$b = 4860$, and $C = 5°$. Then

$c^2 = a^2 + b^2 - 2ab\cos C$, $c = 978.0$ miles.

To find θ : $\dfrac{360°}{360 \text{ minutes}} = \dfrac{\theta}{5 \text{ minutes}}$

$\theta = 5°$

41. Use the figure for Exercise **36**, with the half-diagonals 6 and 7.5 inches this time. Then $b = 6, c = 7.5$, and $A_1 = 63.7°$, $a_1^{\,2} = b^2 + c^2 - 2bc\cos A_1$, $a_1 = 7.2$ inches. Also $b = 6, c = 7.5$, and $A_2 = 116.3°$, $a_2^{\,2} = b^2 + c^2 - 2bc\cos A_2$, $a_2 = 11.5$ inches.

43. Clearly $a = $ distance from B to $C = 11.27 + 8.23 = 19.5$. Similarly $b = 8.23 + 13 = 21.23$, $c = 11.27 + 13 = 24.27$. Apply the Alternate Law of Cosines.

$\cos A = \dfrac{b^2 + c^2 - a^2}{2bc}$, $A = 50.2°$, $\cos B = \dfrac{a^2 + c^2 - b^2}{2ac}$, $B = 56.8°$,

$C = 180° - A - B = 73.0°$.

45. From the Pythagorean theorem, $AD = \sqrt{8^2 - 1^2} = \sqrt{63}$, $BE = \sqrt{7^2 - 1^2} = \sqrt{48}$. Then $\sin ACD = AD/8$, angle $ACD = 82.8°$.

$\qquad \sin BCE = BE/7$, angle $BCE = 81.8°$.

In triangle ABC, apply the Alternate Law of Cosines.

$\cos ACB = \dfrac{8^2 + 7^2 - 9^2}{2 \cdot 8 \cdot 7}$, $\angle ACB = 73.4°$

Then $\angle DCE = 360° - \angle ACD - \angle BCE - \angle ACB = 122°$. Arc $\overset{\frown}{DE}$ is then

$\frac{122}{360} \cdot 2\pi(1) = 2.1$ meters. The length of the rope $= AD + \overset{\frown}{DE} + BE = 16.99$ meters.

8.4 The Law of Sines

1. $b = \dfrac{a \sin B}{\sin A} = \dfrac{6 \sin 22°}{\sin 44°} = 3.2, C = 180° - A - B = 114°, c = \dfrac{a \sin C}{\sin A} = \dfrac{6 \sin 114°}{\sin 44°} = 7.9$

3. $B = 180° - A - C = 30°, \; b = \dfrac{a \sin B}{\sin A} = \dfrac{12 \sin 30°}{\sin 110°} = 6.4, c = \dfrac{a \sin C}{\sin A} = \dfrac{12 \sin 40°}{\sin 110°} = 8.2$

5. $A = 180° - B - C = 86°, \; a = \dfrac{b \sin A}{\sin B} = \dfrac{6 \sin 86°}{\sin 42°} = 8.9, \; c = \dfrac{b \sin C}{\sin B} = \dfrac{6 \sin 52°}{\sin 42°} = 7.1$

7. $C = 180° - A - B = 41.5°, b = \dfrac{a \sin B}{\sin A} = \dfrac{16 \sin 36.2°}{\sin 102.3°} = 9.7,$

$c = \dfrac{a \sin C}{\sin A} = \dfrac{16 \sin 41.5°}{\sin 102.3°} = 10.9$

9. This is the SSA case. $\sin C = \dfrac{c \sin B}{b} = \dfrac{20 \sin 70°}{12} = 1.5662$

$\sin C > 1$ hence no such triangle exists.

11. This is the SSA case. $\sin A = \dfrac{a \sin B}{b} = \dfrac{12 \sin 20°}{15} = .2736$

$A = 25.3°$ or $A = 180° - 23.5° = 154.7°.$

Case 1: $A = 154.7°, \; C = 5.3°, \; c = a \sin C / \sin A = 3.3$

Case 2: $A = 25.3°, \; C = 137.4°, \; c = a \sin C / \sin A = 24.9$

13. This is the SSA case. $\sin C = \dfrac{c \sin A}{a} = \dfrac{12 \sin 102°}{5} = 2.3476$

$\sin C > 1$ hence no such triangle exists.

15. This is the SSA case. $\sin B = \dfrac{b \sin C}{c} = \dfrac{12 \sin 56°}{10} = .8290$

$B = 84.2°$ or $B = 180° - 84.2° = 95.8°.$

Case 1: $B = 95.8°, \; A = 28.2°, \; a = 5.7$

Case 2: $B = 84.2°, \; A = 39.8°, \; a = \dfrac{b \sin A}{\sin B} = 7.7$

17. $b = \dfrac{a \sin B}{\sin A} = \dfrac{5 \sin 6.7°}{\sin 41°} = .89, C = 180° - A - B = 132.3°,$

$c = \dfrac{a \sin C}{\sin A} = \dfrac{5 \sin 132.3°}{\sin 41°} = 5.6$

19. $a^2 = b^2 + c^2 - 2bc\cos A$, $a = 9.8$, $\sin B = \dfrac{b\sin A}{a}$, $B = 23.3°$

$C = 180° - A - B = 81.7°$

21. $\cos A = \dfrac{b^2 + c^2 - a^2}{2bc}$, $A = 18.6°$, $\cos B = \dfrac{a^2 + c^2 - b^2}{2ac}$, $B = 39.6°$

$C = 180° - A - B = 121.8°$

23. $c^2 = a^2 + b^2 - 2ab\cos C$, $c = 13.9$, $\sin A = a\sin C/c$, $A = 60.1°$

$B = 180° - A - C = 72.9°$

25. This is the SSA case. $\sin C = \dfrac{c\sin B}{b} = \dfrac{12.4\sin 62.5°}{17.2} = .6395$

$C = 39.8°$ or $C = 180° - 39.8° = 140.2°$.

Case 1: $C = 39.8°$, $A = 77.7°$, $a = b\sin A/\sin B = 18.9$

Case 2: $C = 140.2°$, impossible, because $B + C = 202.7° > 180°$

27. This is the SSA case. $\sin B = \dfrac{b\sin A}{a} = \dfrac{18.2\sin 50.7°}{10.1} = 1.3944$

$\sin B > 1$ hence no such triangle exists.

29. $a^2 = b^2 + c^2 - 2bc\cos A$, $a = 18.5$, $\cos B = \dfrac{a^2 + c^2 - b^2}{2ac}$, $B = 36.7°$

$C = 180° - A - B = 78.3°$

31. $b = \dfrac{a\sin B}{\sin A} = \dfrac{110\sin 35°}{\sin 19°} = 193.8$, $C = 180° - A - B = 126°$,

$c = \dfrac{a\sin C}{\sin A} = \dfrac{110\sin 126°}{\sin 19°} = 273.3$

33. First determine angle C: $C = 180° - 57° - 42° = 81°$

Then distance $= b = \dfrac{c\sin B}{\sin C} = \dfrac{200\sin 42°}{\sin 81°} = 135.5$ meters.

35. Let A be the angle at the top of the tower. Then

$\sin A = \dfrac{40\sin 57°}{54} = 2.5$, $A = 38.4°$, $B = 180° - 38.4° - 57° = 84.6°$

The required angle $\alpha = 90° - B = 5.4°$.

37. The third angle of the triangle $= 180° - 37.25° - 34.85° = 107.9°$. Clearly the vertical

side of the triangle $= 8 - 5 = 3$. Then $d = \dfrac{3\sin 107.9°}{\sin 34.85°} = 5.0$ feet.

39. Sketch a figure.

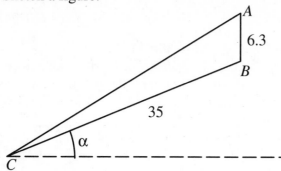

Angle $ACB = 10°$.

$\sin A = \dfrac{35\sin 10°}{6.3}$, $A = 74.7°$

$B = 180° - A - C = 95.3°$.

Then $\alpha = 95.3° - 90° = 5.3°$.

41. Call Harville vertex A, Eastview vertex B, Wellstone vertex C.

$\sin C = \dfrac{18\sin 40°}{20}$, $C = 35.3°$, $B = 180° - A - C = 104.7°$.

Required distance $= b = 20\sin 104.7°/\sin 40° = 30.1$ km.

43. First, $C = 180° - 49° - 128° = 3°\,C$, then using the Law of Sines to get the length of

AB, $AB = \dfrac{144\sin 3°}{\sin 49°} = 9.99$ meters.

45. Label the text figure: $A =$ Ray's position, $B =$ Tom's position, $C =$ plane's position, $D =$ base of tower. Triangle ABD is a right triangle, hence $\cos 8.6° = 5280/c$, $c = 5340$ ft. (Or use the Pythagorean Theorem) $C = 180° - 81° - 67° = 32°$. $b = c\sin B/\sin C = 9761$ ft. The height of the plane is $\sin 81° = h/9761$, $h = 9642$ ft.

47. a. Solve triangle ABC to find $\angle BAC$, the angle at A. $180° - \angle BAC = \angle EAB$.
  Solve triangle ABD to find $\angle ABD$, the angle at B. $180° - \angle ABD = \angle EBA$.
  Now solve triangle EAB using the two angles and included side to find EA.
  b. Follow the steps in part **a.**

$\cos BAC = \dfrac{25^2 + 75^2 - 80^2}{2(25)(75)}$, $\angle BAC = 92.3°$, $\angle EAB = 87.7°$.

$\cos ABD = \dfrac{75^2 + 22^2 - 90^2}{2(75)(22)}$, $\angle ABD = 127.1°$, $\angle EBA = 52.9°$.

$\angle AEB = 180° - 87.7° - 52.9° = 39.4°$

$AE = \dfrac{75\sin 52.9°}{\sin 39.4°} = 94$ ft

49. Sketch a figure.

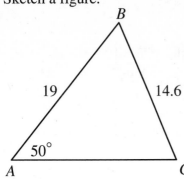

$$\sin C = \frac{19\sin 50°}{14.6} = .9969, \ C = 85.5° \ \text{or} \ C = 94.5°.$$

Case 1: $C = 85.5°, \ B = 44.5°$

Case 2: $C = 94.5°, \ B = 35.5°$ (irrelevant, since the top has already hit the ground in the case 1 situation.)

$$b = \frac{a\sin B}{\sin A} = \frac{14.6\sin 44.5°}{50°} = 13.36 \ \text{meters.}$$

51. a. See example 1 of section 8.3.

b. $\sin C = \dfrac{15\sin 60°}{13} = .9993,$ thus $C = 87.8°,$ or $C = 180° - 87.8° = 92.2°$

c. $A = 27.8°$

8.4.A The Area of a Triangle

1. $\frac{1}{2}ab\sin C = \frac{1}{2}(4)(8)\sin 37° = 9.6$ square units

3. $\frac{1}{2}ab\sin C = \frac{1}{2}(12)(7)\sin 68° = 38.9$

5. Use Heron's formula: $s = \frac{1}{2}(a+b+c) = 22$

$$A = \sqrt{s(s-a)(s-b)(s-c)}$$
$$= \sqrt{22(22-11)(22-15)(22-18)}$$

$= 82.3$ square units

7. Use Heron's formula: $s = \frac{1}{2}(a+b+c) = 13.5$

$$A = \sqrt{s(s-a)(s-b)(s-c)}$$
$$= \sqrt{13.5(13.5-7)(13.5-9)(13.5-11)}$$

$= 31.4$ square units

9. Use the distance formula to find side lengths of $\sqrt{29}, \sqrt{10}, \sqrt{61}$.
Then use Heron's formula: $s = \frac{1}{2}(a+b+c) = 8.1788$

$$A = \sqrt{s(s-a)(s-b)(s-c)}$$

$= 6.5$ square units

11. As in the hint, separate the region into two triangular regions. If this is done by drawing a diagonal from the vertex with the $89°$ angle to the vertex with the $72°$ angle, then the areas of the two triangles are given by the formula Area $= \frac{1}{2}ab\sin C$. The entire area is $\frac{1}{2}(55)(135)\sin 96° + \frac{1}{2}(120)(68.4)\sin 103 = 7691$ square units.

13. From the Law of Sines, the angle A between the 24-foot side and the third side is given by $\sin A = \dfrac{20\sin 44°}{24} = .5789$, $A = 35.3°$. The third angle is then given by $180° - (44° + A) = 100.6°$. The area is then $\frac{1}{2}(20)(24)\sin 100.6° = 235.9$ square feet.

15. Use Heron's formula to find the area of the deck: $s = \frac{1}{2}(65+72+88) = 112.5$.

$$A = \sqrt{s(s-a)(s-b)(s-c)}$$
$$= \sqrt{112.5(112.5-65)(112.5-72)(112.5-88)}$$

$A = 2302.7$ square feet
Divide by 400 square feet per gallon to obtain 5.8 gallons.

17. Let A', B', C' be the points directly below A, B, C.

In right triangle $A'B'B$, $A'B = \sqrt{61}$, $\angle BA'B' = 39.806°$.

In triangle $BA'A$, $\angle BA'A = 90° - 39.806° = 50.194°$, $AB = 6.1$ from the Law of Cosines.

In right triangle $A'C'C$, $A'C = \sqrt{34}$, $\angle CA'C' = 30.964°$.

In triangle $CA'A$, $\angle CA'A = 90° - 30.964° = 59.036°$, $AC = 5.1$.

In right triangle $B'C'C$, $B'C = 5$, $\angle CB'C' = 36.870°$.

In triangle CBB', $\angle CBB' = 90° - 36.870° = 53.130°$, $BC = 4.5$.

Now use Heron's formula to find the area of triangle ABC, with sides $a = 4.5$, $b = 5.1$, $c = 6.1$, obtaining 11.18 square units.

19. a. $A = \dfrac{1}{2}(4)(5)^2 \sin\left(\frac{360°}{4}\right) = 50$

    b. $A = \dfrac{1}{2}(6)(2)^2 \sin\left(\dfrac{360°}{6}\right) = 6\sqrt{3}$

21. Since $12 + 20 < 36$, this violates the Triangle Inequality that states that the sum of two sides of a triangle is greater than the third side. This is not a triangle and the area is undefined.

Chapter 8 Review Exercises

1. $r = \sqrt{x^2 + y^2} = \sqrt{5^2 + 3^2} = \sqrt{34}$

$\sin\theta = \dfrac{3}{\sqrt{34}}, \cos\theta = \dfrac{5}{\sqrt{34}}, \tan\theta = \dfrac{3}{5}$

3. **(d)**

5. Since the hypotenuse is $\sqrt{65}$, **(e)** is true.

7. $b = \sqrt{a^2 + c^2} = \sqrt{10^2 + 14^2} = 2\sqrt{74} \approx 17.2$

$\tan A = \dfrac{10}{14}, A = \tan^{-1}\left(\dfrac{10}{14}\right) \approx 35.5°$

$\angle C = 90° - 35.5 = 54.5°$

9. $\angle A = 90° - \angle C = 90° - 35° = 55°$

$\cos C = \dfrac{a}{b}, \ \cos 35° = \dfrac{12}{b}, \ b = \dfrac{12}{\cos 35°} = 14.65$

$\tan C = \dfrac{c}{a}, \ \tan 35° = \dfrac{c}{12}, \ c = 12\tan 35° = 8.40$

11. Let h be the height of the tower.

$\tan 57.3° = \dfrac{h}{145}, \ h = 145\tan 57.3° = 225.9$ feet

13. $\tan\theta = \dfrac{140}{5280}, \ \theta = \tan^{-1}\dfrac{140}{5280} = 1.52°$

15. $\cos A = \dfrac{b^2 + c^2 - a^2}{2bc}; A = 58.8°; \cos B = \dfrac{a^2 + c^2 - b^2}{2ac}; B = 34.8°$

$C = 180° - A - B = 86.4°$

17. $b^2 = a^2 + c^2 - 2ac\cos B; b = 20.4; \cos A = \dfrac{b^2 + c^2 - a^2}{2bc}; A = 26.4°$

$C = 180° - A - B = 38.6°$

19. Let c be the distance between the trains after 3 hours.
Train 1 has gone $45\cdot3 = 135$ miles $= a$.
Train 2 has gone $70\cdot3 = 210$ miles $= b$.
$c^2 = 135^2 + 210^2 - 2\cdot135\cdot210\cos 120° = 301$ miles

21. $A = 180° - B - C = 16°, a = c\sin A/\sin C = 1.5, b = c\sin B/\sin C = 4.5$

23. This is the SSA case.

$$\sin A = \frac{a \sin C}{c} = \frac{75 \sin 62°}{95} = .6971$$

$A = 44.2°, \ B = 73.8°, b = c \sin B / \sin C = 103.3$

25. This is the SSA case.

$$\sin B = \frac{b \sin A}{a} = \frac{4 \sin 60°}{3.5} = .9897$$

$B = 81.8°$ or $B = 180° - 81.8° = 98.2°$

Case 1: $B = 81.8°, C = 38.2°,$

$\qquad\qquad c = b \sin C / \sin B = 2.5$

Case 2: $B = 98.2°, C = 21.8°,$

$\qquad\qquad c = b \sin C / \sin B = 1.5$

27. $\frac{1}{2} bc \sin A = \frac{1}{2}(24)(15) \sin 55° = 147.4$ square units.

29. Sketch a figure.

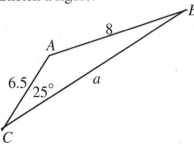

$$\sin B = \frac{6.5 \sin 25°}{8}, \ B = 20.1°, \ A = 134.9°, \ a = \frac{8 \sin 134.9°}{\sin 25°} = 13.4 \text{ km.}$$

(The other value for B is too large for this triangle.)

31. Use the Law of Sines.

$\qquad C = 180° - A - B = 57°, a = b \sin A / \sin B = 38.5, c = b \sin C / \sin B = 43.4$

33. Use the Law of Cosines.

$$b^2 = a^2 + c^2 - 2ac \cos B; b = 12.0; \cos A = \frac{b^2 + c^2 - a^2}{2bc}; A = 75.1°$$

$C = 180° - A - B = 28.9°$

35. Let Alice be vertex A, Joe vertex B, and the flagpole vertex C. Thus $\angle C = 63°$.

$$a = \frac{240 \sin 54°}{\sin 63°} = 217.9 \text{ meters is the distance from Joe to the flagpole.}$$

$$b = \frac{240 \sin 63°}{\sin 63°} = 240 \text{ meters is the distance from Alice to the flagpole.}$$

37. Sketch a figure.

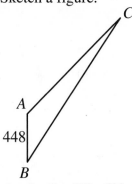

Angle $C = 65° - 62° = 3°$. Angle $B = 90° - 65° = 25°$. Angle $A = 152°$.

a. $b = \dfrac{448 \sin 25°}{\sin 3°} = 3617.65$ feet.

b. $a = \dfrac{448 \sin 152°}{\sin 3°} = 4018.71$ feet.

c. $\sin 65° = h/4019$, $h = 3642.19$ feet.

39. Use the Pythagorean Theorem to obtain the sides of triangle ABC.

$a = \sqrt{18^2 + 10^2} = \sqrt{424}$

$b = \sqrt{18^2 + 12^2} = \sqrt{468}$

$c = \sqrt{10^2 + 12^2} = \sqrt{244}$

Then angle $ABC = \cos^{-1}\left(\dfrac{a^2 + c^2 - b^2}{2ac}\right) = \cos^{-1}\left(\dfrac{424 + 244 - 468}{2\sqrt{424}\sqrt{244}}\right) = 71.89°$.

41. $A = \frac{1}{2}(5)(8)\sin 30° = 10$ square units

43. Use Heron's formula: $s = \frac{1}{2}(7 + 11 + 14) = 16$

$A = \sqrt{16(16 - 7)(16 - 11)(16 - 14)} = 37.95$ square units

Chapter 8 Test

1. **a.** $\sin\theta = \dfrac{2b}{\sqrt{1+4b^2}}$ **b.** $\cos\theta = \dfrac{1}{\sqrt{1+4b^2}}$ **c.** $\tan\theta = 2b$

2. By the drawing, $\sin\theta = \dfrac{7}{160}; \theta = \sin^{-1}\left(\tfrac{7}{160}\right) \approx 2.51°$

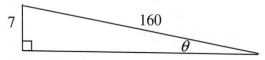

3. $\csc C = 3.5 = \dfrac{b}{c}$

    $3.5 = \dfrac{31.5}{c}; c = 9$

4. By the drawing: $\sin 55° = \dfrac{x}{310}; x = 310\sin 55° \approx 253.94.$ Plus the height of her hand to give: $253.94 + 3 = 256.94$ feet.

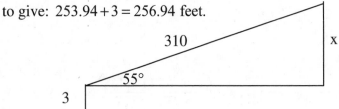

5. $\tan 30° = \dfrac{12}{c}; c = \dfrac{12}{\tan 30°} = \dfrac{12}{\sqrt{3}/3} = 12\sqrt{3}$

6. By the drawing: $\tan 42° = \dfrac{x}{200}; x = 200\tan 42° \approx 180.08$ feet.

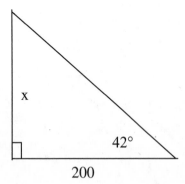

7. $A = 180° - 90° - 37° = 53°$

$$\tan 37° = \frac{c}{3.5}; c = 3.5 \tan 37° \approx 2.64$$

$$\cos 37° = \frac{3.5}{b}; b = \frac{3.5}{\cos 37°} \approx 4.38$$

8. By the drawing: $\theta = 90° - 72° = 18°$. Therefore,

$$\tan 18° = \frac{x}{5500}; x = 5500 \tan 18° \approx 1787.06 \text{ feet.}$$

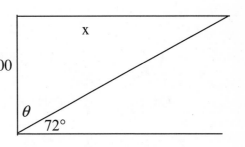

9. $\sin \theta = \frac{149}{250}; \theta = \sin^{-1}\left(\frac{149}{250}\right) \approx 36.58°$.   10. $\cos C = \frac{9}{10}; C = \cos^{-1}\left(\frac{9}{10}\right) \approx 25.84°$

$$A = 90° - 26° = 64.16°$$

11. Since the straight path forms a right triangle with the horizon, the top angle of bottom triangle will be $180° - 90° - 17° = 73°$. Then the angle supplementary with the straight path and the bottom of the statue is $180° - 73° = 107°$. Then by the Law of Cosines:

$$x = \sqrt{120^2 + 180^2 - 2(120)(180)\cos 107°} = 243.78 \text{ feet.}$$

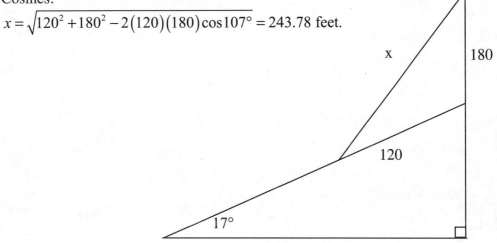

12. $A = 180° - 93.5° - 49.5° = 37°$

$$\frac{\sin 93.5°}{6} = \frac{\sin 46.5°}{c}; c = \frac{6\sin 46.5°}{\sin 93.5°} \approx 4.57$$

$$\frac{\sin 93.5°}{6} = \frac{\sin 37°}{a}; a = \frac{6\sin 37°}{\sin 93.5°} \approx 3.62$$

13. $b = \sqrt{6.9^2 + 16.5^2 - 2(6.9)(16.5)\cos 26.6°} = 10.78$

$\dfrac{\sin 26.6°}{10.78} = \dfrac{\sin A}{6.9}; A = \sin^{-1}\left(\dfrac{6.9\sin 26.6°}{10.78}\right) \approx 16.65°$

$C = 180° - 26.6° - 16.65° = 136.75°$

14. $b = \sqrt{22^2 + 15.3^2 - 2(22)(15.3)\cos 74°} = 23.08$

$\dfrac{\sin 74°}{23.08} = \dfrac{\sin A}{22}; A = \sin^{-1}\left(\dfrac{22\sin 74°}{23.08}\right) \approx 66.41°$

$C = 180° - 74° - 66.41° = 39.59°$

15. $A = \cos^{-1}\left(\dfrac{5.4^2 - 7.2^2 - 12^2}{-2(7.2)(12)}\right) \approx 15.29°$

$\dfrac{\sin 15.29°}{5.4} = \dfrac{\sin B}{7.2}; B = \sin^{-1}\left(\dfrac{7.2\sin 15.29°}{5.4}\right) \approx 20.59°$

$C = 180° - 15.29° - 20.59° = 144.11°$

16. $a = \sqrt{15.2^2 + 19.3^2 - 2(15.2)(19.3)\cos 44°} = 13.47$

$\dfrac{\sin 44°}{13.47} = \dfrac{\sin B}{15.2}; B = \sin^{-1}\left(\dfrac{15.2\sin 44°}{13.47}\right) \approx 51.61°$

$C = 180° - 44° - 51.61° = 84.39°$

17. $\dfrac{\sin 37.2°}{12.3} = \dfrac{\sin C}{20.2}; C = \sin^{-1}\left(\dfrac{20.2\sin 37.2°}{12.3}\right) \approx 83.18°$

Now, $180° - 83.18° = 96.82°$, thus $180° - 96.82° - 37.2° = 45.98°$

$C_2 = 83.18°$ and $C_1 = 96.82°$

$A_2 = 180° - 83.18° - 37.2° = 59.62°$

$A_1 = 180° - 96.82° - 37.2° = 45.98°$

$\dfrac{\sin 37.2°}{12.3} = \dfrac{\sin 59.62°}{a_2}; a_2 = \dfrac{12.3\sin 59.62°}{\sin 37.2°} \approx 17.55$

$\dfrac{\sin 37.2°}{12.3} = \dfrac{\sin 45.98°}{a_1}; a_1 = \dfrac{12.3\sin 45.98°}{\sin 37.2°} \approx 14.63$

18. Let $C = (1,-1), B = (5,1), A = (-8,5),$ then the triangle has sides of length

$$\overline{BC} = \sqrt{(5-1)^2 + (1-(-1))^2} = \sqrt{16+4} = 2\sqrt{5}$$

$$\overline{AC} = \sqrt{(-8-1)^2 + (5-(-1))^2} = \sqrt{81+36} = 3\sqrt{13}$$

$$\overline{AB} = \sqrt{(-8-5)^2 + (5-1)^2} = \sqrt{169+16} = \sqrt{185}$$

Using the Law of Cosines and Law of Sines gives:

$$C = \cos^{-1}\left(\frac{\sqrt{185}^2 - (2\sqrt{5})^2 - (3\sqrt{13})^2}{-2(2\sqrt{5})(3\sqrt{13})}\right) \approx 119.74°$$

$$\frac{\sin 119.74°}{\sqrt{185}} = \frac{\sin B}{3\sqrt{13}}; B = \sin^{-1}\left(\frac{3\sqrt{13}\sin 119.74°}{\sqrt{185}}\right) \approx 43.67°$$

$$C = 180° - 119.74° - 43.67° = 16.59°$$

19. Since the plane flew 425 mph for 1 hour and 42 minutes, the total distance traveled was $425\left(1+\frac{42}{60}\right) = 722.5,$ and the distance for the second part of the flight was $375\left(2+\frac{33}{60}\right) = 956.25.$ Then by the Law of Cosines:

$$x = \sqrt{722.5^2 + 956.25^2 - 2(722.5)(956.25)\cos 155°} \approx 1639.74 \text{ miles}$$

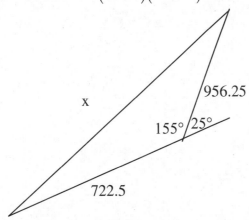

20. Since the straight path forms a right triangle with the horizon, the top angle of bottom triangle will be $180° - 90° - 7° = 83°.$ Then the angle supplementary with the straight path and the bottom of the statue is $180° - 83° = 97°.$ Thus, the top angle of the statue with the shadow is $57°.$ Then by the Law of Sines:

$$\frac{\sin 26°}{x} = \frac{\sin 57°}{9}; x = \frac{9\sin 26°}{\sin 57°} \approx 4.7 \text{ meters.}$$

21. From the drawing, $\tan 54° = \dfrac{h}{x}$ and $\tan 67° = \dfrac{h}{300-x}$. Solving both equations for x

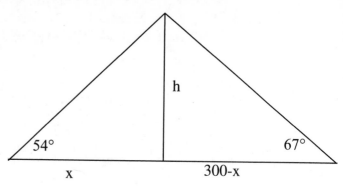

gives $x = \dfrac{h}{\tan 54°}$ and $x = -\dfrac{h}{\tan 67°} + 300$

Giving $\dfrac{h}{\tan 54°} = -\dfrac{h}{\tan 67°} + 300$

$\dfrac{h}{\tan 54°} + \dfrac{h}{\tan 67°} = 300$

$h\left(\dfrac{1}{\tan 54°} + \dfrac{1}{\tan 67°}\right) = 300$

$\therefore h = \dfrac{300}{\dfrac{1}{\tan 54°} + \dfrac{1}{\tan 67°}} \approx 260.64$ feet.

22. a. $s = \dfrac{1}{2}(4+6+9) = 9.5.$ So by Heron's Formula:

$S = \sqrt{9.5(9.5-4)(9.5-6)(9.5-9)} = 9.56$ square units

b. $S = \dfrac{1}{2}(7)(9)\sin 43° = 21.48$ square units

Chapter 9
Applications of Trigonometry

9.1 The Complex Plane and Polar Form for Complex Numbers

1-7. (graphs)

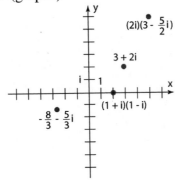

5. $(1+i)(1-i) = 1-i^2$
$$= 1-(-1) = 2$$

7. $2i\left(3-\frac{5}{2}i\right) = 6i - 5i^2$
$$= 6i - 5(-1)$$
$$= 5 + 6i$$

9. $|5-12i| = \sqrt{5^2 + (-12)^2} = 13$

11. $|1+\sqrt{2}i| = \sqrt{1^2 + (\sqrt{2})^2} = \sqrt{3}$

13. $|-12i| = \sqrt{0^2 + (-12)^2} = 12$

15. Let $z=1$, $w=i$.
Then $|z+w| = |1+i| = \sqrt{2}$
but $|z|+|w| = |1|+|i|$
$$= 1+1 = 2.$$

17.

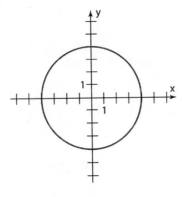

19. The graph consists of all points that are 10 units away from $(1,0)$.

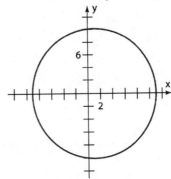

21. The graph consists of all points that are 4 units away from $2i$ or $(0,2)$.

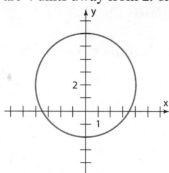

23. The graph consists of all points with x-coordinate equal to 2.

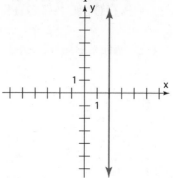

25.
$$2\left(\cos\frac{\pi}{4}+i\sin\frac{\pi}{4}\right)=2\left(\frac{\sqrt{2}}{2}+i\frac{\sqrt{2}}{2}\right)$$
$$=\sqrt{2}+i\sqrt{2}$$

27.
$$\cos\frac{\pi}{2}+i\sin\frac{\pi}{2}=0+i$$
$$=i$$

29.
$$5\left(\cos\frac{2\pi}{3}+i\sin\frac{2\pi}{3}\right)=5\left(-\frac{1}{2}+i\frac{\sqrt{3}}{2}\right)$$
$$=-\frac{5}{2}+i\frac{5\sqrt{3}}{2}$$

31.
$$1.5\left(\cos\frac{\pi}{6}+i\sin\frac{\pi}{6}\right)=1.5\left(\frac{\sqrt{3}}{2}+i\frac{1}{2}\right)$$
$$=\frac{3\sqrt{3}}{4}+\frac{3}{4}i$$

33. $2(\cos 4+i\sin 4)=2(-.6536-.7568i)$
$$=-1.3073-1.5136i$$

35. $4(\cos 2+i\sin 2)=4(-.4161+.9093i)$
$$=-1.6646+3.6372i$$

37. $r=\sqrt{3^2+3^2}=3\sqrt{2},\tan\theta=\frac{3}{3}=1,z$ lies in the first quadrant, hence
$$\theta=\tan^{-1}1=\frac{\pi}{4}\Rightarrow 3+3i=3\sqrt{2}\left(\cos\frac{\pi}{4}+i\sin\frac{\pi}{4}\right).$$

39. $r=\sqrt{2^2+\left(2\sqrt{3}\right)^2}=4,\tan\theta=\frac{2\sqrt{3}}{2}=\sqrt{3},z$ lies in the first quadrant, hence
$$\theta=\tan^{-1}\sqrt{3}=\frac{\pi}{3}\Rightarrow 2+2\sqrt{3}i=4\left(\cos\frac{\pi}{3}+i\sin\frac{\pi}{3}\right).$$

41. $r=\sqrt{\left(3\sqrt{3}\right)^2+(-3)^2}=6,\tan\theta=\frac{-3}{3\sqrt{3}}=-\frac{1}{\sqrt{3}},z$ lies in the fourth quadrant, hence
$$\theta=2\pi+\tan^{-1}\left(-\frac{1}{\sqrt{3}}\right)=\frac{11\pi}{6}\Rightarrow 3\sqrt{3}-3i=6\left(\cos\frac{11\pi}{6}+i\sin\frac{11\pi}{6}\right).$$

43. $r=\sqrt{\left(-\sqrt{3}\right)^2+\left(-\sqrt{3}\right)^2}=\sqrt{6},\tan\theta=\frac{-\sqrt{3}}{-\sqrt{3}}=1,z$ lies in the third quadrant, hence
$$\theta=\pi+\tan^{-1}1=\frac{5\pi}{4}\Rightarrow -\sqrt{3}-\sqrt{3}i=\sqrt{6}\left(\cos\frac{5\pi}{4}+i\sin\frac{5\pi}{4}\right).$$

45. $r = \sqrt{3^2 + 4^2} = 5, \tan \theta = \dfrac{4}{3}, z$ lies in the first quadrant, hence

$\theta = \tan^{-1} \dfrac{4}{3} = .9273 \Rightarrow 3 + 4i = 5\left(\cos(.9273) + i\sin(.9273)\right).$

47. $r = \sqrt{5^2 + (-12)^2} = 13, \tan \theta = \dfrac{-12}{5}, z$ lies in the fourth quadrant, hence

$\theta = 2\pi + \tan^{-1}\left(-\dfrac{12}{5}\right) = 5.1072 \Rightarrow 5 - 12i = 13\left(\cos(5.1072) + i\sin(5.1072)\right).$

49. $r = \sqrt{1^2 + 2^2} = \sqrt{5}, \tan \theta = \dfrac{2}{1} = 2, z$ lies in the first quadrant, hence

$\theta = \tan^{-1} 2 = 1.1071 \Rightarrow 1 + 2i = \sqrt{5}\left(\cos(1.1071) + i\sin(1.1071)\right).$

51. $r = \sqrt{\left(\dfrac{-5}{2}\right)^2 + \left(\dfrac{7}{2}\right)^2} = \sqrt{\dfrac{74}{4}}, \tan \theta = \dfrac{7}{2} \div \left(-\dfrac{5}{2}\right) = -\dfrac{7}{5}, z$ lies in the second quadrant,

hence $\theta = \pi + \tan^{-1}\left(-\dfrac{7}{5}\right) = 2.191 \Rightarrow -\dfrac{5}{2} + \dfrac{7}{2}i = \dfrac{\sqrt{74}}{2}\left(\cos(2.191) + i\sin(2.191)\right).$

53. $\left(\cos\dfrac{\pi}{2} + i\sin\dfrac{\pi}{2}\right) \cdot (\cos\pi + i\sin\pi) = \cos\dfrac{3\pi}{2} + i\sin\dfrac{3\pi}{2} = 0 + i(-1) = -i$

55. $4\left(\cos\dfrac{\pi}{4} + i\sin\dfrac{\pi}{4}\right) \cdot 3\left(\cos\dfrac{\pi}{12} + i\sin\dfrac{\pi}{12}\right) = 12\left(\cos\dfrac{\pi}{3} + i\sin\dfrac{\pi}{3}\right) = 12\left(\dfrac{1}{2} + i\dfrac{\sqrt{3}}{2}\right)$

$$= 6 + 6i\sqrt{3}$$

57. $3\left(\cos\dfrac{\pi}{8} + i\sin\dfrac{\pi}{8}\right) \cdot 12\left(\cos\dfrac{3\pi}{8} + i\sin\dfrac{3\pi}{8}\right) = 36\left(\cos\dfrac{\pi}{2} + i\sin\dfrac{\pi}{2}\right) = 36(0 + i(1)) = 36i$

59. $\dfrac{\cos\pi + i\sin\pi}{\cos\dfrac{2\pi}{3} + i\sin\dfrac{2\pi}{3}} = \cos\dfrac{\pi}{3} + i\sin\dfrac{\pi}{3} = \dfrac{1}{2} + \dfrac{\sqrt{3}}{2}i$

61. $\dfrac{8\left(\cos\dfrac{4\pi}{3} + i\sin\dfrac{4\pi}{3}\right)}{4\left(\cos\dfrac{7\pi}{6} + i\sin\dfrac{7\pi}{6}\right)} = 2\left(\cos\dfrac{\pi}{6} + i\sin\dfrac{\pi}{6}\right) = 2\left(\dfrac{\sqrt{3}}{2} + \dfrac{1}{2}i\right) = \sqrt{3} + i$

63. $\dfrac{6\left(\cos\dfrac{7\pi}{20}+i\sin\dfrac{7\pi}{20}\right)}{4\left(\cos\dfrac{\pi}{10}+i\sin\dfrac{\pi}{10}\right)}=\dfrac{3}{2}\left(\cos\dfrac{\pi}{4}+i\sin\dfrac{\pi}{4}\right)=\dfrac{3}{2}\left(\dfrac{\sqrt{2}}{2}+\dfrac{\sqrt{2}}{2}i\right)=\dfrac{3\sqrt{2}}{4}+\dfrac{3\sqrt{2}}{4}i$

65. $(1+i)\left(1+\sqrt{3}i\right)=\sqrt{2}\left(\cos\dfrac{\pi}{4}+i\sin\dfrac{\pi}{4}\right)\cdot 2\left(\cos\dfrac{\pi}{3}+i\sin\dfrac{\pi}{3}\right)=2\sqrt{2}\left(\cos\dfrac{7\pi}{12}+i\sin\dfrac{7\pi}{12}\right)$

67. $\dfrac{1+i}{1-i}=\dfrac{\sqrt{2}\left(\cos\dfrac{\pi}{4}+i\sin\dfrac{\pi}{4}\right)}{\sqrt{2}\left(\cos\left(-\dfrac{\pi}{4}\right)+i\sin\left(-\dfrac{\pi}{4}\right)\right)}=\cos\dfrac{\pi}{2}+i\sin\dfrac{\pi}{2}$

69. $3i\left(2\sqrt{3}+2i\right)=3\left(\cos\dfrac{\pi}{2}+i\sin\dfrac{\pi}{2}\right)\cdot 4\left(\cos\dfrac{\pi}{6}+i\sin\dfrac{\pi}{6}\right)=12\left(\cos\dfrac{2\pi}{3}+i\sin\dfrac{2\pi}{3}\right)$

71. $i(i+1)\left(-\sqrt{3}+i\right)=\left(\cos\dfrac{\pi}{2}+i\sin\dfrac{\pi}{2}\right)\cdot\sqrt{2}\left(\cos\dfrac{\pi}{4}+i\sin\dfrac{\pi}{4}\right)\cdot 2\left(\cos\dfrac{5\pi}{6}+i\sin\dfrac{5\pi}{6}\right)$

$$=2\sqrt{2}\left(\cos\left(\dfrac{\pi}{2}+\dfrac{\pi}{4}+\dfrac{5\pi}{6}\right)+i\sin\left(\dfrac{\pi}{2}+\dfrac{\pi}{4}+\dfrac{5\pi}{6}\right)\right)$$

$$=2\sqrt{2}\left(\cos\dfrac{19\pi}{12}+i\sin\dfrac{19\pi}{12}\right)$$

73. Since $z=r(\cos\theta+i\sin\theta), i=\cos\dfrac{\pi}{2}+i\sin\dfrac{\pi}{2}, iz=r\left(\cos\left(\theta+\dfrac{\pi}{2}\right)+i\sin\left(\theta+\dfrac{\pi}{2}\right)\right)$ the

argument of iz is $90°$ more than that of z, hence iz is equivalent to rotating z through $90°$ in the complex plane.

75. a. slope $= \dfrac{b-0}{a-0} = \dfrac{b}{a}$ **b.** slope $= \dfrac{d-0}{c-0} = \dfrac{d}{c}$

c. The line L has slope $= \dfrac{d}{c}$; from the point-slope form, the equation is

$$y - b = \dfrac{d}{c}(x-a).$$

d. The line M has slope $= \dfrac{b}{a}$; from the point-slope form, the equation is

$$y - d = \dfrac{b}{a}(x-c).$$

e.

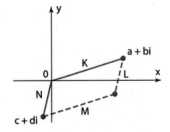

f. Substituting $(a+c, b+d)$ in the equation of line L yields:

$$y - b = \dfrac{d}{c}(x-a)$$

$$b + d - b = \dfrac{d}{c}(a+c-a)$$

$$d = d$$

Substituting $(a+c, b+d)$ in the equation of line M yields:

$$y - d = \dfrac{b}{a}(x-c)$$

$$b + d - d = \dfrac{b}{a}(a+c-c)$$

$$b = b$$

Thus $(a+c, b+d)$ lies on both lines L and M. The conclusion follows as stated in the text.

77. a. $r_2\left(\cos\theta_2 + i\sin\theta_2\right)\left(\cos\theta_2 - i\sin\theta_2\right) = r_2\left(\cos^2\theta_2 + \sin^2\theta_2\right) = r_2 \cdot 1 = r_2$

b. $r_1\left(\cos\theta_1 + i\sin\theta_1\right)\left(\cos\theta_2 - i\sin\theta_2\right)$

$= r_1\left(\cos\theta_1\cos\theta_2 - i\cos\theta_1\sin\theta_2 + i\sin\theta_1\cos\theta_2 - i^2\sin\theta_1\sin\theta_2\right)$

$= r_1\left(\cos\theta_1\cos\theta_2 + \sin\theta_1\sin\theta_2 + i\left(\sin\theta_1\cos\theta_2 - \cos\theta_1\sin\theta_2\right)\right)$

$= r_1\left(\cos(\theta_1 - \theta_2) + i\sin(\theta_1 - \theta_2)\right)$

9.2 DeMoivre's Theorem and nth Roots of Complex Numbers

1. $\left(\cos\dfrac{\pi}{12}+i\sin\dfrac{\pi}{12}\right)^6 = \cos\dfrac{6\pi}{12}+i\sin\dfrac{6\pi}{12} = \cos\dfrac{\pi}{2}+i\sin\dfrac{\pi}{2} = 0+1i \text{ or } i$

3. $\left[2\left(\cos\dfrac{\pi}{24}+i\sin\dfrac{\pi}{24}\right)\right]^8 = 2^8\left(\cos\dfrac{8\pi}{24}+i\sin\dfrac{8\pi}{24}\right) = 256\left(\cos\dfrac{\pi}{3}+i\sin\dfrac{\pi}{3}\right)$

$$= 256\left(\dfrac{1}{2}+i\dfrac{\sqrt{3}}{2}\right) = 128+128i\sqrt{3}$$

5. $\left[3\left(\cos\dfrac{7\pi}{30}+i\sin\dfrac{7\pi}{30}\right)\right]^5 = 3^5\left(\cos\dfrac{35\pi}{30}+i\sin\dfrac{35\pi}{30}\right) = 243\left(\cos\dfrac{7\pi}{6}+i\sin\dfrac{7\pi}{6}\right)$

$$= 243\left(-\dfrac{\sqrt{3}}{2}-i\dfrac{1}{2}\right) = -\dfrac{243\sqrt{3}}{2}-\dfrac{243}{2}i$$

7. $\left(\dfrac{1}{2}+\dfrac{\sqrt{3}}{2}i\right)^3 = \left(\cos\dfrac{\pi}{3}+i\sin\dfrac{\pi}{3}\right)^3 = \cos\pi+i\sin\pi = -1+0i = -1$

9. $(1-i)^{12} = \left[\sqrt{2}\left(\cos\dfrac{7\pi}{4}+i\sin\dfrac{7\pi}{4}\right)\right]^{12} = \left(\sqrt{2}\right)^{12}\left(\cos\dfrac{84\pi}{4}+i\sin\dfrac{84\pi}{4}\right)$

$$= 64(\cos 21\pi+i\sin 21\pi) = 64(-1+0i) = -64+0i = -64$$

11. $\left(\dfrac{\sqrt{3}}{2}+\dfrac{1}{2}i\right)^{10} = \left(\cos\dfrac{\pi}{6}+i\sin\dfrac{\pi}{6}\right)^{10} = \cos\dfrac{10\pi}{6}+i\sin\dfrac{10\pi}{6} = \cos\dfrac{5\pi}{3}+i\sin\dfrac{5\pi}{3} = \dfrac{1}{2}-i\dfrac{\sqrt{3}}{2}$

13. $\left(-\dfrac{1}{\sqrt{2}}+\dfrac{i}{\sqrt{2}}\right)^{14} = \left(\cos\dfrac{3\pi}{4}+i\sin\dfrac{3\pi}{4}\right)^{14} = \cos\dfrac{42\pi}{4}+i\sin\dfrac{42\pi}{4}$

$$= \cos\dfrac{21\pi}{2}+i\sin\dfrac{21\pi}{2} = 0+1i = i$$

15. Apply the roots-of-unity formula with $n = 4$, $k = 0,1,2,3$.
    $k = 0$: $\cos 0+i\sin 0 = 1$
    $k = 1$: $\cos 2\pi/4+i\sin 2\pi/4 = i$
    $k = 2$: $\cos 4\pi/4+i\sin 4\pi/4 = -1$
    $k = 3$: $\cos 6\pi/4+i\sin 6\pi/4 = -i$

17. Apply the roots-of-unity formula with $n = 2, k = 0, 1$.

$$k = 0; \quad \sqrt{36}\left[\cos\left(\frac{\pi/3}{2}\right) + i\sin\left(\frac{\pi/3}{2}\right)\right] = 6\left(\cos\frac{\pi}{6} + i\sin\frac{\pi}{6}\right)$$

$$k = 1; \quad \sqrt{36}\left[\cos\left(\frac{\pi/3 + 2\pi}{2}\right) + i\sin\left(\frac{\pi/3 + 2\pi}{2}\right)\right] = 6\left(\cos\frac{7\pi}{6} + i\sin\frac{7\pi}{6}\right)$$

19. Apply the roots-of-unity formula with $n = 3, \ k = 0, 1, 2$.

$$k = 0: \quad \sqrt[3]{64}\left[\cos\left(\frac{\pi/5}{3}\right) + i\sin\left(\frac{\pi/5}{3}\right)\right] = 4\left(\cos\frac{\pi}{15} + i\sin\frac{\pi}{15}\right)$$

$$k = 1: \quad \sqrt[3]{64}\left[\cos\left(\frac{\pi/5 + 2\pi}{3}\right) + i\sin\left(\frac{\pi/5 + 2\pi}{3}\right)\right] = 4\left(\cos\frac{11\pi}{15} + i\sin\frac{11\pi}{15}\right)$$

$$k = 2: \quad \sqrt[3]{64}\left[\cos\left(\frac{\pi/5 + 4\pi}{3}\right) + i\sin\left(\frac{\pi/5 + 4\pi}{3}\right)\right] = 4\left(\cos\frac{7\pi}{5} + i\sin\frac{7\pi}{5}\right)$$

21. Apply the roots-of-unity formula with $n = 4, \ k = 0, 1, 2, 3$.

$$k = 0: \quad \sqrt[4]{81}\left[\cos\left(\frac{\pi/12}{4}\right) + i\sin\left(\frac{\pi/12}{4}\right)\right] = 3\left(\cos\frac{\pi}{48} + i\sin\frac{\pi}{48}\right)$$

$$k = 1: \sqrt[4]{81}\left[\cos\left(\frac{\pi/12 + 2\pi}{4}\right) + i\sin\left(\frac{\pi/12 + 2\pi}{4}\right)\right] = 3\left(\cos\frac{25\pi}{48} + i\sin\frac{25\pi}{48}\right)$$

$$k = 2: \sqrt[4]{81}\left[\cos\left(\frac{\pi/12 + 4\pi}{4}\right) + i\sin\left(\frac{\pi/12 + 4\pi}{4}\right)\right] = 3\left(\cos\frac{49\pi}{48} + i\sin\frac{49\pi}{48}\right)$$

$$k = 3: \sqrt[4]{81}\left[\cos\left(\frac{\pi/12 + 6\pi}{4}\right) + i\sin\left(\frac{\pi/12 + 6\pi}{4}\right)\right] = 3\left(\cos\frac{73\pi}{48} + i\sin\frac{73\pi}{48}\right)$$

23. Apply the roots-of-unity formula with $n = 5, \ k = 0, 1, 2, 3, 4$ to
$-1 = 1(\cos\pi + i\sin\pi)$

$$k = 0: \quad \sqrt[5]{1}\left[\cos\frac{\pi}{5} + i\sin\frac{\pi}{5}\right] = \cos\frac{\pi}{5} + i\sin\frac{\pi}{5}$$

$$k = 1: \sqrt[5]{1}\left[\cos\left(\frac{\pi + 2\pi}{5}\right) + i\sin\left(\frac{\pi + 2\pi}{5}\right)\right] = \cos\frac{3\pi}{5} + i\sin\frac{3\pi}{5}$$

$$k = 2: \sqrt[5]{1}\left[\cos\left(\frac{\pi + 4\pi}{5}\right) + i\sin\left(\frac{\pi + 4\pi}{5}\right)\right] = \cos\pi + i\sin\pi$$

$$k = 3: \sqrt[5]{1}\left[\cos\left(\frac{\pi + 6\pi}{5}\right) + i\sin\left(\frac{\pi + 6\pi}{5}\right)\right] = \cos\frac{7\pi}{5} + i\sin\frac{7\pi}{5}$$

$$k = 4: \sqrt[5]{1}\left[\cos\left(\frac{\pi + 8\pi}{5}\right) + i\sin\left(\frac{\pi + 8\pi}{5}\right)\right] = \cos\frac{9\pi}{5} + i\sin\frac{9\pi}{5}$$

25. Apply the roots-of-unity formula with $n = 5$, $k = 0, 1, 2, 3, 4$ to

$$i = 1\left(\cos\frac{\pi}{2} + i\sin\frac{\pi}{2}\right)$$

$$k = 0: \sqrt[5]{1}\left[\cos\left(\frac{\pi/2}{5}\right) + i\sin\left(\frac{\pi/2}{5}\right)\right] = \cos\frac{\pi}{10} + i\sin\frac{\pi}{10}$$

$$k = 1: \sqrt[5]{1}\left[\cos\left(\frac{\pi/2 + 2\pi}{5}\right) + i\sin\left(\frac{\pi/2 + 2\pi}{5}\right)\right] = \cos\frac{\pi}{2} + i\sin\frac{\pi}{2}$$

$$k = 2: \sqrt[5]{1}\left[\cos\left(\frac{\pi/2 + 4\pi}{5}\right) + i\sin\left(\frac{\pi/2 + 4\pi}{5}\right)\right] = \cos\frac{9\pi}{10} + i\sin\frac{9\pi}{10}$$

$$k = 3: \sqrt[5]{1}\left[\cos\left(\frac{\pi/2 + 6\pi}{5}\right) + i\sin\left(\frac{\pi/2 + 6\pi}{5}\right)\right] = \cos\frac{13\pi}{10} + i\sin\frac{13\pi}{10}$$

$$k = 4: \sqrt[5]{1}\left[\cos\left(\frac{\pi/2 + 8\pi}{5}\right) + i\sin\left(\frac{\pi/2 + 8\pi}{5}\right)\right] = \cos\frac{17\pi}{10} + i\sin\frac{17\pi}{10}$$

27. Apply the roots-of-unity formula with $n = 2$, $k = 0, 1$ to $1 + i = \sqrt{2}\left(\cos\pi/4 + i\sin\pi/4\right)$ to obtain

$$k = 0: \sqrt[4]{2}\left[\cos\left(\frac{\pi/4}{2}\right) + i\sin\left(\frac{\pi/4}{2}\right)\right] = \sqrt[4]{2}\left(\cos\frac{\pi}{8} + i\sin\frac{\pi}{8}\right)$$

$$k = 1: \sqrt[4]{2}\left[\cos\left(\frac{\pi/4 + 2\pi}{2}\right) + i\sin\left(\frac{\pi/4 + 2\pi}{2}\right)\right] = \sqrt[4]{2}\left(\cos\frac{9\pi}{8} + i\sin\frac{9\pi}{8}\right)$$

29. Apply the roots-of-unity formula with $n = 4$, $k = 0, 1, 2, 3$ to

$$8\sqrt{3} + 8i = 16\left(\cos\frac{\pi}{6} + i\sin\frac{\pi}{6}\right) \text{ to obtain}$$

$$k = 0: \sqrt[4]{16}\left[\cos\left(\frac{\pi/6}{4}\right) + i\sin\left(\frac{\pi/6}{4}\right)\right] = 2\left(\cos\frac{\pi}{24} + i\sin\frac{\pi}{24}\right)$$

$$k = 1: \sqrt[4]{16}\left[\cos\left(\frac{\pi/6 + 2\pi}{4}\right) + i\sin\left(\frac{\pi/6 + 2\pi}{4}\right)\right] = 2\left(\cos\frac{13\pi}{24} + i\sin\frac{13\pi}{24}\right)$$

$$k = 2: \sqrt[4]{16}\left[\cos\left(\frac{\pi/6 + 4\pi}{4}\right) + i\sin\left(\frac{\pi/6 + 4\pi}{4}\right)\right] = 2\left(\cos\frac{25\pi}{24} + i\sin\frac{25\pi}{24}\right)$$

$$k = 3: \sqrt[4]{16}\left[\cos\left(\frac{\pi/6 + 6\pi}{4}\right) + i\sin\left(\frac{\pi/6 + 6\pi}{4}\right)\right] = 2\left(\cos\frac{37\pi}{24} + i\sin\frac{37\pi}{24}\right)$$

31. The solutions are the six sixth roots of -1. Write $-1 = \cos\pi + i\sin\pi$ and apply the root formula to obtain

$$\cos\frac{\pi}{6} + i\sin\frac{\pi}{6} = \frac{\sqrt{3}}{2} + i\frac{1}{2}, \quad \cos\frac{7\pi}{6} + i\sin\frac{7\pi}{6} = -\frac{\sqrt{3}}{2} - i\frac{1}{2}, \quad \cos\frac{\pi}{2} + i\sin\frac{\pi}{2} = i$$

$$\cos\frac{3\pi}{2} + i\sin\frac{3\pi}{2} = -i, \quad \cos\frac{5\pi}{6} + i\sin\frac{5\pi}{6} = -\frac{\sqrt{3}}{2} + i\frac{1}{2}, \quad \cos\frac{11\pi}{6} + i\sin\frac{11\pi}{6} = \frac{\sqrt{3}}{2} - i\frac{1}{2}$$

33. The solutions are the three cube roots of i. Write $i = \cos(\pi/2) + i\sin(\pi/2)$ and apply the root formula to obtain:

$$\cos\frac{\pi}{6} + i\sin\frac{\pi}{6} = \frac{\sqrt{3}}{2} + i\frac{1}{2}, \quad \cos\frac{5\pi}{6} + i\sin\frac{5\pi}{6} = -\frac{\sqrt{3}}{2} + \frac{1}{2}i, \quad \cos\frac{3\pi}{2} + i\sin\frac{3\pi}{2} = -i.$$

35. The solutions are the three cube roots of $-27i$. Write $-27i = 27\left(\cos\frac{3\pi}{2} + i\sin\frac{3\pi}{2}\right)$ and apply the root formula to obtain: $3(\cos(\pi/2) + i\sin(\pi/2)) = 3i$,

$$3\left(\cos\frac{7\pi}{6} + i\sin\frac{7\pi}{6}\right) = -\frac{3\sqrt{3}}{2} - \frac{3i}{2}, \quad 3\left(\cos\frac{11\pi}{6} + i\sin\frac{11\pi}{6}\right) = \frac{3\sqrt{3}}{2} - \frac{3i}{2}.$$

37. The solutions are the five fifth roots of $243i$. Write $243i = 243\left(\cos\frac{\pi}{2} + i\sin\frac{\pi}{2}\right)$ and apply the root formula to obtain $3\left[\cos\left(\frac{\pi + 4k\pi}{10}\right) + i\sin\left(\frac{\pi + 4k\pi}{10}\right)\right], k = 0,1,2,3,4$

39. The solutions are the four fourth roots of $-1 + \sqrt{3}i$. Write $-1 + \sqrt{3}i = 2(\cos(2\pi/3) + i\sin(2\pi/3))$ and apply the root formula to obtain

$$\sqrt[4]{2}\left(\cos\frac{\pi}{6} + i\sin\frac{\pi}{6}\right) = \sqrt[4]{2}\left(\frac{\sqrt{3}}{2} + i\frac{1}{2}\right), \quad \sqrt[4]{2}\left(\cos\frac{2\pi}{3} + i\sin\frac{2\pi}{3}\right) = \sqrt[4]{2}\left(-\frac{1}{2} + i\frac{\sqrt{3}}{2}\right),$$

$$\sqrt[4]{2}\left(\cos\frac{7\pi}{6} + i\sin\frac{7\pi}{6}\right) = \sqrt[4]{2}\left(-\frac{\sqrt{3}}{2} - i\frac{1}{2}\right), \quad \sqrt[4]{2}\left(\cos\frac{5\pi}{3} + i\sin\frac{5\pi}{3}\right) = \sqrt[4]{2}\left(\frac{1}{2} - i\frac{\sqrt{3}}{2}\right)$$

41. With calculator in parametric mode, set the values $0 \le t \le 2\pi, t$-step $\approx .067, -1.5 \le x \le 1.5, -1 \le y \le 1$, and graph the unit circle $x = \cos t, y = \sin t$. Then reset the t-step to $2\pi/7$ and graph again. Using the trace feature, find the seven vertices to be $1, .6235 \pm .7818i, -.2225 \pm .9749i, -.9010 \pm .4339i$.

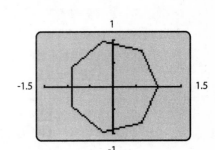

43. With calculator in parametric mode, set the values $0 \le t \le 2\pi$, t-step $\approx .067$, $-1.5 \le x \le 1.5, -1 \le y \le 1$, and graph the unit circle $x = \cos t$, $y = \sin t$. Then reset the t-step to $2\pi/8 = \pi/4$ and graph again. Using the trace feature, find the eight vertices to be $\pm 1, \pm i, .7071 \pm .7071i, -.7071 \pm .7071i$.

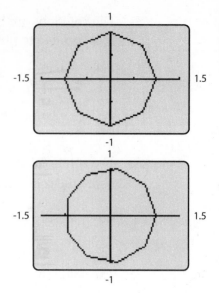

45. With calculator in parametric mode, set the values $0 \le t \le 2\pi$, t-step $\approx .067$, $-1.5 \le x \le 1.5, -1 \le y \le 1$, and graph the unit circle $x = \cos t$, $y = \sin t$. Then reset the t-step to $2\pi/9$ and graph again. Using the trace feature, find the nine vertices to be 1, $.7660 \pm .6428i, .1736 \pm .9848i, -.5 \pm .8660i, -.9397 \pm .3420i$.

47. As suggested, the solutions are the solutions of $x^4 - 1 = 0$ with the exception of 1 itself, that is : $-1, i, -i$.

49. As suggested, the solutions are the solutions of $x^6 - 1 = 0$ with the exception of 1 itself, that is : $\left(-1 \pm i\sqrt{3}\right)/2$, $\left(1 \pm i\sqrt{3}\right)/2$, and -1.

51. Since the argument of v is 12 times the argument of w, when w has increased by 2π, v has increased by 24π, making 12 trips around the circle.

53. Since the u_i are distinct, clearly, the vu_i are distinct. Since v is a solution of $z^n = r(\cos\theta + i\sin\theta)$, we can write $v^n = r(\cos\theta + i\sin\theta)$ and similarly $u_i^n = 1$. Then $\left(vu_i\right)^n = v^n u_i^n = r(\cos\theta + i\sin\theta) \cdot 1 = r(\cos\theta + i\sin\theta)$. Thus the vu_i are the solutions of the equation.

9.3 Vectors in the Plane

1. $\overrightarrow{PQ} = \langle 5-2, 9-3 \rangle = \langle 3,6 \rangle;\ \|\langle 3,6 \rangle\| = \sqrt{3^2+6^2} = 3\sqrt{5}$

3. $\overrightarrow{PQ} = \langle -4-(-7), -5-0 \rangle = \langle 3,-5 \rangle;\ \|\langle 3,-5 \rangle\| = \sqrt{3^2+(-5)^2} = \sqrt{34}$

5. $\overrightarrow{PQ} = \langle 7-1, 11-5 \rangle = \langle 6,6 \rangle$

7. $\overrightarrow{PQ} = \langle -10-(-4), 2-(-8) \rangle = \langle -6,10 \rangle$

9. $\overrightarrow{PQ} = \langle \frac{17}{5}-\frac{4}{5}, -\frac{12}{5}-(-2) \rangle = \langle \frac{13}{5}, -\frac{2}{5} \rangle$

11. $\mathbf{u}+\mathbf{v} = \langle -2,4 \rangle + \langle 6,1 \rangle = \langle 4,5 \rangle,\quad \mathbf{v}-\mathbf{u} = \langle 6,1 \rangle - \langle -2,4 \rangle = \langle 6-(-2), 1-4 \rangle = \langle 8,-3 \rangle$
$2\mathbf{u}-3\mathbf{v} = 2\langle -2,4 \rangle - 3\langle 6,1 \rangle = \langle -4,8 \rangle - \langle 18,3 \rangle = \langle -22,5 \rangle$

13. $\mathbf{u}+\mathbf{v} = \langle 3, 3\sqrt{2} \rangle + \langle 4\sqrt{2}, 1 \rangle = \langle 3+4\sqrt{2}, 3\sqrt{2}+1 \rangle$
$\mathbf{v}-\mathbf{u} = \langle 4\sqrt{2}, 1 \rangle - \langle 3, 3\sqrt{2} \rangle = \langle 4\sqrt{2}-3, 1-3\sqrt{2} \rangle$
$2\mathbf{u}-3\mathbf{v} = 2\langle 3, 3\sqrt{2} \rangle - 3\langle 4\sqrt{2}, 1 \rangle = \langle 6, 6\sqrt{2} \rangle - \langle 12\sqrt{2}, 3 \rangle = \langle 6-12\sqrt{2}, 6\sqrt{2}-3 \rangle$

15. $\mathbf{u}+\mathbf{v} = 2\langle -2,5 \rangle + \frac{1}{4}\langle -7,12 \rangle = \langle -4,10 \rangle + \langle -\frac{7}{4}, 3 \rangle = \langle -\frac{23}{4}, 13 \rangle$
$\mathbf{v}-\mathbf{u} = \frac{1}{4}\langle -7,12 \rangle - 2\langle -2,5 \rangle = \langle -\frac{7}{4}, 3 \rangle - \langle -4,10 \rangle = \langle \frac{9}{4}, -7 \rangle$
$2\mathbf{u}-3\mathbf{v} = 2\langle -4,10 \rangle - 3\langle -\frac{7}{4}, 3 \rangle = \langle -8,20 \rangle - \langle -\frac{21}{4}, 9 \rangle = \langle -\frac{11}{4}, 11 \rangle$

17. $\mathbf{u}+\mathbf{v} = 8\mathbf{i} + 2(3\mathbf{i}-2\mathbf{j}) = 8\mathbf{i}+6\mathbf{i}-4\mathbf{j} = 14\mathbf{i}-4\mathbf{j}$
$\mathbf{v}-\mathbf{u} = 2(3\mathbf{i}-2\mathbf{j}) - 8\mathbf{i} = 6\mathbf{i}-4\mathbf{j}-8\mathbf{i} = -2\mathbf{i}-4\mathbf{j}$
$2\mathbf{u}-3\mathbf{v} = 2(8\mathbf{i}) - 3\cdot2(3\mathbf{i}-2\mathbf{j}) = 16\mathbf{i}-18\mathbf{i}+12\mathbf{j} = -2\mathbf{i}+12\mathbf{j}$

19. $\mathbf{u}+\mathbf{v} = -(2\mathbf{i}+\frac{3}{2}\mathbf{j}) + \frac{3}{4}\mathbf{i} = -2\mathbf{i}-\frac{3}{2}\mathbf{j}+\frac{3}{4}\mathbf{i} = -\frac{5}{4}\mathbf{i}-\frac{3}{2}\mathbf{j}$
$\mathbf{v}-\mathbf{u} = \frac{3}{4}\mathbf{i} + (2\mathbf{i}+\frac{3}{2}\mathbf{j}) = \frac{3}{4}\mathbf{i}+2\mathbf{i}+\frac{3}{2}\mathbf{j} = \frac{11}{4}\mathbf{i}+\frac{3}{2}\mathbf{j}$
$2\mathbf{u}-3\mathbf{v} = -2(2\mathbf{i}+\frac{3}{2}\mathbf{j}) - 3(\frac{3}{4}\mathbf{i}) = -4\mathbf{i}-\frac{6}{2}\mathbf{j}-\frac{9}{4}\mathbf{i} = -\frac{25}{4}\mathbf{i}-3\mathbf{j}$

21. $\mathbf{u}+2\mathbf{w} = \mathbf{i}-2\mathbf{j}+2(-4\mathbf{i}+\mathbf{j}) = \mathbf{i}-2\mathbf{j}-8\mathbf{i}+2\mathbf{j} = -7\mathbf{i}\quad \langle -7,0 \rangle$

23. $\frac{1}{2}\mathbf{w} = \frac{1}{2}(-4\mathbf{i}+\mathbf{j}) = -2\mathbf{i}+\frac{1}{2}\mathbf{j}\quad \langle -2, \frac{1}{2} \rangle$

25. $\frac{1}{4}(8\mathbf{u}+4\mathbf{v}-\mathbf{w})=2\mathbf{u}+\mathbf{v}-\frac{1}{4}\mathbf{w}=2(\mathbf{i}-2\mathbf{j})+(3\mathbf{i}+\mathbf{j})-\frac{1}{4}(-4\mathbf{i}+\mathbf{j})$
$$=2\mathbf{i}-4\mathbf{j}+3\mathbf{i}+\mathbf{j}+\mathbf{i}-\frac{1}{4}\mathbf{j}=6\mathbf{i}-\frac{13}{4}\mathbf{j}\quad\left\langle 6,-\frac{13}{4}\right\rangle$$

27. $\mathbf{v}=\langle\|\mathbf{v}\|\cos\theta,\|\mathbf{v}\|\sin\theta\rangle=\langle 4\cos 0°,4\sin 0°\rangle=\langle 4,0\rangle$

29. $\mathbf{v}=\langle\|\mathbf{v}\|\cos\theta,\|\mathbf{v}\|\sin\theta\rangle=\langle 10\cos 225°,10\sin 225°\rangle=\left\langle -5\sqrt{2},-5\sqrt{2}\right\rangle$

31. $\mathbf{v}=\langle\|\mathbf{v}\|\cos\theta,\|\mathbf{v}\|\sin\theta\rangle=\langle 6\cos 40°,6\sin 40°\rangle=\langle 4.5963,3.8567\rangle$

33. $\mathbf{v}=\langle\|\mathbf{v}\|\cos\theta,\|\mathbf{v}\|\sin\theta\rangle=\left\langle\frac{1}{2}\cos 250°,\frac{1}{2}\sin 250°\right\rangle=\langle -.1710,-.4698\rangle$

35. $\|\mathbf{v}\|=\sqrt{4^2+4^2}=4\sqrt{2}\quad\tan\theta=4/4=1,\ \theta=45°\ (\text{quadrant I})$

37. $\|\mathbf{v}\|=\sqrt{(-8)^2+0^2}=8\quad\tan\theta=0/8=0,\ \theta=180°\ (\text{negative }x\text{-axis})$

39. $\|6\mathbf{j}\|=\sqrt{0^2+6^2}=6\quad\tan\theta=6/0,\text{ undefined},\ \theta=90°\ (\text{positive }y\text{-axis})$

41. $\|-2\mathbf{i}+8\mathbf{j}\|=\sqrt{(-2)^2+8^2}=2\sqrt{17}\quad\tan\theta=8/-2=-4,\ \theta=104.04°\ (\text{quadrant II})$

43. $\|\mathbf{v}\|=\sqrt{4^2+(-5)^2}=\sqrt{41},\quad\text{unit vector}=\frac{1}{\sqrt{41}}\langle 4,-5\rangle=\left\langle\frac{4}{\sqrt{41}},-\frac{5}{\sqrt{41}}\right\rangle$

45. $\|\mathbf{v}\|=\sqrt{5^2+10^2}=5\sqrt{5},\quad\text{unit vector}=\frac{1}{5\sqrt{5}}(5\mathbf{i}+10\mathbf{j})=\frac{1}{\sqrt{5}}\mathbf{i}+\frac{2}{\sqrt{5}}\mathbf{j}=\left\langle\frac{1}{\sqrt{5}},\frac{2}{\sqrt{5}}\right\rangle$

47. $\mathbf{u}=\langle 30\cos 0°,30\sin 0°\rangle=\langle 30,0\rangle,\ \mathbf{v}=\langle 90\cos 60°,90\sin 60°\rangle=\left\langle 45,45\sqrt{3}\right\rangle$

$\text{Resultant}=\mathbf{u}+\mathbf{v}=\left\langle 75,45\sqrt{3}\right\rangle,\ |\mathbf{u}+\mathbf{v}|=\sqrt{75^2+\left(45\sqrt{3}\right)^2}=30\sqrt{13}=108.2\text{ pounds}$

$\tan\theta=\dfrac{45\sqrt{3}}{75},\ \theta=46.1°\ (\text{quadrant I})$

49. $\mathbf{u}=\langle 12\cos 130°,12\sin 130°\rangle=\langle -7.7135,9.1925\rangle$
$\mathbf{v}=\langle 20\cos 250°,20\sin 250°\rangle=\langle -6.8404,-18.7939\rangle$
$\text{Resultant}=\mathbf{u}+\mathbf{v}=\langle -14.5539,-9.6013\rangle$
$|\mathbf{u}+\mathbf{v}|=\sqrt{(-14.5539)^2+(-9.6013)^2}=17.4356\text{ kg}$
$\tan\theta=-9.6013/-14.5539=.6597,\ \theta=213.4132°\ (\text{quadrant III})$

51. $\mathbf{u}_1+\mathbf{u}_2+\mathbf{u}_3=\langle -8,-2\rangle,\ \mathbf{u}_1+\mathbf{u}_2+\mathbf{u}_3+\mathbf{v}=0,$
$\mathbf{v}=-\mathbf{u}_1-\mathbf{u}_2-\mathbf{u}_3=-\langle 2,5\rangle-\langle -6,1\rangle-\langle -4,-8\rangle=\langle 8,2\rangle$

53. $\mathbf{v}+\mathbf{0}=\langle c,d \rangle + \langle 0,0 \rangle = \langle c,d \rangle = \mathbf{v}$ and $\mathbf{0}+\mathbf{v}=\langle 0,0 \rangle + \langle c,d \rangle = \langle c,d \rangle = \mathbf{v}$

55. $\begin{aligned}r(\mathbf{u}+\mathbf{v}) &= r(\langle a,b \rangle + \langle c,d \rangle) \\ &= r\langle a+c, b+d \rangle \\ &= \langle ra+rc, rb+rd \rangle \\ &= \langle ra, rb \rangle + \langle rc+rd \rangle \\ &= r\langle a,b \rangle + r\langle c,d \rangle \\ &= r\mathbf{u}+r\mathbf{v}\end{aligned}$

57. $\begin{aligned}(rs)\mathbf{v} &= rs\langle c,d \rangle \\ &= \langle rsc, rsd \rangle \\ &= r\langle sc, sd \rangle = r(s\mathbf{v}) \\ \langle rsc, rsd \rangle &= \langle src, srd \rangle \\ &= s\langle rc, rd \rangle = s(r\mathbf{v})\end{aligned}$

59. Write the forces as
$\langle 30\cos 0°, 30\sin 0° \rangle = \langle 30,0 \rangle$, $\langle 20\cos 28°, 20\sin 28° \rangle = \langle 17.6590, 9.3894 \rangle$.
Resultant $= \langle 30,0 \rangle + \langle 17.6590, 9.3894 \rangle = \langle 47.6590, 9.3894 \rangle$.
The magnitude of the resultant $= \sqrt{47.6590^2 + 9.3894^2} = 48.575$ pounds .

61. Following the example, write
$$\frac{\left\|\overrightarrow{TP}\right\|}{50} = \sin 40°, \quad \left\|\overrightarrow{TP}\right\| = 50\sin 40° = 32.1 \text{ pounds parallel to plane.}$$
$$\frac{\left\|\overrightarrow{TQ}\right\|}{50} = \cos 40°, \quad \left\|\overrightarrow{TQ}\right\| = 50\cos 40° = 38.3 \text{ pounds perpendicular to plane.}$$

63. Here, $\dfrac{\left\|\overrightarrow{TP}\right\|}{150} = \dfrac{60}{150} = \cos\theta, \quad \theta = \cos^{-1}\left(\dfrac{60}{150}\right) = 66.4°$.

In Exercises **65-69**, navigation angles are measured in degrees clockwise from north. First they are converted to standard position and measured clockwise. Use $\mathbf{p}$ for the plane's speed and direction; $\mathbf{w}$ for the wind's speed and direction.

65. The plane's direction is $60°$, convert to $30°$ in standard position.
$\mathbf{p} = \langle 250\cos 30°, 250\sin 30° \rangle$
The wind is blowing from $330°$ to $150°$; convert to $-60°$ in standard position.
$\mathbf{w} = \langle 40\cos(-60°), 40\sin(-60°) \rangle$
$\mathbf{p} + \mathbf{w} = \langle 250\cos 30° + 40\cos(-60°), 250\sin 30° + 40\sin(-60°) \rangle = \langle 236.5, 90.4 \rangle$
This is a vector with magnitude $\sqrt{236.5^2 + 90.4^2} = 253.2$ mph and direction $\tan^{-1}(90.4/236.5) = 20.9°$. This represents a navigation angle of $90° - 20.9° = 69.1°$.

67. The plane's direction is $300°$, convert to $150°$ in standard position.
$\mathbf{p} = \langle 300\cos 150°, 300\sin 150°\rangle$
The wind's direction is $30°$, convert to $60°$ in standard position.
$\mathbf{w} = \langle 50\cos 60°, 50\sin 60°\rangle$
$\mathbf{p} + \mathbf{w} = \langle 300\cos 150° + 50\cos 60°, 300\sin 150° + 50\sin 60°\rangle = \langle -234.8, 193.3\rangle$
This is a vector with magnitude $\sqrt{(-234.8)^2 + 193.3^2} = 304.1$ mph and direction
(quadrant II) $180° + \tan^{-1}(193.3/-234.8) = 140.5°$.
This represents a navigation angle of $309.5°$.

69. The plane's course is $70°$, convert to $20°$ in standard position. Hence
$\mathbf{p} + \mathbf{w} = \langle 400\cos 20°, 400\sin 20°\rangle$. The wind blowing south is represented by
$\mathbf{w} = \langle 0, -60\rangle$. $\mathbf{p} + \mathbf{w} - \mathbf{w} = \langle 400\cos 20°, 400\sin 20° + 60\rangle = \langle 375.9, 196.8\rangle$.
This is a vector with magnitude $\sqrt{375.9^2 + 196.8^2} = 424.3$ mph and direction
$\tan^{-1}(196.8/375.9) = 27.6°$. This represents a navigation angle of $62.4°$.

71. Represent the current by $\mathbf{c} = \langle -1, 0\rangle$ and the desired direction by $\mathbf{r} = \langle 0, r\rangle$. Then she
should swim in the direction $\mathbf{s}$ given by $\mathbf{s} + \mathbf{c} = \mathbf{r}$, $\mathbf{s} = \mathbf{r} - \mathbf{c} = \langle 0, r\rangle - \langle -1, 0\rangle = \langle 1, r\rangle$.
Since $|\mathbf{s}| = 2.8 = \|\langle 1, r\rangle\| = \sqrt{1 + r^2}$, $r = 2.62$. Then the direction of $\mathbf{s}$ is given by
$\tan^{-1}(r/1) = \tan^{-1}(2.62) = 69.08°$.

73. Following the hint,
$\mathbf{v} = c\cos 65°\mathbf{i} + c\sin 65°\mathbf{j}$, $\quad \mathbf{u} = d\cos 148°\mathbf{i} + d\sin 148°\mathbf{j}$
$\mathbf{u} + \mathbf{v} = 0\mathbf{i} + 400\mathbf{j}$ (to counteract gravity)
Hence: $c\cos 65° + d\cos 148° = 0$
$\quad\quad c\sin 65° + d\sin 148° = 400$
Solve to obtain $c = 341.77$ lbs on $\mathbf{v}$, $d = 170.32$ lbs on $\mathbf{u}$.

75. Represent the force in the $28°$ rope by $\mathbf{u} = a\cos 152°\mathbf{i} + a\sin 152°\mathbf{j}$.
Represent the force in the $38°$ rope by $\mathbf{v} = b\cos 38°\mathbf{i} + b\sin 38°\mathbf{j}$.
$\mathbf{u} + \mathbf{v} = 0\mathbf{i} + 600\mathbf{j}$
Hence: $a\cos 152° + b\cos 38° = 0$, $a\sin 152° + b\sin 38° = 600$
Solve to obtain $a = 517.55$ lb on the $28°$ rope, $b = 579.90$ lbs on the $38°$ rope.

77. a. $\mathbf{v} = \langle x_2 - x_1, y_2 - y_1 \rangle$; $k\mathbf{v} = \langle kx_2 - kx_1, ky_2 - ky_1 \rangle$

   b. $\|\mathbf{v}\| = \sqrt{(x_2 - x_1)^2 + (y_2 - y_1)^2}$; $\|k\mathbf{v}\| = \sqrt{(kx_2 - kx_1)^2 + (ky_2 - ky_1)^2}$

   c. $\|k\mathbf{v}\| = \sqrt{k^2(x_2 - x_1)^2 + k^2(y_2 - y_1)^2} = \sqrt{k^2}\sqrt{(x_2 - x_1)^2 + (y_2 - y_1)^2}$

   $\quad = |k|\sqrt{(x_2 - x_1)^2 + (y_2 - y_1)^2} = |k|\|\mathbf{v}\|$

   d. $\tan\theta = \dfrac{y_2 - y_1}{x_2 - x_1} = \dfrac{k}{k}\dfrac{y_2 - y_1}{x_2 - x_1} = \dfrac{ky_2 - ky_1}{kx_2 - kx_1} = \tan\beta$

   Since the angles have the same tangent, they can only differ by a multiple of π, and so are parallel. They have either the same or the opposite direction.

   e. If $k>0$, the signs of the components of $k\mathbf{v}$ are the same as those of $\mathbf{v}$, so the two vectors lie in the same quadrant. Therefore, they must have the same direction. If $k<0$, then the components of $k\mathbf{v}$ and the components of $\mathbf{v}$ must have opposite signs and the two vectors do not lie in the same quadrant. Therefore, they do not have the same direction and must have opposite directions.

79. a. Since $\mathbf{u} - \mathbf{v} = \langle a - c, b - d \rangle$, $\|\mathbf{u} - \mathbf{v}\| = \sqrt{(a - c)^2 + (b - d)^2}$. The magnitude of $\mathbf{w}$ is given by the distance betwwen the points (a, b) and (c, d). Hence, $\|\mathbf{u} - \mathbf{v}\| = \|\mathbf{w}\|$.

   b. $\mathbf{u} - \mathbf{v}$ lies on the straight line through $(0,0)$ and $(a - c, b - d)$ which has slope $\dfrac{(b - d) - 0}{(a - c) - 0} = \dfrac{b - d}{a - c}$. $\mathbf{w}$ lies on the line joining (a, b) and (c, d). This also has slope $\dfrac{b - d}{a - c}$. Since the slopes are the same, the vectors, $\mathbf{u} - \mathbf{v}$ and $\mathbf{w}$ are parallel. Therefore, they either point in the same direction or in opposite directions. We can see that they have the same direction by considering the signs of the components of $\mathbf{u} - \mathbf{v}$. If $a - c > 0$ and $b - d > 0$, then $a > c$, $b > d$ and $\mathbf{u} - \mathbf{v}$ and $\mathbf{w}$ both point up and right. If $a - c > 0$ and $b - d < 0$, both vectors will point left and up. If $a - c < 0$ and $b - d < 0$, both vectors will point left and down. In any case they point in the same direction.

9.4 The Dot Product

1. $\mathbf{u} \bullet \mathbf{v} = \langle 3,4 \rangle \bullet \langle -5,2 \rangle = 3(-5) + 4 \cdot 2 = -7,\quad \mathbf{u} \bullet \mathbf{u} = \langle 3,4 \rangle \bullet \langle 3,4 \rangle = 3 \cdot 3 + 4 \cdot 4 = 25$
$\mathbf{v} \bullet \mathbf{v} = \langle -5,2 \rangle \bullet \langle -5,2 \rangle = (-5)(-5) + 2 \cdot 2 = 29$

3. $\mathbf{u} \bullet \mathbf{v} = (2\mathbf{i}+\mathbf{j}) \bullet 3\mathbf{i} = 2 \cdot 3 + 1 \cdot 0 = 6,\quad \mathbf{u} \bullet \mathbf{u} = (2\mathbf{i}+\mathbf{j}) \bullet (2\mathbf{i}+\mathbf{j}) = 2 \cdot 2 + 1 \cdot 1 = 5$
$\mathbf{v} \bullet \mathbf{v} = 3\mathbf{i} \bullet 3\mathbf{i} = 3 \cdot 3 = 9$

5. $\mathbf{u} \bullet \mathbf{v} = (3\mathbf{i}+2\mathbf{j}) \bullet (2\mathbf{i}+3\mathbf{j}) = 3 \cdot 2 + 2 \cdot 3 = 12$
$\mathbf{u} \bullet \mathbf{u} = (3\mathbf{i}+2\mathbf{j}) \bullet (3\mathbf{i}+2\mathbf{j}) = 3 \cdot 3 + 2 \cdot 2 = 13$
$\mathbf{v} \bullet \mathbf{v} = (2\mathbf{i}+3\mathbf{j}) \bullet (2\mathbf{i}+3\mathbf{j}) = 2 \cdot 2 + 3 \cdot 3 = 13$

7. $\mathbf{u} \bullet (\mathbf{v}+\mathbf{w}) = \langle 4,3 \rangle \bullet (\langle -5,2 \rangle + \langle 4,-1 \rangle) = \langle 4,3 \rangle \bullet \langle -1,1 \rangle = -4 + 3 = -1$

9. $(\mathbf{u}+\mathbf{v}) \bullet (\mathbf{v}+\mathbf{w}) = (\langle 4,3 \rangle + \langle -5,2 \rangle) \bullet (\langle -5,2 \rangle + \langle 4,-1 \rangle) = \langle -1,5 \rangle \bullet \langle -1,1 \rangle = 1 + 5 = 6$

11. $(3\mathbf{u}+\mathbf{v}) \bullet (2\mathbf{w}) = (3\langle 4,3 \rangle + \langle -5,2 \rangle) \bullet (2\langle 4,-1 \rangle) = (\langle 12,9 \rangle + \langle -5,2 \rangle) \bullet \langle 8,-2 \rangle$
$= \langle 7,11 \rangle \bullet \langle 8,-2 \rangle = 56 - 22 = 34$

13. $\cos\theta = \dfrac{\langle 4,-3 \rangle \bullet \langle 1,2 \rangle}{\|\langle 4,-3 \rangle\|\|\langle 1,2 \rangle\|} = \dfrac{-2}{5\sqrt{5}};\quad \theta = \cos^{-1}\dfrac{-2}{5\sqrt{5}} = 1.75065 \text{ rad}$

15. $\cos\theta = \dfrac{(2\mathbf{i}-3\mathbf{j}) \bullet (-\mathbf{i})}{|2\mathbf{i}-3\mathbf{j}||-\mathbf{i}|} = \dfrac{-2}{\sqrt{13}(1)} = \dfrac{-2}{\sqrt{13}};\quad \theta = \cos^{-1}\dfrac{-2}{\sqrt{13}} = 2.1588 \text{ rad}$

17. $\cos\theta = \dfrac{(\sqrt{2}\mathbf{i}+\sqrt{2}\mathbf{j}) \bullet (\mathbf{i}-\mathbf{j})}{|\sqrt{2}\mathbf{i}+\sqrt{2}\mathbf{j}||\mathbf{i}-\mathbf{j}|} = \dfrac{0}{2\sqrt{2}} = 0;\quad \theta = \dfrac{\pi}{2} \text{ rad}$

19. $\langle 2,6 \rangle \bullet \langle 3,-1 \rangle = 6 - 6 = 0$. Orthogonal

21. $\langle 9,-6 \rangle \bullet \langle -6,4 \rangle = -54 - 24 = -78$. Not orthogonal
However, $\langle 9,-6 \rangle = -\frac{3}{2}\langle -6,4 \rangle$, hence the vectors are parallel.

23. $(2\mathbf{i}-2\mathbf{j}) \bullet (5\mathbf{i}+8\mathbf{j}) = 10 - 16 = -6$. Not orthogonal
Since $2\mathbf{i}-2\mathbf{j} \neq k(5\mathbf{i}+8\mathbf{j})$ the vectors are also not parallel.

25. The vectors will be orthogonal if their dot product is zero.
$(2\mathbf{i}+3\mathbf{j}) \bullet (3\mathbf{i}-k\mathbf{j}) = 6 - 3k = 0$ if $k = 2$

27. The vectors will be orthogonal if their dot product is zero.

$(\mathbf{i} - \mathbf{j}) \bullet \left(k\mathbf{i} + \sqrt{2}\mathbf{j}\right) = k - \sqrt{2} = 0$ if $k = \sqrt{2}$.

29. $proj_{\mathbf{u}}\mathbf{v} = \left(\dfrac{\mathbf{u} \bullet \mathbf{v}}{\|\mathbf{u}\|^2}\right)\mathbf{u} = \dfrac{(3\mathbf{i} - 5\mathbf{j}) \bullet (6\mathbf{i} + 2\mathbf{j})}{\|3\mathbf{i} - 5\mathbf{j}\|^2}(3\mathbf{i} - 5\mathbf{j}) = \dfrac{8}{34}(3\mathbf{i} - 5\mathbf{j}) = \dfrac{12}{17}\mathbf{i} - \dfrac{20}{17}\mathbf{j}$

$proj_{\mathbf{v}}\mathbf{u} = \left(\dfrac{\mathbf{u} \bullet \mathbf{v}}{\|\mathbf{v}\|^2}\right)\mathbf{v} = \dfrac{(3\mathbf{i} - 5\mathbf{j}) \bullet (6\mathbf{i} + 2\mathbf{j})}{\|6\mathbf{i} + 2\mathbf{j}\|^2}(6\mathbf{i} + 2\mathbf{j}) = \dfrac{8}{40}(6\mathbf{i} + 2\mathbf{j}) = \dfrac{6}{5}\mathbf{i} + \dfrac{2}{5}\mathbf{j}$

31. $proj_{\mathbf{u}}\mathbf{v} = \left(\dfrac{\mathbf{u} \bullet \mathbf{v}}{\|\mathbf{u}\|^2}\right)\mathbf{u} = \dfrac{(\mathbf{i} + \mathbf{j}) \bullet (\mathbf{i} - \mathbf{j})}{\|\mathbf{i} + \mathbf{j}\|^2}(\mathbf{i} + \mathbf{j}) = 0(\mathbf{i} + \mathbf{j}) = \mathbf{0}$

$proj_{\mathbf{v}}\mathbf{u} = \left(\dfrac{\mathbf{u} \bullet \mathbf{v}}{\|\mathbf{v}\|^2}\right)\mathbf{v} = \mathbf{0}$ (similarly)

33. $comp_{\mathbf{v}}\mathbf{u} = \dfrac{\mathbf{u} \bullet \mathbf{v}}{\|\mathbf{v}\|} = \dfrac{(10\mathbf{i} + 4\mathbf{j}) \bullet (3\mathbf{i} - 2\mathbf{j})}{\|3\mathbf{i} - 2\mathbf{j}\|} = \dfrac{22}{\sqrt{13}}$

35. $comp_{\mathbf{v}}\mathbf{u} = \dfrac{\mathbf{u} \bullet \mathbf{v}}{\|\mathbf{v}\|} = \dfrac{(3\mathbf{i} + 2\mathbf{j}) \bullet (-\mathbf{i} + 3\mathbf{j})}{\|-\mathbf{i} + 3\mathbf{j}\|} = \dfrac{3}{\sqrt{10}}$

37. $\mathbf{u} \bullet (\mathbf{v} + \mathbf{w}) = \langle a, b\rangle \bullet (\langle c, d\rangle + \langle r, s\rangle)$

$= \langle a, b\rangle \bullet \langle c + r, d + s\rangle$

$= a(c + r) + b(d + s)$

$= (ac + bd) + (ar + bs)$

$= \langle a, b\rangle \bullet \langle c, d\rangle + \langle a, b\rangle \bullet \langle r, s\rangle$

$= \mathbf{u} \bullet \mathbf{v} + \mathbf{u} \bullet \mathbf{w}$

39. $\mathbf{0} \bullet \mathbf{u} = \langle 0, 0\rangle \bullet \langle a, b\rangle = 0a + 0b = 0$

41. If $\theta = 0$, $\mathbf{v} = k\mathbf{u}$ for some positive k.   If $\theta = \pi$, $\mathbf{v} = k\mathbf{u}$ for some negative k.

$\cos\theta = 1$, then $\mathbf{u} \bullet \mathbf{v} = k\mathbf{u} \bullet \mathbf{u}$ $\cos\theta = -1$, then $\mathbf{u} \bullet \mathbf{v} = k\mathbf{u} \bullet \mathbf{u}$

$\qquad\qquad = k\|\mathbf{u}\|^2 \qquad\qquad\qquad\qquad\qquad = k\|\mathbf{u}\|^2 = \mathbf{u} \bullet k\mathbf{u}$

$\qquad\qquad = k\|\mathbf{u}\|\|\mathbf{u}\| \bullet 1 \qquad\qquad\qquad\qquad = \|\mathbf{u}\|(-k\|\mathbf{u}\|)$

$\qquad\qquad = \|\mathbf{u}\|\|\mathbf{v}\|\cos\theta \qquad\qquad\qquad\qquad = -\|\mathbf{u}\|\|\mathbf{v}\|$

$\qquad\qquad\qquad\qquad\qquad\qquad\qquad\qquad\qquad = \|\mathbf{u}\|\|\mathbf{v}\|\cos\theta$

43. The side containing $(1, 2)$ and $(3, 4)$ has vector description $\langle 3 - 1, 4 - 2\rangle = \langle 2, 2\rangle$.

The side containing $(3, 4)$ and $(5, 2)$ has vector description $\langle 5 - 3, 2 - 4\rangle = \langle 2, -2\rangle$.

Since these two vectors are orthogonal, the line segments form a right angle.

45. Many possible answers: One is $\mathbf{u} = \langle 1,0 \rangle$, $\mathbf{v} = \langle 1,1 \rangle$, and $\mathbf{w} = \langle 1,-1 \rangle$.

47. As in Example **7**, the force has magnitude
$$\left| comp_v \, \mathbf{F} \right| = \left\| \mathbf{F} \right\| \cos(90° - 30°) = \left| 600 \cos 60° \right| = 300 \text{ pounds.}$$

49. $\overrightarrow{PQ} = 4\mathbf{i} + \mathbf{j}$, Work $= \mathbf{F} \bullet \overrightarrow{PQ} = (2\mathbf{i} + 5\mathbf{j}) \bullet (4\mathbf{i} + \mathbf{j}) = 13$

51. $\overrightarrow{PQ} = 3\mathbf{i} + 6\mathbf{j}$, Work $= \mathbf{F} \bullet \overrightarrow{PQ} = (2\mathbf{i} + 3\mathbf{j}) \bullet (3\mathbf{i} + 6\mathbf{j}) = 24$

53. The force is a vector $30\cos(-60°)\mathbf{i} + 30\sin(-60°)\mathbf{j} = 15\mathbf{i} - 15\sqrt{3}\mathbf{j} = \mathbf{F}$.
The displacement is a vector $75\mathbf{i} = \mathbf{d}$.
The work $= \mathbf{F} \bullet \mathbf{d} = \left(15\mathbf{i} - 15\sqrt{3}\mathbf{j} \right) \bullet (75\mathbf{i}) = 1125$ ft-lb.

55. Following the hint, the work is
$$-\mathbf{F} \bullet \mathbf{d} = -(0\mathbf{i} - 40\mathbf{j}) \bullet (100\cos 20°\mathbf{i} + 100\sin 20°\,\mathbf{j}) = 4000\sin 20° = 1368 \text{ ft-lb.}$$

Chapter 9 Review Exercises

1. $\left|i(4+2i)\right|+\left|3-i\right|=\left|4i-2\right|+\left|3-i\right|=\sqrt{4^2+(-2)^2}+\sqrt{3^2+(-1)^2}=\sqrt{20}+\sqrt{10}$

3.

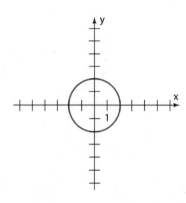

5. $r=\sqrt{1^2+\sqrt{3}^2}=2.$ $\tan\theta=\sqrt{3}/1;\ \theta=\pi/3$ (quadrant I)
$1+\sqrt{3}i=2(\cos\pi/3+i\sin\pi/3)$

7. $2\left(\cos\dfrac{\pi}{12}+i\sin\dfrac{\pi}{12}\right)\cdot4\left(\cos\dfrac{\pi}{6}+i\sin\dfrac{\pi}{6}\right)=8\left(\cos\dfrac{\pi}{4}+i\sin\dfrac{\pi}{4}\right)$

$$=8\left(\dfrac{\sqrt{2}}{2}+i\dfrac{\sqrt{2}}{2}\right)=4\sqrt{2}+4i\sqrt{2}$$

9. $\dfrac{12\left(\cos\dfrac{7\pi}{12}+i\sin\dfrac{7\pi}{12}\right)}{3\left(\cos\dfrac{5\pi}{12}+i\sin\dfrac{5\pi}{12}\right)}=4\left(\cos\dfrac{\pi}{6}+i\sin\dfrac{\pi}{6}\right)=4\left(\dfrac{\sqrt{3}}{2}+\dfrac{1}{2}i\right)=2\sqrt{3}+2i$

11. $\left[\sqrt[3]{3}\left(\cos\dfrac{5\pi}{36}+i\sin\dfrac{5\pi}{36}\right)\right]^{12}=\left(3^{\frac{1}{3}}\right)^{12}\left(\cos\dfrac{5\pi}{3}+i\sin\dfrac{5\pi}{3}\right)$

$$=81\left(\dfrac{1}{2}+\left(-\dfrac{\sqrt{3}}{2}\right)i\right)$$

$$=\dfrac{81}{2}-\dfrac{81\sqrt{3}}{2}i$$

13. Apply the roots-of-unity formula with $n = 6$, $k = 0,1,2,3,4,5$.

$$k = 0: \cos 0 + i\sin 0$$

$$k = 1: \cos\frac{2\pi}{6} + i\sin\frac{2\pi}{6} = \cos\frac{\pi}{3} + i\sin\frac{\pi}{3}$$

$$k = 2: \cos\frac{4\pi}{6} + i\sin\frac{4\pi}{6} = \cos\frac{2\pi}{3} + i\sin\frac{2\pi}{3}$$

$$k = 3: \cos\frac{6\pi}{6} + i\sin\frac{6\pi}{6} = \cos\pi + i\sin\pi$$

$$k = 4: \cos\frac{8\pi}{6} + i\sin\frac{8\pi}{6} = \cos\frac{4\pi}{3} + i\sin\frac{4\pi}{3}$$

$$k = 5: \cos\frac{10\pi}{6} + i\sin\frac{10\pi}{6} = \cos\frac{5\pi}{3} + i\sin\frac{5\pi}{3}$$

15. The solutions are the four fourth roots of $i = \cos\frac{\pi}{2} + i\sin\frac{\pi}{2}$, with $n = 4$, $k = 0,1,2,3$.

$$k = 0: \cos\left(\frac{\pi/2}{4}\right) + i\sin\left(\frac{\pi/2}{4}\right) = \cos\frac{\pi}{8} + i\sin\frac{\pi}{8}$$

$$k = 1: \cos\left(\frac{\pi/2 + 2\pi}{4}\right) + i\sin\left(\frac{\pi/2 + 2\pi}{4}\right) = \cos\frac{5\pi}{8} + i\sin\frac{5\pi}{8}$$

$$k = 2: \cos\left(\frac{\pi/2 + 4\pi}{4}\right) + i\sin\left(\frac{\pi/2 + 4\pi}{4}\right) = \cos\frac{9\pi}{8} + i\sin\frac{9\pi}{8}$$

$$k = 3: \cos\left(\frac{\pi/2 + 6\pi}{4}\right) + i\sin\left(\frac{\pi/2 + 6\pi}{4}\right) = \cos\frac{13\pi}{8} + i\sin\frac{13\pi}{8}$$

17. $\mathbf{u} + \mathbf{v} = \langle 2,-5\rangle + \langle 6,1\rangle = \langle 8,-4\rangle$

19. $\|2\mathbf{v} - 4\mathbf{u}\| = \|2\langle 6,1\rangle - 4\langle 2,-5\rangle\| = \|\langle 12,2\rangle - \langle 8,-20\rangle\| = \|\langle 4,22\rangle\| = \sqrt{4^2 + 22^2} = 10\sqrt{5}$

21. $4\mathbf{u} - \mathbf{v} = 4(-3\mathbf{i} + \mathbf{j}) - (2\mathbf{i} - 5\mathbf{j}) = -12\mathbf{i} + 4\mathbf{j} - 2\mathbf{i} + 5\mathbf{j} = -14\mathbf{i} + 9\mathbf{j}$

23. $\|\mathbf{u} + \mathbf{v}\| = \|-3\mathbf{i} + \mathbf{j} + 2\mathbf{i} - 5\mathbf{j}\| = \|-\mathbf{i} - 4\mathbf{j}\| = \sqrt{(-1)^2 + (-4)^2} = \sqrt{17}$

25. $\mathbf{v} = \langle\|\mathbf{v}\|\cos\theta, \|\mathbf{v}\|\sin\theta\rangle = \langle 5\cos 45°, 5\sin 45°\rangle = \langle 5\sqrt{2}/2, 5\sqrt{2}/2\rangle$

27. A vector in the opposite direction is given by $-3\mathbf{i} + 6\mathbf{j}$. Since the magnitude of this vector is given by $\sqrt{(-3)^2 + 6^2} = 3\sqrt{5}$, the required unit vector is $\frac{1}{3\sqrt{5}}(-3\mathbf{i} + 6\mathbf{j}) = -\frac{1}{\sqrt{5}}\mathbf{i} + \frac{2}{\sqrt{5}}\mathbf{j}$.

29. Let the velocity of the plane be $\mathbf{v} = \langle 300\cos(-30°), 300\sin(-30°) \rangle = \langle 150\sqrt{3}, -150 \rangle$. Let the velocity of the wind be $\mathbf{w} = \langle 0, -40 \rangle$. Hence the course of the plane can be represented by $\mathbf{v} + \mathbf{w} = \langle 150\sqrt{3}, -150 \rangle + \langle 0, -40 \rangle = \langle 150\sqrt{3}, -190 \rangle$. Then

$$\|\mathbf{v} + \mathbf{w}\| = \sqrt{\left(150\sqrt{3}\right)^2 + (-190)^2} = \sqrt{103,600} = 321.87 \text{ mph.}$$

$$\tan\theta = \frac{-190}{150\sqrt{3}}, \quad \theta = -36.18° \text{ which represents a course of } 126.18°.$$

31. $\mathbf{u} \bullet \mathbf{v} = \langle 4, -3 \rangle \bullet \langle -1, 6 \rangle = 4(-1) + (-3)6 = -22$

33. $(\mathbf{u} + \mathbf{v}) \bullet \mathbf{w} = (\langle 4, -3 \rangle + \langle -1, 6 \rangle) \bullet \langle 5, 0 \rangle = \langle 3, 3 \rangle \bullet \langle 5, 0 \rangle = 3 \cdot 5 + 3 \cdot 0 = 15$

35. $\cos\theta = \dfrac{(5\mathbf{i} - 2\mathbf{j}) \bullet (3\mathbf{i} + \mathbf{j})}{\|5\mathbf{i} - 2\mathbf{j}\|\|3\mathbf{i} + \mathbf{j}\|} = \dfrac{13}{\sqrt{29}\sqrt{10}}; \quad \theta = \cos^{-1}\dfrac{13}{\sqrt{29}\sqrt{10}} = .70 \text{ rad}$

37. $proj_\mathbf{v}\mathbf{u} = \left(\dfrac{\mathbf{u} \bullet \mathbf{v}}{\|\mathbf{v}\|^2}\right)\mathbf{v} = \dfrac{(4\mathbf{i} - 3\mathbf{j}) \bullet (2\mathbf{i} + \mathbf{j})}{\|2\mathbf{i} + \mathbf{j}\|^2}(2\mathbf{i} + \mathbf{j}) = \dfrac{5}{\left(\sqrt{5}\right)^2}(2\mathbf{i} + \mathbf{j}) = 2\mathbf{i} + \mathbf{j}$

39. $(\mathbf{u} + \mathbf{v}) \bullet (\mathbf{u} - \mathbf{v}) = \mathbf{u} \bullet \mathbf{u} - \mathbf{u} \bullet \mathbf{v} + \mathbf{v} \bullet \mathbf{u} - \mathbf{v} \bullet \mathbf{v} = \|\mathbf{u}\|^2 - \mathbf{u} \bullet \mathbf{v} + \mathbf{v} \bullet \mathbf{u} - \|\mathbf{v}\|^2 = \|\mathbf{u}\|^2 - \|\mathbf{v}\|^2$
If $\mathbf{u}$ and $\mathbf{v}$ have the same magnitude, this expression equals zero, hence $\mathbf{u} + \mathbf{v}$ and $\mathbf{u} - \mathbf{v}$ are orthogonal.

41. Represent the force of gravity by $\mathbf{F} = \langle 0, -3500 \rangle$ and let $\mathbf{v}$ be a unit vector pointing down the ramp. The force needed to hold the car against the force of gravity is $-proj_\mathbf{v}\mathbf{F}$. The magnitude of this force is
$$\|\mathbf{F}\|\cos\theta = 3500\cos 60° = 3500\left(\tfrac{1}{2}\right) = 1750 \text{ pounds.}$$

Chapter 9 Test

1. $\dfrac{3i}{2}(6+2i)=9i+3i^2=9i+3(-1)=-3+9i$

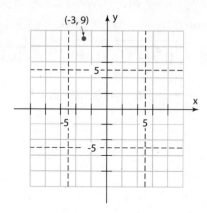

2. First express $3+3i$ in polar form; $r=\sqrt{3^2+3^2}=\sqrt{9+9}=\sqrt{18}=3\sqrt{2}.$ Since

$\cos\theta=\dfrac{3}{3\sqrt{2}}=\dfrac{1}{\sqrt{2}}=\dfrac{\sqrt{2}}{2}=\sin\theta,\theta=\dfrac{\pi}{4}.$ Thus $3+3i=3\sqrt{2}\left(\cos\dfrac{\pi}{4}+i\sin\dfrac{\pi}{4}\right).$ Then by

DeMoivre's Theorem, $(3+3i)^6=\left(3\sqrt{2}\right)^6\left(\cos\left(6\bullet\dfrac{\pi}{4}\right)+i\sin\left(6\bullet\dfrac{\pi}{4}\right)\right)$

$$=5832\left(\cos\left(\dfrac{3\pi}{2}\right)+i\sin\left(\dfrac{3\pi}{2}\right)\right)$$

$$=5832\left(0+i(-1)\right)$$

$$=-5832i$$

3. $|3+2i|=\sqrt{3^2+2^2}=\sqrt{9+4}=\sqrt{13}$

4. To find the cube roots of $64\left(\cos\dfrac{\pi}{16}+i\sin\dfrac{\pi}{16}\right)$ needs to be written in the form

$s^3\left(\cos\beta+i\sin\beta\right).$ Thus, $s=4$ and $\beta=\dfrac{\pi/16+2k\pi}{3}=\dfrac{\pi}{48}+\dfrac{2k\pi}{3},$ for $k=0,1,2.$

$k=0;\quad z=4\left(\cos\dfrac{\pi}{48}+i\sin\dfrac{\pi}{48}\right)$

$k=1;\quad z=4\left(\cos\left(\dfrac{\pi}{48}+\dfrac{2\pi}{3}\right)+i\sin\left(\dfrac{\pi}{48}+\dfrac{2\pi}{3}\right)\right)=4\left(\cos\left(\dfrac{33\pi}{48}\right)+i\sin\left(\dfrac{33\pi}{48}\right)\right)$

$$=4\left(\cos\left(\dfrac{11\pi}{16}\right)+i\sin\left(\dfrac{11\pi}{16}\right)\right)$$

$k=2;\quad z=4\left(\cos\left(\dfrac{\pi}{48}+\dfrac{4\pi}{3}\right)+i\sin\left(\dfrac{\pi}{48}+\dfrac{4\pi}{3}\right)\right)=4\left(\cos\left(\dfrac{65\pi}{48}\right)+i\sin\left(\dfrac{65\pi}{48}\right)\right)$

5. **a.** $\left|\left(\sqrt{5^2+(-3)^2}\right)^2\right|=\left|\left(\sqrt{25+9}\right)^2\right|=\left|\left(\sqrt{34}\right)^2\right|=34$

b. $(5-3i)(5+3i)=25+15i-15i-9i^2=25-9(-1)=34$

6. First write $1+\sqrt{3}i$ in polar form; $r=\sqrt{1^2+\left(\sqrt{3}\right)^2}=\sqrt{1+3}=2.$ Then $\cos\theta=\dfrac{1}{2}$ and

$\sin\theta=\dfrac{\sqrt{3}}{2}$, so $\theta=\dfrac{\pi}{3}$, thus $2\left(\cos\dfrac{\pi}{3}+i\sin\dfrac{\pi}{3}\right)$. To find the cube roots of

$2\left(\cos\dfrac{\pi}{3}+i\sin\dfrac{\pi}{3}\right)$ needs to be written in the form $s^3\left(\cos\beta+i\sin\beta\right)$.

Thus, $s=\sqrt[3]{2}$ and $\beta=\dfrac{\pi/3+2k\pi}{3}=\dfrac{\pi}{9}+\dfrac{2k\pi}{3}$, for $k=0,1,2.$

$k=0;\quad z=\sqrt[3]{2}\left(\cos\dfrac{\pi}{9}+i\sin\dfrac{\pi}{9}\right)$

$k=1;\quad z=\sqrt[3]{2}\left(\cos\left(\dfrac{\pi}{9}+\dfrac{2\pi}{3}\right)+i\sin\left(\dfrac{\pi}{9}+\dfrac{2\pi}{3}\right)\right)=\sqrt[3]{2}\left(\cos\left(\dfrac{7\pi}{9}\right)+i\sin\left(\dfrac{7\pi}{9}\right)\right)$

$k=2;\quad z=\sqrt[3]{2}\left(\cos\left(\dfrac{\pi}{9}+\dfrac{4\pi}{3}\right)+i\sin\left(\dfrac{\pi}{9}+\dfrac{4\pi}{3}\right)\right)=\sqrt[3]{2}\left(\cos\left(\dfrac{13\pi}{9}\right)+i\sin\left(\dfrac{13\pi}{9}\right)\right)$

7. The graph consists of all points 5 units away from $(3,0)$

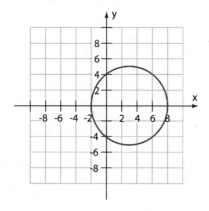

8.   The solutions are the four fourth roots of $-i = \cos\left(\dfrac{3\pi}{2}\right) + i\sin\left(\dfrac{3\pi}{2}\right)$. They are given

by $\cos\left(\dfrac{3\pi}{8} + \dfrac{k\pi}{2}\right) + i\sin\left(\dfrac{3\pi}{8} + \dfrac{k\pi}{2}\right), k = 0,1,2,3.$

$k = 0; \quad z = \left(\cos\dfrac{3\pi}{8} + i\sin\dfrac{3\pi}{8}\right)$

$k = 1; \quad z = \left(\cos\left(\dfrac{3\pi}{8} + \dfrac{\pi}{2}\right) + i\sin\left(\dfrac{3\pi}{8} + \dfrac{\pi}{2}\right)\right) = \left(\cos\left(\dfrac{7\pi}{8}\right) + i\sin\left(\dfrac{7\pi}{8}\right)\right)$

$k = 2; \quad z = \left(\cos\left(\dfrac{3\pi}{8} + \dfrac{\pi}{2}\right) + i\sin\left(\dfrac{3\pi}{8} + \dfrac{\pi}{2}\right)\right) = \left(\cos\left(\dfrac{11\pi}{8}\right) + i\sin\left(\dfrac{11\pi}{8}\right)\right)$

$k = 3; \quad z = \left(\cos\left(\dfrac{3\pi}{8} + \dfrac{3\pi}{2}\right) + i\sin\left(\dfrac{3\pi}{8} + \dfrac{3\pi}{2}\right)\right) = \left(\cos\left(\dfrac{15\pi}{8}\right) + i\sin\left(\dfrac{15\pi}{8}\right)\right)$

9. $3\left(\cos\dfrac{3\pi}{8} + i\sin\dfrac{3\pi}{8}\right)\cdot 13\left(\cos\dfrac{\pi}{8} + i\sin\dfrac{\pi}{8}\right) = 39\left(\cos\left(\dfrac{3\pi}{8} + \dfrac{\pi}{8}\right) + i\sin\left(\dfrac{3\pi}{8} + \dfrac{\pi}{8}\right)\right)$

$$= 39\left(\cos\left(\dfrac{\pi}{2}\right) + i\sin\left(\dfrac{\pi}{2}\right)\right)$$

$$= 39\left(0 + i(1)\right)$$

$$= 39i$$

10. The solutions are the four fourth roots of unit of

$$-648 + 648\sqrt{3}i = 1296\left(\cos\frac{2\pi}{3} + i\sin\frac{2\pi}{3}\right). \text{They are given by}$$

$$6\left(\cos\left(\frac{\pi}{6} + \frac{k\pi}{2}\right) + i\sin\left(\frac{\pi}{6} + \frac{k\pi}{2}\right)\right), k = 0, 1, 2, 3.$$

$$k = 0; \quad z = 6\left(\cos\left(\frac{\pi}{6}\right) + i\sin\left(\frac{\pi}{6}\right)\right) = 6\left(\frac{\sqrt{3}}{2} + i\left(\frac{1}{2}\right)\right) = 3\sqrt{3} + 3i$$

$$k = 1; \quad z = 6\left(\cos\left(\frac{\pi}{6} + \frac{\pi}{2}\right) + i\sin\left(\frac{\pi}{6} + \frac{\pi}{2}\right)\right) = 6\left(\cos\left(\frac{2\pi}{3}\right) + i\sin\left(\frac{2\pi}{3}\right)\right) = 6\left(-\frac{1}{2} + i\left(\frac{\sqrt{3}}{2}\right)\right)$$

$$= -3 + 3i\sqrt{3}$$

$$k = 2; \quad z = 6\left(\cos\left(\frac{\pi}{6} + \frac{2\pi}{2}\right) + i\sin\left(\frac{\pi}{6} + \frac{2\pi}{2}\right)\right) = \left(\cos\left(\frac{7\pi}{6}\right) + i\sin\left(\frac{7\pi}{6}\right)\right) = 6\left(-\frac{\sqrt{3}}{2} + i\left(-\frac{1}{2}\right)\right)$$

$$= -3\sqrt{3} - 3i$$

$$k = 3; \quad z = 6\left(\cos\left(\frac{\pi}{6} + \frac{3\pi}{2}\right) + i\sin\left(\frac{\pi}{6} + \frac{3\pi}{2}\right)\right) = \left(\cos\left(\frac{5\pi}{3}\right) + i\sin\left(\frac{5\pi}{3}\right)\right) = 6\left(\frac{1}{2} + i\left(-\frac{\sqrt{3}}{2}\right)\right)$$

$$= 3 - 3i\sqrt{3}$$

11. $-10i = 10\left(\cos\left(\frac{3\pi}{2}\right) + i\sin\left(\frac{3\pi}{2}\right)\right)$ and $\sqrt{3} + i = 2\left(\cos\frac{\pi}{6} + i\sin\frac{\pi}{6}\right).$ Then

$$\frac{10\left(\cos\left(\frac{3\pi}{2}\right) + i\sin\left(\frac{3\pi}{2}\right)\right)}{2\left(\cos\frac{\pi}{6} + i\sin\frac{\pi}{6}\right)} = 5\left(\cos\left(\frac{3\pi}{2} - \frac{\pi}{6}\right) + i\sin\left(\frac{3\pi}{2} - \frac{\pi}{6}\right)\right)$$

$$= 5\left(\cos\frac{4\pi}{3} + i\sin\frac{4\pi}{3}\right)$$

12. $(x^3 + x^2 + x + 1)(x - 1) = x^4 - 1$, so the solutions are the solutions of $x^4 - 1$ with the exception of 1 itself, that is the solutions are $-1, i, -i$.

$$x^4 = -1; -1 = \cos(\pi) + i\sin(\pi)$$

So, $\cos\left(\pi + \dfrac{k\pi}{2}\right) + i\sin\left(\pi + \dfrac{k\pi}{2}\right), k = 0, 1, 2, 3$

$k = 0$: $\cos(\pi) + i\sin(\pi) = -1$

$k = 1$: $\cos\left(\pi + \dfrac{\pi}{2}\right) + i\sin\left(\pi + \dfrac{\pi}{2}\right) = \cos\left(\dfrac{3\pi}{2}\right) + i\sin\left(\dfrac{3\pi}{2}\right) = -i$

$k = 2$: $\cos\left(\pi + \dfrac{2\pi}{2}\right) + i\sin\left(\pi + \dfrac{2\pi}{2}\right) = \cos(2\pi) + i\sin(2\pi) = 1$

$k = 3$: $\cos\left(\pi + \dfrac{3\pi}{2}\right) + i\sin\left(\pi + \dfrac{3\pi}{2}\right) = \cos\left(\dfrac{5\pi}{2}\right) + i\sin\left(\dfrac{5\pi}{2}\right) = i$

13. $\langle -8 - (-1), -9 - 4 \rangle = \langle -7, -13 \rangle$ is a vector with initial point at the origin.

14. $(\langle 3, 5 \rangle + 2\langle -2, 1 \rangle) \cdot (3\langle 3, 5 \rangle + \langle 3, -1 \rangle) = (\langle 3, 5 \rangle + \langle -4, 2 \rangle) \cdot (\langle 9, 15 \rangle + \langle 3, -1 \rangle)$

$$= (\langle -1, 7 \rangle) \cdot (\langle 12, 14 \rangle)$$
$$= (-1)(12) + (7)(14)$$
$$= 86$$

15. a. $\sqrt{11}\mathbf{i} + \sqrt{5}\mathbf{j}$ **b.** $-\sqrt{11}\mathbf{i} + \sqrt{5}\mathbf{j}$ **c.** $-5\sqrt{11}\mathbf{i} + 6\sqrt{5}\mathbf{j}$

16. $\cos\theta = \dfrac{\langle 9, 3 \rangle \cdot \langle 0, -2 \rangle}{\|\langle 9, 3 \rangle\| \|\langle 0, -2 \rangle\|}$

$\theta = \cos^{-1}\left(\dfrac{\langle 9, 3 \rangle \cdot \langle 0, -2 \rangle}{\|\langle 9, 3 \rangle\| \|\langle 0, -2 \rangle\|}\right) = \cos^{-1}\left(\dfrac{-6}{(3\sqrt{10})(2)}\right) = \cos^{-1}\left(\dfrac{-1}{\sqrt{10}}\right)$

$\approx 108.4349° = 1.8925$ rad

17. $\mathbf{v} = 3\cos 210°\mathbf{i} + 3\sin 210°\mathbf{j}$

$$\mathbf{v} = 3\left(-\dfrac{\sqrt{3}}{2}\right)\mathbf{i} + 3\left(-\dfrac{1}{2}\right)\mathbf{j}$$

$$\mathbf{v} = -\dfrac{3\sqrt{3}}{2}\mathbf{i} - \dfrac{3}{2}\mathbf{j} = \left\langle -\dfrac{3\sqrt{3}}{2}, -\dfrac{3}{2} \right\rangle$$

18. If the vectors are orthogonal, then their dot product is 0, thus;

$$(3k\mathbf{i} - 6\mathbf{j}) \cdot (-7\mathbf{i} + \mathbf{j}) = 0$$

$$(3k)(-7) + (-6)(1) = 0$$

$$-21k - 6 = 0$$

$$k = -\frac{6}{21} = -\frac{2}{7}$$

19. $\mathbf{u} = \dfrac{-7\mathbf{i} - 6\mathbf{j}}{\|-7\mathbf{i} - 6\mathbf{j}\|} = \dfrac{-7\mathbf{i} - 6\mathbf{j}}{\sqrt{49 + 36}} = \dfrac{-7\mathbf{i} - 6\mathbf{j}}{\sqrt{85}} = -\dfrac{7\mathbf{i}}{\sqrt{85}} - \dfrac{6\mathbf{j}}{\sqrt{85}}$

20. $\text{comp}_v \mathbf{u} = \dfrac{(\mathbf{i} - 2\mathbf{j}) \cdot (3\mathbf{i} + \mathbf{j})}{\|(3\mathbf{i} + \mathbf{j})\|} = \dfrac{3 + (-2)}{\sqrt{10}} = \dfrac{1}{\sqrt{10}}$

21. a. $\langle 1, 6 \rangle + \langle 7, -3 \rangle + \langle -2, 0 \rangle + \langle -9, 8 \rangle = \langle -3, 11 \rangle$ **b.** $\langle -3, 11 \rangle + \langle v_1, v_2 \rangle = \mathbf{0}$

$$\langle v_1, v_2 \rangle = \langle 3, -11 \rangle$$

22. $\overrightarrow{PQ} = 6\mathbf{i} - 6\mathbf{j}, \; W = \mathbf{F} \cdot \overrightarrow{PQ} = (8\mathbf{i} + \mathbf{j}) \cdot (6\mathbf{i} - 6\mathbf{j}) = 48 + (-6) = 42$

23. Let $\mathbf{F}_1 = \langle 3.9 \cos 255°, 3.9 \sin 255° \rangle$ be the force of the swimmer, and let
$\mathbf{F}_2 = \langle x \cos 0°, x \sin 0° \rangle$ be the force of the river. Since the swimmer reached the bank south of the starting point, the horizontal component of the resultant $\mathbf{F}_1 + \mathbf{F}_2$ must equal 0. Thus, $3.9 \cos 255° + x \cos 0° = \mathbf{0}$

$$x = -\frac{3.9 \cos 255°}{\cos 0°} \approx 1.01 \text{ mph.}$$

24. $W = \|\mathbf{F}\| \|\mathbf{d}\| \cos \theta = 17(200) \cos 20° = 3194.95 \text{ ft-lb.}$

Chapter 10
Analytic Geometry

10.1 Circles and Ellipses

Note: Throughout this chapter, windows have been chosen on the TI-83, wherever possible, to show vertices as often as possible without distorting the shapes of the graphs unreasonably.

1. Circle; $x^2 + (y-3)^2 = 4$

3. Ellipse; $x^2 + 6y^2 = 18$

5. Circle; $(x-4)^2 + (y-3)^2 = 4$

7. Complete the square:
$$x^2 + y^2 + 8x - 6y - 15 = 0$$
$$x^2 + 8x + 16 + y^2 - 6y + 9 = 15 + 16 + 9$$
$$(x+4)^2 + (y-3)^2 = 40$$
Center: $(-4, 3)$, radius $2\sqrt{10}$

9. Complete the square:
$$x^2 + y^2 + 6x - 4y - 15 = 0$$
$$x^2 + 6x + 9 + y^2 - 4y + 4 = 15 + 9 + 4$$
$$(x+3)^2 + (y-2)^2 = 28$$
Center: $(-3, 2)$, radius $\sqrt{28} = 2\sqrt{7}$

11. Complete the square:
$$x^2 + y^2 + 25x + 10y = -12$$
$$x^2 + 25x + \frac{625}{4} + y^2 + 10y + 25$$
$$= -12 + \frac{625}{4} + 25$$
$$\left(x + \frac{25}{2}\right)^2 + (y+5)^2 = \frac{677}{4}$$
Center: $\left(-\frac{25}{2}, -5\right)$, radius $\frac{\sqrt{677}}{2} \approx 13.01$

13. Ellipse; center $(0,0)$; vertices $(-5,0)$ and $(5,0)$; foci $\left(-\sqrt{21}, 0\right)$ and $\left(\sqrt{21}, 0.\right)$.

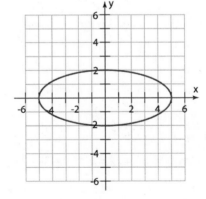

15. Ellipse; center $(0,0)$; vertices $(0,2)$ and $(0,-2)$; foci $(0,-1)$ and $(0,1)$

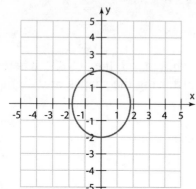

17. Ellipse; center $(0,0)$; vertices $(-9,0)$ and $(9,0)$; foci $\left(-\sqrt{32},0\right)$ and $\left(\sqrt{32},0\right)$

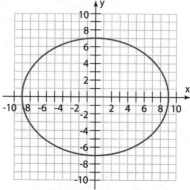

19. Circle; center $(0,0)$; radius ½

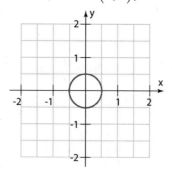

21. $a = 7$, $b = 2$, major axis horizontal

$$\frac{x^2}{7^2} + \frac{y^2}{2^2} = 1$$

$$\frac{x^2}{49} + \frac{y^2}{4} = 1$$

23. $2a = 12$, $2b = 8$, so $a = 6$, $b = 4$, major axis horizontal

$$\frac{x^2}{6^2} + \frac{y^2}{4^2} = 1$$

$$\frac{x^2}{36} + \frac{y^2}{16} = 1$$

25. $a = 7$, $b = 3$, major axis vertical

$$\frac{x^2}{3^2} + \frac{y^2}{7^2} = 1 \text{ or } \frac{x^2}{9} + \frac{y^2}{49} = 1$$

27. $a = 4$, $b = 2$

$$\pi ab = \pi \cdot 4 \cdot 2 = 8\pi$$

29. Rewrite: $\dfrac{x^2}{4} + \dfrac{y^2}{3} = 1$

$$a = 2, \ b = \sqrt{3}$$

$$\pi ab = \pi \cdot 2 \cdot \sqrt{3} = 2\pi\sqrt{3}$$

31. Rewrite: $\dfrac{6x^2}{14} + \dfrac{2y^2}{14} = 1$

$$\frac{x^2}{7/3} + \frac{y^2}{7} = 1$$

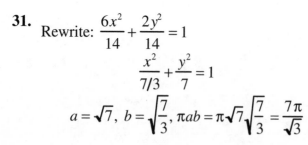

$$a = \sqrt{7}, \ b = \sqrt{\frac{7}{3}}, \ \pi ab = \pi\sqrt{7}\sqrt{\frac{7}{3}} = \frac{7\pi}{\sqrt{3}}$$

33. Ellipse; center $(1,5)$; vertices $(1,2)$, **35.** Ellipse; center $(-1,4)$; vertices $(1,8)$; foci $\left(0,-\sqrt{5}+5\right),\left(0,\sqrt{5}+5\right)$

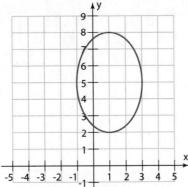

35. Ellipse; center $(-1,4)$; vertices $(-5,4)$,

$(3,4)$;   foci

$\left(-2\sqrt{2}-1,4\right),\left(2\sqrt{2}-1,4\right)$

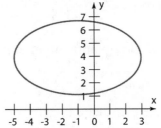

37. Ellipse; center $(-3,1)$; vertices $(-3,4)$,

$(-3,-2)$; foci $\left(-3,\sqrt{5}+1\right),\left(-3,-\sqrt{5}+1\right)$

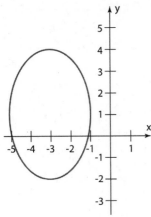

39. Complete the square to get:

$x^2+y^2+6x-8y+5=0$

$\left(x^2+6x+9\right)+\left(y^2-8y+16\right)=-5+9+16$

$\left(x+3\right)^2+\left(y-4\right)^2=20$

Which is the equation of a circle with center at $(-3,4)$.

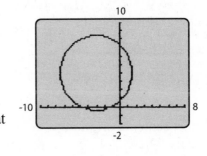

41. Complete the square to get:

$$4x^2 + y^2 + 24x - 4y + 36 = 0$$

$$4\left(x^2 + 6x + 9\right) + \left(y^2 - 4y + 4\right) = -36 + 36 + 4$$

$$4\left(x + 3\right)^2 + \left(y - 2\right)^2 = 4$$

$$\frac{\left(x + 3\right)^2}{1} + \frac{\left(y - 2\right)^2}{4} = 1$$

Which is the equation of an ellipse with center at $(-3, 2)$.

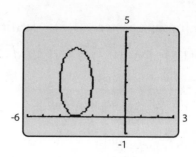

43. Complete the square to get:

$$9x^2 + 25y^2 - 18x + 50y = 191$$

$$9\left(x^2 - 2x + 1\right) + 25\left(y^2 + 2y + 1\right) = 191 + 9 + 25$$

$$9\left(x - 1\right)^2 + 25\left(y + 1\right)^2 = 225$$

$$\frac{\left(x - 1\right)^2}{25} + \frac{\left(y + 1\right)^2}{9} = 1$$

Which is the equation of an ellipse with center at $(1, -1)$.

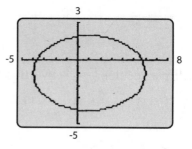

45. $2a = 7 - (-1),\ a = 4$
$2b = 4 - 0,\ b = 2$
major axis vertical
$$\frac{\left(x - 2\right)^2}{4} + \frac{\left(y - 3\right)^2}{16} = 1$$

47. $2a = 12,\ a = 6,\ 2b = 5,\ b = 5/2$
major axis vertical
$$\frac{\left(x - 7\right)^2}{25/4} + \frac{\left(y + 4\right)^2}{36} = 1$$

49. $a = 9 - 3 = 6,\ b = (-2) - (-6) = 4$
major axis horizontal
$$\frac{\left(x - 3\right)^2}{36} + \frac{\left(y + 2\right)^2}{16} = 1$$

51. $2a = 14,\ a = 7,\ 2b = 8,\ b = 4$
$$\frac{\left(x + 5\right)^2}{49} + \frac{\left(y - 3\right)^2}{16} = 1$$
or $\dfrac{\left(x + 5\right)^2}{16} + \dfrac{\left(y - 3\right)^2}{49} = 1$

53. Circle, center at origin: $2x^2 + 2y^2 - 8 = 0$

55. Ellipse, major axis vertical, center in first quadrant: $2x^2 + y^2 - 8x - 6y + 9 = 0$

57. Ellipse, major axis vertical, center in third quadrant: $\dfrac{\left(x + 3\right)^2}{4} + \dfrac{\left(y + 3\right)^2}{8} = 1$

59. The ellipse has y-intercepts $(0, 4)$ and $(0, -4)$. Therefore, the circle, which has center at the origin, has radius 4. The equation of the circle must be $x^2 + y^2 = 16$.

61. $2a = 477,736$ $a = 238,868$ $2b = 477,078$ $b = 238,539$

$c = \sqrt{a^2 - b^2} = \sqrt{(238,868)^2 - (238,539)^2} = 12,533$

The minimum distance is $a - c = 238,868 - 12,533 = 226,335$ miles.

The maximum distance is $a + c = 238,868 + 12,533 = 251,401$ miles.

63. The figure shows an ellipse with major axis $2a = 100$, minor axis $b = 30$. Then $a = 50$, hence the equation is $\dfrac{x^2}{50^2} + \dfrac{y^2}{30^2} = 1$. The foci of the ellipse are given by calculation $c^2 = 50^2 - 30^2, c = \sqrt{1600} = 40$. Then the distance between the two people is $2c = 80$ feet.

65. In an appropriate coordinate system, the ellipse has equation $\dfrac{x^2}{50^2} + \dfrac{y^2}{20^2} = 1$.

When $x = 25$, $\dfrac{25^2}{50^2} + \dfrac{y^2}{20^2} = 1$. Solve to obtain $y = 10\sqrt{3} = 17.3$ feet.

67. $a^2 = 100$ $b^2 = 99$ $c = \sqrt{a^2 - b^2} = \sqrt{100 - 99} = 1$, eccentricity $= \frac{c}{a} = \frac{1}{10} = .1$

69. $a^2 = 40$ $b^2 = 10$ $c = \sqrt{a^2 - b^2} = \sqrt{40 - 10} = \sqrt{30}$

eccentricity $= \dfrac{c}{a} = \dfrac{\sqrt{30}}{\sqrt{40}} = \dfrac{\sqrt{3}}{2} = .87$

71. The closer the eccentricity is to zero, the more the ellipse resembles a circle. The closer the eccentricity is to 1, the more elongated the ellipse becomes.

73. The minimum distance from the center is $a - c = 6540 + 6400 = 12,940$.

The maximum distance from the center is $a + c = 22,380 + 6400 = 28,780$.

Solving these two equations yields $a = 20,860$, $c = 7920$.

Hence, eccentricity $= \dfrac{c}{a} = \dfrac{7920}{20,860} = .38$

75. If $a = b$, the equation becomes $\dfrac{x^2}{a^2} + \dfrac{y^2}{a^2} = 1$ or $x^2 + y^2 = a^2$.

This is the equation of a circle with center at the origin, radius a.

77. The guest can be anywhere inside an ellipse with foci 50 feet apart: $c = 25$, and sum of distances to foci: $150 - 50 = 100$: $a = 50$. The fence is an ellipse with major axis 100 ft, minor axis $2\left(\sqrt{50^2 - 25^2}\right) = 50\sqrt{3}$ ft. ≈ 86.6 ft.

The area is $\pi a b = \pi \cdot 50 \cdot 25\sqrt{3} = 1250\pi\sqrt{3}$ sq ft. ≈ 6801.7 sq ft.

10.2 Hyperbolas

1. Ellipse, center at origin, major axis horizontal: $x^2 + 4y^2 = 1$.

3. Hyperbola, focal axis on x-axis: $2x^2 - y^2 = 8$.

5. Ellipse: center at origin, major axis vertical: $6x^2 + 2y^2 = 18$.

7. Hyperbola, vertices $(\pm 2, 0)$, asymptotes $y = \pm\dfrac{1}{2}x$, foci $\left(\pm\sqrt{5}, 0\right)$

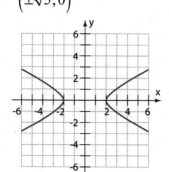

9. Write: $\dfrac{y^2}{5} - \dfrac{x^2}{3} = 1$

    Hyperbola, vertices $\left(0, \pm\sqrt{5}\right)$,

    asymptotes $y = \pm\sqrt{\dfrac{3}{5}}x$,

    foci $\left(0, \pm\sqrt{8}\right)$

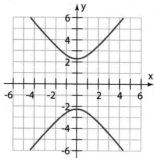

11. Hyperbola, vertices $(0, \pm 3)$, asymptotes $y = \pm\dfrac{3}{4}x$, foci $(0, \pm 5)$

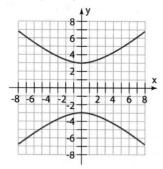

13. Write: $\dfrac{x^2}{1} - \dfrac{y^2}{1/4} = 1$

    $\dfrac{x^2}{1^2} - \dfrac{y^2}{(1/2)^2} = 1$

    $a = 1,\ b = \frac{1}{2}$. Hyperbola, vertices

    $(\pm 1, 0)$, asymptotes $y = \pm\dfrac{1}{2}x$,

    foci $\left(\pm\dfrac{\sqrt{5}}{2}, 0\right)$

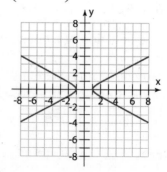

15. Center $(0,0); a = 2; b = 4$

$$\frac{x^2}{4} - \frac{y^2}{16} = 1$$

17. Center $(0,0); a = 5; b = 2$

$$\frac{y^2}{25} - \frac{x^2}{4} = 1$$

19. Center $(-2,-1); a = 3; b = 2$

$$\frac{(x+2)^2}{9} - \frac{(y+1)^2}{4} = 1$$

21. Since the x-intercepts are ± 3, $a = 3$.

Since an asymptote is $y = 2x$, $\dfrac{b}{a} = 2$, $b = 2a$, $b = 6$. Equation: $\dfrac{x^2}{9} - \dfrac{y^2}{36} = 1$.

23. Since a vertex is $(2,0)$, $a = 2$. The equation must be of form $\dfrac{x^2}{2^2} - \dfrac{y^2}{b^2} = 1$. Since

$\left(4, \sqrt{3}\right)$ is on the graph, its coordinates must satisfy the equation. Hence

$$\frac{4^2}{2^2} - \frac{\left(\sqrt{3}\right)^2}{b^2} = 1$$

$$b^2 = 1$$

The equation is: $\dfrac{x^2}{4} - \dfrac{y^2}{1} = 1$.

25. Hyperbola, center $(-1,-3)$, vertices $(-1,-8), (-1,2)$, foci $\left(-1, -3 \pm \sqrt{41}\right)$, asymptotes $y + 3 = \pm\dfrac{5}{4}(x+1)$.

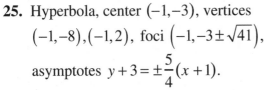

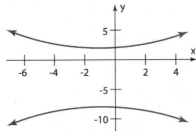

27. Hyperbola, center $(-3,2)$, vertices $(-4,2),(-2,2)$, foci $\left(-3 \pm \sqrt{5}, 2\right)$, asymptotes $y - 2 = \pm 2(x+3)$.

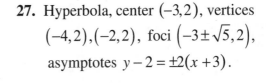

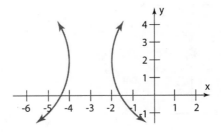

29. $(y+4)^2 - 8(x-1)^2 = 8$

$$\frac{(y+4)^2}{8} - \frac{(x-1)^2}{1} = 1$$

Hyperbola, center $(1,-4)$, vertices $\left(1, -4 \pm \sqrt{8}\right)$, foci $(1,-7),(1,-1)$, asymptotes $y + 4 = \pm\sqrt{8}(x-1)$.

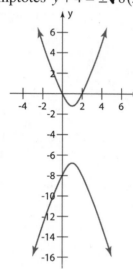

31. $4y^2 - x^2 + 6x - 24y + 11 = 0$

$4\left(y^2 - 6y + 9\right) - \left(x^2 - 6x + 9\right)$
$$= -11 + 36 - 9$$

$4(y-3)^2 - (x-3)^2 = 16$

$$\frac{(y-3)^2}{4} - \frac{(x-3)^2}{16} = 1$$

Hyperbola, center $(3,3)$, vertices $(3,1),(3,5)$, foci $\left(3, 3 \pm \sqrt{20}\right)$

asymptotes $y - 3 = \pm\dfrac{1}{2}(x-3)$.

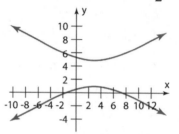

33. $2x^2 + 2y^2 - 12x - 16y + 26 = 0$

$2\left(x^2 - 6x + 9\right) + 2\left(y^2 - 8y + 16\right)$
$$= -26 + 18 + 32$$

$(x-3)^2 + (y-4)^2 = 12$

Circle, center $(3,4)$, radius $\sqrt{12}$.

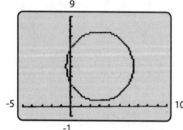

35. $2x^2 + 3y^2 - 12x - 24y + 54 = 0$

$2\left(x^2 - 6x + 9\right) + 3\left(y^2 - 8y + 16\right)$
$$= -54 + 18 + 48$$

$$\frac{(x-3)^2}{6} + \frac{(y-4)^2}{4} = 1$$

Ellipse, center $(3,4)$, major axis horizontal.

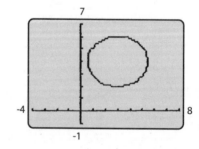

37. $x^2 - 3y^2 + 4x + 12y = 20$

$(x^2 + 4x + 4) - 3(y^2 - 4y + 4) = 20 + 4 - 12$

$\dfrac{(x+2)^2}{12} - \dfrac{(y-2)^2}{4} = 1$

Hyperbola, center $(-2, 2)$,
focal axis on horizontal line
$y = 2$, $a = \sqrt{12}$,

$b = 2$, asymptotes $y - 2 = \pm\dfrac{1}{\sqrt{3}}(x+2)$.

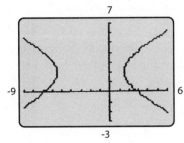

39. Since the center is $(-2, 3)$ and the focal axis is vertical, the equation must be of form:

$\dfrac{(y-3)^2}{a^2} - \dfrac{(x+2)^2}{b^2} = 1.$

Since the distance from center to vertex is $3 - 1 = 2$, $a = 2$; $\dfrac{(y-3)^2}{4} - \dfrac{(x+2)^2}{b^2} = 1$

Since $(-2 + 3\sqrt{10}, 11)$ is on the graph, its coordinates satisfy the equation.

$\dfrac{(11-3)^2}{4} - \dfrac{(-2+3\sqrt{10}+2)^2}{b^2} = 1$, $b^2 = 6$. The equation is $\dfrac{(y-3)^2}{4} - \dfrac{(x+2)^2}{6} = 1.$

41. Since the center is $(4, 2)$ and the focal axis is horizontal, the equation must be of

form: $\dfrac{(x-4)^2}{a^2} - \dfrac{(y-2)^2}{b^2} = 1.$

Since the distance from center to vertex is $7 - 4 = 3$, $a = 3$; $\dfrac{(x-4)^2}{9} - \dfrac{(y-2)^2}{b^2} = 1$

Since an asymptote has slope $\dfrac{4}{3}$, $\dfrac{b}{a} = \dfrac{4}{3}$, $b = 4$.

The equation is $\dfrac{(x-4)^2}{9} - \dfrac{(y-2)^2}{16} = 1.$

43. Center at origin, focal axis is vertical: $y^2 - 2x^2 = 6$

45. Center in first quadrant: $\dfrac{(y-2)^2}{4} - \dfrac{(x-3)^2}{9} = 1$

47. Center in third quadrant: $\dfrac{(x+3)^2}{3} - \dfrac{(y+3)^2}{4} = 1$

49. The graphs are shown below on one set of coordinate axes.

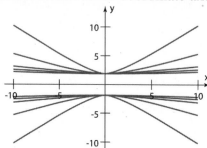

The "curviest" curve has $b = 2$; the flattest curve has $b = 20$. Clearly, as b increases, the graphs get flatter. However, although the hyperbola may look like two horizontal lines for very large b, it is still a hyperbola, with asymptotes $y = \pm 2x/b$ that have non-zero slopes.

51. The asymptotes of $\dfrac{x^2}{a^2} - \dfrac{y^2}{a^2} = 1$ are $y = \pm \dfrac{a}{a} x$ or $y = \pm x$.

Since they have slopes 1 and –1 and $1(-1) = -1$, they are perpendicular.

53. The data require the position (x, y) of the explosion to satisfy: The distances of (x, y) from two points 5280 feet apart differ by 2200 feet. This defines a hyperbola with $2c = 5280$, $2a = 2200$, thus $a = 1100$, $c = 2640$, $b^2 = 2640^2 - 1100^2 = 5,759,600$.

If the points are $(\pm 1100, 0)$, then the equation must be: $\dfrac{x^2}{1,210,000} - \dfrac{y^2}{5,759,600} = 1$.

The exact location, however, cannot be determined from only the given information.

55. From the diagram, we see that $(5, 0)$ is an x-intercept, thus $a = 5$.

Since the asymptotes are $y = \pm \dfrac{3}{4} x$, the point on the asymptote directly above the x-intercept is $\left(5, \dfrac{15}{4}\right)$, which tells us that $b = \dfrac{15}{4}$.

Therefore, the equation of the particle's path is $\dfrac{x^2}{5^2} - \dfrac{y^2}{(15/4)^2} = 1$ or $\dfrac{x^2}{25} - \dfrac{16y^2}{225} = 1$.

57. $a^2 = 10$ $b^2 = 40$ $c^2 = a^2 + b^2 = 50$ $e = \frac{c}{a} = \frac{\sqrt{50}}{\sqrt{10}} = \sqrt{5} \approx 2.24$

59. Rewrite: $\dfrac{(y-2)^2}{3} - \dfrac{(x+2)^2}{6} = 1$; $a^2 = 3$ $b^2 = 6$ $c^2 = a^2 + b^2 = 9$ $e = \frac{c}{a} = \frac{3}{\sqrt{3}} = \sqrt{3}$

61. Rewrite: $4(x^2 - 4x + 4) - 5(y^2 + 10y + 25) = -71 + 16 - 125$ or $\dfrac{(y+5)^2}{36} - \dfrac{(x-2)^2}{45} = 1$

$a^2 = 36$ $b^2 = 45$ $c^2 = a^2 + b^2 = 81$ $e = \frac{c}{a} = \frac{9}{6} = \frac{3}{2}$

10.3 Parabolas

1. Parabola, opens to the right: $6x = y^2$

3. Hyperbola, focal axis horizontal: $2x^2 - y^2 = 8$

5. Ellipse, major axis horizontal: $x^2 + 6y^2 = 18$

7. $p = 4$

$y^2 = 4(4)x$

$y^2 = 16x$

9. $p = -8$

$x^2 = 4(-8)y$

$x^2 = -32y$

11. Parabola, opens upward.

Rewrite: $x^2 = \frac{1}{3}y$

$4p = \frac{1}{3}$

$p = \frac{1}{12}$

Focus: $\left(0, \frac{1}{12}\right)$, Directrix $y = -\frac{1}{12}$

13. Parabola, opens upward.

Rewrite: $x^2 = 4y$

$4p = 4$

$p = 1$

Focus: $(0,1)$, Directrix $y = -1$

15. Parabola, opens to the left.

Rewrite: $y^2 = -8x$

$4p = -8$

$p = -2$

Focus: $(-2,0)$, Directrix $x = 2$

17. Rewrite: $x - 2 = y^2$. $4p = 1$; $p = \frac{1}{4}$. Opens to the right.

Vertex: $(2,0)$ Focus: $\left(2\frac{1}{4},0\right)$ Directrix: $x = 2 - \frac{1}{4} = \frac{7}{4}$

19. Rewrite: $-1(x - 2) = (y + 1)^2$. $4p = -1$; $p = -\frac{1}{4}$. Opens to the left.

Vertex: $(2,-1)$ Focus: $\left(\frac{7}{4},-1\right)$ Directrix: $x = 2 + \frac{1}{4} = \frac{9}{4}$

21. Rewrite: $3\left(x - \frac{2}{3}\right) = (y + 3)^2$. $4p = 3$; $p = \frac{3}{4}$. Opens to the right.

Vertex: $\left(\frac{2}{3},-3\right)$ Focus: $\left(\frac{2}{3} + \frac{3}{4},-3\right) = \left(\frac{17}{12},-3\right)$ Directrix: $x = \frac{2}{3} - \frac{3}{4} = -\frac{1}{12}$

23. Complete the square: $x = y^2 - 9y$

$$x + \frac{81}{4} = y^2 - 9y + \frac{81}{4}$$

$$x + \frac{81}{4} = \left(y - \frac{9}{2}\right)^2$$

$4p = 1$; $p = \frac{1}{4}$. Opens to the right.

Vertex: $\left(-\frac{81}{4}, \frac{9}{2}\right)$

Focus: $\left(-\frac{81}{4} + \frac{1}{4}, \frac{9}{2}\right) = \left(-20, \frac{9}{2}\right)$

Directrix: $x = -\frac{81}{4} - \frac{1}{4} = -\frac{41}{2}$

25. Complete the square: $y = 3x^2 + x - 4$

$$y + 4 + \frac{3}{36} = 3\left(x^2 + \frac{1}{3}x + \frac{1}{36}\right)$$

$$\frac{1}{3}\left(y + \frac{49}{12}\right) = \left(x + \frac{1}{6}\right)^2$$

$4p = \frac{1}{3}$; $p = \frac{1}{12}$. Opens upward.

Vertex: $\left(-\frac{1}{6}, -\frac{49}{12}\right)$

Focus: $\left(-\frac{1}{6}, -\frac{49}{12} + \frac{1}{12}\right) = \left(-\frac{1}{6}, -4\right)$

Directrix: $x = -\frac{49}{12} - \frac{1}{12} = -\frac{25}{6}$

27. Complete the square: $y = -3x^2 + 4x + 5$

$$y - 5 - 4/3 = -3\left(x^2 - 4/3 x + 4/9\right)$$

$$-1/3(y - 19/3) = (x - 2/3)^2$$

$4p = -\frac{1}{3}$; $p = -\frac{1}{12}$. Opens downward.

Vertex: $\left(\frac{2}{3}, \frac{19}{3}\right)$, Focus: $\left(\frac{2}{3}, \frac{19}{3} - \frac{1}{12}\right) = \left(\frac{2}{3}, \frac{25}{4}\right)$, Directrix: $y = \frac{19}{3} + \frac{1}{12} = \frac{77}{12}$

29. The latus rectum is the line segment made of the points directly above and below, or directly left and right of the focus.

For, $x^2 = -8y$, the focus is $(0, -2)$. To find the points directly above and below, solve

$x^2 = 4py$, when $y = p$

$x^2 = 4py = 4p^2 = 16$

$x = \pm 4$

Therefore the latus rectum is the line segment between $(-4, -2)$ and $(4, -2)$.

31. For, $y^2 = 20x$, the focus is $(5, 0)$. To find the points directly left and right, solve

$y^2 = 4px$, when $x = p$

$y^2 = 4px = 4p^2 = 100$

$y = \pm 10$

Therefore the latus rectum is the line segment between $(5, -10)$ and $(5, 10)$.

33. For, $x^2 - 4y = 0$, the focus is $(0, 1)$. To find the points directly above and below, solve

$x^2 = 4py$, when $y = p$

$x^2 = 4py = 4p^2 = 4$

$x = \pm 2$

Therefore the latus rectum is the line segment between $(-2, 1)$ and $(2, 1)$.

35. Vertex: (1, 2)

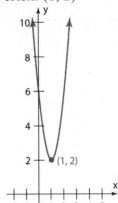

37. Vertex: (0, 2)

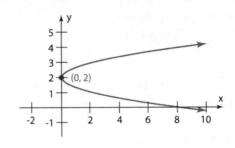

39. Vertex: (2, –5)

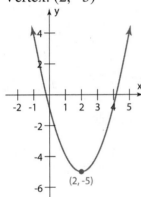

41. Vertex: (–1, –1)

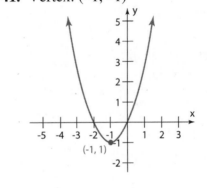

43. The equation has form $x^2 = 4py$. Since (2, 12) is on the graph, its coordinates satisfy the equation, hence $2^2 = 4p \cdot 12$, $4p = \frac{1}{3}$.
$x^2 = \frac{1}{3}y$ or $y = 3x^2$

45. The equation has form $(x-1)^2 = 4py$. Since (2, 13) is on the graph, its coordinates satisfy the equation, hence $(2-1)^2 = 4p \cdot 13$, $4p = \frac{1}{13}$.
$(x-1)^2 = \frac{1}{13}y$ or $y = 13(x-1)^2$

47. The equation has form $(y-1)^2 = 4p(x-2)$. Since (5,0) is on the graph, its coordinates satisfy the equation, hence $(0-1)^2 = 4p(5-2)$, $4p = \frac{1}{3}$.
$(y-1)^2 = \frac{1}{3}(x-2)$ or $(x-2) = 3(y-1)^2$

49. Since the focus is to the right of the vertex a distance $\frac{1}{16}$, $p = \frac{1}{16}$, and the parabola opens to the right. The equation is $4(\frac{1}{16})(x+3) = (y+2)^2$ or $(x+3) = 4(y+2)^2$.

51. Since the focus is above the vertex a distance $\frac{1}{8}$, $p = \frac{1}{8}$, and the parabola opens upward. The equation is $4(\frac{1}{8})(y-1) = (x-1)^2$ or $(y-1) = 2(x-1)^2$.

53. A rough sketch (not shown) indicates that the parabola must open to the right. The equation has form $(y-3)^2 = 4p(x+1)$. Since $(8,0)$ is on the graph, its coordinates satisfy the equation, hence

$$(0-3)^2 = 4p(8+1)$$
$$4p = 1$$

The equation is $(y-3)^2 = (x+1)$. Confirm that $(0,4)$ is on the graph, since its coordinates also satisfy the equation.

55. Complete the square:
$$x^2 = 6x - y - 5$$
$$x^2 - 6x + 9 = -y - 5 + 9$$
$$(x-3)^2 = -(y-4)$$
Parabola, vertex $(3,4)$

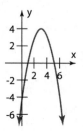

57. Complete the square:
$$3y^2 - 2y = x - 1$$
$$3\left(y^2 - \tfrac{2}{3}y + \tfrac{1}{9}\right) = x - 1 + \tfrac{1}{3}$$
$$\left(y - \tfrac{1}{3}\right)^2 = \tfrac{1}{3}\left(x - \tfrac{2}{3}\right)$$
Parabola, vertex $\left(\tfrac{2}{3}, \tfrac{1}{3}\right)$

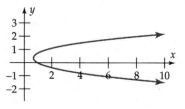

59. Complete the square:
$$3x^2 + 3y^2 - 6x - 12y - 6 = 0$$
$$3\left(x^2 - 2x + 1\right) + 3\left(y^2 - 4y + 4\right) = 6 + 3 + 12$$
$$(x-1)^2 + (y-2)^2 = 7$$
Circle, center $(1,2)$

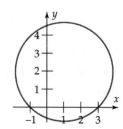

61. Complete the square:
$$2x^2 - y^2 + 16x + 4y + 24 = 0$$
$$2\left(x^2 + 8x + 16\right) - \left(y^2 - 4y + 4\right) = -24 + 32 - 4$$
$$\frac{(x+4)^2}{2} - \frac{(y-2)^2}{4} = 1$$
Hyperbola, center $(-4,2)$,
vertices $\left(-4 \pm \sqrt{2}, 2\right)$

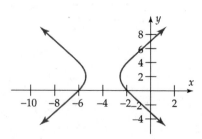

63. Parabola, vertex in first quadrant, opens upward: $y = (x-4)^2 + 2$

65. Parabola, vertex in third quadrant, opens downward: $y = -x^2 - 8x - 18$

67. Parabola, vertex in third quadrant, opens upward: $y = (x+5)^2 - 3$

69. Clearly, 0. The vertex of $y = x^2 + c$ is at $(0, c)$.

71. Replace x with 9 and solve $4y^2 + 4y = 5 \cdot 9 - 12$ to obtain
$y = \dfrac{-1 \pm \sqrt{34}}{2}$: $\left(9, \dfrac{-1 \pm \sqrt{34}}{2}\right)$ are the points.

73. a. The latus rectum is the line segment made of the points directly above and below, or directly left and right of the focus. To find the points directly above and below, solve $y^2 = 4px$, when $x = p$

$$y^2 = 4px = 4p^2$$
$$y = \pm 2p$$

Therefore the latus rectum is the line segment between $(p, -2p)$ and $(p, 2p)$.

b. To find the points directly left and right, solve $x^2 = 4py$, when $y = p$

$$x^2 = 4py = 4p^2$$
$$x = \pm 2p$$

Therefore the latus rectum is the line segment between $(-2p, p)$ and $(2p, p)$.

75. By appropriate choice of axes the parabola can be viewed as opening upward with vertex at the origin. The equation is then $x^2 = 4py$. When $x = \frac{1}{2}(4) = 2$, $y = 1.5$, hence $2^2 = 4p(1.5)$

$$p = \tfrac{2}{3}$$

The receiver should be placed at the focus, $\frac{2}{3}$ feet or 8 inches from the vertex.

77. If the parabola is viewed as opening upward with vertex at the origin, its equation is of form $x^2 = 4py$, with p given as 130, hence, $x^2 = 520y$. When $x = 150$, then

$$y = \frac{150^2}{520} = 43.3 \text{ feet. This is the depth of the dish.}$$

79. Choose the origin at the bottom of the mirror. The equation of the parabola is then of the form $x^2 = 4py$, with p given as 16.75, hence when $y = .096$, then

$$x = \pm\sqrt{67(.096)} = \pm 2.536, \text{ so the diameter is } 2.536 - (-2.536) = 5.072 \text{ meters.}$$

81. Choose the origin at the center of the bridge. The equation of the parabola is then $x^2 = 4py$. When $x = \frac{1}{2}(420) = 210$, $y = 100$, hence

$$210^2 = 4p \cdot 100$$
$$4p = 441$$

The equation is then $x^2 = 441y$. One hundred feet from one of the towers, take

$$x = 210 - 100 = 110. \text{ Then } y = \frac{110^2}{441} = 27.44 \text{ feet.}$$

10.3.A Parametric Equations for Conic Sections

1. $x^2 + y^2 = 25$ is an equation of a circle. In parametric form:
$x = 5\cos t, y = 5\sin t$

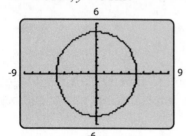

3. $(x-4)^2 + (y+2)^2 = 9$ is an equation of a circle. In parametric form:
$x = 3\cos t + 4, y = 3\sin t - 2$

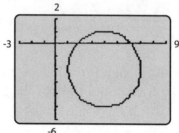

5. $(x+1)^2 = 5 - y^2$ is an equation of a circle. In parametric form:
$x = \sqrt{5}\cos t - 1, y = \sqrt{5}\sin t$

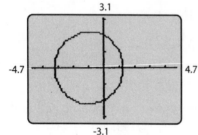

7. $\dfrac{x^2}{10} - 1 = -\dfrac{y^2}{36}$ is an equation of an ellipse. In parametric form:
$x = \sqrt{10}\cos t, y = 6\sin t$

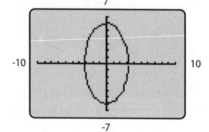

9. $4x^2 + 4y^2 = 1$ is an equation of an ellipse. In parametric form:
$x = \frac{1}{2}\cos t, y = \frac{1}{2}\sin t$

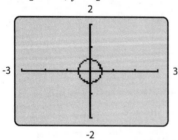

11. $\dfrac{(x-1)^2}{4} + \dfrac{(y-5)^2}{9} = 1$ is an equation of an ellipse. In parametric form:
$x = 2\cos t + 1, y = 3\sin t + 5$

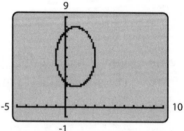

13. $\dfrac{(x+1)^2}{16}+\dfrac{(y-4)^2}{8}=1$ is an equation of an ellipse. In parametric form:
$x=4\cos t-1,\ y=\sqrt{8}\sin t+4$

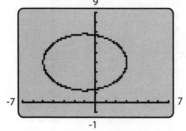

15. $\dfrac{x^2}{10}-\dfrac{y^2}{36}=1$ is an equation of a hyperbola. In parametric form:
$x=\dfrac{\sqrt{10}}{\cos t},\quad y=6\tan t$

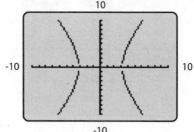

17. $x^2-4y^2=1$ is an equation of a hyperbola. In parametric form:
$x=\dfrac{1}{\cos t},\ y=\dfrac{1}{2}\tan t$

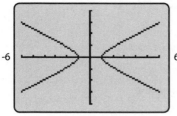

19. $\dfrac{(x+3)^2}{25}-\dfrac{(y+1)^2}{16}=1$ is an equation of a hyperbola. In parametric form:
$x=4\tan t-1,\ y=\dfrac{5}{\cos t}-3$

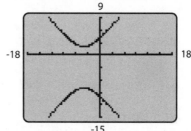

21. $\dfrac{(x+3)^2}{1}-\dfrac{(y-2)^2}{4}=1$ is an equation of a hyperbola. In parametric form:
$x=\dfrac{1}{\cos t}-3,\ y=2\tan t+2$

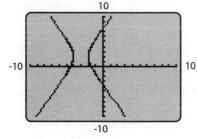

23. $8x=2y^2$ is an equation of a parabola. In parametric form:
$x=t^2/4\quad y=t$

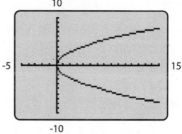

25. $y = 4(x-1)^2 + 2$ is an equation of a parabola. In parametric form:
$x = t$ $y = 4(t-1)^2 + 2$

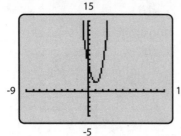

27. $x = 2(y-2)^2$ is an equation of a parabola. In parametric form:
$x = 2(t-2)^2$ $y = t$

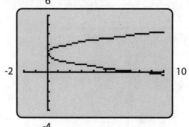

29. This is an equation of a circle with center $(0,-5)$ and $r = 3$.

31. This is an equation of an ellipse with center $(4,0)$.

33. This is an equation of a hyperbola with center $(2,4)$.

35. This is an equation of a hyperbola with center $(0,-3)$.

37. This is an equation of a parabola with vertex $(4,-3)$.

39. a. $x^2 + y^2 = \cos^2(.5t) + \sin^2(.5t) = 1$

   b. $x^2 + y^2 = \cos^2(.5t) + \left(-\sin^2(.5t)\right) = \cos^2(.5t) + \sin^2(.5t) = 1$

   c. Since $\cos(.5t)$ and $\sin(.5t)$ have a period of 4π, these parametric equations trace our half of a circle from $0 \le x \le 2\pi$.

For 41 and 43, use a viewing window of $-4.7 \le x \le 4.7, -3.1 \le y \le 3.1$.

41. The parametric equations are:
$x = 3\cos t, y = 3\sin t$ for the face
$x = .5\cos t + 1.3, y = .5\sin t + 1.3$ for the right eye
$x = .5\cos t - 1.3, y = .5\sin t + 1.3$ for the left eye
$x = 1.5\cos(.5t), y = 1.5\sin(.5t) - .5$ for the frown

43. The parametric equations are:
$x = 2\cos t, y = 3\sin t$ for the face
$x = .3\cos t + 1, y = .4\sin t + 1.4$ for the right eye
$x = .3\cos t - 1, y = .4\sin t + 1.4$ for the left eye
$x = \cos(.5t), y = -1.2\sin(.5t) - .9$ for the smile

10.4 Rotations and Second-Degree Equations

1. $A = 1, B = -2, C = 3$. $B^2 - 4AC = (-2)^2 - 4(1)(3) = -8 < 0$. Ellipse

3. $A = 1, B = 2, C = 1$. $B^2 - 4AC = 2^2 - 4(1)(1) = 0$. Parabola

5. $A = 17, B = -48, C = 31$. $B^2 - 4AC = (-48)^2 - 4(17)(31) = 196 > 0$. Hyperbola

7. $A = 9, B = 0, C = 4$.
$B^2 - 4AC = 0^2 - 4(9)(4) = -144 < 0$.
Ellipse. Solve for y to obtain
$y = \frac{1}{2}\left(2 \pm 3\sqrt{-5 - 6x - x^2}\right)$.
Use window $-6 \le x \le 3, -2 \le y \le 4$

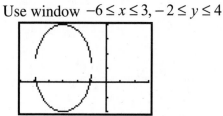

9. $A = -1, B = 0, C = 4$.
$B^2 - 4AC = 0^2 - 4(-1)(4) = 16 > 0$
Hyperbola Solve for y to obtain
$y = \frac{1}{2}\left(6 \pm \sqrt{25 - 6x + x^2}\right)$.
Use window $-7 \le x \le 13, -3 \le y \le 9$

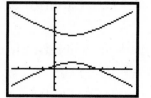

11. $A = 0, B = 0, C = 3$.
$B^2 - 4AC = 0^2 - 4(0)(3) = 0$. Parabola
Solve for y to obtain
$y = \frac{1}{3}\left(1 \pm \sqrt{3x - 2}\right)$.
Use window $-1 \le x \le 8, -3 \le y \le 3$

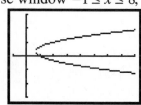

13. $A = 41, B = -24, C = 34$.
$B^2 - 4AC = (-24)^2 - 4(41)(34)$
$= -5000 < 0$
Ellipse. Solve for y to obtain
$y = \frac{1}{34}\left(12x \pm 5\sqrt{34 - 50x^2}\right)$.
Use window $-1.5 \le x \le 1.5, -1 \le y \le 1$

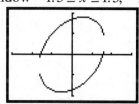

15. $A = 17, B = -48, C = 31.$

$B^2 - 4AC = (-48)^2 - 4(17)(31)$

$= 196 > 0$

Hyperbola Solve for y to obtain

$y = \frac{1}{31}\left(24x \pm 7\sqrt{x^2 - 31}\right)$. Use window

$-15 \le x \le 15, -10 \le y \le 10$

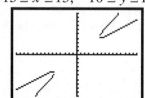

17. $A = 9, B = 24, C = 16.$

$B^2 - 4AC = 24^2 - 4(9)(16) = 0$.

Parabola. Solve for y to obtain

$y = \frac{1}{16}\left(65 - 12x \pm 5\sqrt{169 - 120x}\right)$.

Use window $-19 \le x \le 2, -1 \le y \le 13$

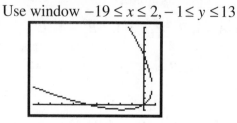

19. $A = 23, B = 26\sqrt{3}, C = -3.$

$B^2 - 4AC = \left(26\sqrt{3}\right)^2 - 4(23)(-3)$

$= 2304 > 0$

Hyperbola. Solve for y to obtain

$y = \frac{1}{3}\left(8\sqrt{3} + 13\sqrt{3}x \pm 24\sqrt{1 + x + x^2}\right)$.

Use window

$-15 \le x \le 15, -15 \le y \le 15$

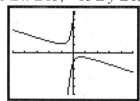

21. $A = 17, B = -12, C = 8.$

$B^2 - 4AC = (-12)^2 - 4(17)(8)$

$= -400 < 0$

Ellipse. Solve for y to obtain

$y = \frac{1}{4}\left(3x \pm \sqrt{160 - 25x^2}\right)$.

Use window $-6 \le x \le 6, -4 \le y \le 4$

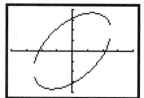

23. $A = 3, B = 2\sqrt{3}, C = 1.$

$B^2 - 4AC = \left(2\sqrt{3}\right)^2 - 4(3)(1) = 0$.

Parabola.

Solve for y to obtain

$y = (2 - x)\sqrt{3} \pm 2\sqrt{7 - 4x}$.

Use window $-9 \le x \le 4, -2 \le y \le 10$

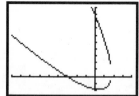

10.4.A Rotation of Axes

1. $x = 3, y = 2$
Using the rotation equations, write:
$3 = u\cos 45° - v\sin 45° = (u - v)/\sqrt{2}$

$2 = u\sin 45° + v\cos 45° = (u + v)/\sqrt{2}$

Solving, we obtain $u = \dfrac{5\sqrt{2}}{2}, v = -\dfrac{\sqrt{2}}{2}$.

$\left(\dfrac{5\sqrt{2}}{2}, -\dfrac{\sqrt{2}}{2}\right)$

3. $x = 1, y = 0$
Using the rotation equations, write:
$1 = u\cos 30° - v\sin 30° = (u\sqrt{3} - v)/2$

$0 = u\sin 30° + v\cos 30° = (u + v\sqrt{3})/2$

Solving, we obtain $u = \dfrac{\sqrt{3}}{2}, v = -\dfrac{1}{2}$.

$\left(\dfrac{\sqrt{3}}{2}, -\dfrac{1}{2}\right)$

5. Using the rotation equations, write: $x = u\cos 45° - v\sin 45° = (u - v)/\sqrt{2}$
$y = u\sin 45° + v\cos 45° = (u + v)/\sqrt{2}$

Substituting, we obtain $\dfrac{(u - v)}{\sqrt{2}}\dfrac{(u + v)}{\sqrt{2}} = 1$.

$u^2 - v^2 = 2$ or $\dfrac{u^2}{2} - \dfrac{v^2}{2} = 1$

7. Using the rotation equations, write:
$x = u\cos 30° - v\sin 30° = (u\sqrt{3} - v)/2$, $y = u\sin 30° + v\cos 30° = (u + v\sqrt{3})/2$
Substituting, we obtain

$7\left(\dfrac{u\sqrt{3} - v}{2}\right)^2 - 6\sqrt{3}\left(\dfrac{u\sqrt{3} - v}{2}\right)\left(\dfrac{u + v\sqrt{3}}{2}\right) + 13\left(\dfrac{u + v\sqrt{3}}{2}\right)^2 - 16 = 0.$

$7\left(3u^2 - 2\sqrt{3}uv + v^2\right) - 6\sqrt{3}\left(\sqrt{3}u^2 + 3uv - uv - \sqrt{3}v^2\right) + 13\left(u^2 + 2\sqrt{3}uv + 3v^2\right) = 4\cdot 16$

$16u^2 + 64v^2 = 4\cdot 16$

$u^2 + 4v^2 = 4$ or $\dfrac{u^2}{4} + v^2 = 1$

9. $A = 41 \quad B = -24 \quad C = 34$

$\cot 2\theta = \dfrac{A - C}{B} = \dfrac{41 - 34}{-24} = -\dfrac{7}{24}, \cos 2\theta = -\dfrac{7}{25}$

$\sin\theta = \sqrt{\dfrac{1 - \cos 2\theta}{2}} = \sqrt{\dfrac{1 - (-7/25)}{2}} = \sqrt{\dfrac{1 + 7/25}{2}} = \dfrac{4}{5}, \cos\theta = \dfrac{3}{5}, \theta = \sin^{-1}\dfrac{4}{5} \approx 53.13°$

The rotation equations are: $x = \dfrac{3u - 4v}{5}, \quad y = \dfrac{4u + 3v}{5}$

11. $A = 17$, $B = -48$, $C = 31$, so $\cot 2\theta = \dfrac{A-C}{B} = \dfrac{17-31}{-48} = \dfrac{7}{24}$ and $\cos 2\theta = \dfrac{7}{25}$.

$\sin \theta = \sqrt{\dfrac{1-\cos 2\theta}{2}} = \sqrt{\dfrac{1-7/25}{2}} = \dfrac{3}{5}$, $\cos \theta = \dfrac{4}{5}$, $\theta = 36.87°$

The rotation equations are: $x = \dfrac{4u-3v}{5}$, $y = \dfrac{3u+4v}{5}$

13. a. Substitute the rotation equations: $x = u\cos\theta - v\sin\theta$, $y = u\sin\theta + v\cos\theta$

to obtain: $A(u\cos\theta - v\sin\theta)^2 + B(u\cos\theta - v\sin\theta)(u\sin\theta + v\cos\theta) +$

$C(u\sin\theta + v\cos\theta)^2 + D(u\cos\theta - v\sin\theta) + E(u\sin\theta + v\cos\theta) + F = 0$

Multiply out and collect terms to obtain: $(A\cos^2\theta + B\sin\theta\cos\theta + C\sin^2\theta)u^2 +$

$(-2A\sin\theta\cos\theta + B(\cos^2\theta - \sin^2\theta) + 2C\sin\theta\cos\theta)uv +$

$(A\sin^2\theta - B\sin\theta\cos\theta + C\cos^2\theta)v^2 + (D\cos\theta + E\sin\theta)u +$

$(-D\sin\theta + E\cos\theta)v + F = 0$

b. Indeed. $B' = 2(C-A)\sin\theta\cos\theta + B(\cos^2\theta - \sin^2\theta)$

c. Immediately upon applying the double-angle identities,
$B' = (C-A)\sin 2\theta + B\cos 2\theta$.

d. Then if $\cot 2\theta = (A-C)/B$ we have $\cos 2\theta / \sin 2\theta = (A-C)/B$ or
$(C-A)\sin 2\theta + B\cos 2\theta = 0$. Hence $B' = 0$.

15. a. Using the results of Exercise 13(a), write

$B'^2 - 4A'C' = \left[2(C-A)\sin\theta\cos\theta + B(\cos^2\theta - \sin^2\theta)\right]^2 -$

$4(A\cos^2\theta + B\sin\theta\cos\theta + C\sin^2\theta)(A\sin^2\theta - B\sin\theta\cos\theta + C\cos^2\theta)$

Expanding and collecting terms yields

$B'^2 - 4A'C' = A^2(4\cos^2\theta\sin^2\theta - 4\cos^2\theta\sin^2\theta) +$

$B^2(\cos^4\theta - 2\cos^2\theta\sin^2\theta + \sin^4\theta + 4\cos^2\theta\sin^2\theta) +$

$C^2(4\cos^2\theta\sin^2\theta - 4\cos^2\theta\sin^2\theta) +$

$AB(-4\cos^3\theta\sin\theta + 4\cos\theta\sin^3\theta + 4\cos^3\theta\sin\theta - 4\cos\theta\sin^3\theta) +$

$AC(-8\sin^2\theta\cos^2\theta - 4\cos^4\theta - 4\sin^4\theta) +$

$BC(4\cos^3\theta\sin\theta - 4\cos\theta\sin^3\theta - 4\cos^3\theta\sin\theta + 4\cos\theta\sin^3\theta)$

The coefficients of A^2, C^2, AB, and BC are seen to reduce to zero, and the right side reduces to: $B'^2 - 4A'C' = B^2(\cos^4\theta + 2\cos^2\theta\sin^2\theta + \sin^4\theta) -$

$4AC(\cos^4\theta + 2\cos^2\theta\sin^2\theta + \sin^4\theta)$

$= (B^2 - 4AC)(\cos^2\theta + \sin^2\theta) = B^2 - 4AC$

b. $B^2 - 4AC = B'^2 - 4A'C' = -4A'C'$ if $B' = 0$.

Therefore if $B^2 - 4AC < 0$, $-4A'C' < 0$, $A'C' > 0$, and the graph is an ellipse.
Similarly if $B^2 - 4AC = 0$, $-4A'C' = 0$, $A'C' = 0$, and the graph is a parabola.
And if $B^2 - 4AC > 0$, $-4A'C' > 0$, $A'C' < 0$, and the graph is a hyperbola.

10.5 Plane Curves and Parametric Equations

1. $-5 \le x \le 6, -2 \le y \le 2$

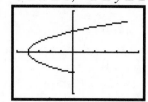

3. $-3 \le x \le 4, -2 \le y \le 3$

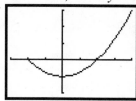

5. $-3 \le x \le 3, -2 \le y \le 2$

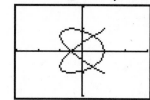

7. $-5 \le x \le 4, -4 \le y \le 5$

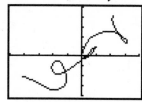

9. $0 \le x \le 14, -15 \le y \le 0$

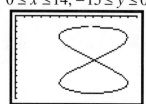

11. $-2 \le x \le 20, -11 \le y \le 11$

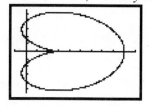

13. $-12 \le x \le 12, -12 \le y \le 12$

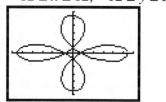

15. $-2 \le x \le 20, -20 \le y \le 4$

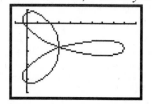

17. $-25 \le x \le 22, -25 \le y \le 26$

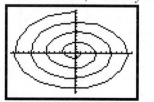

19. $x = t-3, y = 2t+1 \quad t \geq 0$

$t = x+3$

$y = 2(x+3)+1 = 2x+7 \quad x > -3$

21. $x = -2+t^2, y = 1+2t^2$

$t^2 = x+2$

$y = 1+2(x+2)$

$y = 2x+5 \quad x \geq -2$

23. $x = t^3 - 3t^2 + 2t, y = t-1$

$t = y+1$

$x = (y+1)^3 - 3(y+1)^2 + 2(y+1)$

$x = y^3 - y$

25. $x = \sqrt{t}, y = t^4 - 1 \quad t \geq 0$

$t = x^2$

$y = (x^2)^4 - 1$

$y = x^8 - 1$

27. $x = e^t, y = t$

$x = e^y$

$(\text{or } y = \ln x)$

29. $x = 3\cos t, y = 3\sin t$

$x^2 + y^2 = (3\cos t)^2 + (3\sin t)^2$

$x^2 + y^2 = 9$

31. The graphs look identical on the page (and hence are not shown). Both give a straight line segment between $P = (-4,7)$ and $Q = (2,-5)$. However, the graph in (a) moves from P to Q, while the graph in (b) moves from Q to P.

33. $(x-7)^2 + (y+4)^2 = 6^2$ is a circle.

Parametrize as:

$x-7 = 6\cos t, x = 6\cos t + 7$

$y+4 = 6\sin t, y = 6\sin t - 4$

$0 \leq t \leq 2\pi, -3 \leq x \leq 17, -11 \leq y \leq 3$

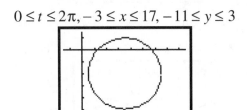

35. Complete the square to obtain $(x-2)^2 + (y+2)^2 = 9$. Then parametrization as

$x = 3\cos t + 2, \ y = 3\sin t - 2$

37. $\dfrac{x^2}{25} + \dfrac{y^2}{18} = 1$ is an ellipse. Parameterize as

$x = 5\cos t, \quad y = \sqrt{18}\sin t.$

$-5 \leq x \leq 5, -5 \leq y \leq 5$

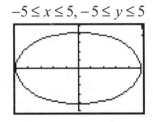

39. $\dfrac{x^2}{9} - \dfrac{y^2}{15} = 1$ is a hyperbola. Parameterize as

$$x = \dfrac{3}{\cos t}, \; y = \sqrt{15}\tan t$$

$-10 \le x \le 10, -10 \le y \le 10$

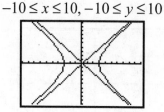

41. $\dfrac{(y-2)^2}{36} + \dfrac{(x-5)^2}{24} = 1$ is a hyperbola.

Parameterize as $x = \sqrt{24}\tan t + 5$,

$$y = \dfrac{6}{\cos t} + 2.$$

$-10 \le x \le 10, -10 \le y \le 10$

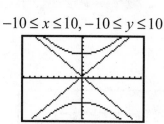

43. a. $\dfrac{d-b}{c-a}$ **b.** $y - b = \dfrac{d-b}{c-a}(x-a)$ or $\dfrac{y-b}{d-b} = \dfrac{x-a}{c-a}$

c. Solving both equations for t, we obtain $\dfrac{x-a}{c-a} = t$, $\dfrac{y-b}{d-b} = t$.

Thus $\dfrac{y-b}{d-b} = \dfrac{x-a}{c-a}$ results.

45. In Exercise **44**, the parametrization
$x = a + (c-a)t, y = b + (d-b)t, 0 \le t \le 1$
was developed. Here:
$x = -6 + (12 - (-6))t$ $y = 12 + (-10 - 12)t$
$x = 18t - 6$ $y = 12 - 22t, \; 0 \le t \le 1$

Use window
$-8 \le x \le 14, -12 \le y \le 14$

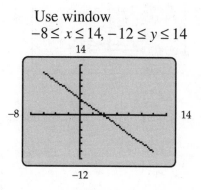

47. In Exercise **44**, the parametrization
$x = a + (c-a)t, y = b + (d-b)t, 0 \le t \le 1$
was developed. Here:
$x = 18 + (-16 - 18)t$ $y = 4 + (14 - 4)t$
$x = 18 - 34t$ $y = 10t + 4, \; 0 \le t \le 1$

Use window
$-20 \le x \le 20, 0 \le y \le 18$

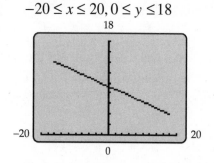

49. a. $k = 1$

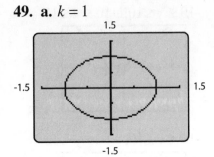

$k = 2$

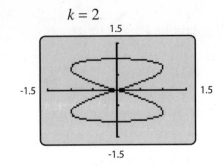

$k = 3$

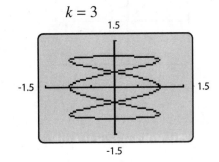

$k = 4$

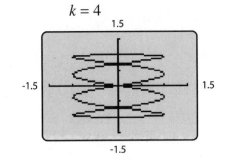

b. $k = 5$

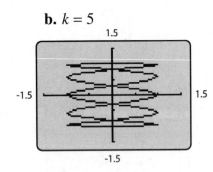

$k = 6$

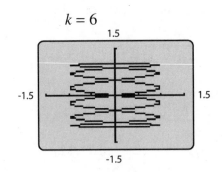

51. a. – d. Are shown on the same graph with a window of $-20 \leq x \leq 20, -1 \leq y \leq 8$.

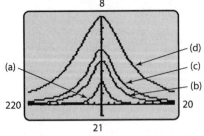

e. The shape will be the same as the graphs above, but the peak will occur at the point $(0, 6)$.

53. Graph using the window
$-15 \le x \le 5, 0 \le y \le 20$.

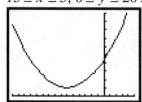

There appears to be a local minimum at $(-6, 2)$.
Since $y = 2$ must be the minimum value for y,
and this corresponds to $t = 0$, therefore $x = -6$,
this is confirmed.

55. Graph using the window
$0 \le x \le 8, 0 \le y \le 6$.

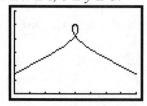

There appears to be a local maximum at $(4, 5)$.
Since $y = 5$ must be the maximum value for y,
and this corresponds to $t = 0$, therefore $x = 4$,
this is confirmed.

57. a. Apply the text formula for projectile
motion to obtain parametric equations
$x = (110 \cos 28°)t$, $y = (110 \sin 28°)t - 16t^2$.

Graph using the window
$0 \le t \le 3.5, 0 \le x \le 350, 0 \le y \le 50$
(calculator in degree mode).

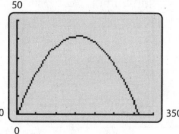

b. Find that $y = 0$ when $t = 3.23$. This is the time that the skeet is in the air.

c. The highest point occurs midway in the flight when $t = 1.61$ seconds; this yields
on substitution $y = 41.67$ feet.

59. a. Apply the text formula for projectile motion to obtain parametric equations
$x = (88 \cos 48°)t$, $y = (88 \sin 48°)t - 16t^2 + 4$
Graph using the window $0 \le t \le 4.5, 0 \le x \le 250, 0 \le y \le 100$.

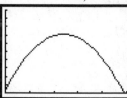

b. Trace to find that when $x = 200$, y is greater than 40, so the arrow will go over
the wall.

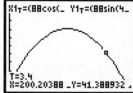

61. Apply the text formula for projectile motion to obtain parametric equations

$$x = (v\cos 30°)t \quad y = (v\sin 30°)t - 16t^2$$

$$x = v\frac{\sqrt{3}}{2}t \qquad y = \frac{v}{2}t - 16t^2$$

When $x = 300$, $y = 0$, hence $300 = v\frac{\sqrt{3}}{2}t \quad 0 = \frac{v}{2}t - 16t^2$

Thus $t = \dfrac{600}{v\sqrt{3}}$. Substituting into the second equation gives $0 = \dfrac{v}{2}\left(\dfrac{600}{v\sqrt{3}}\right) - 16\left(\dfrac{600}{v\sqrt{3}}\right)^2$.

The positive solution of this equation is $v = 105.29$ ft/sec.

63. a. The four paths are shown in the window $0 \le t \le 8, 0 \le x \le 350, 0 \le y \le 200$.

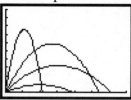

 b. The ball went farthest for the 40° angle.
 c. The ball would go farthest of all for a 45° angle. The 40° and 45° angle paths are
    shown below.

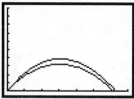

65. a. The diagram shows that $\dfrac{\pi}{2} + \theta = t$, $\theta = t - \dfrac{\pi}{2}$.

  In right triangle PQC,

  $\cos\theta = CQ/3$ $\sin\theta = PQ/3$

  $CQ = 3\cos\theta$ $PQ = 3\sin\theta$

  $x = OT - CQ = 3t - 3\cos(t - \pi/2)$ $y = CT + PQ = 3 + 3\sin(t - \pi/2)$.

 b. $\cos(t - \pi/2) = \cos t\cos\pi/2 + \sin t\sin\pi/2 = \sin t$

  $\sin(t - \pi/2) = \sin t\cos\pi/2 - \cos t\sin\pi/2 = -\cos t$

  Hence $x = 3t - 3\sin t = 3(t - \sin t)$

  $\qquad\quad y = 3 - 3\cos t = 3(1 - \cos t)$.

67. a. The diagram shows that $\dfrac{3\pi}{2}+\theta=t$, $\theta=t-\dfrac{3\pi}{2}$.

In right triangle PQC,
$\cos\theta=CQ/3$ $\qquad\qquad$ $\sin\theta=PQ/3$
$CQ=3\cos\theta$ $\qquad\qquad$ $PQ=3\sin\theta$
$x=OT+CQ=3t+3\cos(t-3\pi/2)$ $\qquad$ $y=CT-PQ=3-3\sin(t-3\pi/2)$.

b. $\cos(t-3\pi/2)=\cos t\cos(3\pi/2)+\sin t\sin(3\pi/2)=-\sin t$
$\sin(t-3\pi/2)=\sin t\cos(3\pi/2)-\cos t\sin(3\pi/2)=\cos t$

Hence $x=3t-3\sin t=3(t-\sin t)$, $y=3-3\cos t=3(1-\cos t)$.

69. a. Graphs A and B do not collide. A passes through the point of intersection when $t=1.2$ and B passes through the point when $t=0.95$.

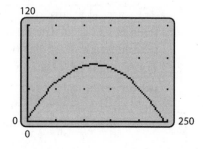

b. A and B are closest together near $t=1.10$.

c. A and C very nearly collide near $t=1.13$.

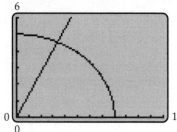

d. The given function is obtained by applying the distance formula to the points $(8\cos t,5\sin t)$ from A and $(3t,4t)$ from C. In function mode, graph

$$y=\sqrt{(8\cos x-3x)^2+(5\sin x-4x)^2}\text{ and}$$

use trace or the minimum routine to show that there is a minimum at $x=1.1322$, however, the minimum is not zero.

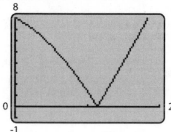

71. a. Examine the figures below:

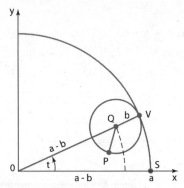
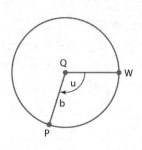

The center Q of the small circle is always at distance $a - b$ from the origin. Let t be the angle that the line from the origin to Q makes with the x-axis. Then the coordinates of Q are $x = (a - b)\cos t$, $y = (a - b)\sin t$. The change in x-coordinate on the small circle from Q to P is $b\cos u$, where u is the angle that PQ makes with the positive x-axis. Similarly, the change in y-coordinate from Q to P is $-b\sin u$. Therefore, the coordinates of P are

$$x = (a - b)\cos t + b\cos u$$

$$y = (a - b)\sin t - b\sin u$$

Since the inner circle rolls without slipping, the arc length that P moves around the inner circle from P to W must equal the arc length that the inner circle has moved along the circumference of the larger circle from S to V. Since the length of a circular arc is the radius times the angle, this means $bu = (a - b)t$, or $u = (a - b)t / b$. Therefore

$$x = (a - b)\cos t + b\cos\left(\frac{a - b}{b}t\right)$$

$$y = (a - b)\sin t - b\sin\left(\frac{a - b}{b}t\right)$$

b. In this case

$$x = 4\cos t + \cos 4t$$

$$y = 4\sin t - \sin 4t$$

$$0 \le t \le 2\pi, -6 \le x \le 7, -5 \le y \le 5$$

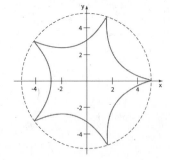

c. In this case

$$x = 3\cos t + 2\cos\tfrac{3}{2}t$$

$$y = 3\sin t - 2\sin\tfrac{3}{2}t$$

$$0 \le t \le 4\pi, -6 \le x \le 7, -5 \le y \le 5$$

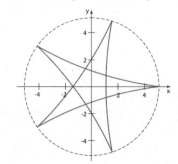

10.6 Polar Coordinates

1. $P - \left(2, \dfrac{\pi}{4}\right), Q - \left(3, \dfrac{2\pi}{3}\right), R - (5, \pi), S - \left(7, \dfrac{7\pi}{6}\right), T - \left(4, \dfrac{3\pi}{2}\right), U - \left(6, -\dfrac{\pi}{3}\right)$
$V - (7, 0)$

3.

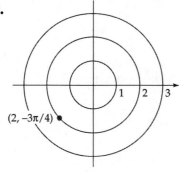

5.

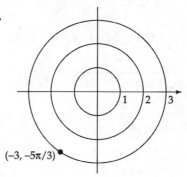

7. $\left(3, \dfrac{7\pi}{3}\right), \left(3, -\dfrac{5\pi}{3}\right), \left(-3, \dfrac{4\pi}{3}\right), \left(-3, -\dfrac{2\pi}{3}\right)$

9. $\left(2, \dfrac{4\pi}{3}\right), \left(2, -\dfrac{8\pi}{3}\right), \left(-2, \dfrac{\pi}{3}\right), \left(-2, -\dfrac{5\pi}{3}\right)$

11. $\left(\sqrt{3}, \dfrac{11\pi}{4}\right), \left(\sqrt{3}, -\dfrac{5\pi}{4}\right), \left(-\sqrt{3}, \dfrac{7\pi}{4}\right), \left(-\sqrt{3}, -\dfrac{\pi}{4}\right)$

13. $r = 3, \ \theta = \dfrac{\pi}{3}, \ x = 3\cos\dfrac{\pi}{3} = \dfrac{3}{2}, \ y = 3\sin\dfrac{\pi}{3} = \dfrac{3\sqrt{3}}{2}, \ \left(\dfrac{3}{2}, \dfrac{3\sqrt{3}}{2}\right)$

15. $\left(-1\cos\dfrac{5\pi}{6}, -1\sin\dfrac{5\pi}{6}\right) = \left(\dfrac{\sqrt{3}}{2}, -\dfrac{1}{2}\right)$ **17.** $(1.5\cos 5, 1.5\sin 5) = (.4255, -1.438)$

19. $\left(-4\cos(-\dfrac{\pi}{7}), -4\sin\left(-\dfrac{\pi}{7}\right)\right) = (-3.604, 1.736)$

21. $r = \sqrt{x^2 + y^2} = \sqrt{\left(3\sqrt{3}\right)^2 + (-3)^2} = 6, \ \tan\theta = \dfrac{y}{x} = \dfrac{-3}{3\sqrt{3}} = -\dfrac{1}{\sqrt{3}}$
$\theta = -\dfrac{\pi}{6}$ (Quadrant IV), $\left(6, -\dfrac{\pi}{6}\right)$

23. $r = \sqrt{1^2 + 1^2} = \sqrt{2}$, $\tan\theta = \dfrac{1}{1} = 1$, $\theta = \dfrac{\pi}{4}$ (Quadrant I), $\left(\sqrt{2}, \dfrac{\pi}{4}\right)$

25. $r = \sqrt{3^2 + \left(3\sqrt{3}\right)^2} = 6$, $\tan\theta = \dfrac{3\sqrt{3}}{\sqrt{3}} = 3$, $\theta = \dfrac{\pi}{3}$ (Quadrant I), $\left(6, \dfrac{\pi}{3}\right)$

27. $r = \sqrt{2^2 + 4^2} = 2\sqrt{5}$, $\tan\theta = \dfrac{4}{2} = 2$, $\theta = \tan^{-1} 2$ (Quadrant I),

$\left(2\sqrt{5}, \tan^{-1} 2\right) = \left(2\sqrt{5}, 1.107\right)$

29. $r = \sqrt{(-5)^2 + (2.5)^2} = \sqrt{31.25} = \dfrac{5}{2}\sqrt{5}$, $\tan\theta = \dfrac{2.5}{-5} = -\dfrac{1}{2}$ (Quadrant II)

$\theta = \pi + \tan^{-1}\left(-\dfrac{1}{2}\right)$, $\left(\dfrac{5}{2}\sqrt{5}, \pi + \tan^{-1}\left(-\dfrac{1}{2}\right)\right) = (5.59, 2.6679)$

31. $r = \sqrt{0^2 + (-2)^2} = 2$, $\tan\theta = \dfrac{-2}{0} =$ undefined, $\theta = \dfrac{3\pi}{2}$ $\left(2, \dfrac{3\pi}{2}\right)$

33. $r = \sqrt{(-2)^2 + 4^2} = \sqrt{20} = 2\sqrt{5}$, $\tan\theta = \dfrac{4}{-2} = 7$, Quadrant II

$\theta = \pi + \tan^{-1}(-2)$, $\left(2\sqrt{5}, \pi + \tan^{-1}(-2)\right) = (4.47, 2.0344)$

35. $x^2 + y^2 = 25$, $r = 5^2$, $r = 5$

37. $x = 12$
$r\cos\theta = 12$
$r = \dfrac{12}{\cos\theta} = 12\sec\theta$

39. $y = 2x + 1$
$r\sin\theta = 2r\cos\theta + 1$
$r(\sin\theta - 2\cos\theta) = 1$
$r = \dfrac{1}{\sin\theta - 2\cos\theta}$

41. $r = 3$
$r^2 = 9$
$x^2 + y^2 = 9$

43. $\theta = \pi/6$
$\tan\theta = \tan(\pi/6)$
$\dfrac{\sin\theta}{\cos\theta} = \dfrac{1}{\sqrt{3}}$
$\dfrac{y}{x} = \dfrac{1}{\sqrt{3}}$
$y = \dfrac{x}{\sqrt{3}}$

45. $r = \sec\theta$
$r = \dfrac{1}{\cos\theta}$
$r\cos\theta = 1$
$x = 1$

47. $r^2 = \tan\theta$
$x^2 + y^2 = \dfrac{y}{x}$
$x^3 + xy^2 = y$

49. $r = 2\sin\theta$

$r^2 = 2r\sin\theta$

$x^2 + y^2 = 2y$

51. $r = \dfrac{4}{1 + \sin\theta}$

$r(1 + \sin\theta) = 4$

$r + r\sin\theta = 4$

$\sqrt{x^2 + y^2} + y = 4$

$x^2 + y^2 = (4 - y)^2$

$x^2 = 16 - 8y$

53.

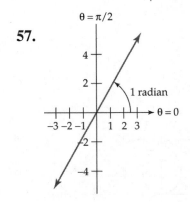

55.

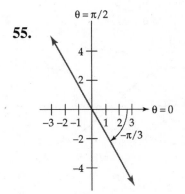

57.

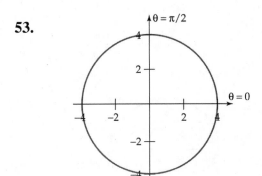

59.

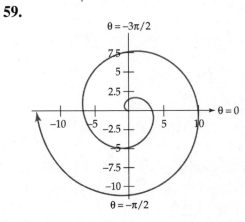

61.

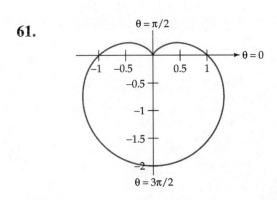

63.

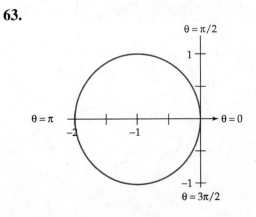

65.

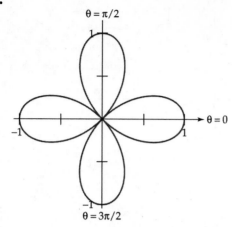

67.

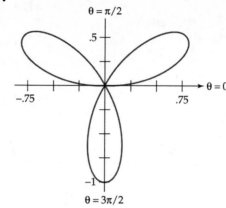

69.

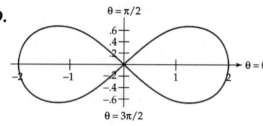

71.

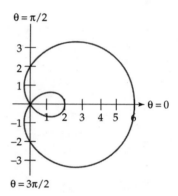

73.

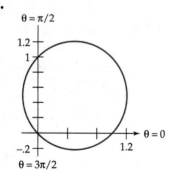

75.

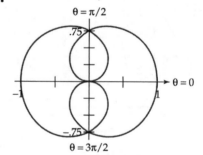

77.

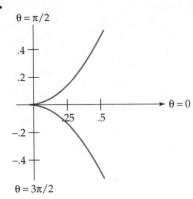

79.

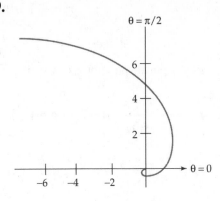

81.

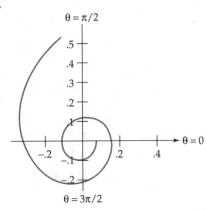

83. a.

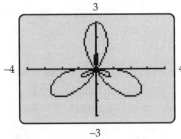

b.

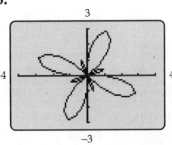

c.

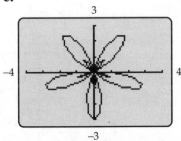

85. Following the hint, write
$$r = a\sin\theta + b\cos\theta$$

$$r^2 = ar\sin\theta + br\cos\theta$$

$$x^2 + y^2 = ay + bx$$

$$x^2 - bx + \frac{b^2}{4} + y^2 - ay + \frac{a^2}{4} = \frac{a^2 + b^2}{4}$$

$$\left(x - \frac{b}{2}\right)^2 + \left(y - \frac{a}{2}\right)^2 = \frac{a^2 + b^2}{4}$$

This is the equation of a circle.

87. Following the hint, the distance from (r, θ) to (s, β) is given by the Law of Cosines.
$$d^2 = r^2 + s^2 - 2rs\cos(\theta - \beta)$$

$$d = \sqrt{r^2 + s^2 - 2rs\cos(\theta - \beta)}$$

10.7 Polar Equations of Conics

1. $e = 1$, parabola, vertex at $\theta = \pi$. Graph **(d)**

3. $r = \dfrac{3}{1 - 2\sin\theta}$: $e = 2$, hyperbola, vertices at $\theta = \dfrac{3\pi}{2}, \dfrac{\pi}{2}$. Graph **(c)**

5. $r = \dfrac{2}{1 - \frac{2}{3}\sin\theta}$: $e = \dfrac{2}{3}$, ellipse, vertices at $\theta = \dfrac{\pi}{2}, \dfrac{3\pi}{2}$. Graph **(a)**

7. $r = \dfrac{4}{1 + \frac{4}{3}\sin\theta}$: $e = \dfrac{4}{3}$, hyperbola.   9. $r = \dfrac{8/3}{1 + \sin\theta}$: $e = 1$, parabola.

11. $r = \dfrac{1/3}{1 - \frac{2}{3}\cos\theta}$: $e = \dfrac{2}{3}$, ellipse.

13. $a^2 = 100$, $b^2 = 99$, $c^2 = a^2 - b^2 = 1$, $e = c/a = 1/10 = .1$

15. $a^2 = 10$, $b^2 = 40$, $c^2 = a^2 + b^2 = 50$, $e = c/a = \sqrt{50}/\sqrt{10} = \sqrt{5}$

17. Complete the square: $16(x^2 - 2x + 1) - 9(y^2 - 4y + 4) = -124 + 16 - 36$

    $\dfrac{(y-2)^2}{16} - \dfrac{(x-1)^2}{9} = 1$; $a^2 = 16$, $b^2 = 9$, $c^2 = a^2 + b^2 = 25$; $e = \dfrac{c}{a} = \dfrac{5}{4}$

19. **a.**

    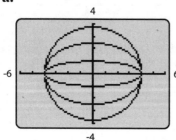

    b. The eccentricities are:

    $\dfrac{\sqrt{15}}{4} = .9682$

    $\dfrac{\sqrt{10}}{4} = .7906$

    $\dfrac{\sqrt{2}}{4} = .3536$

    c. As the eccentricity approaches 0, the ellipse becomes more circular. An "ellipse" with eccentricity 0 is actually a circle.

21. The vertex of the parabola is $(4, \pi)$. Graph using the window $-5 \le x \le 5$, $-10 \le y \le 10$.

    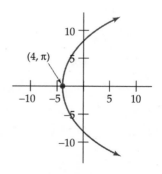

23. The vertices of the hyperbola are $(-2,0)$ and $\left(\frac{2}{3},\pi\right)$. Graph using the window $-3 \le x \le 1, -3 \le y \le 3$.

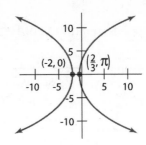

25. The vertices of the ellipse are $\left(10,\frac{\pi}{2}\right)$ and $\left(\frac{10}{7},\frac{3\pi}{2}\right)$. Graph using the window $-10 \le x \le 10, -2 \le y \le 10$.

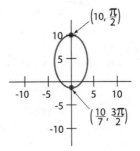

27. The vertices of the ellipse are $(15,0)$ and $(3,\pi)$. Graph using the window $-5 \le x \le 15, -10 \le y \le 10$.

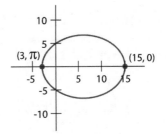

29. The vertex of the parabola is $\left(\frac{3}{2},\frac{\pi}{2}\right)$. Graph using the window $-10 \le x \le 10, -5 \le y \le 2$.

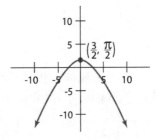

31. The vertices of the hyperbola are $\left(2,\frac{\pi}{2}\right)$ and $\left(-10,\frac{3\pi}{2}\right)$. Graph using the window $-10 \le x \le 10, -5 \le y \le 15$.

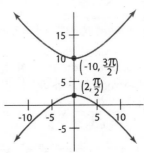

33. The polar equation is of form $r = \dfrac{d}{1-\cos\theta}$. Since $(3,\pi)$ satisfies the equation,

$$3 = \frac{d}{1-\cos\pi}, \quad d=6, \quad r = \frac{6}{1-\cos\theta}$$

35. The polar equation is of form $r = \dfrac{ed}{1+e\sin\theta}$. Since $\left(2,\dfrac{\pi}{2}\right)$ and $\left(8,\dfrac{3\pi}{2}\right)$ satisfy the

equation, $2 = \dfrac{ed}{1+e\sin\pi/2}$ $8 = \dfrac{ed}{1+e\sin 3\pi/2}$

$$2 + 2e = 8 - 8e$$

$$e = 3/5$$

$$ed = 16/5$$

$$r = \dfrac{16/5}{1+\frac{3}{5}\sin\theta} = \dfrac{16}{5+3\sin\theta}$$

37. The polar equation is of form $r = \dfrac{ed}{1+e\cos\theta}$. Since $(1,0)$ and $(-3,\pi)$ satisfy the

equation, $1 = \dfrac{ed}{1+e\cos 0}$ $-3 = \dfrac{ed}{1+e\cos\pi}$

$$1 + e = -3 + 3e$$

$$e = 2$$

$$ed = 3$$

$$r = \dfrac{3}{1+2\cos\theta}$$

39. Rewrite the equation of the directrix as $r = -2/\cos\theta$, $r\cos\theta = -2$, $x = -2$.

The directrix is a vertical line 2 units to the left of the pole. The equation is of form

$r = \dfrac{ed}{1-e\cos\theta}$ with $e = 4, d = 2$.

$$r = \dfrac{8}{1-4\cos\theta}$$

41. Rewrite the equation of the directrix as $r = -3/\sin\theta$, $r\sin\theta = -3$, $y = -3$.

The directrix is a horizontal line 3 units below the pole. The equation is of form

$r = \dfrac{ed}{1-e\sin\theta}$ with $e = 1, d = 3$.

$$r = \dfrac{3}{1-\sin\theta}$$

43. Rewrite the equation of the directrix as $r = 2/\cos\theta$, $r\cos\theta = 2$, $x = 2$.

The directrix is a vertical line 2 units to the right of the pole. The equation is of form

$r = \dfrac{ed}{1+e\cos\theta}$ with $e = \dfrac{1}{2}, d = 2$.

$$r = \dfrac{1}{1+\frac{1}{2}\cos\theta} = \dfrac{2}{2+\cos\theta}$$

45. Since the directrix is left of the pole, the equation is of form $r = \dfrac{2d}{1 - 2\cos\theta}$.

Since $(1, 2\pi/3)$ is on the graph,

$$1 = \frac{2d}{1 - 2\cos(2\pi/3)}$$

$$d = 1.$$

$$r = \frac{2}{1 - 2\cos\theta}$$

47. Since $a^2 = \dfrac{e^2 d^2}{\left(1 - e^2\right)^2}$ and $b^2 = \dfrac{e^2 d^2}{1 - e^2}$, $\dfrac{a^2}{b^2} = \dfrac{1}{1 - e^2}$.

Since $e < 1$, $0 < e^2 < 1$, $1 > 1 - e^2 > 0$, $\dfrac{1}{1 - e^2} > 1$.

Therefore $\dfrac{a^2}{b^2} > 1$ and $a^2 > b^2$. $a > b$ since both are positive.

49. Since the orbit is a parabola, $e = 1$.

$$r = \frac{d}{1 - \cos\theta}$$

If $(60, \pi/3)$ is on the graph, then

$$60 = \frac{d}{1 - \cos(\pi/3)}$$

$$d = 30$$

$$r = \frac{30 \cdot 10^7}{1 - \cos\theta}$$

Chapter 10 Review Exercises

1. $a^2 = 20$, $b^2 = 16$, $c^2 = 20 - 16 = 4$: Ellipse, vertices $\left(0, \pm 2\sqrt{5}\right)$, foci $(0, \pm 2)$.

3. $a^2 = 16$, $b^2 = 7$, $c^2 = 16 - 7 = 9$: Ellipse, center $(1, 3)$,
vertices $(1, 3 \pm 4)$, that is, $(1, -1)$ and $(1, 7)$, foci $(1, 3 \pm 3)$, that is, $(1, 0)$ and $(1, 6)$.

5. Rewrite the equation as $x^2 = 10/7\, y$. $4p = 10/7$, $p = 5/14$.
The axis of the parabola is vertical and it opens upward.
Focus: $(0, 5/14)$ Directrix: $y = -5/14$

7. Ellipse

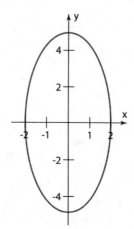

9. Ellipse

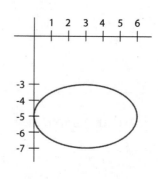

11. $a = 5$, $b = 2$. Hyperbola

Asymptotes $y + 4 = \pm \dfrac{5}{2}(x - 1)$

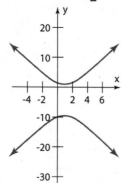

13. Complete the square:
$$x^2 - 10x + 25 + 4y^2 = -9 + 25$$
$$(x - 5)^2 + 4y^2 = 16$$
$$\frac{(x - 5)^2}{16} + \frac{y^2}{4} = 1$$

Ellipse

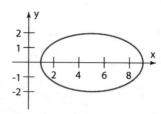

15. Rewrite as $y = 2(x-3)^2 + 3$. Parabola

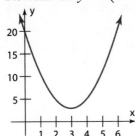

17. Complete the square to obtain $x - 1 = (y+1)^2$. Parabola

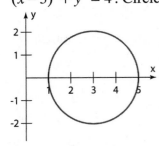

19. Complete the square to obtain $(x-3)^2 + y^2 = 4$. Circle

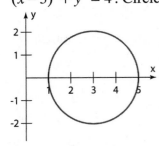

21. Complete the square: $x^2 + 8x + 16 + y^2 + 10y + 25 = -33 + 16 + 25$

$(x+4)^2 + (y+5)^2 = 8$. Center: $(-4,-5)$, radius: $2\sqrt{2}$

23. Complete the square: $4(x^2 - 8x + 16) + 3(y^2 + 12y + 36) = -124 + 64 + 108$

$4(x-4)^2 + 3(y+6)^2 = 48$. Center: $(4,-6)$

25. Since the distance from center to vertex is 2, $a = 2$. The equation has form

$\dfrac{(x-3)^2}{4} + \dfrac{(y-1)^2}{b^2} = 1$. Since $\left(2, 1 + \sqrt{3/2}\right)$ is on the ellipse,

$\dfrac{(2-3)^2}{4} + \dfrac{\left(1 + \sqrt{3/2} - 1\right)^2}{b^2} = 1$

$\dfrac{1}{4} + \dfrac{3}{2b^2} = 1$

$\dfrac{1}{b^2} = \dfrac{1}{2}$. The equation is $\dfrac{(x-3)^2}{4} + \dfrac{(y-1)^2}{2} = 1$.

27. The equation has form $\dfrac{y^2}{a^2} - \dfrac{(x-3)^2}{b^2} = 1$. Since the distance from center to vertex

is 2, write $\dfrac{y^2}{4} - \dfrac{(x-3)^2}{b^2} = 1$. Since $\left(1, \sqrt{5}\right)$ is on the hyperbola,

$$\dfrac{\left(\sqrt{5}\right)^2}{4} - \dfrac{(1-3)^2}{b^2} = 1$$

$$\dfrac{5}{4} - \dfrac{4}{b^2} = 1$$

$$\dfrac{1}{b^2} = \dfrac{1}{16}. \text{ The equation is } \dfrac{y^2}{4} - \dfrac{(x-3)^2}{16} = 1.$$

29. Since the axis is horizontal and the parabola opens to the left, the equation has form
$\left(y - \left(-\tfrac{1}{2}\right)\right)^2 = 4p\left(x - \tfrac{3}{2}\right)$. Since $(-3, 1)$ is on the parabola,

$$\left(1 + \tfrac{1}{2}\right)^2 = 4p\left(-3 - \tfrac{3}{2}\right)$$

$$-\tfrac{1}{2} = 4p$$

$$\left(y + \tfrac{1}{2}\right)^2 = -\tfrac{1}{2}\left(x - \tfrac{3}{2}\right).$$

31. Circle; parametric equations
$x = 3\cos t + 5,\ y = 3\sin t - 2$

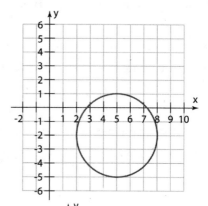

33. Ellipse; parametric equations
$x = 2\cos t,\ y = 5\sin t$

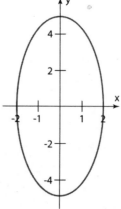

35. Ellipse; parametric equations
$$x = 3\cos t + 3, \, y = 2\sin t - 5$$

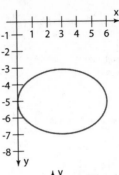

37. Hyperbola; parametric equations
$$x = 2\tan t + 1, \, y = \frac{5}{\cos t - 4} - 4$$

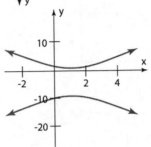

39. Parabola; parametric equations
$$x = (t+4)^2 - 2, \, y = t$$

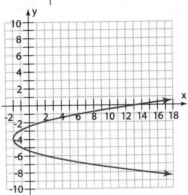

41. Following the hint, write the equation of the ellipse as $\dfrac{x^2}{25} + \dfrac{y^2}{100} = 1$. Then the points

with y-coordinate 8 satisfy $\dfrac{x^2}{25} + \dfrac{8^2}{100} = 1$, $x = \pm 3$. The distance between these points

is $3 - (-3) = 6$ feet.

43. By appropriate choice of axes the parabola can be viewed as opening upward with vertex at the origin. The equation is then $x^2 = 4py$. When

$x = \dfrac{1}{2}(3) = 1.5, \, y = 1$, hence $1.5^2 = 4p(1)$

$$p = \frac{9}{16}$$

The receiver should be placed at the focus, $\dfrac{9}{16}$ feet or 6.75 inches from the vertex.

45. $A = 3, \, B = 2\sqrt{2}, \, C = 2, \, B^2 - 4AC = \left(2\sqrt{2}\right)^2 - 4(3)(2) = -16 < 0.$ Ellipse

47. $A = -3, B = 4, C = 0, B^2 - 4AC = 4^2 - 4(-3)(0) = 16 > 0.$ Hyperbola

49. Solve for y to obtain
$$y = \frac{x \pm \sqrt{24 - 3x^2}}{2}.$$
Graph using $-6 \le x \le 6, -4 \le y \le 4$.

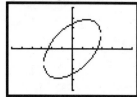

51. Solve for y to obtain $y = \dfrac{2 - x^2}{x}$.
Graph using $-9 \le x \le 9, -6 \le y \le 6$.

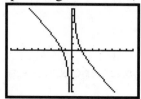

53. Solve for y to obtain
$$y = \frac{-2\sqrt{2} - 3x \pm \sqrt{8 + 20x\sqrt{2} + 5x^2}}{2}.$$
Graph using
$-15 \le x \le 10, -10 \le y \le 20$.

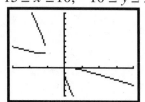

55. $x = u\cos 60° - v\sin 60° = \dfrac{u - v\sqrt{3}}{2}$

$y = u\sin 60° + v\cos 60° = \dfrac{u\sqrt{3} + v}{2}.$

57. $A = 1, B = 1, C = 1$
$$\cot 2\theta = \frac{A - C}{B} = \frac{1 - 1}{1} = 0$$
$$2\theta = 90°$$
$$\theta = 45°$$

59. $-15 \le x \le 15, -10 \le y \le 10$

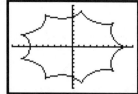

61. $-35 \le x \le 32, -2 \le y \le 16$

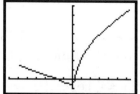

63. Graph using the window
$-10 \le x \le 10, -10 \le y \le 10$

$t = 2 - y$

$x = 2(2 - y) - 1$

$x = 3 - 2y$

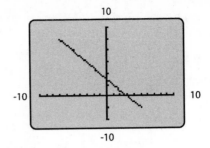

65. Graph using the window
$-1 \le x \le 1, 0 \le y \le 2$

$x^2 = \cos^2 t \quad \dfrac{y}{2} = \sin^2 t$

$x^2 + \dfrac{y}{2} = 1$

$y = 2 - 2x^2$

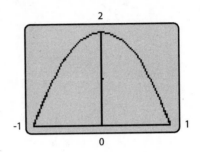

67. (b) and **(c)** only yield a portion of the parabola and therefore are not parametrizations of the curve.

69.

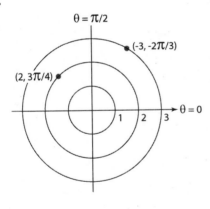

71.

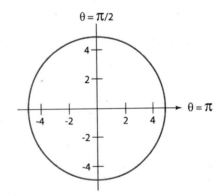

73.

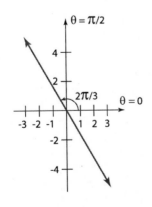

75.

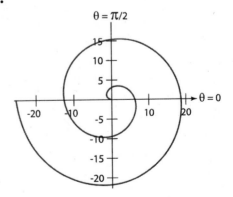

77.

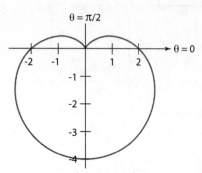

79.

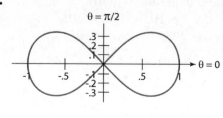

81. $x = r\cos\theta = 3\cos\left(-\dfrac{2\pi}{3}\right) = 3\left(-\dfrac{1}{2}\right) = -\dfrac{3}{2}$

$y = r\sin\theta = 3\sin\left(-\dfrac{2\pi}{3}\right) = 3\left(-\dfrac{\sqrt{3}}{2}\right) = -\dfrac{3\sqrt{3}}{2}$

$\left(-\dfrac{3}{2}, -\dfrac{3\sqrt{3}}{2}\right)$

83. Rewrite the equation as

$\dfrac{x^2}{28} + \dfrac{y^2}{84} = 1.$

$a^2 = 84, b^2 = 28, c^2 = 56$

$e = \dfrac{c}{a} = \dfrac{\sqrt{56}}{\sqrt{84}} = \dfrac{\sqrt{6}}{3} \approx 0.8165$

85. The vertices of the ellipse are $\left(12, \frac{\pi}{2}\right)$ and $\left(4, \frac{3\pi}{2}\right)$.
Graph using the window $-15 \le x \le 15, -5 \le y \le 15$.

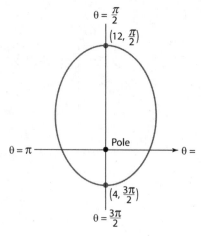

87. The vertices of the hyperbola are $(4,0)$ and $(-2,\pi)$.
Graph using the window $0 \le x \le 6, -8 \le y \le 8$.

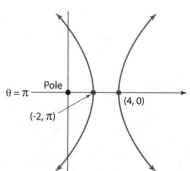

89. The polar equation is of form $r = \dfrac{ed}{1 + e\cos\theta}$. Since $(4,0)$ and $(6,\pi)$ satisfy the

equations $4 = \dfrac{ed}{1+e}$ and $6 = \dfrac{ed}{1-e}$, solve for e.

$$4 + 4e = 6 - 6e$$
$$e = 1/5$$
$$ed = 24/5$$
$$r = \dfrac{24/5}{1 + \frac{1}{5}\cos\theta} = \dfrac{24}{5 + \cos\theta}$$

91. Rewrite the equation of the directrix as $r = \dfrac{2}{\cos\theta}$, $r\cos\theta = 2$, $x = 2$. The directrix is

a vertical line 2 units to the right of the pole. $d = 2, e = 1$

The polar equation is $r = \dfrac{2}{1 + \cos\theta}$.

Chapter 10 Test

1. a. $y - .2x^2 = 0$ Thus, the focus is $\left(0, \frac{5}{4}\right)$ and the directrix is $y = -\frac{5}{4}$.

$x^2 = 5y$

$x^2 = 4(1.25)y$

b. Since the focus is $(0,9)$ and the directrix is $y = -9$, then $p = 9$. Therefore, the equation of the parabola is $x^2 = 4(p)y = 36y$.

2. a. $36y^2 - 9x^2 = 324$

$$\frac{y^2}{9} - \frac{x^2}{36} = 1$$

This is an equation of a hyperbola, with $a = 3$, $b = 6$, and $c = \sqrt{3^2 + 6^2} = 3\sqrt{5}$.

b. The center is $(0,0)$, the vertices are $(0,3)$ and $(0,-3)$, and the foci are $\left(0, \sqrt{45}\right)$ and $\left(0, -\sqrt{45}\right)$.

c.

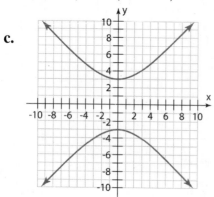

3. $x^2 + 3y^2 + 2x - 18y = 8$

$\left(x^2 + 2x + 1\right) + 3\left(y^2 - 6y + 9\right) = 8 + 1 + 27$

$(x+1)^2 + 3(y-3)^2 = 36$

$$\frac{(x+1)^2}{36} + \frac{(y-3)^2}{12} = 1$$

This an equation of an ellipse with center $(-1,3)$.

4. a. $y^2 + 4y + x + 2 = 0$

$y^2 + 4y + 4 = -x - 2 + 4$

$(y+2)^2 = -x + 2$

$(y+2)^2 = -(x-2)$

$(y+2)^2 = 4\left(-\frac{1}{4}\right)(x-2)$

The vertex is $(2,-2)$.

b.

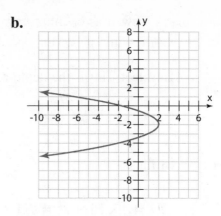

5.   Since the center is $(-6,1), h = -6$ and $k = 1.$ Also since a vertex is $(-4,1), a = 2.$ Now, substitute the point $\left(-2, 1 - 4\sqrt{3}\right)$ into the equation and solve for b.

$$\frac{(x+6)^2}{2^2} - \frac{(y-1)^2}{b^2} = 1$$

$$\frac{(-2+6)^2}{2^2} - \frac{\left(1-4\sqrt{3}-1\right)^2}{b^2} = 1$$

$$\frac{16}{4} - \frac{\left(-4\sqrt{3}\right)^2}{b^2} = 1$$

$$4 - \frac{48}{b^2} = 1$$

$$-\frac{48}{b^2} = -3$$

$$-48 = -3b^2$$

$$16 = b^2$$

Thus the equation of the hyperbola is $\dfrac{(x+6)^2}{4} - \dfrac{(y-1)^2}{16} = 1.$

6.   The center is $(0,0),$ so $h = k = 0.$ The vertices are $(0,\pm 2),$ so $a = 2.$ Also from the graph $b = 3,$ therefore the equation of the hyperbola is $\dfrac{y^2}{4} - \dfrac{x^2}{9} = 1.$

7. a. $A = 2, B = 4, C = 2;\ B^2 - 4AC = 4^2 - 4(2)(2) = 0$ Parabola

   b. $A = 2, B = -3, C = -2;\ B^2 - 4AC = (-3)^2 - 4(2)(-2) = 25 > 0$ Hyperbola

8. a. Elipse   **b.** Center $(1,-2)$; Vertices $(-3,-2),(5,-2)$

9. If the parabola is viewed as opening upward with vertex at the origin, its equation is of form $x^2 = 4py$, with p given as 110, hence $x^2 = 440y$. When $x = 135$, then
$$y = \frac{1355^2}{440} = 41.42 \text{ feet. This is the depth of the dish.}$$

10. **a.** $7x^2 + 4y^2 = 28$

$$\frac{x^2}{4} + \frac{y^2}{7} = 1 \text{ Ellipse}$$

$$a = \sqrt{7}; b = 2; c = \sqrt{\left(\sqrt{7}\right)^2 + 2^2} = \sqrt{11}$$

b. Center $(0,0)$

Vertices $\left(0, \pm\sqrt{7}\right)$

Foci $\left(0, \pm\sqrt{11}\right)$

c.

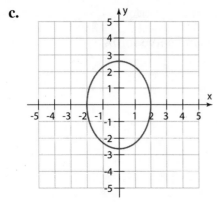

11. $4x^2 + 6xy + 2y^2 - 8y - 1 = 0$

$A = 4, B = 6, C = 2$

$B^2 - 4AC = 6^2 - 4(4)(2) = 4 > 0$

Hyperbola
Use a viewing window of
$-15 \le x \le 15, -200 \le y \le 75$

12. $S = \left(7, \frac{7\pi}{6}\right)$ $T = \left(5, \frac{3\pi}{2}\right)$

13. $x = 5\cos t - 3$, $y = 5\sin t + 4$

14. $x = -3(y-4)^2 - 5$

$$(y-4)^2 = -\frac{1}{3}(x+5)$$

$$(y-4)^2 = 4\left(-\frac{4}{3}\right)(x+5)$$

In parametric form:

$$x = \frac{(t-4)^2}{4\left(-\frac{4}{3}\right)} - 5 = -3(t-4)^2 - 5$$

$$y = t$$

15.

a. $x = 6\cos\dfrac{\pi}{4} = 6\left(\dfrac{\sqrt{2}}{2}\right) = 3\sqrt{2}$

    $y = 6\sin\dfrac{\pi}{4} = 6\left(\dfrac{\sqrt{2}}{2}\right) = 3\sqrt{2}$

    $\left(3\sqrt{2}, 3\sqrt{2}\right)$

b. $r = \sqrt{\left(9\sqrt{3}\right)^2 + \left(-9\right)^2} = 18$

    $\tan\theta = \dfrac{-9}{9\sqrt{3}} = -\dfrac{1}{\sqrt{3}}$

    $\theta = \tan^{-1}\left(-\dfrac{1}{\sqrt{3}}\right) = -\dfrac{\pi}{6}$

    $\left(18, -\dfrac{\pi}{6}\right)$

16. $t = y - 1; x = (y-1)^2 - (y-1)$

    $x = y^2 - 2y + 1 - y + 1$

    $x = y^2 - 3y + 2$

17.

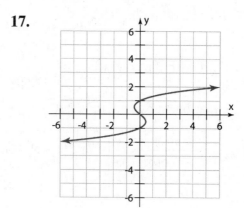

18. a. Apply the text formula for a projectile motion to obtain parametric equations
$$x = (80\cos 45°)t = 40\sqrt{2}, \quad y = (80\sin 45° t)t - 16t^2 + 5 = 40\sqrt{2}t - 16t^2 + 5$$

b. The ball will hit the ground when the y coordinate is zero, so solve the quadratic equation: $\quad 0 = -16t^2 + (80\sin 45°)t + 5 = -16t^2 + 40\sqrt{2}t + 5$. By the quadratic formula $t \approx 3.62$ seconds.

19. $x = \dfrac{3}{\cos t}, \quad y = 4\tan t$

20. $x^2 + y^2 + 8x - 4y + 16 = 0$

    $\left(x^2 + 8x + 16\right) + \left(y^2 - 4y + 4\right) = -16 + 16 + 4$

    $(x+4)^2 + (y-2)^2 = 4$

    $\dfrac{(x+4)^2}{4} + \dfrac{(y-2)^2}{4} = 1$

The parametric form of an ellipse is $x = 2\cos t - 4$, $y = 2\sin t + 2$.

21.

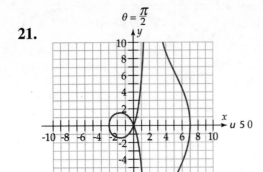

22. Start with the equation $r = \dfrac{ed}{1 + e\cos\theta}$. Since the coordinates of the vertices satisfy the

equation, substitute in to get $3 = \dfrac{ed}{1 + e\cos 0} = \dfrac{ed}{1+e}$ or $3(1+e) = ed$ and

$5 = \dfrac{ed}{1 + e\cos\pi} = \dfrac{ed}{1-e}$ or $5(1-e) = ed$. Therefore, $3(1+e) = 5(1-e)$

$$3 + 3e = 5 - 5e$$
$$8e = 2$$
$$e = 1/4$$

Substituting e into one of the above equations gives d, $3\left(1+\frac{1}{4}\right) = \frac{1}{4}d; d = 15$.

Thus, $r = \dfrac{\left(\frac{1}{4}\right)(15)}{1 + \frac{1}{4}\cos\theta} = \dfrac{15}{4 + \cos\theta}$.

Chapter 11
Systems of Equations

11.1 Systems of Linear Equations in Two Variables

1. Substitute:

$$2(-1)+3=1? \quad 1=1 \text{ True}$$

$$-3(-1)+2(3)=9? \quad 9=9 \text{ True}$$

The values are a solution.

3. Substitute:

$$\tfrac{1}{3}(2)+\tfrac{1}{2}(-1)=\tfrac{1}{6}? \quad \tfrac{1}{6}=\tfrac{1}{6} \text{ True}$$

$$\tfrac{1}{2}(2)+\tfrac{1}{3}(-1)=\tfrac{2}{3}? \quad \tfrac{2}{3}=\tfrac{2}{3} \text{ True}$$

The values are a solution.

5. Since the last equation requires $z = 1$, values for which $z = -1$ cannot be a solution.

7.
$$x+3y=6$$
$$3x-y=2$$
$$x=6-3y$$
$$3(6-3y)-y=2$$
$$y=-8/5$$
$$x=6-3(8/5)$$
$$x=6/5$$

9.
$$3x-2y=4$$
$$2x+y=-1$$
$$y=-2x-1$$
$$3x-2(-2x-1)=4$$
$$x=2/7$$
$$y=-2(2/7)-1$$
$$y=-11/7$$

11.
$$r+s=0$$
$$r-s=5$$
$$r=-s$$
$$-s-s=5$$
$$s=-5/2$$
$$r=5/2$$

13.
$$x+y=c+d$$
$$x-y=2c-d$$
$$x=y+2c-d$$
$$y+2c-d+y=c+d$$
$$y=d-\tfrac{1}{2}c$$
$$x=d-\tfrac{1}{2}c+2c-d$$
$$x=\tfrac{3}{2}c$$

15.
$$2x-2y=12$$
$$\underline{-2x+3y=10}$$
$$y=22$$
$$2x-2(22)=12$$
$$x=28$$

17. $x + 3y = -1$
$2x - y = 5$
Multiply the second
equation by 3.
$x + 3y = -1$
$\underline{6x - 3y = 15}$
$7x \quad = 14$
$x = 2$
$2(2) - y = 5$
$y = -1$

19. $2x + 3y = 15$
$8x + 12y = 40$
Multiply the first
equation by -4.
$-8x - 12y = -60$
$\underline{8x + 12y = 40}$
$0 = -20$
Inconsistent

21. $4x - 5y = 2$
$12x - 15y = 6$
Multiply the first equation by -3.
$-12x + 15y = -6$
$\underline{12x - 15y = 6}$
$0 = 0$
$x = b$, where b is any real number
$y = \dfrac{4b - 2}{5}$

23. $\dfrac{x}{3} - \dfrac{y}{2} = -3$
$\dfrac{2x}{5} + \dfrac{y}{5} = -2$
Clear fractions.
$2x - 3y = -18$
$2x + y = -10$
Multiply the first
equation by -1.
$-2x + 3y = 18$
$\underline{2x + y = -10}$
$4y = 8$
$y = 2$
$2x + 2 = -10$
$x = -6$

25. $\dfrac{x+y}{4}-\dfrac{x-y}{3}=1$

$\dfrac{x+y}{4}+\dfrac{x-y}{2}=9$

Clear fractions and collect terms   to obtain:

$-x+7y=12$

$3x-y=36$

Multiply the first

equation by 3.

$-3x+21y=36$

$\underline{\ \ \ 3x-y=36\ \ \ }$

$20y=72$

$y=3.6$

$-x+7(3.6)=12$

$x=13.2$

27. $\dfrac{1}{x}-\dfrac{3}{y}=2$

$\dfrac{2}{x}+\dfrac{1}{y}=3$

Multiply the second

equation by 3.

$\dfrac{1}{x}-\dfrac{3}{y}=2$

$\dfrac{6}{x}+\dfrac{3}{y}=9$

$\overline{\dfrac{7}{x}\qquad=11}$

$\dfrac{1}{x}=\dfrac{11}{7}$

$x=\dfrac{7}{11}$

$\dfrac{11}{7}-\dfrac{3}{y}=2$

$y=-7$

29. $\dfrac{2}{x}+\dfrac{3}{y}=8$

$\dfrac{3}{x}-\dfrac{1}{y}=1$

Multiply the second

equation by 3.

$\dfrac{2}{x}+\dfrac{3}{y}=8$

$\dfrac{9}{x}-\dfrac{3}{y}=3$

$\overline{\dfrac{11}{x}\qquad=11}$

$x=1$

$\dfrac{2}{1}+\dfrac{3}{y}=8$

$y=\dfrac{1}{2}$

31. a. Solve each equation for y, and observe that a positive slope represents an increasing population; a negative slope represents a decreasing population.

Philadelphia: $y = -7.96x + 1588.47$

Houston: $y = 26.67x + 1644.64$

b. Solve the system of equations form part (a)

$$26.67x + 1644.64 = -7.96x + 1588.47$$

$$x = -1.62$$

Which corresponds to the year 1988.

33. Solve the system of equations $y = -135x + 31,065$

$$y = 9.77x + 14,302.67$$

$$-135x + 31,065 = 9.77x + 14,302.67$$

$$x = 115.7$$

Which corresponds to the year 2115.

35. Multiply the first equation by d and the second by $-b$.

$$adx + bdy = rd$$

$$\underline{-bcx - bdy = -sb}$$

$$x(ad - bc) = rd - sb$$

$$x = \frac{rd - sb}{ad - bc}$$

Multiply the first equation by $-c$ and the second by a.

$$-acx - bcy = -rc$$

$$\underline{acx + ady = as}$$

$$(ad - bc)y = as - rc$$

$$y = \frac{as - rc}{ad - bc}$$

37. Since the two lines have slopes $-\frac{1}{2}$ and 2, they are perpendicular and intersect in one point, regardless of c. Therefore there is exactly one solution for the system.

39. Substitute: $4d = 2$

$2c + 16d = 2$ Solve to obtain $d = \frac{1}{2},\ c = -3$.

41. Substitute to obtain $.85x = 40 - 1.15x$. Solve to obtain $x = 20$. Then $p = .85(20) = 17$.

Price: $17; Quantity: 20,000.

43. Substitute to obtain $300 - 30x = 80 + 25x$. Solve to obtain $x = 4$.

Then $p = 300 - 30(4) = 180$. Price: $180; Quantity: 4000.

45. Let $x =$ the number of adults and $y =$ the number of children.

$$x + y = 200$$

$$8x + 5y = 1435$$

Solve to obtain $x = 145$ adults and $y = 55$ children.

47. Let $x =$ the speed of the boat and $y =$ the speed of the current.
Then the speed upstream $= x - y$ and the speed downstream $= x + y$.
$\left(15 \text{ minutes} = \frac{1}{4} \text{ hour}, \ 12 \text{ minutes} = \frac{1}{5} \text{ hour}\right)$

$$\frac{1}{4}(x - y) = 4$$

$$\frac{1}{5}(x + y) = 4$$

Solve to obtain $x = 18$ mph, $y = 2$ mph.

49. Let $x =$ the number of pounds of cashews, and $y =$ the number of pounds of peanuts.
Then solve: $x + y = 6$

$$4.4x + 1.2y = 12$$

To obtain $x = 1.5$ pounds of cashews and $y = 4.5$ pounds of peanuts.

51. Let x be the amount of 50% alloy and y be the amount of 75% alloy. Then
$$x + y = 40 \quad \text{(amounts of alloy add up)}$$

$$.5x + .75y = .6(40) \quad \text{(amounts of silver add up)}$$

Clear the decimals to obtain: $\quad x + y = 40$
$$50x + 75y = 2400$$

Solve to obtain $x = 24$ grams of 50% alloy, $y = 16$ grams of 75% alloy.

53. Note that $80 per month equals $960 per year, and $9.50 per month is $114 per year.
a. Electric: $y = 2000 + 960x$. Solar: $y = 14,000 + 114x$.
b. Electric: $y = 2000 + 960(5) = \$6800$. Solar: $y = 14,000 + 114(5) = \$14,570$.
c. Solve the system of equations in part **a** by substitution:
$$2000 + 960x = 14,000 + 114x$$

$$846x = 12,000$$

$$x = 14.2$$

The costs will be the same in the fifteenth year. Electric heating will be cheaper before that: solar afterwards.

55. a. $R = 250,500x$ **b.** $C = 1,295,000 + 206,500x$ **c.** $P = R - C = 44,000x - 1,295,000$
d. $R = 109,000x$, $C = 440,000 + 82,000x$, $P = R - C = 27,000x - 440,000$
e. Set the two expressions for profit equal and solve for x,
$$27,000x - 440,000 = 44,000x - 1,295,000$$

$$855,000 = 17,000x$$

$$x = 50.3 \text{ weeks}$$

$$P = \$918,000$$

f. It would be better to open off Broadway if a run of 50 weeks or less is expected.

57. a. $R = 60x$

  b. Set $R = C$.
  $$60x = 45x + 6000$$
  $$15x = 6000$$
  $$x = 400 \text{ hedge trimmers}$$

59. Set $R = C$.
$$95x = 80x + 7500$$
$$15x = 7500$$
$$x = 500$$
Since the break-even point is above the maximum number that can be sold, the product should not be produced.

61. Let x = amount at 9% and y = amount at 11%. Then
$$y = x - 8000$$
$$.09x + .11y = 2010$$
Solve by substitution: $.09x + .11(x - 8000) = 2010$, $x = \$14,450$, $y = \$6,450$.

63. a.   **b.**

$$f(x) = .373x + 27.02$$

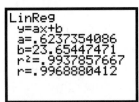

$$f(x) = .624x + 23.65$$

c.

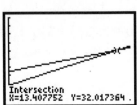

Plotting the two functions and applying the intersection routine yields the year 2003.

65. Let x = how many bowls and y = how many plates. Then
$$3x + 2y = 480 \quad \text{(time)}$$
$$.25x + .20y = 44 \quad \text{(costs)}$$
Solve the system to obtain $x = 80$ bowls and $y = 120$ plates.

11.1.A Systems of Non-Linear Equations

1. Solve the first equation for y in terms of x to obtain $y = x^2$. Substitute into the second equation to obtain:
$$-2x + x^2 = 3$$
$$x^2 - 2x - 3 = 0$$
$$x = 3 \quad x = -1$$
If $x = 3$, $y = 9$. If $x = -1$, $y = 1$.

3. Solve the first equation for y in terms of x to obtain $y = x^2$. Substitute into the second equation to obtain:
$$x + 2x^2 = 5$$
$$2x^2 + x - 5 = 0$$
$$x = \frac{-1 \pm \sqrt{41}}{4}$$
If $x = \dfrac{-1 - \sqrt{41}}{4}$, $y = \dfrac{21 + \sqrt{41}}{8}$.

If $x = \dfrac{-1 + \sqrt{41}}{4}$, $y = \dfrac{21 - \sqrt{41}}{8}$.

5. Solve the first equation for y in terms of x to obtain $y = 10 - x$.
Substitute into the second equation to obtain:
$$x(10 - x) = 21$$
$$x^2 - 10x + 21 = 0$$
$$x = 7 \quad x = 3$$
If $x = 7$, $y = 3$. If $x = 3$, $y = 7$.

7. Solve the second equation for x in terms of y to obtain $x = 2y + 6$.
Substitute into the first equation to obtain: $(2y + 6)y + y^2 = 9$
$$3y^2 + 6y - 9 = 0$$
$$y = -3 \quad y = 1$$
If $y = 1$, $x = 8$. If $y = -3$, $x = 0$.

9. Solve the second equation for x in terms of y to obtain $x = y + 2$.
Substitute into the first equation to get:
$$(y + 2)^2 + y^2 - 4(y + 2) - 4y = -4$$
$$y^2 + 4y + 4 + y^2 - 4y - 8 - 4y = -4$$
$$2y^2 - 4y = 0$$
$$y = 0 \quad y = 2$$
If $y = 0$, $x = 2$. If $y = 2$, $x = 4$.

11. Solve the second equation for x^2 in terms of y to obtain $x^2 = 19 - y$.
Substitute into the first equation to obtain:
$$19 - y + y^2 = 25$$
$$y^2 - y - 6 = 0$$
$$y = 3 \quad y = -2$$
If $y = 3$, $x = \pm 4$.
If $y = -2$, $x = \pm\sqrt{21}$. (Four solutions.)

13. Graph both equations using the window $-3 \le x \le 5$, $-15 \le y \le 10$, and apply the intersection routine to obtain $(-1.6237, -8.1891)$, $(1.3163, 1.0826)$, and $(2.8073, 2.4814)$. (Only one of the calculator screens is shown.)

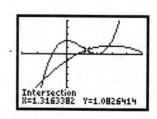

Intersection
X=1.3163382   Y=1.0826414

15. Graph both equations using the window $-3 \le x \le 3$, $-1 \le y \le 5$, and apply the intersection routine to obtain $(-1.9493, .4412)$, $(.3634, .9578)$, and $(1.4184, .5986)$. (Only one of the calculator screens is shown.)

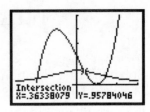

17. Graph both equations using the window $-\pi \le x \le \pi$, $-2 \le y \le 2$, and apply the intersection routine to obtain $(-.9519, -.8145)$.

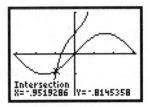

19. $25x^2 - 16y^2 = 400$

$-9x^2 + 4y^2 = -36$

Multiply the second equation by 4.

$25x^2 - 16y^2 = 400$

$-36x^2 + 16y^2 = -144$

$\overline{-11x^2 \qquad = 256}$

This equation has no real solutions, and therefore the system has no solutions.

21. Solve the second equation for y in terms of x to obtain $y = x - 3$. Substitute into the first equation to obtain $3x^2 + 4(x-3)^2 - 18x + 14(x-3) = 1$. Solve this equation using the quadratic formula to obtain $y = 2 \pm \sqrt{5}$. The solutions are therefore: $x = 2 + \sqrt{5}, y = -1 + \sqrt{5}$ and $x = 2 - \sqrt{5}, y = -1 - \sqrt{5}$.

23. Solve the second equation for y in terms of x to obtain $y = x^2 + x + 1$. Substitute into the first to get: $x^2 + 4x(x^2 + x + 1) + 4(x^2 + x + 1)^2 - 30x - 90(x^2 + x + 1) + 450 = 0$
Graph the left side of this equation using the window $-7 \le x \le 7, -150 \le y \le 500$ and apply the zero (root) routine to obtain four roots: $-4.8093, -3.1434, 2.1407$ and 2.8120. Since $y = x^2 + x + 1$ the solutions are therefore:
$(-4.8093, 19.3201), (-3.1434, 7.7374), (2.1407, 7.7230)$ and $(2.8120, 11.7195)$.

25. Solve the second equation for y in terms of x to obtain $y = \pm\sqrt{64 - 4x^2}$. Solve the

first equation for y in terms of x to obtain: $y = \dfrac{3x - 5 \pm \sqrt{x^2 - 24x + 37}}{2}$.

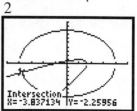

Intersection
X=-3.837134 Y=-2.25956

Graph these four equations using the
window $-5 \le x \le 5, -10 \le y \le 10$ and
apply the intersection routine to obtain
$(-3.8371, -2.2596)$, and $(-.9324, -7.7796)$.
(Only one of the calculator screens is shown.)

27. Solve the equations for y in terms of x to obtain

$y = \dfrac{-3x \pm \sqrt{5x^2 + 8}}{2}$ and $y = \dfrac{5x \pm \sqrt{84 - 11x^2}}{6}$.

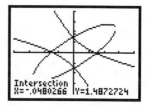

Intersection
X=-.0480266 Y=1.4872724

Graph these four equations using the window
$-4 \le x \le 4, -4 \le y \le 4$ and apply the
intersection routine to obtain $(\pm 1.4873, \mp .0480)$
and $(\pm .0480, \mp 1.4873)$. (Only one of the
calculator screens is shown.)

29. As in example 5, the distance from (x, y) to $(2.7, 15.9)$ is

$\sqrt{(x - 2.7)^2 + (y - 15.9)^2} = 12.7$. Similarily, the distance from (x, y) to $(13.9, 9.9)$ is

$\sqrt{(x - 13.9)^2 + (y - 9.9)^2} = 13.3$. Solving both equations for y in terms of x to obtain

$y = \pm\sqrt{(12.7)^2 - (x - 2.7)^2} + 15.9$ and $y = \pm\sqrt{(13.3)^2 - (x - 13.9)^2} + 9.9$. Graph these

four equations using a calculator and the intersection routine to obtain $(2.4, 3.2)$.

31. Let x be the length and width of the square ends and y be the third dimension. Then
$x^2y = 16$ (volume) and $2x^2 + 4xy = 40$ (surface area).
Solve the first equation for y in terms of x to obtain $y = 16/x^2$, then substitute into
the second equation to obtain: $2x^2 + 4x(16/x^2) = 40$ or $2x^3 - 40x + 64 = 0$.
Solve graphically or algebraically to obtain $x = 2$ and $x = 3.1231$ as the two positive
roots. If $x = 2$ meters, $y = 4$ meters, so the dimensions will be $2 \times 2 \times 4$ meters.
If $x = 3.1231$ meters, $y = 1.6404$ meters so the dimensions will be
$3.1213 \times 3.1213 \times 1.6404$ meters.

33. Let x and y be the numbers. Then
$$x + y = -17$$
$$xy = 52$$
Solve the first equation for y in terms of x to obtain $y = -17 - x$. Substitute into the second equation to obtain: $x(-17 - x) = 52$
$$x^2 + 17x + 52 = 0$$
$$x = -4 \quad x = -13$$
The two numbers are $x = -4$ and $y = -13$ (or the other way around).

35. Let x and y be the numbers. Solve $x - y = 25.75$, $xy = 127.5$ to obtain $x = 30$ and $y = 4.25$ or $x = -4.25$ and $y = -30$.

37. Let x and y be the numbers. Then
$$y = x + 1$$
$$xy = 4.16$$
Substitute the expression for y from the first equation into the second to obtain: $x(x + 1) = 4.16$
$$x^2 + x - 4.16 = 0$$
$$(x - 1.6)(x + 2.6) = 0$$
The only positive solution is $x = 1.6$, $y = 2.6$.

39. Let x be the length and y be the width. Solve $2x + 2y = 53$, $xy = 165$ to obtain $x = 16.5$ feet and $y = 10$ feet (or the other way around).

41. Let x and y be the lengths. Solve $x^2 + y^2 = 35^2$, $\frac{1}{2}xy = 225$ to obtain $x = 32.1$ cm and $y = 14.0$ cm (or the other way around).

11.2 Large Systems of Linear Equations

1.
$$\begin{pmatrix} 2 & -3 & 4 & | & 1 \\ 1 & 2 & -6 & | & 0 \\ 3 & -7 & 4 & | & -3 \end{pmatrix}$$

3.
$$\begin{pmatrix} 2 & -\frac{5}{2} & \frac{2}{3} & | & 0 \\ 1 & -\frac{1}{4} & 4 & | & 0 \\ 0 & -3 & \frac{1}{2} & | & 0 \end{pmatrix}$$

5. $3x - 5y = 4$
$\qquad 9x + 7y = 2$

7. $x \quad + \quad z \quad = 1$
$\quad x \; - \; y + 4z - 2w = 3$
$\quad 4x + 2y + 5z \quad = 2$

9. $x = 3/2$
$\quad y = 5$
$\quad z = -2$
$\quad w = 0$

11 $x + 2w = 3$
$\quad y + 3w = 5$
$\qquad z = 2$
Yields: $x = 3 - 2w$, $y = 5 - 3w$, $z = 2$, $w = t$ any real number

13. Original:
$$\begin{array}{rcl} -x + 3y + 2z & = & 0 \quad [A] \\ 2x - y - z & = & 3 \quad [B] \\ x + 2y + 3z & = & 0 \quad [C] \end{array}$$

Eliminate x:
$$\begin{array}{rcl} x + 2y + 3z & = & 0 \quad [C] \\ 5y + 5z & = & 0 \quad [A+C]=[D] \\ -5y - 7z & = & 3 \quad [B-2C]=[E] \end{array}$$

Then eliminate y:
$$\begin{array}{rcl} x + 2y + 3z & = & 0 \quad [C] \\ y + z & = & 0 \quad [D]\div 5 \\ -2z & = & 3 \quad [E+D] \end{array}$$

Then $z = -3/2$. Back substitution yields $y = 3/2$, $x = 3/2$.

15. Write the augmented matrix
$$\begin{pmatrix} 1 & 1 & 1 & | & 1 \\ 1 & -2 & 2 & | & 4 \\ 2 & -1 & 3 & | & 5 \end{pmatrix}$$

Obtain zeros in Column 1:
$$\begin{pmatrix} 1 & 1 & 1 & | & 1 \\ 0 & -3 & 1 & | & 3 \\ 0 & 3 & -1 & | & -3 \end{pmatrix} \begin{matrix} \\ B-A \\ C-2A \end{matrix}$$

Get a zero in Col. 2, eliminating Row 3:
$$\begin{pmatrix} 1 & 1 & 1 & | & 1 \\ 0 & -3 & 1 & | & 3 \\ 0 & 0 & 0 & | & 0 \end{pmatrix} \begin{matrix} \\ \\ C+B \end{matrix}$$

This is the augmented matrix of:
$x + y + z = 1$
$-3y + z = 3$
Set $z = t$. Then
$y = -1 + \frac{1}{3}t$, $x = 2 - \frac{4}{3}t$.
Where t is any real number.

17. Write the augmented matrix:

$$\begin{pmatrix} 1 & -2 & 4 & 6 \\ 1 & 1 & 13 & 6 \\ -2 & 6 & -1 & -10 \end{pmatrix}$$

First, obtain zeros in Col. 1:

$$\begin{pmatrix} 1 & -2 & 4 & 6 \\ 0 & 3 & 9 & 0 \\ 0 & 2 & 7 & 2 \end{pmatrix} \begin{matrix} \\ B-A \\ C+2A \end{matrix}$$

Second, obtain a 1 as the leading coefficient in Row 2, and 0's above and below:

$$\begin{pmatrix} 1 & 0 & 10 & 6 \\ 0 & 1 & 3 & 0 \\ 0 & 0 & 1 & 2 \end{pmatrix} \begin{matrix} A+\frac{2}{3}B \\ \frac{1}{3}B \\ C-\frac{2}{3}B \end{matrix}$$

Last, obtain zeros in Col. 3:

$$\begin{pmatrix} 1 & 0 & 0 & -14 \\ 0 & 1 & 0 & -6 \\ 0 & 0 & 1 & 2 \end{pmatrix} \begin{matrix} -10C+A \\ -3C+B \\ \end{matrix}$$

The solution is read directly:
$x=-14, y=-6, z=2$.

Details of the reduction to reduced row-echelon form are not supplied from here on; the rref routine on the T-83 should be used.

19. Write the augmented matrix:

$$\begin{pmatrix} 1 & 1 & 1 & 200 \\ 1 & -2 & 0 & 0 \\ 2 & 3 & 5 & 600 \\ 2 & -1 & 1 & 200 \end{pmatrix}$$

Reduce to:

$$\begin{pmatrix} 1 & 0 & 0 & 100 \\ 0 & 1 & 0 & 50 \\ 0 & 0 & 1 & 50 \\ 0 & 0 & 0 & 0 \end{pmatrix}$$

The solution is
$x=100, y=50, z=50$.

21. Write the augmented matrix:

$$\begin{pmatrix} 11 & 10 & 9 & 5 \\ 1 & 2 & 3 & 1 \\ 3 & 2 & 1 & 1 \end{pmatrix}$$

Start by exchanging Rows 1 and 2:

$$\begin{pmatrix} 1 & 2 & 3 & 1 \\ 11 & 10 & 9 & 5 \\ 3 & 2 & 1 & 1 \end{pmatrix}$$

In reduced row-echelon form this becomes:
$$\begin{pmatrix} 1 & 0 & -1 & 0 \\ 0 & 1 & 2 & \frac{1}{2} \\ 0 & 0 & 0 & 0 \end{pmatrix}$$

The solutions are:
$z=t, y=-2t+\frac{1}{2}, x=t$.
Where t is any real number.

23. Write the augmented matrix:

$$\begin{pmatrix} 5 & -1 & 7 \\ 1 & 1 & 5 \\ 4 & -2 & 2 \end{pmatrix}$$

In reduced row-echelon form this becomes:

$$\begin{pmatrix} 1 & 0 & 2 \\ 0 & 1 & 3 \\ 0 & 0 & 0 \end{pmatrix}$$

The solution is $x = 2$, $y = 3$.

25. Write the augmented matrix:

$$\begin{pmatrix} 1 & -4 & -13 & 4 \\ 1 & -2 & -3 & 2 \\ -3 & 5 & 4 & 2 \end{pmatrix}$$

In reduced row-echelon form this becomes:

$$\begin{pmatrix} 1 & 0 & 7 & 0 \\ 0 & 1 & 5 & 0 \\ 0 & 0 & 0 & 1 \end{pmatrix}$$

The last row represents the equation $0 = 1$, hence the system has no solution.

27. Write the augmented matrix:

$$\begin{pmatrix} 4 & 1 & 3 & 7 \\ 1 & -1 & 2 & 3 \\ 3 & 2 & 1 & 4 \end{pmatrix}$$

In reduced row-echelon form this becomes:

$$\begin{pmatrix} 1 & 0 & 1 & 2 \\ 0 & 1 & -1 & -1 \\ 0 & 0 & 0 & 0 \end{pmatrix}$$

The solutions are
$z = t$, $y = t - 1$, $x = -t + 2$.
Where t is any real number.

28. Write the augmented matrix:

$$\begin{pmatrix} 2 & 5 & 3 & -5 \\ 5 & -8 & -2 & -2 \\ 1 & -18 & -8 & 8 \end{pmatrix}$$

In reduced row-echelon form this becomes:

$$\begin{pmatrix} 1 & 0 & \frac{14}{41} & \frac{50}{41} \\ 0 & 1 & \frac{19}{41} & -\frac{21}{41} \\ 0 & 0 & 0 & 0 \end{pmatrix}$$

The solutions are
$z = t$, $y = \frac{-21 - 19t}{41}$, $x = \frac{-50 - 14t}{41}$.
Where t is any real number.

29. Write the augmented matrix:

$$\begin{pmatrix} 1 & 1 & 1 & 0 \\ 3 & -1 & 1 & 0 \\ 5 & -1 & 1 & 0 \end{pmatrix}$$

In reduced row-echelon form this becomes:

$$\begin{pmatrix} 1 & 0 & 0 & 0 \\ 0 & 1 & 0 & 0 \\ 0 & 0 & 1 & 0 \end{pmatrix}$$

The solution is
$x = y = z = 0$.

31. Write the augmented matrix:

$$\begin{pmatrix} 2 & 1 & 3 & -2 & -6 \\ 4 & 3 & 1 & -1 & -2 \\ 1 & 1 & 1 & 1 & -5 \\ -2 & -2 & 2 & 2 & -10 \end{pmatrix}$$

In reduced row-echelon form this becomes:

$$\begin{pmatrix} 1 & 0 & 0 & 0 & -1 \\ 0 & 1 & 0 & 0 & 1 \\ 0 & 0 & 1 & 0 & -3 \\ 0 & 0 & 0 & 1 & -2 \end{pmatrix}$$

The solution is
$x = -1$, $y = 1$, $z = -3$, $w = -2$.

33. Write the augmented matrix:

$$\begin{pmatrix} 1 & -2 & -1 & -3 & -3 \\ -1 & 1 & 1 & 0 & 0 \\ 0 & 4 & 3 & -2 & -1 \\ 2 & -2 & 0 & 1 & 1 \end{pmatrix}$$

In reduced row-echelon form this becomes:

$$\begin{pmatrix} 1 & 0 & 0 & 0 & \frac{7}{31} \\ 0 & 1 & 0 & 0 & \frac{6}{31} \\ 0 & 0 & 1 & 0 & \frac{1}{31} \\ 0 & 0 & 0 & 1 & \frac{29}{31} \end{pmatrix}$$

The solution is $x = 7/31$, $y = 6/31$, $z = 1/31$, $w = 29/31$.

35. Following the hint, write the system as:

$$\begin{aligned} 3u &- v &+ 4w &= -13 \\ u &+ 2v &- w &= 12 \\ 4u &- v &+ 3w &= -7 \end{aligned}$$

Write the augmented matrix:

$$\begin{pmatrix} 3 & -1 & 4 & -13 \\ 1 & 2 & -1 & 12 \\ 4 & -1 & 3 & -7 \end{pmatrix}$$

In reduced row-echelon form:

$$\begin{pmatrix} 1 & 0 & 0 & 2 \\ 0 & 1 & 0 & 3 \\ 0 & 0 & 1 & -4 \end{pmatrix}$$

Thus $u = 2$, $v = 3$, $w = -4$, hence $x = 1/2$, $y = 1/3$, $z = -1/4$.

37. Adding the first two equations yields $5x = 3$. Thus $x = 3/5$, $y = 1/5$ is the only possible solution. Check that this satisfies all four equations.

39. Write $x + 2y = 3$ as $-2x - 4y = -6$ and add to the second equation to obtain $-y = -2$, $y = 2$, $x = -1$, as the only possible solution. Check that this satisfies all four equations.

41. Multiply both sides of the equation by $(x-1)(x+3)$ to get

$$4x = A(x+3) + B(x-1) = (A+B)x + (3A-B).$$

Equating the coefficients gives the system of equations: $A + B = 4$
$$3A - B = 0$$

which has solution $A = 1$, $B = 3$.

43. Multiply both sides of the equation by $(x+2)(x-3)^2$ to get
$$2x + 1 = A(x^2 - 6x + 9) + B(x^2 - x - 6) + C(x+2) \text{ or}$$
$$2x + 1 = (A+B)x^2 + (-6A - B + C)x + (9A - 6B + 2C). \text{ Equate coefficients to get:}$$

$$\begin{aligned} A &+ B & &= 0 \\ -6A &- B &+ C &= 2 \\ 9A &- 6B &+ 2C &= 1 \end{aligned}$$

which has solution $A = -3/25$, $B = 3/25$, $C = 7/5$.

45. Multiply both sides of the equation by $(x+1)(x^2-x+1)$ to get

$5x^2+1=A(x^2-x+1)+(Bx+C)(x+1)$ or

$5x^2+1=(A+B)x^2+(-A+B+C)x+(A+C)$. Equate coefficients to get:

$$A+B\phantom{{}+C}=5$$
$$-A+B+C=0$$
$$A\phantom{{}+B}+C=1$$

which has solution $A=2, B=3, C=-1$.

47. Let x = amount borrowed from friend, y = amount borrowed from bank, z = amount borrowed from insurance company.   Then:
$$x+y+z=10{,}000$$
$$y=2x$$
$$.08x+.09y+.05z=830$$

Form the augmented matrix:

$$\begin{pmatrix} 1 & 1 & 1 & 10000 \\ -2 & 1 & 0 & 0 \\ .08 & .09 & .05 & 830 \end{pmatrix}$$

In reduced row-echelon form this becomes:

$$\begin{pmatrix} 1 & 0 & 0 & 3000 \\ 0 & 1 & 0 & 6000 \\ 0 & 0 & 1 & 1000 \end{pmatrix}$$

$3000 from friend, $6000 from bank, $1000 from insurance company.

49. Let x = amount in fund, y = amount in bonds, z = amount in food.
Then:
$$x+y+z=70{,}000$$
$$y=2x$$
$$.02x+.1y+.06z=4800$$

Form the augmented matrix:

$$\begin{pmatrix} 1 & 1 & 1 & 70000 \\ -2 & 1 & 0 & 0 \\ .02 & .1 & .06 & 4800 \end{pmatrix}$$

In reduced row-echelon form:

$$\begin{pmatrix} 1 & 0 & 0 & 15000 \\ 0 & 1 & 0 & 30000 \\ 0 & 0 & 1 & 25000 \end{pmatrix}$$

$15,000 in fund, $30,000 in bonds, $25,000 in the fast food franchise.

51. Let x = number of cups of Progresso, y = number of cups of H.C., z = number of cups of Campbell. Then
$$100x+130y+130z=2030 \quad \text{(calories)}$$
$$970x+480y+880z=11{,}900 \quad \text{(sodium)}$$
$$6x+8y+8z=124 \quad \text{(protein)}$$

Form the augmented matrix:

$$\begin{pmatrix} 100 & 130 & 130 & 2030 \\ 970 & 480 & 880 & 11900 \\ 6 & 8 & 8 & 124 \end{pmatrix}$$

In reduced row-echelon form this becomes:

$$\begin{pmatrix} 1 & 0 & 0 & 6 \\ 0 & 1 & 0 & 9 \\ 0 & 0 & 1 & 2 \end{pmatrix}$$

6 cups of Progresso, 9 cups of Healthy Choice, 2 cups of Campbell's, the serving size is 1.7 cups.

53. Let x = number of bedroom models, y = number of living room models, z = numbe of whole-house models. Then: $10x + 20y + 60z = 440$
$$8x + 8y + 28z = 248$$

Form the augmented matrix: In reduced row-echelon form this becomes:

$$\begin{pmatrix} 10 & 20 & 60 & 440 \\ 8 & 8 & 28 & 248 \end{pmatrix} \qquad \begin{pmatrix} 1 & 0 & 1 & 18 \\ 0 & 1 & \frac{5}{2} & 13 \end{pmatrix}$$

Thus if $z = t$, $y = 13 - \frac{5}{2}t$, $x = 18 - t$. t must be positive, even, and less than 5. This yields 3 solutions.

$t = 0$: 18 bedroom models, 13 living-room models, 0 whole-house models
$t = 2$: 16 bedroom models, 8 living-room models, 2 whole-house models
$t = 4$: 14 bedroom models, 3 living-room models, 4 whole-house models

55. a. X-ray 1: $a + b$. X-ray 2: $a + c$.
  b. The system is: Form the augmented matrix: In reduced row-echelon form:

$$\begin{array}{l} a + b = .75 \\ a + c = .60 \\ b + c = .54 \end{array} \qquad \begin{pmatrix} 1 & 1 & 0 & .75 \\ 1 & 0 & 1 & .60 \\ 0 & 1 & 1 & .54 \end{pmatrix} \qquad \begin{pmatrix} 1 & 0 & 0 & .405 \\ 0 & 1 & 0 & .345 \\ 0 & 0 & 1 & .195 \end{pmatrix}$$

$a = .405, b = .345, c = .195$

  c. a is bone, b is tumorous, c is healthy.

57. Let x = number of chairs, y = number of chests, z = number of tables. Then
$.2x + .5y + .3z = 1950$ (cutting)

$.3x + .4y + .1z = 1490$ (assembly)

$.1x + .6y + .4z = 2160$ (finishing)

Form the augmented matrix: In reduced row-echelon form this becomes:

$$\begin{pmatrix} .2 & .5 & .3 & 1950 \\ .3 & .4 & .1 & 1490 \\ .1 & .6 & .4 & 2160 \end{pmatrix} \qquad \begin{pmatrix} 1 & 0 & 0 & 2000 \\ 0 & 1 & 0 & 1600 \\ 0 & 0 & 1 & 2500 \end{pmatrix}$$

Produce 2000 chairs, 1600 chests, and 2500 tables.

59. a. Following the hint, obtain :

Intersection B: $x + y = 700 + 400$

Intersection C: $y + z = 200 + 500$

Intersection D: $z + t = 300 + 300$

b. Form the augmented matrix:

$$\begin{pmatrix} 1 & 0 & 0 & 1 & 1000 \\ 1 & 1 & 0 & 0 & 1100 \\ 0 & 1 & 1 & 0 & 700 \\ 0 & 0 & 1 & 1 & 600 \end{pmatrix}$$

In reduced row-echelon form this becomes:

$$\begin{pmatrix} 1 & 0 & 0 & 1 & 1000 \\ 0 & 1 & 0 & -1 & 100 \\ 0 & 0 & 1 & 1 & 600 \\ 0 & 0 & 0 & 0 & 0 \end{pmatrix}$$

Solution: $z = 600 - t$, $y = 100 + t$, $x = 1000 - t$.

c. The solution requires $0 \le t \le 600$. Thus the smallest number of cars leaving A on 4th avenue is $t = 0$. The largest is 600. Thus the smallest number of cars leaving A on Euclid is $x = 400$. The largest is 1000. Thus the smallest number of cars leaving C on 5th avenue is 100. The largest is 700. Thus the smallest number of cars leaving C on Chester is $z = 0$. The largest is 600.

11.3 Matrix Methods for Square Systems

1. A is a 2×3 matrix. B is a 3×4 matrix. AB is a 2×4 matrix, but BA is not defined.

3. A is a 3×2 matrix. B is a 2×3 matrix. AB is a 3×3 matrix. BA is a 2×2 matrix.

5. A is a 3×2 matrix. B is a 2×2 matrix. AB is a 3×2 matrix, but BA is not defined.

7. AB is a 2×3 matrix.
$$\begin{pmatrix} 1(-1)+2(7) & 1(2)+2(1) & 1(3)+2(0) \\ -4(-1)+3(7) & -4(2)+3(1) & -4(3)+3(0) \end{pmatrix} = \begin{pmatrix} 13 & 4 & 3 \\ 25 & -5 & -12 \end{pmatrix}$$

9. AB is a 3×2 matrix.
$$\begin{pmatrix} 1(1)+0(1)+-2(0) & 1(1)+0(0)+-2(1) \\ 0(1)+3(1)+-1(0) & 0(1)+3(0)+-1(1) \\ 2(1)+4(1)+0(0) & 2(1)+4(0)+0(1) \end{pmatrix} = \begin{pmatrix} 1 & -1 \\ 3 & -1 \\ 6 & 2 \end{pmatrix}$$

11. AB is a 4×4 matrix.
$$\begin{pmatrix} 2(1)+0(1)+(-1)1 & 2(0)+0(1)+(-1)1 & 2(1)+0(0)+(-1)1 & 2(1)+0(1)+(-1)0 \\ 1(1)+1(1)+2(1) & 1(0)+1(1)+2(1) & 1(1)+1(0)+2(1) & 1(1)+1(1)+2(0) \\ 0(1)+2(1)+(-3)1 & 0(0)+2(1)+(-3)1 & 0(1)+2(0)+(-3)1 & 0(1)+2(1)+(-3)0 \\ 2(1)+3(1)+0(1) & 2(0)+3(1)+0(1) & 2(1)+3(0)+0(1) & 2(1)+3(1)+0(0) \end{pmatrix}$$
$$= \begin{pmatrix} 1 & -1 & 1 & 2 \\ 4 & 3 & 3 & 2 \\ -1 & -1 & -3 & 2 \\ 5 & 3 & 2 & 5 \end{pmatrix}$$

13. $AB = \begin{pmatrix} 7 & 8 \\ 14 & 11 \end{pmatrix}$ $BA = \begin{pmatrix} -1 & 2 \\ 8 & 19 \end{pmatrix}$ These are not equal.

15. $AB = \begin{pmatrix} 8 & 24 & -8 \\ 2 & -2 & 6 \\ -3 & -21 & 15 \end{pmatrix}$ $BA = \begin{pmatrix} 19 & 9 & 8 \\ -10 & 2 & 0 \\ 0 & 0 & 0 \end{pmatrix}$
These are not equal.

17. Form the augmented matrix:
$$\left(\begin{array}{cc|cc} 1 & 2 & 1 & 0 \\ 3 & 4 & 0 & 1 \end{array}\right)$$

In reduced row-echelon form this is:
$$\left(\begin{array}{cc|cc} 1 & 0 & -2 & 1 \\ 0 & 1 & 3/2 & -1/2 \end{array}\right)$$

Therefore $A^{-1} = \begin{pmatrix} -2 & 1 \\ 3/2 & -1/2 \end{pmatrix}$

19. The matrix is not reducible by row reduction to the identity. Therefore the matrix has no inverse.

21. The matrix is not reducible by row reduction to the identity. Therefore the matrix has no inverse.

23. Form the augmented matrix:
$$\left(\begin{array}{ccc|ccc} 5 & 0 & 2 & 1 & 0 & 0 \\ 2 & 2 & 1 & 0 & 1 & 0 \\ -3 & 1 & -1 & 0 & 0 & 1 \end{array}\right)$$

In reduced row-echelon form this becomes:
$$\left(\begin{array}{ccc|ccc} 1 & 0 & 0 & -3 & 2 & -4 \\ 0 & 1 & 0 & -1 & 1 & -1 \\ 0 & 0 & 1 & 8 & -5 & 10 \end{array}\right)$$

Therefore the inverse matrix is:
$$\begin{pmatrix} -3 & 2 & -4 \\ -1 & 1 & -1 \\ 8 & -5 & 10 \end{pmatrix}$$

25. Form the matrix equation $AX = B$ with
$$A = \begin{pmatrix} -1 & 1 & 0 \\ -1 & 0 & 1 \\ 6 & -2 & -3 \end{pmatrix}, \quad X = \begin{pmatrix} x \\ y \\ z \end{pmatrix}, \quad B = \begin{pmatrix} 1 \\ -2 \\ 3 \end{pmatrix}$$

The solution is given by :
$$X = A^{-1}B = \begin{pmatrix} -1 \\ 0 \\ -3 \end{pmatrix}$$
$$x = -1, y = 0, z = -3$$

27. Form the matrix equation $AX = B$ with
$$A = \begin{pmatrix} 2 & 1 & 0 \\ -4 & -1 & -3 \\ 3 & 1 & 2 \end{pmatrix}, \quad X = \begin{pmatrix} x \\ y \\ z \end{pmatrix}, \quad B = \begin{pmatrix} 0 \\ 1 \\ 2 \end{pmatrix}$$

The solution is given by :
$$X = A^{-1}B = \begin{pmatrix} -8 \\ 16 \\ 5 \end{pmatrix}$$
$$x = -8, y = 16, z = 5$$

29. Form the matrix equation $AX = B$ with

$$A = \begin{pmatrix} 1 & 1 & 0 & 2 \\ 2 & -1 & 1 & -1 \\ 3 & 3 & 2 & -2 \\ 1 & 2 & 1 & 0 \end{pmatrix}, \; X = \begin{pmatrix} x \\ y \\ z \\ w \end{pmatrix}, \; B = \begin{pmatrix} 3 \\ 5 \\ 0 \\ 2 \end{pmatrix}$$

The solution is given by :

$$X = A^{-1}B = \begin{pmatrix} -.5 \\ -2.1 \\ 6.7 \\ 2.8 \end{pmatrix}$$

31. Form the matrix equation $AX = B$ with:

$$A = \begin{pmatrix} 1 & 1 & 6 & 2 & 0 \\ 1 & 0 & 5 & 2 & -3 \\ 3 & 2 & 17 & 6 & -4 \\ 4 & 3 & 21 & 7 & -2 \\ -6 & -5 & -36 & -12 & 3 \end{pmatrix}, \; X = \begin{pmatrix} x \\ y \\ z \\ v \\ w \end{pmatrix}, \; B = \begin{pmatrix} 1.5 \\ 2 \\ 2.5 \\ 3 \\ 3.5 \end{pmatrix}$$

The solution is given by :

$$X = A^{-1}B = \begin{pmatrix} 10.5 \\ 5 \\ -13 \\ 32 \\ 2.5 \end{pmatrix}$$

Thus $x = 10.5, y = 5, z = -13, v = 32, w = 2.5$.

33. Form the matrix equation $AX = B$ with

$$A = \begin{pmatrix} 1 & 2 & 2 & -2 \\ 4 & 4 & -1 & 5 \\ -2 & 5 & 6 & 4 \\ 5 & 13 & 7 & 12 \end{pmatrix}, \; X = \begin{pmatrix} x \\ y \\ z \\ w \end{pmatrix}, \; B = \begin{pmatrix} -23 \\ 7 \\ 0 \\ -7 \end{pmatrix}$$

The solution is (given in exact fractional form):

$$X = A^{-1}B = \begin{pmatrix} -1149/161 \\ 426/161 \\ -1124/161 \\ 579/161 \end{pmatrix}$$

35. Form the matrix equation $AX = B$ as in previous problems. Since A^{-1} exists, $A^{-1}B$ is defined and

$$X = A^{-1}B = \begin{pmatrix} 0 \\ 0 \\ 0 \\ 0 \\ 0 \end{pmatrix}$$

gives the solution $x = y = z = v = w = 0$.

37. Since the matrix A of the system has no inverse, the inverse method fails. Write the augmented matrix:

$$\left(\begin{array}{cccc|c} 1 & 2 & 3 & 0 & 1 \\ 3 & 2 & 4 & 0 & -1 \\ 2 & 6 & 8 & 1 & 3 \\ 2 & 0 & 2 & -2 & 3 \end{array} \right)$$

In reduced row-echelon form this becomes:

$$\left(\begin{array}{cccc|c} 1 & 0 & 0 & 1 & 0 \\ 0 & 1 & 0 & 5/2 & 0 \\ 0 & 0 & 1 & -1 & 0 \\ 0 & 0 & 0 & 0 & 1 \end{array} \right)$$

Since the last row is equivalent to $0 = 1$ the system is inconsistent and has no solution.

39. Since the matrix A of the system has no inverse, the inverse method fails. Write the augmented matrix:

$$\begin{pmatrix} 1 & 0 & 0 & 3 & | & -2 \\ 1 & -4 & -1 & 3 & | & -7 \\ 0 & 4 & 1 & 0 & | & 5 \\ -1 & 12 & 3 & -3 & | & 17 \end{pmatrix}$$

In reduced row-echelon form this becomes:

$$\begin{pmatrix} 1 & 0 & 0 & 3 & | & -2 \\ 0 & 1 & 1/4 & 0 & | & 5/4 \\ 0 & 0 & 0 & 0 & | & 0 \\ 0 & 0 & 0 & 0 & | & 0 \end{pmatrix}$$

The solutions can be written as $w=t$, $z=s$, $y=\frac{5}{4}-\frac{1}{4}s$, $x=-2-3t$, where s and t are any real numbers.

41. Write the equation of the parabola as $y=ax^2+bx+c$. Since the three given points must satisfy the equation, solve the system.

$$\begin{array}{rrrrr} 9a & - & 3b & + & c & = & 15 \\ a & + & b & + & c & = & -7 \\ 25a & + & 5b & + & c & = & 111 \end{array}$$

Form the matrix equation $AX=B$ with

$$A=\begin{pmatrix} 9 & -3 & 1 \\ 1 & 1 & 1 \\ 25 & 5 & 1 \end{pmatrix}, X=\begin{pmatrix} a \\ b \\ c \end{pmatrix}, B=\begin{pmatrix} 15 \\ -7 \\ 111 \end{pmatrix}$$

The solution is given by :

$$X=A^{-1}B=\begin{pmatrix} 35/8 \\ 13/4 \\ -117/8 \end{pmatrix}$$

The equation of the parabola is therefore $y=\dfrac{35}{8}x^2+\dfrac{13}{4}x-\dfrac{117}{8}$.

43. a. The points $(0,315),(21,337)$ and $(43,371)$ must satisfy the equation $y=ax^2+bx+c$. Solve the system:

$$\begin{array}{rrrrr} & & & c & = & 315 \\ 441a & + & 21b & + & c & = & 337 \\ 1849a & + & 43b & + & c & = & 371 \end{array}$$

to obtain $a=\dfrac{115}{9933}, b=\dfrac{7991}{9933}, c=315$.

b. Use the table feature to obtain :

$$\begin{array}{lll} 1983 — x=25 & y=342.35 \text{ ppm} \\ 1993 — x=35 & y=357.34 \text{ ppm} \\ 2003 — x=45 & y=374.65 \text{ ppm} \end{array}$$

45. a. The points $(2,169), (6,260)$ and $(10,320)$ must satisfy the equation $y = ax^2 + bx + c$. Solve the system:

$$
\begin{array}{rcrcrcl}
4a & + & 2b & + & c & = & 169 \\
36a & + & 6b & + & c & = & 260 \\
100a & + & 10b & + & c & = & 320
\end{array}
$$

to obtain $y = -\dfrac{31}{32}x^2 + \dfrac{61}{2}x + \dfrac{895}{8}$.

b. Use the table feature to obtain :

$$
\begin{array}{lll}
1993 - x = 3 & y = \$194.66 \\
1998 - x = 8 & y = \$293.88 \\
2002 - x = 12 & y = \$338.38
\end{array}
$$

47. Note first that $e^{\ln x} = x$ and $e^{-\ln x} = \dfrac{1}{x}$. Substitute to obtain the system:

$$
\begin{array}{rcrcrcl}
a & + & b & + & c & = & -2 \\
2a & + & \frac{1}{2}b & + & c & = & 1 \\
4a & + & \frac{1}{4}b & + & c & = & 4
\end{array}
$$

Solve the system to obtain $a = 1, b = -4, c = 1, y = e^x - 4e^{-x} + 1$.

49. Let x = price of one pair of jeans, y = price of one jacket , z = price of one sweater, w = price of one shirt. Solve the system $AX = B$ with

$$
A = \begin{pmatrix} 3000 & 3000 & 2200 & 4200 \\ 2700 & 2500 & 2100 & 4300 \\ 5000 & 2000 & 1400 & 7500 \\ 7000 & 1800 & 600 & 8000 \end{pmatrix}, \ X = \begin{pmatrix} x \\ y \\ z \\ w \end{pmatrix}, \ B = \begin{pmatrix} 507,650 \\ 459,075 \\ 541,225 \\ 571,500 \end{pmatrix}
$$

to obtain $x = \$34.50, \ y = \$72, \ z = \$44, \ w = \$21.75$.

51. Let x = number of box A, y = number of box B , z = number of box C. Solve the system:

$$
\begin{array}{rcrcrcll}
.6x & + & .3y & + & .5z & = & 41,400 & (\text{chocolates}) \\
.4x & + & .4y & + & .3z & = & 29,400 & (\text{mints}) \\
& & .3y & + & .2z & = & 16,200 & (\text{caramels})
\end{array}
$$

to obtain $x = 15,000, \ y = 18,000, \ z = 54,000$ boxes.

Chapter 11 Review Exercises

1.

$$
\begin{aligned}
-5x + 3y &= 4 \quad [A] \\
2x - y &= -3 \quad [B] \\
6x - 3y &= -9 \quad 3[B] \\
x &= -5 \quad [A]+3[B]
\end{aligned}
$$

Substitute $x=-5$ into equation $[B]$ to obtain $y=-7$.

3.

$$
\begin{aligned}
3x + 4y &= 7 \quad [A] \\
2x - 3y &= -1 \quad [B] \\
6x + 8y &= 14 \quad 2[A] \\
-6x + 9y &= 3 \quad -3[B] \\
17y &= 17 \quad 2[A]-3[B]
\end{aligned}
$$

$y=1$. Substitute into equation $[B]$ to obtain $x=1$.

5. Multiply the first equation by 2 and add the two equations.

$$
\begin{aligned}
21.78x + 2y &= 369.84 \\
\underline{16.06x - 2y} &= \underline{32.04} \\
37.84x &= 401.88 \\
x &= 10.62
\end{aligned}
$$

$$y = 184.92 - 10.89(10.62) = 69.26$$

10.62 corresponds to 2000. 69 days were found.

7. Let x = amount of 40% alloy, y = amount of 80% alloy. Then

$$
\begin{aligned}
x + y &= 50 \quad [A] \\
.4x + .8y &= .75(50) \quad [B] \\
-4x - 4y &= -200 \quad -4[A] \\
\underline{4x + 7y} &= \underline{375} \quad 10[B] \\
4y &= 175 \\
y &= 43\tfrac{3}{4} \text{ lbs of 80\% alloy} \\
x &= 6\tfrac{1}{4} \text{ lbs of 40\% alloy}
\end{aligned}
$$

9. Solve the second equation for y in terms of x to obtain $y = 2x + 3$.
Substitute into the first equation to obtain:

$$
\begin{aligned}
x^2 - (2x + 3) &= 0 \\
x^2 - 2x - 3 &= 0 \\
x = 3 \text{ or } x &= -1
\end{aligned}
$$

If $x = 3$, $y = 9$. If $x = -1$, $y = 1$.

11. Solve the second equation for y in terms of x to obtain $y = 2 - x$. Substitute into the first equation to obtain:

$$x^2 + (2 - x)^2 = 16$$
$$x^2 + 4 - 4x + x^2 = 16$$
$$x^2 - 2x - 6 = 0$$
$$x = 1 \pm \sqrt{7}$$

If $x = 1 + \sqrt{7}$, $y = 1 - \sqrt{7}$. If $x = 1 - \sqrt{7}$, $y = 1 + \sqrt{7}$.

13. The augmented matrix is:

$$\begin{pmatrix} 3 & -5 & 2 & -9 \\ 3 & 2 & 0 & 0 \\ 1 & 1 & 3 & 19 \end{pmatrix}$$

15. The matrix of the system is:

$$\begin{pmatrix} 1 & 7 & -1 & 2 & 13 \\ 3 & -5 & -1 & -1 & -15 \\ 2 & -2 & -3 & 0 & -19 \end{pmatrix}$$

17. The augmented matrix is:

$$\begin{pmatrix} 2 & 5 & 1 & 8 \\ 1 & 0 & 2 & 9 \\ -2 & -2 & -2 & -13 \end{pmatrix}$$

In reduced row-echelon form this is:

$$\begin{pmatrix} 1 & 0 & 0 & -1 \\ 0 & 1 & 0 & 1 \\ 0 & 0 & 1 & 5 \end{pmatrix}$$

The solution is
$x = -1$, $y = 1$, $z = 5$.

19. The augmented matrix is:

$$\begin{pmatrix} 4 & 3 & -3 & 2 \\ 5 & -3 & 2 & 10 \\ 2 & -2 & 3 & 14 \end{pmatrix}$$

In reduced row-echelon form this is:

$$\begin{pmatrix} 1 & 0 & 0 & 2 \\ 0 & 1 & 0 & 4 \\ 0 & 0 & 1 & 6 \end{pmatrix}$$

The solution is
$x = 2$, $y = 4$, $z = 6$.

21. The augmented matrix is:

$$\begin{pmatrix} 1 & -2 & -3 & 1 \\ 0 & 5 & 10 & 0 \\ 8 & -6 & -4 & 8 \end{pmatrix}$$

In reduced row-echelon form this is:

$$\begin{pmatrix} 1 & 0 & 1 & 1 \\ 0 & 1 & 2 & 0 \\ 0 & 0 & 0 & 0 \end{pmatrix}$$

The solution is $z = t$, $y = -2t$, $x = 1 - t$, t any real number.

23. Since each equation is a multiple of $2x - y = 3$, only **(c)** is true.

25. Multiply both sides of the equation by $(x-3)(x+2)$ to get

$4x-7 = A(x+2)+B(x-3)$, $4x-7=(A+B)x+(2A-3B)$. Equate coefficients to obtain the system of equations: $A+B=4$
$$2A-3B=-7$$
Multiply the first equation by -2 and add to get $B=3$. Use this value to find $A=1$.

27. $\begin{pmatrix} -1(2)+0(4) & -1(-3)+0(1) \\ 0(2)+(-1)(4) & 0(-3)+(-1)(1) \end{pmatrix} = \begin{pmatrix} -2 & 3 \\ -4 & -1 \end{pmatrix}$

29. Since A is 2×2 and E is 3×2 the product AE does not exist.

31. Form the augmented matrix:

$$\begin{pmatrix} -5 & 2 & | & 1 & 0 \\ 7 & -3 & | & 0 & 1 \end{pmatrix}$$

In reduced row-echelon form this is:

$$\begin{pmatrix} 1 & 0 & | & -3 & -2 \\ 0 & 1 & | & -7 & -5 \end{pmatrix}$$

The inverse matrix is: $\begin{pmatrix} -3 & -2 \\ -7 & -5 \end{pmatrix}$

33. Form the augmented matrix:

$$\begin{pmatrix} 3 & 2 & 6 & | & 1 & 0 & 0 \\ 1 & 1 & 2 & | & 0 & 1 & 0 \\ 2 & 2 & 5 & | & 0 & 0 & 1 \end{pmatrix}$$

In reduced row-echelon form this is:

$$\begin{pmatrix} 1 & 0 & 0 & | & 1 & 2 & -2 \\ 0 & 1 & 0 & | & -1 & 3 & 0 \\ 0 & 0 & 1 & | & 0 & -2 & 1 \end{pmatrix}$$

The inverse matrix is:

$$\begin{pmatrix} 1 & 2 & -2 \\ -1 & 3 & 0 \\ 0 & -2 & 1 \end{pmatrix}$$

35. Form the matrix equation $AX = B$ with

$$A=\begin{pmatrix} 1 & 0 & 2 & 6 \\ 3 & 4 & -2 & -1 \\ 5 & 0 & 2 & -5 \\ 4 & -4 & 2 & 3 \end{pmatrix}, X=\begin{pmatrix} x \\ y \\ z \\ w \end{pmatrix}, B=\begin{pmatrix} 2 \\ 0 \\ -4 \\ 1 \end{pmatrix}$$

$x = -1/85$, $y = -14/85$, $z = -21/34$, $w = 46/85$

The solution is given by :

$$X = A^{-1}B = \begin{pmatrix} -1/85 \\ -14/85 \\ -21/34 \\ 46/85 \end{pmatrix}$$

37. Write the equation of the parabola as $y = ax^2 + bx + c$. Since the three given points must satisfy the equation, solve the system.

$$
\begin{array}{rrrrr}
9a & - & 3b & + & c & = & 52 \\
4a & + & 2b & + & c & = & 17 \\
64a & + & 8b & + & c & = & 305
\end{array}
$$

Form the matrix equation $AX = B$ with   The solution is given by :

$$
A = \begin{pmatrix} 9 & -3 & 1 \\ 4 & 2 & 1 \\ 64 & 8 & 1 \end{pmatrix}, \; X = \begin{pmatrix} a \\ b \\ c \end{pmatrix}, \; B = \begin{pmatrix} 52 \\ 17 \\ 305 \end{pmatrix} \qquad X = A^{-1}B = \begin{pmatrix} 5 \\ -2 \\ 1 \end{pmatrix}
$$

The equation of the parabola is therefore $y = 5x^2 - 2x + 1$.

39. a. The points $(9,58), (12,70)$ and $(14,77)$ must satisfy the equation

$y = ax^2 + bx + c$. Solve the system:

$$
\begin{array}{rrrrr}
81a & + & 9b & + & c & = & 58 \\
144a & + & 12b & + & c & = & 70 \\
196a & + & 14b & + & c & = & 77
\end{array}
$$

to obtain $y = -.1x^2 + 6.1x + 11.2$

b. Use the table feature to obtain : $2000 - x = 10$ $y = 62.2$ hours
$\qquad\qquad\qquad\qquad\qquad\qquad\quad$ $2015 - x = 25$ $y = 101.2$ hours

41. Let x = number of students, y = number of faculty, z = number of others. Solve the system:

$$
\begin{array}{rrrrr}
x & + & y & + & z & = & 460 \\
x & - & 3y & & & = & 0 \\
x & + & 1.5y & + & 2z & = & 570
\end{array}
$$

to obtain (among other information) $y = 100$ faculty.

43. Let x = pounds of corn, y = pounds of soybeans, z = pounds of by-products. Solve the system:

$$
\begin{array}{rrrrrl}
10x & + & 20y & + & 30x & = & 1800 & (\text{fiber}) \\
30x & + & 20y & + & 40z & = & 2800 & (\text{fat}) \\
20x & + & 40y & + & 25z & = & 2200 & (\text{protein})
\end{array}
$$

to obtain $x = 30$ pounds of corn, $y = 15$ pounds of soybeans, $z = 40$ pounds of by-products.

Chapter 11 Test

1. Let x = Basho Bites and y = Health Nuggets. Then solve the system of equations:
$$x + y = 15$$
$$2.25x + 2.50y = 35.25$$
Solving the first equation for x and then substituting gives:
$$x = 15 - y$$
$$2.25(15 - y) + 2.50y = 35.25$$
$$33.75 - 2.25y + 2.50y = 35.25$$
$$.25y = 1.5$$
$$y = 6$$
Thus, $x = 15 - 6 = 9$.

2. Multiplying both sides by $(x - 4)(x^2 + 3)$ gives
$$5x^2 - 8x - 10 = A(x^2 + 3) + (7Bx + C)(x - 4)$$
$$= Ax^2 + 3A + 7Bx^2 - 28Bx + Cx - 4C$$
$$= x^2(A + 7B) - x(28B - C) - (4C - 3A)$$
Therefore, $5 = A + 7B$
$$8 = 28B - C$$
$$10 = 4C - 3A$$
Solving the first equation for B, and the third for C and substituting into the second equation gives:
$$B = \frac{5 - A}{7}; C = \frac{10 + 3A}{4}$$
$$8 = 28\left(\frac{5 - A}{7}\right) - \frac{10 + 3A}{4}$$
$$8 = 4(5 - A) - \frac{10 + 3A}{4}$$
$$32 = 16(5 - A) - 10 - 3A$$
$$32 = 70 - 19A$$
$$-38 = -19A$$
$$A = 2$$
Then, $B = \frac{5 - 2}{7} = \frac{3}{7}$, and $C = \frac{10 + 3(2)}{4} = 4$.

3. When $x = 0$, $c = 3$. When $x = 1$: $4 = a + b + 3$; $1 = a + b$. When $x = 2$:
$3 = 4a + 2b + 3$; $0 = 4a + 2b$. Solve the system of equations to find a and b:

$1 = a + b$

$0 = 4a + 2b$

Solving the first equation for a and substituting:

$a = 1 - b$

$0 = 4(1 - b) + 2b$

$0 = 4 - 4b + 2b$

$\overline{-4 = -2b}$

$b = 2$

Hence, $a = 1 - 2 = -1$.

4. From the graph the solution is $(5,3)$.

5. a. $x + 2y = 1$

$3x + 4y = 4$

Solving the first equation for x
and substituting;

$x = 1 - 2y$

$3(1 - 2y) + 4y = 4$

$3 - 6y + 4y = 4$

$-2y = 1$

$y = -\dfrac{1}{2}$

Thus, $x = 1 - 2\left(-\dfrac{1}{2}\right) = 1 + 1 = 2$.

c. $\begin{aligned} x^2 + 3\ln y = -2 \\ y = e^x \end{aligned}$

Substituting y into the first
equation gives

$x^2 + 3\ln(e^x) = -2$

$x^2 + 3x = -2$

$x^2 + 3x + 2 = 0$

$(x + 2)(x + 1) = 0$

$x = -2$ or $x = -1$

$y = e^{-2}$ or $y = e^{-1}$

b. $x - y = -5.0$

$3x + 3y = 7.8$

Solving the first equation for x
and substituting;

$x = y - 5.0$

$3(y - 5.0) + 3y = 7.8$

$3y - 15.0 + 3y = 7.8$

$6y = 22.8$

$y = 3.8$

Thus, $x = 3.8 - 5.0 = -1.2$.

6. Graphing each system of equations gives the number of solutions to be:
   a. 1   **b.** Infinite (same line)   **c.** 0   **d.** 4

7. Let x be the first number and y be the second number. Then solve the system of equations:

$$x + y = 100$$

$$xy = 2343.75$$

Solving for x and substituting into the second equation gives

$$x = 100 - y$$

$$(100 - y)y = 2343.75$$

$$100y - y^2 = 2343.75$$

$$0 = y^2 - 100y + 2343.75$$

By the quadratic equation

$$y = \frac{100 \pm \sqrt{(-100)^2 - 4(1)(2343.75)}}{2(1)}$$

$$y = 37.5 \text{ or } y = 62.5$$

Thus, $x = 100 - 37.5 = 62.5$.

8. **a.** $6x - y - 4z = -4$

   $$y + z = 0$$

   $$6z = 12$$

   The last equation gives $z = 2$, substituting that value into the second equation gives $y + 2 = 0$; $y = -2$.

   Substituting the values for y and z into the first equation gives

   $$6x - (-2) - 4(2) = -4$$

   $$6x + 2 - 8 = -4$$

   $$6x = 2$$

   $$x = \frac{1}{3}.$$

Continued on next page

8. continued

b. As a matrix, the system of equations is $\begin{pmatrix} 1 & 1 & 1 & | & 8 \\ 10 & -3 & -5 & | & 4 \\ -2 & -4 & 2 & | & 7 \end{pmatrix}$

To get zeros in the first column take $-10A + B \rightarrow B$ and $2A + C \rightarrow C$

$\begin{pmatrix} 1 & 1 & 1 & | & 8 \\ 0 & -13 & -15 & | & -76 \\ 0 & -2 & 4 & | & 23 \end{pmatrix}$. To get zeros in the second column take $-2B + 13C \rightarrow C$

$\begin{pmatrix} 1 & 1 & 1 & | & 8 \\ 0 & -13 & -15 & | & -76 \\ 0 & 0 & 82 & | & 451 \end{pmatrix}$. Divide last row by 82 to get $\begin{pmatrix} 1 & 1 & 1 & | & 8 \\ 0 & -13 & -15 & | & -76 \\ 0 & 0 & 1 & | & 5.5 \end{pmatrix}$.

To get zeros in the third column take $15C + B \rightarrow B$ and $-C + A \rightarrow A$

$\begin{pmatrix} 1 & 1 & 0 & | & 2.5 \\ 0 & -13 & 0 & | & 6.5 \\ 0 & 0 & 1 & | & 5.5 \end{pmatrix}$. Divide the second row by -13 to get $\begin{pmatrix} 1 & 1 & 0 & | & 2.5 \\ 0 & 1 & 0 & | & -.5 \\ 0 & 0 & 1 & | & 5.5 \end{pmatrix}$.

Finally, get a zero in the first row take $-B + A \rightarrow A$ $\begin{pmatrix} 1 & 0 & 0 & | & 3 \\ 0 & 1 & 0 & | & -.5 \\ 0 & 0 & 1 & | & 5.5 \end{pmatrix}$.

The solutions are $x = 3$, $y = -\dfrac{1}{2}$, $z = \dfrac{11}{2}$

9. a. $AB = \begin{pmatrix} 6 & 5 & 1 \\ 1 & 3 & 3 \\ -2 & 3 & 3 \end{pmatrix}$; $BA = \begin{pmatrix} 3 & 6 & -3 \\ 0 & 8 & 2 \\ -2 & -2 & 1 \end{pmatrix}$

b. AB is not defined; $BA = \begin{pmatrix} -3 & 9 & 1 \\ -12 & 1 & 3 \end{pmatrix}$

c. $AB = \begin{pmatrix} 1 & 0 \\ 0 & 1 \end{pmatrix}$; $BA = \begin{pmatrix} 1 & 0 \\ 0 & 1 \end{pmatrix}$

Chapter 12
Discrete Algebra

12.1 Sequences and Series

1. $a_1 = 2 \cdot 1 + 6 = 8$, $a_2 = 2 \cdot 2 + 6 = 10$, $a_3 = 2 \cdot 3 + 6 = 12$, $a_4 = 2 \cdot 4 + 6 = 14$, $a_5 = 2 \cdot 5 + 6 = 16$

3. $a_1 = \dfrac{1}{1^3} = 1$, $a_2 = \dfrac{1}{2^3} = \dfrac{1}{8}$, $a_3 = \dfrac{1}{3^3} = \dfrac{1}{27}$, $a_4 = \dfrac{1}{4^3} = \dfrac{1}{64}$, $a_5 = \dfrac{1}{5^3} = \dfrac{1}{125}$

5. $a_1 = \dfrac{1}{2^1} = \dfrac{1}{2}$

$a_2 = \dfrac{2}{2^2} = \dfrac{1}{2}$

$a_3 = \dfrac{3}{2^3} = \dfrac{3}{8}$

$a_4 = \dfrac{4}{2^4} = \dfrac{1}{4}$

$a_5 = \dfrac{5}{2^5} = \dfrac{5}{32}$

7. $a_1 = (-1)^1 \sqrt{1+2} = -\sqrt{3}$

$a_2 = (-1)^2 \sqrt{2+2} = 2$

$a_3 = (-1)^3 \sqrt{3+2} = -\sqrt{5}$

$a_4 = (-1)^4 \sqrt{4+2} = \sqrt{6}$

$a_5 = (-1)^5 \sqrt{5+2} = -\sqrt{7}$

9. $a_1 = 4 + (-.1)^1 = 3.9$

$a_2 = 4 + (-.1)^2 = 4.01$

$a_3 = 4 + (-.1)^3 = 3.999$

$a_4 = 4 + (-.1)^4 = 4.0001$

$a_5 = 4 + (-.1)^5 = 3.99999$

11. $a_1 = (-1)^1 + 3 \cdot 1 = 2$

$a_2 = (-1)^2 + 3 \cdot 2 = 7$

$a_3 = (-1)^3 + 3 \cdot 3 = 8$

$a_4 = (-1)^4 + 3 \cdot 4 = 13$

$a_5 = (-1)^5 + 3 \cdot 5 = 14$

13. $a_1 = 3$, $a_2 = 1$, $a_3 = 4$, $a_4 = 1$, $a_5 = 5$

15. $a_n = (-1)^n$

17. $a_n = \dfrac{n}{n+1}$

19. The terms differ by 5, starting with $a_1 = 2$, gives the sequence:
$a_2 = 7 = 2 + 5 \bullet 1$; $a_3 = 12 = 2 + 5 \bullet 2$; $a_4 = 17 = 2 + 5 \bullet 3$; etc.
In general, $a_n = 2 + (n-1) \bullet 5 = 2 + 5n - 5 = 5n - 3$.

21. Rewrite a $3 \bullet 2^0, 3 \bullet 2^1, 3 \bullet 2^2, \ldots$ to see $a_n = 3 \cdot 2^{n-1}$

23. Rewrite as $4\sqrt{1}, 4\sqrt{2}, 4\sqrt{3}, \ldots$ to see $a_n = 4\sqrt{n}$.

469

25. $a_1 = 4$

$a_2 = 2a_1 + 3 = 2 \cdot 4 + 3 = 11$

$a_3 = 2a_2 + 3 = 2 \cdot 11 + 3 = 25$

$a_4 = 2a_3 + 3 = 2 \cdot 25 + 3 = 53$

$a_5 = 2a_4 + 3 = 2 \cdot 53 + 3 = 109$

27. $a_1 = -16$

$a_2 = \dfrac{a_1}{2} = \dfrac{-16}{2} = -8$

$a_3 = \dfrac{a_2}{2} = \dfrac{-8}{2} = -4$

$a_4 = \dfrac{a_3}{2} = \dfrac{-4}{2} = -2$

$a_5 = \dfrac{a_4}{2} = \dfrac{-2}{2} = -1$

29. $a_1 = 2$

$a_2 = 2 - a_1 = 2 - 2 = 0$

$a_3 = 3 - a_2 = 3 - 0 = 3$

$a_4 = 4 - a_3 = 4 - 3 = 1$

$a_5 = 5 - a_4 = 5 - 1 = 4$

31. $a_1 = 1$

$a_2 = -2$

$a_3 = 3$

$a_4 = a_3 + a_2 + a_1 = 3 + (-2) + 1 = 2$

$a_5 = a_4 + a_3 + a_2 = 2 + 3 + (-2) = 3$

33. $a_0 = 2$

$a_1 = 3$

$a_2 = a_1\left(\frac{1}{2}a_0\right) = 3\left(\frac{1}{2} \cdot 2\right) = 3$

$a_3 = a_2\left(\frac{1}{2}a_1\right) = 3\left(\frac{1}{2} \cdot 3\right) = \frac{9}{2}$

$a_4 = a_3\left(\frac{1}{2}a_2\right) = \frac{9}{2}\left(\frac{1}{2} \cdot 3\right) = \frac{27}{4}$

35. $\displaystyle\sum_{i=1}^{11} i$

37. $\displaystyle\sum_{i=1}^{7} \frac{1}{2^{i+6}}$ or $\displaystyle\sum_{i=7}^{13} \frac{1}{2^i}$

39. $\displaystyle\sum_{i=1}^{7} \frac{2^i}{i}$

41. $\displaystyle\sum_{k=1}^{7} k = 1 + 2 + 3 + 4 + 5 + 6 + 7 = 28$

43. $\displaystyle\sum_{i=1}^{4} (i^2 + 1) = (1^2 + 1) + (2^2 + 1) + (3^2 + 1) + (4^2 + 1) = 34$

45. $\displaystyle\sum_{i=1}^{5} 3i = 3 + 6 + 9 + 12 + 15 = 45$

47. $\displaystyle\sum_{n=1}^{16} (2n - 3) = \left[2(1) - 3\right] + \left[2(2) - 3\right] + \left[2(3) - 3\right] + \ldots + \left[2(16) - 3\right]$

$= (-1) + (1) + (3) + \ldots + (29) = 224$

49. $\displaystyle\sum_{n=15}^{36}\left(n^2-8\right)=\left(15^2-8\right)+\left(16^2-8\right)+\left(17^2-8\right)+\ldots+\left(36^2-8\right)=15,015$

51. $a_1=1^2-5\cdot1+2=-2$

$a_2=2^2-5\cdot2+2=-4$

$a_3=3^2-5\cdot3+2=-4$

$a_4=4^2-5\cdot4+2=-2$

$a_5=5^2-5\cdot5+2=2$

$a_6=6^2-5\cdot6+2=8$

$a_1+a_2+a_3=-10$

$a_1+a_2+a_3+a_4+a_5+a_6=-2$

53. $a_1=(-1)^{1+1}\cdot5=5$

$a_2=(-1)^{2+1}\cdot5=-5$

$a_3=(-1)^{3+1}\cdot5=5$

$a_4=(-1)^{4+1}\cdot5=-5$

$a_5=(-1)^{5+1}\cdot5=5$

$a_6=(-1)^{6+1}\cdot5=-5$

$a_1+a_2+a_3=5$

$a_1+a_2+a_3+a_4+a_5+a_6=0$

55. $\displaystyle\sum_{n=1}^{6}\frac{1}{2n+1}\approx.9551$

57. $\displaystyle\sum_{n=1}^{5}\frac{(-1)^{n+1}n}{n+7}\approx.2558$

59. $a_{12}=\left(1+\dfrac{1}{12}\right)^{12}=2.613035$

61. $a_{102}=\dfrac{102^3-102^2+5(102)}{3(102)^2+2(102)-1}=33.465$

63. $\displaystyle\sum_{k=1}^{14}\frac{1}{k^2}=\frac{1}{1^2}+\frac{1}{2^2}+\ldots+\frac{1}{14^2}=1.5759958$

65. After 1 month the interest is $\frac{.05}{12}(9500)$. After paying \$218.78, the principal is $9500+\frac{.05}{12}(9500)-218.78=9500\left(1+\frac{.05}{12}\right)-218.78$. Each month, then, the previous month's principal is multiplied by $\left(1+\frac{.05}{12}\right)$ and 218.78 is subtracted. Thus

$a_0=9500,\ a_n=a_{n-1}\left(1+\dfrac{.05}{12}\right)-218.78$

Using the table feature of a calculator in sequence mode;

$a_{12}=\$7299.70,\ a_{36}=\$2555.50$

67. a. The approximate number of bachelor's degrees awarded in 2004 was

$a_4=25.3\cdot4+1250=1351.2,$ or 1,351,200.

The approximate number of bachelor's degrees awarded in 2007 was

$a_7=25.3\cdot7+1250=1427.1,$ or 1,427,100.

b. The approximate number of bachelor's degrees awarded between 2004 and 2007

was $\displaystyle\sum_{n=4}^{7}(25.3n+1250)=8486.7,$ or 8,486,700.

69. a. The spending per person in 2005 was $c_5 = .16(5)^2 + 2.7(5) + 25.1 = \$42.60$. The spending per person in 2007 was $c_7 = .16(7)^2 + 2.7(7) + 25.1 = \$51.84$.

b. The total spent per person from 2000 to 2007 was

$$\sum_{n=0}^{7} (.16n^2 + 2.7n + 25.1) = \$298.80.$$

71. b. $a_{17} = 59, a_{18} = 61, a_{19} = 67, a_{20} = 71$

73. $a_1 = 2^2 = 4, a_2 = 3^2 = 9, a_3 = 5^2 = 25, a_4 = 7^2 = 49, a_5 = 11^2 = 121$

75. a_1 is the largest prime integer less than 5, thus $a_1 = 3$.

a_2 is the largest prime integer less than 10, thus $a_2 = 7$.

a_3 is the largest prime integer less than 15, thus $a_3 = 13$.

a_4 is the largest prime integer less than 20, thus $a_4 = 19$.

a_5 is the largest prime integer less than 25, thus $a_5 = 23$.

77. a. 1, 1, 2, 3, 5, 8, 13, 21, 34, 55

b. 1, 2, 4, 7, 12, 20, 33, 54, 88, 143

c. The nth partial sum is 1 less than the $(n + 2)$nd term, i.e. $a_{n+2} - 1$.

79. $5(1)^2 + 4(-1)^1 = 1 = 1^2$

$5(1)^2 + 4(-1)^2 = 9 = 3^2$

$5(2)^2 + 4(-1)^3 = 16 = 4^2$

$5(3)^2 + 4(-1)^4 = 49 = 7^2$

$5(5)^2 + 4(-1)^5 = 121 = 11^2$

$5(8)^2 + 4(-1)^6 = 324 = 18^2$

$5(13)^2 + 4(-1)^7 = 841 = 29^2$

$5(21)^2 + 4(-1)^8 = 2209 = 47^2$

$5(34)^2 + 4(-1)^9 = 5776 = 76^2$

$5(55)^2 + 4(-1)^{10} = 15129 = 123^2$

81. Since

$a_n = a_{n-1} + a_{n-2}, \ a_n - a_{n-1} = a_{n-2}.$

Thus $a_3 - a_2 = a_1$

$a_4 - a_3 = a_2$

$a_5 - a_4 = a_3$

$\cdots\cdots\cdots\cdots$

$\underline{a_{k+2} - a_{k+1} = a_k}$

Adding: $a_{k+2} - a_2 = \displaystyle\sum_{n=1}^{k} a_n$

Since $a_2 = 1$, $a_{k+2} - 1 = \displaystyle\sum_{n=1}^{k} a_n$.

12.2 Arithmetic Sequences

1. $3-1=5-3=7-5=9-7=2$. Therefore, the sequence is arithmetic, with common difference of 2.

3. $2-1=1; 4-2=2$. Therefore, the sequence is not arithmetic.

5. $\log 2 - \log 1 = \log\left(\dfrac{2}{1}\right) = \log 2$; $\log 4 - \log 2 = \log\left(\dfrac{4}{2}\right) = \log 2$;

$\log 8 - \log 4 = \log\left(\dfrac{8}{4}\right) = \log 2$; $\log 16 - \log 8 = \log\left(\dfrac{16}{8}\right) = \log 2$

Therefore, the sequence is arithmetic with a common difference of $\log 2$.

7. $-\dfrac{4}{6} - \dfrac{1}{3} = -\dfrac{15}{9} - \left(-\dfrac{4}{6}\right) = -1$ and so on. Therefore the sequence is arithmetic with a common difference of -1.

9. $a_1 = 5 + 4 \cdot 1 = 9$
$a_2 = 5 + 4 \cdot 2 = 13$
$a_3 = 5 + 4 \cdot 3 = 17$
$a_4 = 5 + 4 \cdot 4 = 21$
$a_5 = 5 + 4 \cdot 5 = 25$
The sequence is arithmetic, with common difference $13 - 9 = 4$.

11. $c_1 = (-1)^1 = -1$
$c_2 = (-1)^2 = 1$
$c_3 = (-1)^3 = -1$
$c_4 = (-1)^4 = 1$
$c_5 = (-1)^5 = -1$
Since $c_2 - c_1 = 2$ but $c_3 - c_2 = -2$, the sequence is not arithmetic.

13. $a_1 = 2 + (-1)^1 \cdot 1 = 1$
$a_2 = 2 + (-1)^2 \cdot 2 = 4$
$a_3 = 2 + (-1)^3 \cdot 3 = -1$
$a_4 = 2 + (-1)^4 \cdot 4 = 6$
$a_5 = 2 + (-1)^5 \cdot 5 = -3$
Since $c_2 - c_1 = 3$, but $c_3 - c_2 = -5$, the sequence is not arithmetic.

15. $a_1 = e^{\ln(1)} = 1$
$a_2 = e^{\ln(2)} = 2$
$a_3 = e^{\ln(3)} = 3$
$a_4 = e^{\ln(4)} = 4$
$a_5 = e^{\ln(5)} = 5$
The sequence is arithmetic with a common difference of 1.

17. $a_n = 3 - 2n, \quad a_{n+1} = 3 - 2(n+1)$

$a_{n+1} - a_n = [3 - 2(n+1)] - (3 - 2n) = -2$

The sequence is arithmetic with
a common difference -2.

19. $a_n = 4 + \dfrac{n}{3}, \quad a_{n+1} = 4 + \dfrac{n+1}{3}$

$a_{n+1} - a_n = \left(4 + \dfrac{n+1}{3}\right) - \left(4 + \dfrac{n}{3}\right) = \dfrac{1}{3}$

The sequence is arithmetic with
a common difference $\frac{1}{3}$.

21. $a_n = \dfrac{5 + 3n}{2}, \quad a_{n+1} = \dfrac{5 + 3(n+1)}{2}$

$a_{n+1} - a_n = \dfrac{5 + 3(n+1)}{2} - \dfrac{5 + 3n}{2} = \dfrac{3}{2}$

The sequence is arithmetic with
a common difference $\frac{3}{2}$.

23. $a_n = c + 2n \quad a_{n+1} = c + 2(n+1)$

$a_{n+1} - a_n = [c + 2(n+1)] - (c + 2n) = 2$

The sequence is arithmetic with
a common difference 2.

25. $a_5 = a_1 + (5-1)d = 5 + (5-1)2 = 13$

$a_n = a_1 + (n-1)d = 5 + (n-1)2$

$\quad = 2n + 3$

27. $a_5 = a_1 + (5-1)d = 4 + (5-1)\frac{1}{4} = 5$

$a_n = a_1 + (n-1)d = 4 + (n-1)\frac{1}{4}$

$\quad = \dfrac{n + 15}{4}$

29. $a_5 = a_1 + (5-1)d = 10 + (5-1)\left(-\frac{1}{2}\right) = 8$

$a_n = a_1 + (n-1)d = 10 + (n-1)\left(-\frac{1}{2}\right)$

$\quad = \dfrac{21 - n}{2}$

31. $a_5 = a_1 + (5-1)d = 8 + (5-1)(.1) = 8.4$

$a_n = a_1 + (n-1)d = 8 + (n-1)(.1)$

$\quad = 7.9 + .1n$

33. $a_4 = a_1 + (4-1) \cdot d$

$12 = a_1 + (4-1)2$

$a_1 = 6$

$a_n = a_1 + (n-1)d$

$a_n = 6 + (n-1)2$

$a_n = 2n + 4$

35. $a_3 = a_1 + (3-1) \cdot d$

$3 = a_1 + (3-1)5$

$a_1 = -7$

$a_n = a_1 + (n-1)d$

$a_n = -7 + (n-1)5$

$a_n = 5n - 12$

37. $a_2 = a_1 + d = 4$

$a_6 = a_1 + 5d = 32$

Eliminating a_1 yields

$4d = 28$

$d = 7$, hence $a_1 = -3$.

$a_n = a_1 + (n-1)d$

$a_n = -3 + (n-1)(7)$

$a_n = 7n - 10$

39. $a_5 = a_1 + 4d = 0$

$a_9 = a_1 + 8d = 6$

Eliminating a_1 yields

$4d = 6$

$d = \frac{3}{2}$, hence $a_1 = -6$.

$a_n = a_1 + (n-1)d$

$a_n = -6_1 + (n-1)\left(\frac{3}{2}\right)$

$a_n = \dfrac{3n-15}{2}$

41. $\displaystyle\sum_{n=1}^{k} a_n = ka_1 + \frac{k(k-1)}{2}d = 6\cdot 2 + \frac{6(6-1)}{2}5 = 87$

43. $\displaystyle\sum_{n=1}^{k} a_n = ka_1 + \frac{k(k-1)}{2}d = 7\left(\frac{3}{4}\right) + \frac{7(7-1)}{2}\left(-\frac{1}{2}\right) = -\frac{21}{4}$

45. $\displaystyle\sum_{n=1}^{k} a_n = \frac{k}{2}(a_1 + a_k) = \frac{6}{2}((-4)+14) = 30$

47. $\displaystyle\sum_{n=1}^{k} a_n = \frac{k}{2}(a_1 + a_k) = \frac{9}{2}(6+(-24)) = -81$

49. $\displaystyle\sum_{n=1}^{k} a_n = \frac{k}{2}(a_1 + a_k) = \frac{20}{2}((3\cdot 1 + 4)+(3\cdot 20 + 4)) = 10(7+64) = 710$

51. $\displaystyle\sum_{n=1}^{k} a_n = \frac{k}{2}(a_1 + a_k) = \frac{30}{2}\left(\left(3-\frac{1}{2}\right)+\left(3-\frac{30}{2}\right)\right) = -\frac{285}{2}$

53. $\displaystyle\sum_{n=1}^{k} a_n = \frac{k}{2}(a_1 + a_k) = \frac{40}{2}\left(\frac{1+3}{6}+\frac{40+3}{6}\right) = 20\left(\frac{47}{6}\right) = \frac{470}{3}$

55. a. $a_1 = 4154, a_6 = 5864$

$a_6 = a_1 + (6-1)d$

$d = \dfrac{a_6 - a_1}{5} = \dfrac{5864 - 4154}{5} = 342$

$a_n = 4154 + (n-1)342$

$= 3812 + 342n$

b. Use the formula found in part (a):

$2005 - a_7 = 3812 + 342.7 = \$6206$

$2008 - a_{10} = 3812 + 342.10 = \$7232$

57. The first 12 rows form an arithmetic sequence with $a_1 = 6, d = 2$.
The twelfth row has $a_{12} = a_1 + (12-1)d = 6 + 11 \cdot 2 = 28$ seats.

The first twelve rows have $\sum_{n=1}^{n} a_n = \frac{12}{2}(6+28) = 204$ seats.

Then there are 8 more rows of 28 seats each — 224 more seats for a total of 428 seats.

59. The lengths form an arithmetic sequence with $a_1 = 24$ and $a_9 = 18$.

$$a_9 = a_1 + (9-1)\cdot d$$

$$d = \frac{a_9 - a_1}{8} = \frac{18-24}{8} = -\frac{3}{4}$$

Thus, the rungs decrease by $\frac{3}{4}$ of an inch: $23.25, 22.5, 21.75, 21, 20.25, 19.5, 18.75$ inches are the seven intermediate lengths.

61. These form an arithmetic sequence with $a_1 = 2, d = 2$. Find k, the number of terms:

$$100 = a_1 + (k-1)\cdot d$$

$$100 = 2 + (k-1)2$$

Thus, $k = 50$. $\sum_{n=1}^{k} a_n = \frac{k}{2}(a_1 + a_k) = \frac{50}{2}(2+100) = 2550$

63. These form an arithmetic sequence with $a_1 = 1, d = 1$. Clearly, $k = 200$.

Thus, $\sum_{n=1}^{k} a_n = \frac{k}{2}(a_1 + a_k) = \frac{200}{2}(1+200) = 20{,}100.$

65. These form an arithmetic sequence with $a_1 = 10{,}000, d = 7500$. In the tenth year
$a_{10} = a_1 + (10-1)d = 10{,}000 + 9 \cdot 7500 = \$77{,}500$. The total profit is given by

$$\sum_{n=1}^{k} a_n = \frac{k}{2}(a_1 + a_k) = \frac{10}{2}(10{,}000 + 77{,}500) = \$437{,}500.$$

67. a. $a_1 = 136.4, a_{10} = 217.4$

$$a_{10} = a_1 + (10-1)d$$

$$d = \frac{a_{10} - a_1}{9} = \frac{217.4 - 136.4}{9} = 9$$

$$a_n = 136.4 + (n-1)9 = 127.4 + 9n$$

b. Use the formula found in part (a):
$$2005 - a_{12} = 127.4 + 3\cdot 12$$
$$= \$235.4 \text{ billion}$$

c. $a_7 + a_8 + a_9 + a_{10} + a_{11} + a_{12} = \frac{6}{2}(a_7 + a_{10}) = \$1277.4$ billion

69. a. $c_1 = 15,828.80, \; c_{19} = 28,932.80$

$c_{19} = c_1 + (19-1)d$

$d = \dfrac{c_{19} - c_1}{18} = \dfrac{28,932.80 - 15,828.80}{18} = \$728$

$c_n = 15,828.80 + (n-1)728 = 15,100.80 + 728n$

b. Use the above formula, $2002 - n = 25$

$\displaystyle\sum_{n=1}^{25} a_n = \dfrac{25}{2}(c_1 + c_{25}) = \dfrac{25}{2}\left(15,828.80 + 15,828.80 + (25-1)728\right) = \$614,120$

12.3 Geometric Sequences

1. $a_2 - a_1 = a_3 - a_2 = a_4 - a_3 = a_5 - a_4 = 5$. Arithmetic

3. $a_2 \div a_1 = a_3 \div a_2 = a_4 \div a_3 = a_5 \div a_4 = 1/2$. Geometric

5. $a_2 - a_1 = a_3 - a_2 = a_4 - a_3 = -2$. Arithmetic

7. $a_2 \div a_1 = a_3 \div a_2 = a_4 \div a_3 = a_5 \div a_4 = -\frac{1}{2}$. Geometric

9. $a_2 \div a_1 = a_3 \div a_2 = a_4 \div a_3 = a_5 \div a_4 = \sqrt{2}$. Geometric

11. Since this sequence is constant it is both arithmetic ($d = 0$) and geometric ($r = 1$).

13. $a_6 = a_1 r^5 = 5 \cdot 2^5 = 160$
$a_n = a_1 r^{n-1} = 5 \cdot 2^{n-1}$

15. $a_6 = a_1 r^5 = 4\left(\frac{1}{4}\right)^5 = 1/256$
$a_n = a_1 r^{n-1} = 4\left(\frac{1}{4}\right)^{n-1} = 4^{2-n}$

17. $a_6 = a_1 r^5 = 10\left(-\frac{1}{2}\right)^5 = -\frac{5}{16}$
$a_n = a_1 r^{n-1} = 10\left(-\frac{1}{2}\right)^{n-1} = \frac{5(-1)^{n-1}}{2^{n-2}}$

19. $a_2 = a_1 r$
$12 = a_1\left(\frac{1}{3}\right)$
$a_1 = 36$
$a_6 = a_1 r^5 = 36\left(\frac{1}{3}\right)^5 = \frac{4}{27}$
$a_n = a_1 r^{n-1} = 36\left(\frac{1}{3}\right)^{n-1}$

21. $a_4 = a_1 r^3$
$-\frac{4}{5} = a_1\left(\frac{2}{5}\right)^3$
$a_1 = -\frac{25}{2}$
$a_6 = a_1 r^5 = \left(-\frac{25}{2}\right)\left(\frac{2}{5}\right)^5 = -\frac{16}{125}$
$a_n = a_1 r^{n-1} = \left(-\frac{25}{2}\right)\left(\frac{2}{5}\right)^{n-1} = -\frac{2^{n-2}}{5^{n-3}}$

23. $a_n \div a_{n-1} = \left(-\frac{1}{2}\right)^n \div \left(-\frac{1}{2}\right)^{n-1} = -\frac{1}{2}$. The sequence is geometric, with common ratio $-\frac{1}{2}$.

25. $a_n \div a_{n-1} = 5^{n+2} \div 5^{(n-1)+2} = 5$. Thus the sequence is geometric, with common ratio 5.

27. $a_n \div a_{n-1} = \left(\sqrt{5}\right)^n \div \left(\sqrt{5}\right)^{n-1} = \sqrt{5}$. Geometric sequence, with common ratio $\sqrt{5}$.

29. $a_n \div a_{n-1} = 10e^{4n} \div 10e^{4(n-1)} = e^4$. Geometric sequence, with common ratio e^4.

31. $a_2 \div a_1 = -64 \div 256 = -\frac{1}{4} = r$

$\quad a_5 = a_1 r^4 = 256\left(-\frac{1}{4}\right)^4 = 1$

$\quad a_n = a_1 r^{n-1} = 256\left(-\frac{1}{4}\right)^{n-1} = \left(-\frac{1}{4}\right)^{n-5}$

33. $a_2 \div a_1 = 5 \div \frac{1}{2} = 10 = r$

$\quad a_5 = a_1 r^4 = \frac{1}{2}(10)^4 = 5000$

$\quad a_n = a_1 r^{n-1} = \frac{1}{2}(10)^{n-1}$

35. $\qquad a_6 = a_1 r^5 = \frac{1}{16}$

$\qquad a_3 = a_1 r^2 = 4$

$\quad a_6 \div a_3 = r^3 = \frac{1}{16} \div 4 = \frac{1}{64},$

$\quad$ hence $r = \frac{1}{4},\ a_5 = a_6 \div r = \frac{1}{16} \div \frac{1}{4} = \frac{1}{4}$

$\qquad a_5 = \frac{1}{4}$

$\qquad a_1 = a_3 \div r^2 = 4 \div \left(\frac{1}{4}\right)^2 = 64$

$\qquad a_n = a_1 r^{n-1} = 64\left(\frac{1}{4}\right)^{n-1} = 4^{4-n}$

37. $a_7 = a_1 r^6$

$\quad 20 = 5 \cdot r^6$

$\quad r^6 = 4$

$\quad r = \sqrt[6]{4} = \sqrt[3]{2} \quad (\text{since } r > 0)$

$\quad a_5 = 5 \cdot \left(\sqrt[3]{2}\right)^4 = 10\sqrt[3]{2}$

$\quad a_n = 5\left(\sqrt[3]{2}\right)^{n-1}$

39. $\displaystyle\sum_{n=1}^{k} a_n = a_1\left(\frac{1-r^k}{1-r}\right)$

$\displaystyle\sum_{n=1}^{6} a_n = 5\left(\frac{1-\left(\frac{1}{2}\right)^6}{1-\frac{1}{2}}\right) = \frac{315}{32}$

41. $\displaystyle\sum_{n=1}^{k} a_n = a_1\left(\frac{1-r^k}{1-r}\right)$

$\quad a_2 = a_1 r,\ 6 = a_1 \cdot 2,\ a_1 = 3$

$\displaystyle\sum_{n=1}^{7} a_n = 3\left(\frac{1-2^7}{1-2}\right) = 381$

43. This is a geometric sequence with $a_1 = 2,\ r = 2.$

$\displaystyle\sum_{n=1}^{k} a_n = a_1\left(\frac{1-r^k}{1-r}\right)$

$\displaystyle\sum_{n=1}^{7} 2^n = 2\left(\frac{1-2^7}{1-2}\right) = 254$

45. This is a geometric sequence with $a_1 = -\frac{1}{3},\ r = -\frac{1}{3}.$

$\displaystyle\sum_{n=1}^{k} a_n = a_1\left(\frac{1-r^k}{1-r}\right)$

$\displaystyle\sum_{n=1}^{9}\left(-\frac{1}{3}\right)^n = -\frac{1}{3}\left(\frac{1-\left(-\frac{1}{3}\right)^9}{1-\left(-\frac{1}{3}\right)}\right) = -\frac{4921}{19683}$

47. This is a geometric sequence with $a_1 = 4,\ r = \frac{3}{2}.$

$\displaystyle\sum_{j=1}^{6}\left(\frac{3}{2}\right)^{j-1} = 4\left(\frac{1-\left(\frac{3}{2}\right)^6}{1-\left(\frac{3}{2}\right)}\right) = \frac{665}{8}$

49. a. $r = a_2 \div a_1 = 5.687 \div 5.912 = .9619$, $a_n = a_1 r^{n-1} = 5.912(.9619)^{n-1}$

b. 2007 corresponds to $n = 11$.

$a_{11} = 5.912(.9619)^{10} = 4.01$ students per computer

c. Solve $3 = 5.912(.9619)^n$

$n = \dfrac{\ln(3/5.912)}{\ln(.9619)} = 18.46$. This corresponds to 2015.

51. a. $r = a_2 \div a_1 = 4.1672 \div 3.9631 = 1.0515$, $a_n = a_1 r^{n-1} = 3.9631(1.0515)^{n-1}$

b. $2000 - a_{10} = 3.9631(1.0515)^9 = \$6.228$ billion

$2004 - a_{14} = 3.9631(1.0515)^{13} = \$7.613$ billion

$2008 - a_{18} = 3.9631(1.0515)^{18} = \$9.786$ billion

53. a. $b_3 = 232.22$, $b_1 = 204.74$

$b_3 = b_1 r^2$

$r = \sqrt{b_3 \div b_1} = \sqrt{232.22 \div 204.74} = 1.065$

$b_n = b_1 r^{n-1} = 204.74(1.065)^{n-1}$

b. $\displaystyle\sum_{n=1}^{k} b_n = b_1\left(\dfrac{1-r^k}{1-r}\right)$; $k = 9$ in 2009

$\displaystyle\sum_{n=1}^{9} b_n = 204.74\left(\dfrac{1-(1.065)^9}{1-1.065}\right) = \$2401.98$

55. Following Example **7** of the text, during the first bounce the ball travels $4 + 4 = 8$ ft, during the second, half of this, and so on. This is a geometric sequence with $a_1 = 8, r = \frac{1}{2}$.

Total distance $=$ initial drop $+ \displaystyle\sum_{n=1}^{6} a_1 r^n = 8 + a_1\left(\dfrac{1-r^6}{1-r}\right) = 8 + 8\left(\dfrac{1-\left(\frac{1}{2}\right)^6}{1-\frac{1}{2}}\right) = 23.75$ ft.

57. a. Follow the procedure in example 8 to obtain:

$c_n = 20\left[\dfrac{1-e^{-.1155(5)n}}{1-e^{-.1155(5)}}\right] = 20\left[\dfrac{1-e^{-.5775n}}{1-e^{-.5775}}\right]$

b. After the 9th hour - $n = 9$ $c_9 = 20\left[\dfrac{1-e^{-.5775(9)}}{1-e^{-.5775}}\right] = 45.337$ mg.

59. This forms a geometric sequence with $a_1 = .01$, $r = 2$.

$$\sum_{n=1}^{31} a_n = .01\left(\frac{1-2^{31}}{1-2}\right) = \$21,474,836.47$$

61. This forms a geometric sequence with
$a_1 = 8000(.75)$ (value after 1 yr), $r = .75$.
$a_5 = a_1 r^4 = 8000(.75)(.75)^4 = \$1898.44$

63. Since $a_n = a_1 r^{n-1}$, $\log a_n = \log a_1 + (n-1)\log r$.
Therefore $\log a_n - \log a_{n-1} = [\log a_1 + (n-1)\log r] - [\log a_1 + (n-2)\log r] = \log r$.
This shows that the $\{\log a_n\}$ sequence is arithmetic with common difference $\log r$.

65. Each term $a_k = a_1 r^{k-1} = 2^{k-1}$. The sum of the preceding terms is

$$\sum_{n=1}^{k-1} a_n = a_1\left(\frac{1-r^{k-1}}{1-r}\right) = 1\left(\frac{1-2^{k-1}}{1-2}\right) = 2^{k-1} - 1$$

Thus each term 2^{k-1} equals 1 plus the sum of the preceding terms.

67. First determine how many payments are necessary before $1/25$ of the balance falls below \$5. Form a geometric sequence with $a_1 = 200$ and $r = 24/25$. Then $a_n = a_1 r^{n-1} = 200\left(\frac{24}{25}\right)^{n-1}$. Use the Table feature of the calculator to observe that
$a_{12} = 127.65$, $a_{13} = 122.54$ and $a_{14} = 117.64$.

Thus after 13 payments, with a balance of \$122.54, $1/25$ of the outstanding balance is less than \$5. Then it would take $\$122.54 \div \$5$ or 25 more payments of \$5 to complete the payment, for a total of 3 years and 2 months.

12.3.A Infinite Series

1. This is a convergent geometric series with $a_1 = \frac{1}{2}$ and $r = \frac{1}{2}$. Its sum is

$$\frac{a_1}{1-r} = \frac{\frac{1}{2}}{1-\frac{1}{2}} = 1.$$

3. This is a convergent geometric series with $a_1 = .06$ and $r = .06$. Its sum is

$$\frac{a_1}{1-r} = \frac{.06}{1-.06} = \frac{3}{47}.$$

5. This is a convergent geometric series with $a_1 = 500$ and $r = \frac{200}{500} = \frac{2}{5}$. Its sum is

$$\frac{a_1}{1-r} = \frac{500}{1-\frac{2}{5}} = \frac{2500}{3} = 833\frac{1}{3}.$$

7. This is a convergent geometric series with $a_1 = 2$ and $r = \sqrt{2}/2$. Its sum is

$$\frac{a_1}{1-r} = \frac{2}{1-\left(\sqrt{2}/2\right)} = \frac{4}{2-\sqrt{2}} = 4 + 2\sqrt{2}.$$

9. $.2+.02+.002+\ldots$ is a convergent infinite series with $a_1 = .2$ and $r = .1$. Its sum is

$$\frac{a_1}{1-r} = \frac{.2}{1-.1} = \frac{2}{9}.$$

11. $5.4+.027+.00027+.0000027+\ldots$, after the first term, is a convergent infinite series with $a_1 = .027$ and $r = .01$. Its sum is $5.4 + \dfrac{a_1}{1-r} = 5.4 + \dfrac{.027}{1-.01} = \dfrac{54}{10} + \dfrac{27}{990} = \dfrac{597}{110}.$

13. $2.1+.0425+.0000425+\ldots$, after the first term, is a convergent infinite series with $a_1 = .0425$ and $r = .001$. Its sum is $2.1 + \dfrac{a_1}{1-r} = 2.1 + \dfrac{.0425}{1-.001} = \dfrac{21}{10} + \dfrac{425}{9990} = \dfrac{10702}{4995}.$

15. $1.74+.00241+.00000241+\ldots$, after the first term, is a convergent infinite series with $a_1 = .00241$ and $r = .001$.

Its sum is $1.74 + \dfrac{a_1}{1-r} = 1.74 + \dfrac{.00241}{1-.001} = \dfrac{174}{100} + \dfrac{241}{99900} = \dfrac{174,067}{99,900}.$

17. a. Since $\sum_{n=1}^{\infty} 2(1.5)^n = 2(1.5)^1 + 2(1.5)^2 + 2(1.5)^3 + \ldots = 3 + 3(1.5)^1 + 3(1.5)^2 + \ldots$, this is

a geometric series with $a_1 = 3, r = 1.5$.

b. $\sum_{n=1}^{k} a_n = a_1\left(\dfrac{1-r^k}{1-r}\right) = 3\left(\dfrac{1-1.5^k}{1-1.5}\right) = 6(1.5^k - 1)$. Hence, let $f(x) = 6(1.5^x - 1)$

c. Graph the function of part **b** using the window $0 \le x \le 30, 0 \le y \le 10,000$ and observe that the function increases faster and faster as x gets large; the graph does not approach a horizontal line, and the series does not converge.

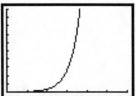

19. $s_1 = \dfrac{\pi}{2} = 1.570796327$

$s_2 = \dfrac{\pi}{2} - \dfrac{(\pi/2)^3}{3!} = .9248322293$

$s_3 = \dfrac{\pi}{2} - \dfrac{(\pi/2)^3}{3!} + \dfrac{(\pi/2)^5}{5!} = 1.004524856$

$s_4 = \dfrac{\pi}{2} - \dfrac{(\pi/2)^3}{3!} + \dfrac{(\pi/2)^5}{5!} - \dfrac{(\pi/2)^7}{7!} = .9998431014$

$s_5 = \dfrac{\pi}{2} - \dfrac{(\pi/2)^3}{3!} + \dfrac{(\pi/2)^5}{5!} - \dfrac{(\pi/2)^7}{7!} + \dfrac{(\pi/2)^9}{9!} = 1.000003543$

At this point, we may conjecture that the sum is 1.

21. The total amount of aminophylline in patient's blood streams after n doses is

$b_n = 20 + 20e^{-.5775} + 20\left[e^{-.5775}\right]^2 + \ldots + 20\left[e^{-.5775}\right]^{n-1}$

Each new dose adds another term to this sum. So if the doses are continued for a long time, the total amount will be approximately given by the geometric seires,

$20 + 20e^{-.5775} + 20\left[e^{-.5775}\right]^2 + \ldots + 20\left[e^{-.5775}\right]^{n-1} + \ldots$, which has $a_1 = 20$ and $r = e^{-.5775}$.

Since $e^{-.1155} < 1$, this series converges and its sum is

$\dfrac{a_1}{1-r} = \dfrac{20}{1-e^{-.5775}} = 45.58923$ mg.

12.4 The Binomial Theorem

1. $6! = 6 \cdot 5 \cdot 4 \cdot 3 \cdot 2 \cdot 1 = 720$

3. $\dfrac{8!}{6!} = \dfrac{8 \cdot 7 \cdot 6 \cdot 5 \cdot 4 \cdot 3 \cdot 2 \cdot 1}{6 \cdot 5 \cdot 4 \cdot 3 \cdot 2 \cdot 1} = 56$

5. $\dfrac{12!}{9!3!} = \dfrac{12 \cdot 11 \cdot 10 \cdot 9!}{9!3 \cdot 2 \cdot 1} = \dfrac{1320}{6} = 220$

7. $\dbinom{6}{2} = \dfrac{6!}{2!4!} = \dfrac{6 \cdot 5 \cdot 4 \cdot 3 \cdot 2 \cdot 1}{2 \cdot 1 \cdot 4 \cdot 3 \cdot 2 \cdot 1} = 15$

9. $\dbinom{100}{99} = \dfrac{100!}{99!1!} = \dfrac{100 \cdot 99!}{99!} = 100$

11. $\dbinom{4}{1}\dbinom{5}{3} = \dfrac{4!}{1!3!} \cdot \dfrac{5!}{3!2!} = \dfrac{4 \cdot 3 \cdot 2 \cdot 1}{1 \cdot 3 \cdot 2 \cdot 1} \cdot \dfrac{5 \cdot 4 \cdot 3 \cdot 2 \cdot 1}{3 \cdot 2 \cdot 1 \cdot 2 \cdot 1} = 40$

13. $\dbinom{5}{3} + \dbinom{5}{2} - \dbinom{6}{3} = \dfrac{5!}{3!2!} + \dfrac{5!}{2!3!} - \dfrac{6!}{3!3!} = \dfrac{5 \cdot 4 \cdot 3!}{3!2 \cdot 1} + \dfrac{5 \cdot 4 \cdot 3!}{2!3!} - \dfrac{6 \cdot 5 \cdot 4 \cdot 3!}{3!3 \cdot 2 \cdot 1} = 10 + 10 - 20 = 0$

15. Display the coefficients on the calculator to show
$\dbinom{6}{0} + \dbinom{6}{1} + \dbinom{6}{2} + \dbinom{6}{3} + \dbinom{6}{4} + \dbinom{6}{5} + \dbinom{6}{6} = 1 + 6 + 15 + 20 + 15 + 6 + 1 = 64$

17. $\dbinom{100}{96} = \dfrac{100!}{96!4!} = \dfrac{100 \cdot 99 \cdot 98 \cdot 97 \cdot 96!}{96!4 \cdot 3 \cdot 2 \cdot 1} = 3,921,225$

19. $(x+1)^4 = \dbinom{4}{0}x^4 + \dbinom{4}{1}x^3 \cdot 1 + \dbinom{4}{2}x^2 \cdot 1^2 + \dbinom{4}{3}x \cdot 1^3 + \dbinom{4}{4} \cdot 1^4$
$= x^4 + 4x^3 + 6x^2 + 4x + 1$

21. $(x+2)^4 = \dbinom{4}{0}x^4 + \dbinom{4}{1}x^3 \cdot 2 + \dbinom{4}{2}x^2 \cdot 2^2 + \dbinom{4}{3}x \cdot 2^3 + \dbinom{4}{4} \cdot 2^4$
$= x^4 + 8x^3 + 24x^2 + 32x + 16$

23. $(x+y)^5 = \dbinom{5}{0}x^5 + \dbinom{5}{1}x^4y + \dbinom{5}{2}x^3y^2 + \dbinom{5}{3}x^2y^3 + \dbinom{5}{4}xy^4 + \dbinom{5}{5}y^5$
$= x^5 + 5x^4y + 10x^3y^2 + 10x^2y^3 + 5xy^4 + y^5$

25. $(a-b)^5 = \left(a+(-b)\right)^5$

$$= \binom{5}{0}a^5 + \binom{5}{1}a^4(-b) + \binom{5}{2}a^3(-b)^2 + \binom{5}{3}a^2(-b)^3 + \binom{5}{4}a(-b)^4 + \binom{5}{5}(-b)^5$$

$$= a^5 - 5a^4b + 10a^3b^2 - 10a^2b^3 + 5ab^4 - b^5$$

27. $(2x+y^2)^5 = \binom{5}{0}(2x)^5 + \binom{5}{1}(2x)^4(y^2) + \binom{5}{2}(2x)^3(y^2)^2 + \binom{5}{3}(2x)^2(y^2)^3$

$$+ \binom{5}{4}(2x)(y^2)^4 + \binom{5}{5}(y^2)^5$$

$$= 32x^5 + 80x^4y^2 + 80x^3y^4 + 40x^2y^6 + 10xy^8 + y^{10}$$

29. $(\sqrt{x}+1)^6 = \binom{6}{0}(\sqrt{x})^6 + \binom{6}{1}(\sqrt{x})^5 1 + \binom{6}{2}(\sqrt{x})^4 1^2 + \binom{6}{3}(\sqrt{x})^3 1^3$

$$+ \binom{6}{4}(\sqrt{x})^2 1^4 + \binom{6}{5}(\sqrt{x})1^5 + \binom{6}{6}1^6$$

$$= x^3 + 6x^2\sqrt{x} + 15x^2 + 20x\sqrt{x} + 15x + 6\sqrt{x} + 1$$

31. $(1-c)^{10} = \binom{10}{0}1^{10} + \binom{10}{1}1^9(-c) + \binom{10}{2}1^8(-c)^2 + \ldots + \binom{10}{9}1(-c)^9 + \binom{10}{10}(-c)^{10}$

$$= 1 - 10c + 45c^2 - 120c^3 + 210c^4 - 252c^5 + 210c^6 - 120c^7 + 45c^8 - 10c^9 + c^{10}$$

33. $(x^{-3}+x)^4 = \binom{4}{0}(x^{-3})^4 + \binom{4}{1}(x^{-3})^3 x + \binom{4}{2}(x^{-3})^2 x^2 + \binom{4}{3}(x^{-3})x^3 + \binom{4}{4}x^4$

$$= x^{-12} + 4x^{-8} + 6x^{-4} + 4 + x^4$$

35. $(1+\sqrt{3})^4 + (1-\sqrt{3})^4 = \binom{4}{0}1^4 + \binom{4}{1}1^3(\sqrt{3}) + \binom{4}{2}1^2(\sqrt{3})^2 + \binom{4}{3}1(\sqrt{3})^3$

$$+ \binom{4}{4}(\sqrt{3})^4 + \binom{4}{0}1^4 + \binom{4}{1}1^3(-\sqrt{3}) + \binom{4}{2}1^2(-\sqrt{3})^2 + \binom{4}{3}1(-\sqrt{3})^3 + \binom{4}{4}(-\sqrt{3})^4$$

$$= 1 + 4\sqrt{3} + 18 + 12\sqrt{3} + 9 + 1 - 4\sqrt{3} + 18 - 12\sqrt{3} + 9 = 56$$

37. $(1+i)^6 = \binom{6}{0}1^6 + \binom{6}{1}1^5 i + \binom{6}{2}1^4 i^2 + \binom{6}{3}1^3 i^3 + \binom{6}{4}1^2 i^4 + \binom{6}{5}1i^5 + \binom{6}{6}i^6$

$$= 1 + 6i + 15i^2 + 20i^3 + 15i^4 + 6i^5 + i^6$$

$$= 1 + 6i - 15 - 20i + 15 + 6i - 1 = -8i$$

39. $g(x) = f(x+1) = (x+1)^3 - 2(x+1)^2 + (x+1) - 4$

$\qquad = x^3 + 3x^2 + 3x + 1 - 2x^2 - 4x - 2 + x + 1 - 4$

$\qquad = x^3 + x^2 - 4$

41. $g(x) = f(x-2) = (x-2)^4 + 3(x-2) + 5$

$\qquad = x^4 - 8x^3 + 24x^2 - 32x + 16 + 3x - 6 + 5$

$\qquad = x^4 - 8x^3 + 24x^2 - 29x + 15$

43. In this term, y has exponent 2, the binomial coefficient is $\binom{5}{2}$, and x has exponent 3.

$\binom{5}{2} x^3 y^2 = 10 x^3 y^2$

45. In this term, $-d$ has exponent 4, the binomial coefficient is $\binom{7}{4}$, and c has

exponent 3. $\binom{7}{4} c^3 (-d)^4 = 35 c^3 d^4$

47. In this term, $\frac{u}{2}$ has exponent 3, the binomial coefficient is $\binom{7}{3}$, and u^{-2} has

exponent 4. $\binom{7}{3}\left(\frac{u}{2}\right)^3 (u^{-2})^4 = 35\frac{u^3}{8}(u^{-8}) = \frac{35}{8u^5}$

49. This is the term with exponent 5 for x, 4 for $-y^2$, and binomial coefficient $\binom{9}{4}$.

$\binom{9}{4}(2x)^5 (-y^2)^4 = 126(32x^5)y^8 = 4032 x^5 y^8$. Coefficient: 4032

51. This term results from $(2x)^3$ and $\left(\frac{1}{x^2}\right)^3$ and thus has binomial coefficient $\binom{6}{3}$.

$\binom{6}{3}(2x)^3\left(\frac{1}{x^2}\right)^3 = 20(8)x^3\left(\frac{1}{x^6}\right) = \frac{160}{x^3}$. Coefficient: 160

53. $f(x) = x^5 + 5x^4 + 10x^3 + 10x^2 + 5x + 1$

$\qquad = \binom{5}{0}x^5 + \binom{5}{1}x^4 + \binom{5}{2}x^3 + \binom{5}{3}x^2 + \binom{5}{4}x + \binom{5}{5}$

$\qquad = (x+1)^5$

55. $16z^4 + 32z^3 + 24z^2 + 8z + 1 = (2z)^4 + 4(2z)^3 + 6(2z)^2 + 4(2z) + 1 = (2z+1)^4$

57. a. $\dbinom{9}{1} = \dfrac{9!}{1!8!} = \dfrac{9 \cdot 8!}{8!} = 9 \qquad \dbinom{9}{8} = \dfrac{9!}{8!1!} = 9$

b. $\dbinom{n}{1} = \dfrac{n!}{1!(n-1)!} = \dfrac{n(n-1)!}{(n-1)!} = n \qquad \dbinom{n}{n-1} = \dfrac{n!}{(n-1)!1!} = \dfrac{n \cdot (n-1)!}{(n-1)!} = n$

59. Since $2 = 1+1$,

$$2^n = (1+1)^n = \dbinom{n}{0}1^n + \dbinom{n}{1}1^{n-1} \cdot 1 + \dbinom{n}{2}1^{n-2} \cdot 1^2 + \ldots + \dbinom{n}{n}1^n = \dbinom{n}{0} + \dbinom{n}{1} + \dbinom{n}{2} + \ldots + \dbinom{n}{n}$$

61. $(\cos\theta + i\sin\theta)^4 = \dbinom{4}{0}(\cos\theta)^4 + \dbinom{4}{1}(\cos\theta)^3(i\sin\theta) + \dbinom{4}{2}(\cos\theta)^2(i\sin\theta)^2$

$$+ \dbinom{4}{3}(\cos\theta)(i\sin\theta)^3 + \dbinom{4}{4}(i\sin\theta)^4$$

$$= \cos^4\theta + 4\cos^3\theta(i\sin\theta) + 6\cos^2\theta(-\sin^2\theta)$$

$$+ 4\cos\theta(-i\sin^3\theta) + \sin^4\theta$$

$$= \cos^4\theta - 6\cos^2\theta\sin^2\theta + \sin^4\theta + i(4\cos^3\theta\sin\theta - 4\cos\theta\sin^3\theta)$$

63. a. $f(x+h) - f(x) = (x+h)^5 - x^5$

$$= \dbinom{5}{0}x^5 + \dbinom{5}{1}x^4h + \dbinom{5}{2}x^3h^2 + \dbinom{5}{3}x^2h^3 + \dbinom{5}{4}xh^4 + \dbinom{5}{5}h^5 - x^5$$

$$= \dbinom{5}{1}x^4h + \dbinom{5}{2}x^3h^2 + \dbinom{5}{3}x^2h^3 + \dbinom{5}{4}xh^4 + \dbinom{5}{5}h^5$$

b. $\dfrac{f(x+h) - f(x)}{h} = \dfrac{h\left[\dbinom{5}{1}x^4 + \dbinom{5}{2}x^3h + \dbinom{5}{3}x^2h^2 + \dbinom{5}{4}xh^3 + \dbinom{5}{5}h^4\right]}{h}$

$$= \dbinom{5}{1}x^4 + \dbinom{5}{2}x^3h + \dbinom{5}{3}x^2h^2 + \dbinom{5}{4}xh^3 + \dbinom{5}{5}h^4$$

c. If h is very close to 0, the above quantity is very close to $\dbinom{5}{1}x^4$ or $5x^4$.

65. a. $f(x+h) - f(x) = (x+h)^{12} - x^{12}$

$$= \binom{12}{0}x^{12} + \binom{12}{1}x^{11}h + \binom{12}{2}x^{10}h^2 + \cdots + \binom{12}{12}h^{12} - x^{12}$$

$$= \binom{12}{1}x^{11}h + \binom{12}{2}x^{10}h^2 + \cdots + \binom{12}{12}h^{12}$$

b. $\dfrac{f(x+h) - f(x)}{h} = \dfrac{h\left[\binom{12}{1}x^{11} + \binom{12}{2}x^{10}h + \cdots + \binom{12}{12}h^{11}\right]}{h}$

$$= \binom{12}{1}x^{11} + \binom{12}{2}x^{10}h + \cdots + \binom{12}{12}h^{11}$$

c. If h is very close to 0, the above quantity is very close to $\binom{12}{1}x^{11}$ or $12x^{11}$.

67. a. $f(x+10)$ adds 10 to each input; hence, if $x = 0$, $f(x+10) = f(10)$, that is, if $x = 0$, the value of the function for $1990 + 10 = 2000$ is calculated.

b. The graph of $g(x) = f(x+10)$ is the graph of $f(x)$ shifted to the left 10 units.

c. $g(x) = f(x+10) = -.052(x+10)^3 + .7(x+10)^2 + 42.14(x+10) + 213.4$

$$= -.052(x^3 + 30x^2 + 300x + 1000) + .7(x^2 + 20x + 100) + 42.14(x+10) + 213.4$$

$$= -.052x^3 - .86x^2 + 40.54x + 652.8$$

69. In general, $a! = a(a-1)(a-2)...1 = a(a-1)!$

a. Thus, $(n-r)! = (n-r)[(n-r)-1]! = (n-r)[n-(r+1)]!$

b. $(n-r)! = (n+1-r-1)! = [(n+1)-(r+1)]!$

c. $\binom{n}{r+1} + \binom{n}{r} = \dfrac{n!}{(r+1)![n-(r+1)]!} + \dfrac{n!}{r!(n-r)!}$

$= \dfrac{(n-r)n!}{(r+1)!(n-r)[n-(r+1)]!} + \dfrac{(r+1)n!}{(r+1)r!(n-r)!}$

$= \dfrac{(n-r)n!}{(r+1)!(n-r)!} + \dfrac{(r+1)n!}{(r+1)!(n-r)!}$

$= \dfrac{[(n-r)+(r+1)]n!}{(r+1)!(n-r)!}$

$= \dfrac{(n+1)n!}{(r+1)![(n+1)-(r+1)]!}$

$= \dfrac{(n+1)!}{(r+1)![(n+1)-(r+1)]!} = \binom{n+1}{r+1}$

d. Since each entry in row n has form $\binom{n}{k}$, the sum of two adjacent entries is

$\binom{n}{r+1} + \binom{n}{r}$. These are the entries above and to the right and left of $\binom{n+1}{r+1}$.

This explains the assertion.

71. First note that all the terms of $(1+.001)^{1000}$ are all positive. Next, notice that the first

two terms are $1^{1000} + \binom{1000}{1} \cdot 1^{999} \cdot (.001) = 1 + 1000(.001) = 1+1 = 2$. Hence, the entire

sum is greater than 2.

12.5 Mathematical Induction

1. When $n = 1$, both sides of the equation are equal to 1, thus the statement is true. Assume that the statement is true for $n = k$. Then
    $$1 + 2 + 2^2 + 2^3 + \cdots + 2^{k-1} = 2^k - 1.$$ Add 2^k to both sides of the equation to get:
    $$1 + 2 + 2^2 + 2^3 + \cdots + 2^{k-1} + 2^k = 2^k - 1 + 2^k = 2\left(2^k\right) - 1 = 2^{k+1} - 1$$
    Thus the statement is true for $n = k + 1$. Therefore, by the Principle of Mathematical Induction the statement is true for all positive integers n.

3. When $n = 1$, both sides of the equation are equal to 1, thus the statement is true. Assume that the statement is true for $n = k$. Then
    $$1 + 3 + 5 + 7 + \cdots + (2k - 1) = k^2.$$ Add the next odd integer, $(2k + 1)$ to both sides:
    $$1 + 3 + 5 + 7 + \cdots + (2k - 1) + (2k + 1) = k^2 + (2k + 1) = (k + 1)^2.$$ Thus the statement is true for $n = k + 1$. Therefore, by the Principle of Mathematical Induction the statement is true for all positive integers n. That is, the sum of the first n odd integers is n^2.

5. When $n = 1$, both sides of the equation are equal to $\frac{1}{2}$, thus the statement is true. Assume that the statement is true for $n = k$. Then
    $$\frac{1}{2} + \frac{1}{4} + \frac{1}{8} + \cdots + \frac{1}{2^k} = 1 - \frac{1}{2^k}.$$ Add $\frac{1}{2^{k+1}}$ to both sides of the equation:
    $$\frac{1}{2} + \frac{1}{4} + \frac{1}{8} + \cdots + \frac{1}{2^k} + \frac{1}{2^{k+1}} = 1 - \frac{1}{2^k} + \frac{1}{2^{k+1}} = 1 + \frac{-2 + 1}{2^{k+1}} = 1 - \frac{1}{2^{k+1}}.$$ Thus the statement is true for $n = k + 1$. Therefore the statement is true for all positive integers n.

7. When $n = 1$, both sides of the equation are equal to $\frac{1}{2}$, thus the statement is true. Assume that the statement is true for $n = k$. Then
    $$\frac{1}{1 \cdot 2} + \frac{1}{2 \cdot 3} + \frac{1}{3 \cdot 4} + \cdots + \frac{1}{k(k+1)} = \frac{k}{k+1}.$$ Add the next term $\dfrac{1}{(k+1)(k+2)}$ to both sides of the equation:
    $$\frac{1}{1 \cdot 2} + \frac{1}{2 \cdot 3} + \frac{1}{3 \cdot 4} + \cdots + \frac{1}{k(k+1)} + \frac{1}{(k+1)(k+2)} = \frac{k}{k+1} + \frac{1}{(k+1)(k+2)}$$
    $$= \frac{(k+1)^2}{(k+1)(k+2)} = \frac{k+1}{k+2}$$
    Thus the statement is true for $n = k + 1$. Therefore the statement is true for all positive integers n.

9. When $n = 1$, we have $1 + 2 > 1$, which is true. Assume that the statement is true for $n = k$. Then $k + 2 > k$. Add 1 to both sides of the inequality to get $k + 2 + 1 > k + 1$, $(k + 1) + 2 > (k + 1)$ and the statement is true for $n = k + 1$. Therefore, the statement $n + 2 > n$ is true for all positive integers n.

11. When $n = 1$, we have $3^1 \geq 3(1)$, which is true. Assume that the statement is true for $n = k$. Then $3^k \geq 3(k)$. Multiply both sides of the inequality by 3 to get $3(3^k) \geq 3(3k), 3^{k+1} \geq 9k, 3^{k+1} \geq 3k + 6k$. Since k is a positive integer, $6k > 3$ and the statement becomes $3^{k+1} \geq 3k + 3, 3^{k+1} \geq 3(k+1)$. Thus the statement is true for $n = k + 1$. Therefore, $3^n \geq 3n$ is true for all positive integers n.

13. When $n = 1$, we have $3(1) > 1 + 1$, which is true. Assume that the statement is true for $n = k$. Then $3k > k + 1$. Add 3 to both sides of the inequality to get $3k + 3 > k + 4$, and since $k + 4 > k + 2$ we have $3k + 3 > k + 2, 3(k+1) > (k+1) + 1$ and the statement is true for $n = k + 1$. Therefore, the statement $3n > n + 1$ is true for all positive integers n.

15. When $n = 1$, $2^{2n+1} + 1 = 2^3 + 1 = 9$ and the statement is true since 3 is a factor of 9. Assume that the statement is true for $n = k$. Then 3 is a factor of $2^{2k+1} + 1$. Thus there is an integer p we can multiply by 3 to get $2^{2k+1} + 1$. That is $3p = 2^{2k+1} + 1$ or $3p - 1 = 2^{2k+1}$. Now consider:
$$2^{2(k+1)+1} + 1 = 2^{2k+3} + 1 = 2^{2k+1}2^2 + 1 = (3p-1)2^2 + 1 = 12p - 3 = 3(4p-1).$$
Thus, 3 is a factor of $2^{2(k+1)+1} + 1$ and the statement is true for $n = k + 1$. Thus, 3 is a factor of $2^{2n+1} + 1$ for all positive integers n.

17. When $n = 1$, $3^{2n+2} - 8n - 9 = 64$ and the statement is true since 64 is a factor of 64. Assume that the statement is true for $n = k$. Then 64 is a factor of $3^{2k+2} - 8k - 9$. Thus there is an integer p we can multiply by 64 to get $3^{2k+2} - 8k - 9$. That is $64p = 3^{2k+2} - 8k - 9$ or $3^{2k+2} = 64p + 8k + 9$. Now consider:
$$3^{2(k+1)+2} - 8(k+1) - 9 = 3^{2k+4} - 8k - 17 = 9(3^{2k+2}) - 8k - 17$$
$$= 9(64p + 8k + 9) - 8k - 17$$
$$= 64(9p + k + 1)$$
Therefore, 64 is a factor of $3^{2(k+1)+2} - 8(k+1) - 9$ and the statement is true for $n = k + 1$. Thus, 64 is a factor of $3^{2n+2} - 8n - 9$ for all positive integers n.

19. When $n = 1$, both sides of the equation are equal to c and the statement is true. Assume that the statement is true for $n = k$. Then
$$c + (c+d) + (c+2d) + (c+3d) + \cdots + + (c + (k-1)d) = \frac{k(2c + (k-1)d)}{2}$$
Add $c + kd$ to both sides of the equation to obtain:
$$c + (c+d) + (c+2d) + \ldots + (c+(k-1)d) + (c+kd) = \frac{k(2c+(k-1)d)}{2} + (c+kd)$$
$$= \frac{2kc + k(k-1)d + 2c + 2kd}{2} = \frac{2c(k+1) + (k+1)kd}{2} = \frac{(k+1)(2c+kd)}{2}$$
and the statement is true for $n = k + 1$. Thus it is true for all positive integers n.

21. a. $x^2 - y^2 = (x - y)(x + y)$

$x^3 - y^3 = (x - y)(x^2 + xy + y^2)$

$x^4 - y^4 = (x - y)(x^3 + x^2 y + xy^2 + y^3)$

b. $x^n - y^n = (x - y)(x^{n-1} + x^{n-2}y + x^{n-3}y^2 + \ldots + y^{n-1})$

From part **a** we see the statement is true for $n = 2$. Assume that the statement is true for $n = k$. Then $x^k - y^k = (x - y)(x^{k-1} + x^{k-2}y + x^{k-3}y^2 + \ldots + y^{k-1})$.

Consider $x^{k+1} - y^{k+1} = xx^k - yy^k = xx^k - xy^k + xy^k - yy^k$

$= x(x^k - y^k) + (x - y)y^k$

$= x(x - y)(x^{k-1} + x^{k-2}y + x^{k-3}y^2 + \ldots + y^{k-1}) + (x - y)y^k$

$= (x - y)\left[x(x^{k-1} + x^{k-2}y + x^{k-3}y^2 + \ldots + y^{k-1}) + y^k\right]$

$= (x - y)(x^k + x^{k-1}y + x^{k-2}y^2 + \ldots + xy^{k-1} + y^k)$

and the statement is true for $n = k + 1$. Thus it is true for all positive integers n.

23. This statement is false. For example 9 is an odd positive integer and it is not a prime.

25. When $n = 1$ the statement is true. Assume that the statement is true for $n = k$. Then $(k + 1)^2 > k^2 + 1$. Then

$(k + 2)^2 = k^2 + 4k + 4 = k^2 + 2k + 1 + 2k + 3 = (k + 1)^2 + (2k + 3)$

$> k^2 + 1 + (2k + 3) = k^2 + 2k + 1 + 3 = (k + 1)^2 + 1 + 3$

$> (k + 1)^2 + 1$ and the statement is true for $n = k + 1$. Thus $(n + 1)^2 > n^2 + 1$ is true for all positive integers n.

27. This statement is false, counterexample $n = 2, n^4 - n + 4 = 18$ and 4 is not a factor of 18.

29. When $n = 5$, we have $2(5) - 4 > 5$, which is true. Assume that the statement is true for $n = k$ where $k \geq 5$. Then $2k - 4 > k$.

And $2(k + 1) - 4 = 2k - 4 + 2 > k + 2 > k + 1$. Thus the statement is true for $n = k + 1$. Therefore by induction the statement is true for all $n \geq 5$.

31. When $n = 2$, we have $2^2 > 2$, which is true. Assume that the statement is true for $n = k$ where $k \geq 2$. Then $k^2 > k$. Then $(k + 1)^2 = k^2 + 2k + 1 > k + 2k + 1 > k + 1$.

Then the statement is true for $n = k + 1$. Thus $n^2 > n$ for all $n \geq 2$.

33. When $n = 4$, we have $3^4 > 2^4 + 10(4)$, which is true. Assume that the statement is true for $n = k$ where $k \geq 4$. Then $3^k > 2^k + 10k$. And
$$3^{k+1} = 3 \cdot 3^k > 3(2^k + 10k) = 3 \cdot 2^k + 30k > 2 \cdot 2^k + 30k > 2^{k+1} + 10k + 10$$
$$= 2^{k+1} + 10(k+1).$$
Therefore the statement is true for $n = k + 1$. Thus $3^n > 2^n + 10n$ for all $n \geq 4$.

35. a. When $n = 2$, you can move the stack in 3 moves. If you start on peg 1, you move the the top ring to peg 2, then the bottom ring to peg 3 and the the top ring on top of the bottom ring.
When $n = 3$, you can move the stack in 7 moves. It takes 3 moves to put the top 2 rings onto peg 2, then the fourth move puts the bottom ring to peg 3 and then 3 more moves to move the top two on top of the bottom. That is $3 + 1 + 3 = 7$.
When $n = 4$, it takes 15 moves, 7 to move the top three rings to peg 2, 1 to move the bottom ring to peg 3 and 7 more to put the top three on the bottom. That is $7 + 1 + 7 = 15$.

b. You can move 1 ring in one move, 2 rings in three moves, 3 rings in seven moves, 4 rings in fifteen moves. Thus we conjecture that it takes $2^n - 1$ moves to move n rings. Clearly this holds true for $n = 1$. Assume that the statement is true for $n = k$. That is, it takes $2^k - 1$ moves to move k rings. Now consider $k + 1$ rings. You can move the top k rings in $2^k - 1$ moves, one move to move the bottom ring, and $2^k - 1$ moves to move the top k rings on the bottom. This gives a total of $2^k - 1 + 1 + 2^k - 1 = 2(2^k) - 1 = 2^{k+1} - 1$ moves. Thus the conjecture is true for $n = k + 1$. It will take $2^n - 1$ moves to move n rings for all positive integers n.

37. The statement is true for $n = 1$. Assume that the statement is true for $n = k$. Then $z^k = r^k[\cos(k\theta) + i\sin(k\theta)]$. And
$$z^{k+1} = z \cdot z^k = z\left[r^k(\cos(k\theta) + i\sin(k\theta))\right] = r(\cos\theta + i\sin\theta)r^k(\cos(k\theta) + i\sin(k\theta))$$
$$= r^{k+1}\left[\cos\theta\cos(k\theta) + i^2\sin\theta\sin(k\theta) + i\cos\theta\sin(k\theta) + i\sin\theta\cos(k\theta)\right]$$
$$= r^{k+1}\left[\cos\theta\cos(k\theta) - \sin\theta\sin(k\theta) + i(\cos\theta\sin(k\theta) + \sin\theta\cos(k\theta))\right]$$
$$= r^{k+1}\left[\cos(\theta + k\theta) + i\sin(\theta + k\theta)\right] = r^{k+1}\left[\cos((k+1)\theta) + i\sin((k+1)\theta)\right]$$
Therefore the statement is true for $n = k + 1$. Hence the statement is true for all positive integers n.

Chapter 12 Review Exercises

1. $a_1 = 2 \cdot 1 - 5 = -3$
$a_2 = 2 \cdot 2 - 5 = -1$
$a_3 = 2 \cdot 3 - 5 = 1$
$a_4 = 2 \cdot 4 - 5 = 3$

3. $a_1 = \left(\frac{-1}{1}\right)^2 = 1$
$a_2 = \left(\frac{-1}{2}\right)^2 = \frac{1}{4}$
$a_3 = \left(\frac{-1}{3}\right)^2 = \frac{1}{9}$
$a_4 = \left(\frac{-1}{4}\right)^2 = \frac{1}{16}$

5. $\left(\dfrac{2}{5}\right)^n$

7. a. $2004 - n = 4; a_4 = \$4.444$ billion
$2006 - n = 6; a_6 = \$5.374$ billion
$2010 - n = 10; a_{10} = \$7.150$ billion

b. Solve $9.6 = -.0035x^2 + .5x + 2.5$
$0 = -.0035x^2 + .5x - 7.1$
$x = 16$
Which corresponds to 2016.

9. $\displaystyle\sum_{n=1}^{9} \frac{n}{n+1}$

11. $\displaystyle\sum_{n=2}^{4}\left(3n^2 - n + 1\right) = \left(3 \cdot 2^2 - 2 + 1\right) + \left(3 \cdot 3^2 - 3 + 1\right) + \left(3 \cdot 4^2 - 4 + 1\right) = 11 + 25 + 45 = 81$

13. This is a geometric sequence with
$a_1 = \frac{1}{9}, r = 3$
$$\sum_{n=1}^{4} a_n = a_1\left(\frac{1 - r^4}{1 - r}\right)$$
$$= \frac{1}{9}\left(\frac{1 - 3^4}{1 - 3}\right) = \frac{40}{9}$$

15. a. 2003-2004 — $n = 14; b_{14} = \$25,472$

b. $\displaystyle\sum_{n=0}^{3} b_n = 11,561.7 + 12,232.8 + 12,942.8 + 13,694.0 \approx \$50,431$

c. $\displaystyle\sum_{n=14}^{17} b_n = 25,472.0 + 26,950.4 + 28,514.6 + 30,169.7 \approx \$111,107$

17. $a_n = a_1 + (n-1)d$

$a_n = 3 + (n-1)(-6)$

$a_n = 9 - 6n$

19. $a_3 = a_1 + 2d$

$d = \dfrac{a_3 - a_1}{2} = \dfrac{7 - (-5)}{2} = 6$

$a_n = a_1 + (n-1)d$

$a_n = -5 + (n-1)6$

$a_n = 6n - 11$

21. $\displaystyle\sum_{n=1}^{k} a_n = ka_1 + \dfrac{k(k-1)}{2}d$

$\displaystyle\sum_{n=1}^{11} a_n = 11\cdot 5 + \dfrac{11\cdot 10}{2}(-2) = -55$

23. $a_n = a_1 r^{n-1}$

$a_n = 2\cdot 3^{n-1}$

25. $a_2 = a_1 r, \ a_7 = a_2 r^5$

$r^5 = \dfrac{a_7}{a_2} = \dfrac{6}{192} = \dfrac{1}{32}$

$r = \dfrac{1}{2}, \ a_1 = \dfrac{a_2}{r} = \dfrac{192}{1/2} = 384$

$a_n = a_1 r^{n-1} = 384\left(\tfrac{1}{2}\right)^{n-1} = \dfrac{3}{2^{n-8}}$

27. $\displaystyle\sum_{n=1}^{5} a_n = a_1\left(\dfrac{1-r^5}{1-r}\right) = \dfrac{1}{4}\left(\dfrac{1-3^5}{1-3}\right) = \dfrac{121}{4}$

29. Since $a_1 = 4, a_5 = 23, a_5 = a_1 + 4D, D = \dfrac{a_5 - a_1}{4} = \dfrac{19}{4}$. (using D = common difference)

Then $b = a_2 = 4 + \dfrac{19}{4} = \dfrac{35}{4}, c = a_3 = \dfrac{35}{4} + \dfrac{19}{4} = \dfrac{27}{2}, d = a_4 = \dfrac{27}{2} + \dfrac{19}{4} = \dfrac{73}{4}$.

31. a. $a_1 = 7.898, a_2 = 8.972$

$r = a_2 \div a_1 = 8.972 \div 7.898 = 1.136$

$a_n = 7.898(1.136)^{n-1}$

b. $2005 - n = 5; a_5 = 7.898(1.136)^4 = \$13.153$ billion

c. $\displaystyle\sum_{n=1}^{6} a_n = 7.898\left(\dfrac{1-1.136^6}{1-1.136}\right) = \$66.736$ billion

33. a. Follow example 8 of section 12.3 to obtain:

$$b_n = 100\left[\frac{1-e^{-1.3863n(2)}}{1-e^{-1.3863(2)}}\right] = 100\left[\frac{1-e^{-2.7726n}}{1-e^{-2.7726}}\right]$$

b. $b_5 = 100\left[\dfrac{1-e^{-2.7726(5)}}{1-e^{-2.7726}}\right] = 106.6665$ mg

c. To find the approximate steady state amount of furocimide when the doses are given every two hours, calculate with larger and larger n values:

$$b_5 = 100\left[\frac{1-e^{-2.7726(5)}}{1-e^{-2.7726}}\right] = 106.6665 \text{ mg}$$

$$b_6 = 100\left[\frac{1-e^{-2.7726(6)}}{1-e^{-2.7726}}\right] = 106.6666 \text{ mg}$$

$$b_{10} = 100\left[\frac{1-e^{-2.7726(10)}}{1-e^{-2.7726}}\right] = 106.6666 \text{ mg}$$

Therefore, the approximate steady state amount of furocimide when the doses are given every two hours is 106.6666 mg.

35. This is a geometric series with $a_1 = 1$ and $r = \frac{1}{2}$. $\dfrac{a_1}{1-r} = \dfrac{1}{1-\frac{1}{2}} = 2$

37. $\displaystyle\binom{15}{12} = \frac{15!}{12!\,3!} = \frac{15\cdot14\cdot13\cdot12!}{12!\,3\cdot2\cdot1} = \frac{15\cdot14\cdot13}{3\cdot2\cdot1} = 455$

39. $\dfrac{20!\,5!}{6!\,17!} = \dfrac{20\cdot19\cdot18\cdot17!\,5!}{6\cdot5!\,17!} = \dfrac{20\cdot19\cdot18}{6} = 1140$

41. $\displaystyle\binom{n+1}{n} = \frac{(n+1)!}{n!\,(n+1-n)!} = \frac{(n+1)!}{n!\,1!} = \frac{(n+1)n!}{n!\cdot1} = n+1$

43. $f(x) = x^3 - 2x^2 + 3x + 1$

$g(x) = f(x-1) = (x-1)^3 - 2(x-1)^2 + 3(x-1) + 1$

$\qquad = x^3 - 3x^2 + 3x - 1 - 2x^2 + 4x - 2 + 3x - 3 + 1$

$\qquad = x^3 - 5x^2 + 10x - 5$

45. This term results from $(2y)^4$ and $(x^2)^1$ and thus has binomial coefficient $\displaystyle\binom{5}{1}$.

$\displaystyle\binom{5}{1}(2y)^4(x^2)^1 = 5\cdot16y^4x^2 = 80x^2y^4$. Coefficient: 80

47. When $n = 1$, both sides of the equation are equal to 1, thus the statement is true. Assume that the statement is true for $n = k$. Then

$1 + 5^1 + 5^2 + \ldots + 5^{k-1} = \dfrac{5^k - 1}{4}$. Add the 5^k to both sides of the equation:

$1 + 5^1 + 5^2 + \ldots + 5^{k-1} + 5^k = \dfrac{5^k - 1}{4} + 5^k = \dfrac{5^k - 1 + 4 \cdot 5^k}{4} = \dfrac{5 \cdot 5^k - 1}{4} = \dfrac{5^{k+1} - 1}{4}$. Thus the

statement is true for $n = k + 1$. And by induction it is true for all positive integers n.

49. Clearly the statement is true for $x = 0$, hence assume $x \neq 0$. Then the statement is true for $n = 1$. Assume that the statement is true for $n = k$, then $\left| x^k \right| < 1$. Multiply both sides of this inequality by $|x|$ to obtain $\left| x^k \right| \cdot |x| < 1 \cdot |x|$. Thus $\left| x^{k+1} \right| = \left| x^k \right| \cdot |x| < |x| < 1$. Thus the statement is true for $n = k + 1$. Thus, it is true for all positive integers n.

51. When $n = 1$, both sides of the equation are equal to 1, thus the statement is true. Assume that the statement is true for $n = k$. Then $1 + 4^1 + 4^2 + \ldots + 4^{k-1} = \dfrac{4^k - 1}{3}$.

Add 4^k to both sides: $1 + 4^1 + 4^2 + \ldots + 4^{k-1} + 4^k = \dfrac{4^k - 1}{3}^k + 4^k = \dfrac{4^{k+1} - 1}{3}$. Thus the

statement is true for $n = k + 1$. And, by induction it is true for all positive integers n.

53. When $n = 1$, $9^n - 8n - 1 = 0$ and the the statement is true since 8 is a factor of 0. Assume that the statement is true for $n = k$. Then 8 is a factor of $9^k - 8k - 1$. That is, there is an integer p that we can multiply by 8 to get $9^k - 8k - 1$. Thus $8p = 9^k - 8k - 1$ or $8p + 8k + 1 = 9^k$. Consider $9^{k+1} - 8(k + 1) - 1$. Rearrange this expression to get $9\left(9^k\right) - 8k - 9 = 9(8p + 8k + 1) - 8k - 9 = 72p + 64k = 8(9p + 8k)$ and so 8 is a factor of $9^{k+1} - 8(k + 1) - 1$. Thus the statement is true for $n = k + 1$. Therefore, by induction the statement is true for all positive integers n.

Chapter 12 Test

1. $a_1 = (-1)^{1+2} - (1+7) = -9$

    $a_2 = (-1)^{2+2} - (2+7) = -8$

    $a_3 = (-1)^{3+2} - (3+7) = -11$

    $a_4 = (-1)^{4+2} - (4+7) = -10$

    $a_5 = (-1)^{5+2} - (5+7) = -13$

2. **a.** \$3782   **b.** $3412 + 3597 + 3782 + 3967 + 4152 + 4337 = \$23,247$

3. **a.** $c_5 = c_1 r^{5-1}$ **b.** Use $c_n = r^{n-1} c_1$ to obtain:

    $$\frac{2}{3} = c_1 \left(-\frac{1}{3} \right)^4 \qquad\qquad \frac{2}{3} = \left(-\frac{1}{3} \right)^{5-1} c_1; \ \frac{2}{3} = \left(-\frac{1}{3} \right)^4 c_1$$

    $$\frac{2}{3} = c_1 \left(\frac{1}{81} \right)^4 \qquad\qquad \frac{2}{3} = \left(\frac{1}{81} \right) c_1$$

    $$c_1 = 54 \qquad\qquad\qquad\quad c_1 = 54$$

    $$c_6 = 54 \left(-\frac{1}{3} \right)^5 = -\frac{2}{9}$$

4. $\displaystyle\sum_{k=1}^{25} \left(2k^2 - 4k + 3 \right) = 9825$

5. The sequence has $a_n = \log 2^{n+1} = (n+1)\log 2$. So $a_n - a_{n-1} = (n+1)\log 2 - n\log 2 = n\log 2 + \log 2 - n\log 2 = \log 2$. Hence, the sequence is arithmetic with common difference of $\log 2$.

6. $\dfrac{4^{2-4}}{4^{1-4}} = \dfrac{4^{-2}}{4^{-3}} = 4; \ \dfrac{4^{3-4}}{4^{2-4}} = 4; \ \dfrac{4^{4-4}}{4^{3-4}} = 4; \dots; \dfrac{4^{n-4}}{4^{(n-1)-4}} = 4$

    Thus the sequence is geometric with a common ratio of 4.

7. $a_1 = 1$

    $a_2 = 3$

    $a_3 = 2a_2 + 3a_1 = 2(3) + 3(1) = 9$

    $a_4 = 2a_3 + 3a_2 = 2(9) + 3(3) = 27$

    $a_5 = 2a_4 + 3a_3 = 2(27) + 3(9) = 81$

8. $a_n - a_{n-1} = (9b + 8cn) - (9b + 8c(n-1)) = 9b + 8cn - 9b - 8cn + 8c = 8c.$ So   the sequence is arithmetic with common difference $8c$.

9. **a.** $320 = r^6 a_1$; $10 = ra_1 \Rightarrow a_1 = \dfrac{10}{r}$ Substituting a_1 into the first equation and solving

for r gives $320 = r^6 \left(\dfrac{10}{r}\right)$; $320 = 10r^5$; $32 = r^5$; $\sqrt[5]{32} = r = 2.$ Then use the value of

r to find $a_1 : 10 = 2a_1$; $a_1 = 5.$ Therefore the formula is $a_n = 2^{n-1}(5).$

b. $a_5 = (5)2^{5-1} = (5)2^4 = 80$

10. The third partial sum is $\displaystyle\sum_{n=1}^{3} \left(2n - 3n^2\right)^2 = 506.$

The sixth partial sum is $\displaystyle\sum_{n=1}^{6} \left(2n - 3n^2\right)^2 = 15{,}547.$

11. $\displaystyle\sum_{n=1}^{8} b_n = \dfrac{8}{2}(0 + 34) = 136$

12. $\displaystyle\sum_{n=1}^{7} c_n = 18\left(\dfrac{1 - \left(\frac{1}{2}\right)^7}{1 - \left(\frac{1}{2}\right)}\right) = \dfrac{1143}{32} \approx 36$

13. $\displaystyle\sum_{k=12}^{16} (-2)^k$

14. $\displaystyle\sum_{n=1}^{40} \dfrac{7 - 8n}{4} = -1570$

15. On the first bounce it rises $10(.44) = 4.4$ feet and falls 4.4 feet for a total of 8.8 feet. On the second bounce it rises and falls 44% of the previous height and hence travels 44% of 8.8 feet. If a_n denotes the distance traveled on the nth bounce, then

$a_1 = 8.8,$ $a_2 = 8.8(.44),$ $a_3 = 8.8(.44)^2$

and in general, $a_n = 8.8(.44)^{n-1}.$ So, the total distance traveled is

$$10 + \sum_{n=1}^{10} a_n = 10 + a_1\left(\dfrac{1 - r^{10}}{1 - r}\right) = 10 + 8.8\left(\dfrac{1 - (.44)^{10}}{1 - .44}\right) = 25.70 \text{ feet.}$$

16. **a.** 2004: $a_7 = 258{,}205(1.07)^7 \approx 414{,}621$ people

2007: $a_{10} = 258{,}205(1.07)^{10} \approx 507{,}928$ people

b. 2010: $a_{13} = 258{,}205(1.07)^{13} \approx 622{,}234$ people

17. $\displaystyle\sum_{n=1}^{100} 8n = 8\sum_{n=1}^{100} n = 8\left[\dfrac{n}{2}(a_1 + a_{100})\right] = 8\left[\dfrac{100}{2}(1 + 100)\right] = 40{,}400$

18. $\displaystyle\sum_{n=1}^{\infty}\left[\frac{1}{2}\left(\frac{1}{5}\right)^{n-1}\right]=\frac{\frac{1}{2}}{1-\frac{1}{5}}=\frac{5}{8}$

19. $\dbinom{5}{0}-\dbinom{5}{1}+\dbinom{5}{2}-\dbinom{5}{3}+\dbinom{5}{4}-\dbinom{5}{5}=1-5+10-10+5-1=0$

20. $f(x)=x^6+x+1$

$g(x)=f(x+1)=(x+1)^6+(x+1)+1$
$=x^6+6x^5+15x^4+20x^3+15x^2+6x+1+x+1+1$
$=x^6+6x^5+15x^4+20x^3+15x^2+7x+3$

21. When $n=1,$ both sides of the equation are equal to 10, thus the statement is true. Assume that the statement is true for $n=k,$ so that

$10+12+14+...+2(k+4)=k^2+9k.$ Add $2(k+5)$ to both sides:

$10+12+14+...+2(k+4)+2(k+5)=k^2+9k+2(k+5)$
$=k^2+9k+2k+10$
$=(k^2+2k+1)+9k+9$
$=(k+1)^2+9(k+1).$

So the statement is true for $n=k+1.$ By induction, the statement is true for every positive integer $n.$

22. $(4u-v^3)^6=\dbinom{6}{0}(4u)^6+\dbinom{6}{1}(4u)^5(-v^3)+\dbinom{6}{2}(4u)^4(-v^3)^2+\dbinom{6}{3}(4u)^3(-v^3)^3$

$+\dbinom{6}{4}(4u)^2(-v^3)^4+\dbinom{6}{5}(4u)(-v^3)^5+\dbinom{6}{6}(-v^3)^6$

$=4096u^6-6114u^5v^3+3840u^4v^6-1280u^3v^9+240u^2v^{12}-24uv^{15}+v^{18}$

23. The fifth term of the expansion is: $\dbinom{7}{4}(\sqrt{x})^3(-\sqrt{10})^4=35\cdot100(\sqrt{x})^3=3500x\sqrt{x}$

24. When $n=1,$ the statement is "4 is a factor of $5^1+3,$" which is true since $5^1+3=8=4\cdot2.$ Assume that the statement is true for $n=k,$ so that 4 is a factor of $5^k+3,$ say $5^k+3=4\cdot R.$ Then

$5^{k+1}+3=5\cdot5^k+3=4\cdot5^k+5^k+3=4+5^k+4R=4(5^k+R).$

Hence, 4 is a factor of $5^{k+1}.$ By induction, 4 is a factor of $5^{k+1}+3$ for every positive integer $n.$

25. By the definition of combination, $\binom{n}{1} = \dfrac{n!}{1!(n-1)!} = \dfrac{n(n-1)!}{(n-1)!} = n,$ and

$$\binom{n}{n-1} = \frac{n!}{(n-1)!(n-(n-1))!} = \frac{n!}{(n-1)!(1)!} = \frac{n(n-1)!}{(n-1)!} = n$$

26. When $n = 3$, $3^2 > 2(3) + 1$, which is true. Assume that the statement is true for $n = k$.

Then $k^2 > 2k + 1$. Then

$$(k+1)^2 = k^2 + 2k + 1 > (2k+1) + (2k+1) > 2k + 1 + 2 = 2(k+1) + 1.$$

Thus the statement is true for $n = k + 1$. By induction, the statement is true for all positive integers $n \geq 3$.

Chapter 13
Limits and Continuity

13.1 Limits of Functions

1. For the function $f(x) = \dfrac{x^6 - 1}{x^4 - 1}$, $f(-1.01) = 1.515$, $f(-1.0001) = 1.50015$, $f(-1.000001) = 1.5000015$. Also $f(-.99) = 1.485$, $f(-.9999) = 1.49985$. The limit seems to be $\frac{3}{2}$.

3. For the function $f(x) = \dfrac{\tan x - x}{x^3}$, $f(.01) = .3333$, $f(.001) = .333333$, $f(.0001) = .3333333$. Also $f(-.01) = .3333$, $f(-.001) = .333333$. The limit seems to be $\frac{1}{3}$.

5. For the function $f(x) = \dfrac{x - \tan x}{x - \sin x}$, $f(.01) = -2.001$, $f(.001) = -2.000001$, $f(.0001) = -1.999988$. Also $f(-.01) = -2.0001$, $f(-.001) = -2.000001$. The limit seems to be -2.

7. For the function $f(x) = \dfrac{x}{\ln|x|}$, $f(.01) = -.002$, $f(.001) = -.0001$, $f(.0001) = -.00001$. Also $f(-.01) = .002$, $f(-.001) = .0001$. The limit seems to be 0.

9. For the function $f(x) = \dfrac{e^x - 1}{\sin x}$, $f(.01) = 1.005$, $f(.001) = 1.0005$, $f(.0001) = 1.00005$. Also $f(-.01) = .995$, $f(-.001) = .9995$. The limit seems to be 1.

11. For the function $f(x) = \dfrac{\sin x}{1 - \cos x}$, choose values approaching π from the left. $f(3.1) = .0208$, $f(3.14) = .0008$, $f(3.141) = .0003$, $f(3.1415) = .00005$. The limit seems to be 0.

13. For the function $f(x) = \sqrt{x}\,\ln x$, choose values approaching 0 from the right. $f(.0001) = -.0921$, $f(.00001) = -.0364$, $f(.000001) = -.0138$, $f(.00000001) = -.0018$. The limit seems to be 0.

15. For the function $f(x) = \dfrac{\sin(6x)}{x}$, choose values approaching 0 from the left. $f(-.1) = 5.6464$, $f(-.01) = 5.9964$, $f(-.001) = 5.999964$, $f(-.0001) = 5.99999964$. The limit seems to be 6.

17. $\lim_{x\to-3} f(x)=-1,\ \lim_{x\to 0}f(x)=1,\ \lim_{x\to 2}f(x)$ does not exist.

19. $\lim_{x\to-3} f(x)=2,\ \lim_{x\to 0}f(x)$ does not exist, $\lim_{x\to 2} f(x)=0.$

21. $\lim_{x\to-3} f(x)=2,\ \lim_{x\to 0}f(x)=1,\ \lim_{x\to 2}f(x)=1.$

23. a. $\lim_{x\to-2^-} f(x)=0.$ **b.** $\lim_{x\to 0^+}f(x)=1.$ **c.** $\lim_{x\to 3^-} f(x)=-1.$ **d.** $\lim_{x\to 3^+} f(x)=-1.$

25. a. $\lim_{x\to-2^-} f(x)=1.$ **b.** $\lim_{x\to 0^+} f(x)=0.$
c. $\lim_{x\to 3^-} f(x)$ does not exist.   **d.** $\lim_{x\to 3^+} f(x)$ does not exist.

27. $\lim_{x\to 3} f(x)=\dfrac{11}{5}=f(3).$

29. $\lim_{x\to 2}\dfrac{x^2-5x+6}{x^2-6x+8}=\lim_{x\to 2}\dfrac{(x-2)(x-3)}{(x-2)(x-4)}=\lim_{x\to 2}\dfrac{(x-3)}{(x-4)}=\dfrac{1}{2}.$ $f(2)$ is not defined.

31. The limit does not exist.

33. $\lim_{x\to 2^+}\sqrt{x^2-4}=0=f(2).$

35. The limit does not exist.

37. For the function $f(x)=\dfrac{x}{e^x-1}$, $f(.01)=.995,\ f(.001)=.9995,\ f(.0001)=.99995.$
Also $f(-.01)=1.005, f(-.001)=1.0005.$ The limit seems to be 1. $f(0)$ is not defined.

39. For the function $f(x)=\dfrac{\ln x}{x-1}$, $f(1.01)=.995,\ f(1.001)=.9995,\ f(1.0001)=.99995.$
Also $f(.99)=1.005, f(.9999)=1.00005.$ The limit seems to be 1. $f(1)$ is not defined.

41. For the function $f(x)=x\ln|x|$, $f(.01)=-.046,\ f(.001)=-.0069,$
$f(.0001)=-.00092.$ Also $f(-.01)=.046, f(-.0001)=.00092.$ The limit seems to be 0. $f(0)$ is not defined.

43. $\lim_{x\to\pi/2} x\cos x=\dfrac{\pi}{2}\cdot 0=0=f\left(\dfrac{\pi}{2}\right).$

45. For the function $f(x)=\dfrac{|x-1|}{x-1}$, $f(.99)=-1,\ f(.999)=-1,\ f(.999)=-1.$

47. The limit does not exist.

49. $\lim\limits_{x\to 1} e^x \sin\dfrac{\pi x}{2} = e \cdot 1 = e = f(1)$.

51. a.

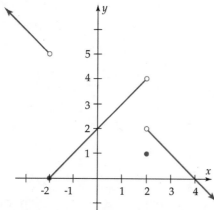

b. $\lim\limits_{x\to -2} f(x)$ does not exist.

c. $\lim\limits_{x\to 1} f(x) = 3$

d. $\lim\limits_{x\to 2} f(x)$ does not exist.

53. a.

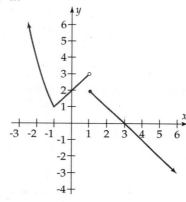

b. $\lim\limits_{x\to -1} g(x) = 1$

c. $\lim\limits_{x\to 0} g(x) = 2$

d. $\lim\limits_{x\to 1} g(x)$ does not exist.

55. The point P, where the terminal side of an angle of t radians in standard position meets the unit circle, has coordinates $(\cos t, \sin t)$. The picture shows that as t gets

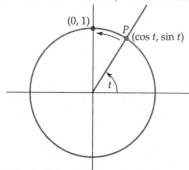

closer and closer to $\pi/2$, the terminal side gets very close to the y-axis and the point P moves along the unit circle toward the point $(0,1)$. Hence the second coordinate of P, $\sin t$, gets closer to 1.

57. For the function $h(x) = [\![x]\!] + [\![-x]\!]$, when x is close to 2, but slightly greater than 2, $[\![x]\!] = 2$ and $[\![-x]\!] = -3$. When x is close to 2, but slightly less than 2, $[\![x]\!] = 1$ and $[\![-x]\!] = -2$. In either case, for all values of x except 2 itself, $h(x) = -1$, so the limit of the function is -1.

59. When x is close to 2, but slightly greater than 2, $[\![x]\!]=2$, hence $\lim\limits_{x\to 2^+}[\![x]\!]=2$. When x is close to 2, but slightly less than 2, $[\![x]\!]=1$, hence $\lim\limits_{x\to 2^-}[\![x]\!]=1$.

61. Write $r(x)=\dfrac{h(x)}{x}$, where $h(x)$ is the function of Exercise **57**. Then:

$$\lim_{x\to 3}r(x)=\frac{\lim\limits_{x\to 3}h(x)}{\lim\limits_{x\to 3}x}=-\frac{1}{3}$$

63. The graph of the function is shown here using the window $-.2\le x\le.2, -1\le y\le 1$.

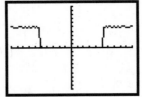

The trace function shows that all values near 0 are mapped to zero, which would suggest a value of the limit as 0 rather than the actual value of 0.5.

13.2 Properties of Limits

1. $\lim_{x \to 4}(f(x) + g(x)) = \lim_{x \to 4} f(x) + \lim_{x \to 4} g(x) = 5 + 0 = 5$

3. $\lim_{x \to 4} \dfrac{f(x)}{g(x)}$ does not exist because $\lim_{x \to 4} g(x) = 0$

5. $\lim_{x \to 4} f(x)g(x) = \lim_{x \to 4} f(x) \lim_{x \to 4} g(x) = 5 \cdot 0 = 0$

7. $\lim_{x \to 4} \dfrac{3h(x)}{2f(x) + g(x)} = \dfrac{3 \lim_{x \to 4} h(x)}{2 \lim_{x \to 4} f(x) + \lim_{x \to 4} g(x)} = \dfrac{3(-2)}{2 \cdot 5 + 0} = -\dfrac{3}{5}$

9. This is a polynomial function, hence $\lim_{x \to 2} f(x) = f(2) = 6 \cdot 2^3 - 2 \cdot 2^2 + 5 \cdot 2 - 3 = 47$

11. This is a rational function, $f(-2)$ is defined, hence

$\lim_{x \to -2} f(x) = f(-2) = \dfrac{3(-2) - 1}{2(-2) + 3} = 7$

13. $\lim_{x \to 3} \dfrac{x^2 - x - 6}{x^2 - 2x - 3} = \lim_{x \to 3} \dfrac{(x - 3)(x + 2)}{(x - 3)(x + 1)} = \dfrac{3 + 2}{3 + 1} = \dfrac{5}{4}$

15. $\lim_{x \to 1} \dfrac{x^3 - 1}{x^2 - 1} = \lim_{x \to 1} \dfrac{(x - 1)(x^2 + x + 1)}{(x - 1)(x + 1)} = \lim_{x \to 1} \dfrac{x^2 + x + 1}{x + 1} = \dfrac{1 + 1 + 1}{1 + 1} = \dfrac{3}{2}$

17. $\lim_{x \to 4^-} \dfrac{x - 4}{x^2 - 16} = \lim_{x \to 4^-} \dfrac{x - 4}{(x - 4)(x + 4)} = \lim_{x \to 4^-} \dfrac{1}{x + 4} = \dfrac{1}{4 + 4} = \dfrac{1}{8}$

19. This rational function is not defined at $x = 3$, and as $x \to 3^+$, values of the function exceed all bounds. The limit does not exist.

21. $\lim_{x \to 1} \sqrt{x^3 + 6x^2 + 2x + 5} = \sqrt{1^3 + 6 \cdot 1^2 + 2 \cdot 1 + 5} = \sqrt{14}$

23. $\lim_{x \to 1^+}\left(\sqrt{x - 1} + 3\right) = \sqrt{1 - 1} + 3 = 3$

25. $\lim_{x \to -2.5^+}\left(\sqrt{5 + 2x} + x\right) = \sqrt{5 + 2(-2.5)} + (-2.5) = -2.5$

27. $\displaystyle\lim_{x\to 3}\frac{\sqrt{x}-\sqrt{3}}{x-3} = \lim_{x\to 3}\frac{\left(\sqrt{x}-\sqrt{3}\right)}{\left(\sqrt{x}-\sqrt{3}\right)\left(\sqrt{x}+\sqrt{3}\right)} = \lim_{x\to 3}\frac{1}{\sqrt{x}+\sqrt{3}} = \frac{1}{2\sqrt{3}}$

29. $\displaystyle\lim_{x\to 0}\frac{\frac{1}{x+5}-\frac{1}{5}}{x} = \lim_{x\to 0}\frac{5-(x+5)}{5x(x+5)} = \lim_{x\to 0}\frac{-1}{5(x+5)} = -\frac{1}{25}$

31. $\displaystyle\lim_{x\to 1}\left[\frac{1}{x-1}-\frac{2}{x^2-1}\right] = \lim_{x\to 1}\frac{x+1-2}{x^2-1} = \lim_{x\to 1}\frac{x-1}{(x-1)(x+1)} = \lim_{x\to 1}\frac{1}{x+1} = \frac{1}{1+1} = \frac{1}{2}$

33. $\displaystyle\lim_{x\to 0}\frac{x^2}{|x|} = \lim_{x\to 0}\frac{|x|^2}{|x|} = \lim_{x\to 0}|x| = |0| = 0$

35. $\displaystyle\lim_{x\to -3^+}\left[\frac{|x+3|}{x+3}+\sqrt{x+3}+1\right] = \lim_{x\to -3^+}\left[\frac{x+3}{x+3}+\sqrt{x+3}+1\right] = \lim_{x\to -3^+}\left[1+\sqrt{x+3}+1\right]$

$$= 1+\sqrt{-3+3}+1 = 2$$

37. $\displaystyle\lim_{x\to 0}\frac{\sqrt{2-x}-\sqrt{2}}{x} = \lim_{x\to 0}\left[\frac{\sqrt{2-x}-\sqrt{2}}{x}\cdot\frac{\sqrt{2-x}+\sqrt{2}}{\sqrt{2-x}+\sqrt{2}}\right] = \lim_{x\to 0}\frac{2-x-2}{x\left(\sqrt{2-x}+\sqrt{2}\right)}$

$$= \lim_{x\to 0}\frac{-1}{\sqrt{2-x}+\sqrt{2}} = \frac{-1}{2\sqrt{2}}$$

39. Note: $\displaystyle\lim_{x\to 0^+}\left[\frac{|x|}{x}-\frac{x}{|x|}\right] = \lim_{x\to 0^+}\left[\frac{x}{x}-\frac{x}{x}\right] = \lim_{x\to 0^+}0 = 0.$

$$\lim_{x\to 0^-}\left[\frac{|x|}{x}-\frac{x}{|x|}\right] = \lim_{x\to 0^-}\left[\frac{-x}{x}-\frac{x}{-x}\right] = \lim_{x\to 0^-}0 = 0.$$

Since the left and right limits exist and are equal to 0, $\displaystyle\lim_{x\to 0}\left[\frac{|x|}{x}-\frac{x}{|x|}\right] = 0.$

41. $\displaystyle\lim_{h\to 0}\frac{f(2+h)-f(2)}{h} = \lim_{h\to 0}\frac{(2+h)^2-2^2}{h} = \lim_{h\to 0}\frac{4h+h^2}{h} = \lim_{h\to 0}(4+h) = 4$

43. $\displaystyle\lim_{h\to 0}\frac{f(2+h)-f(2)}{h} = \lim_{h\to 0}\frac{\left[(2+h)^2+(2+h)+1\right]-\left(2^2+2+1\right)}{h}$

$$= \lim_{h\to 0}\frac{4+4h+h^2+2+h+1-7}{h} = \lim_{h\to 0}\frac{5h+h^2}{h} = \lim_{h\to 0}(5+h) = 5$$

45. $\lim\limits_{h\to 0}\dfrac{f(0+h)-f(0)}{h}=\lim\limits_{h\to 0}\dfrac{|h|-|0|}{h}=\lim\limits_{h\to 0}\dfrac{|h|}{h}$ does not exist, because the left hand limit

is -1 and the right hand limit is 1.

47. $\lim\limits_{h\to 0}\dfrac{f(x+h)-f(x)}{h}=\lim\limits_{h\to 0}\dfrac{[2(x+h)+3]-(2x+3)}{h}=\lim\limits_{h\to 0}\dfrac{2x+2h+3-2x-3}{h}$

$\qquad\qquad = \lim\limits_{h\to 0}\dfrac{2h}{h}=\lim\limits_{h\to 0}2=2$

49. $\lim\limits_{h\to 0}\dfrac{f(x+h)-f(x)}{h}=\lim\limits_{h\to 0}\dfrac{\left[(x+h)^2+(x+h)\right]-\left(x^2+x\right)}{h}$

$\qquad\qquad = \lim\limits_{h\to 0}\dfrac{x^2+2xh+h^2+x+h-x^2-x}{h}$

$\qquad\qquad = \lim\limits_{h\to 0}\dfrac{2xh+h^2+h}{h}=\lim\limits_{h\to 0}(2x+h+1)=2x+1$

51. $\lim\limits_{h\to 0}\dfrac{f(x+h)-f(x)}{h}=\lim\limits_{h\to 0}\dfrac{\sqrt{x+h+1}-\sqrt{x+1}}{h}$

$\qquad\qquad = \lim\limits_{h\to 0}\dfrac{1}{\sqrt{x+h+1}+\sqrt{x+1}}=\dfrac{1}{2\sqrt{x+1}}$

53. $\lim\limits_{h\to 0}\dfrac{f(x+h)-f(x)}{h}=\lim\limits_{h\to 0}\dfrac{\sqrt{(x+h)^2+1}-\sqrt{x^2+1}}{h}$

$\qquad\qquad = \lim\limits_{h\to 0}\dfrac{2x+h}{\sqrt{(x+h)^2+1}+\sqrt{x^2+1}}=\dfrac{2x}{2\sqrt{x^2+1}}=\dfrac{x}{\sqrt{x^2+1}}$

55. $f(x)=\dfrac{|x|}{x}$, $g(x)=\dfrac{-x}{|x|}$, $c=0$

13.2.A The Formal Definition of Limit

1. Given $\varepsilon > 0$, we require a $\delta > 0$ such that if $0 < |x-3| < \delta, |(3x-2)-7| < \varepsilon$.
    Since $|(3x-2)-7| = |3x-9| = 3|(x-3)| < \varepsilon$ if $|x-3| < \varepsilon/3$, choose $\delta = \varepsilon/3$.
    Then if $0 < |x-3| < \delta,\ 0 < |x-3| < \varepsilon/3$
$$0 < 3|x-3| < \varepsilon$$
$$|3x-9| < \varepsilon$$
$$|3x-2-7| < \varepsilon \text{ as required.}$$

3. We require a $\delta > 0$ such that if $0 < |x-5| < \delta,\ |x-5| < \varepsilon$. Clearly $\delta = \varepsilon$ will do.
    Then if $0 < |x-5| < \delta,\ 0 < |x-5| < \varepsilon$ as required.

5. We require a $\delta > 0$ such that if $0 < |x-2| < \delta$, then $|(6x+3)-15| < \varepsilon$.
    Since $|(6x+3)-15| = |6x-12| = 6|x-2| < \varepsilon$ if $|x-2| < \varepsilon/6$, choose $\delta = \varepsilon/6$.
    Then if $0 < |x-2| < \delta,\ 0 < |x-2| < \varepsilon/6$
$$0 < 6|x-2| < \varepsilon$$
$$|6x-12| < \varepsilon$$
$$|(6x+3)-15| < \varepsilon \text{ as required.}$$

7. We require a $\delta > 0$ such that if $0 < |x-1| < \delta,\ |4-4| < \varepsilon$.
    Any positive δ, whether $\delta = 2000$ or 6 or 0.005 will do, since $|4-4| = 0 < \varepsilon$
    for any positive ε whatsoever.

9. We require a $\delta > 0$ such that if $0 < |x-4| < \delta,\ |x-6+2| < \varepsilon$. Clearly $\delta = \varepsilon$ will do.
    Then if $0 < |x-4| < \delta,\ 0 < |x-6+2| < \varepsilon$ as required.

11. Given $\varepsilon > 0$, we require a $\delta > 0$ such that if $0 < |x+2| < \delta, |2x+5-1| < \varepsilon$.
    Since $0 < |2x+5-1| = |2x+4| = 2|x+2| < \varepsilon$ if $|x+2| < \varepsilon/2$, choose $\delta = \varepsilon/2$.
    Then if $0 < |x+2| < \delta,\ 0 < |x+2| < \varepsilon/2$
$$0 < 2|x+2| < \varepsilon$$
$$|2+4x| < \varepsilon$$
$$|3x+5-1| < \varepsilon \text{ as required.}$$

13. Given $\varepsilon > 0$, we require a $\delta > 0$ such that if $0 < |x-0| < \delta, |x^2-0| < \varepsilon$.
    Since $|x^2| < \varepsilon$ exactly when $|x| < \sqrt{\varepsilon}$, choose $\delta = \sqrt{\varepsilon}$.
    Then if $0 < |x-0| < \delta,\ 0 < |x| < \sqrt{\varepsilon},\ |x^2| < \varepsilon$ as required.

15. We require a $\delta > 0$ such that if $0 < |x - c| < \delta$, then $0 < |(f(x) - g(x)) - (L - M)| < \varepsilon$.

   By the triangle inequality,

   $$|(f(x) - g(x)) - (L - M)| = |(f(x) - L) + (M - g(x))|$$

   $$\leq |f(x) - L| + |M - g(x)|$$

   $$\leq |f(x) - L| + |g(x) - M|$$

   Now let ε be any positive number. There are positive numbers δ_1 and δ_2 such that
   if $0 < |x - c| < \delta_1$ then $|f(x) - L| < \varepsilon/2$,

   if $0 < |x - c| < \delta_2$ then $|g(x) - M| < \varepsilon/2$.

   So choose δ to be the smaller of δ_1, δ_2. Then if $0 < |x - c| < \delta$,

   $$|(f(x) - g(x)) - (L - M)| \leq |f(x) - L| + |g(x) - M| \leq \frac{\varepsilon}{2} + \frac{\varepsilon}{2} \leq \varepsilon.$$

   This proves that if $\lim_{x \to c} f(x) = L$ and $\lim_{x \to c} g(x) = M$, then $\lim_{x \to c}(f(x) - g(x)) = L - M$.

17. (a) The function is not defined at $x=0$. Near the origin the graph is similar to the

   graph of $f(x) = \sin\left(\dfrac{\pi}{x}\right)$ with one important exception. In the graph of $g(x)$ the

   height of the waves gets smaller and smaller as x approaches 0. This is difficult to
   see, even with technology (the viewing window with $-.1 < x < .1$
   and $-.1 < y < .1$ gives a rough idea).

   (b) Given $\varepsilon > 0$, we require a $\delta > 0$ such that if $0 < |x - 0| < \delta$, $\left|x \sin\left(\dfrac{\pi}{x}\right) - 0\right| < \varepsilon$.

   Since $\left|x \sin\left(\dfrac{\pi}{x}\right)\right| < \varepsilon$ exactly when $|x| < \varepsilon$, choose $\delta = \varepsilon$.

   Then if $0 < |x - 0| < \delta$, $0 < |x| < \varepsilon$, $\left|x \sin\left(\dfrac{\pi}{x}\right)\right| < \varepsilon$ as required.

13.3 Continuity

1. The function is undefined at 3 and 6, and therefore discontinuous at these points.

3. The function is continuous at –2 and 3, but undefined and thus discontinuous at 0.

5. The function has no limit as x approaches –2, hence it is discontinuous there. It is continuous at 0 and 3.

7. $f(3) = 3^2 + 5(3-2)^7 = 14$, $\lim\limits_{x \to 3} f(x) = \lim\limits_{x \to 3} x^2 + 5\left[\lim\limits_{x \to 3}(x-2)\right]^7 = 14$

    Since $\lim\limits_{x \to 3} f(x) = f(3)$, the function is continuous at $x = 3$.

9. $f(2) = \dfrac{2^2 - 9}{(2^2 - 2 - 6)(2^2 + 6 \cdot 2 + 9)} = \dfrac{1}{20}$

    $\lim\limits_{x \to 2} f(x) = \dfrac{\lim\limits_{x \to 2}(x^2 - 9)}{\lim\limits_{x \to 2}(x^2 - x - 6)\lim\limits_{x \to 2}(x^2 + 6x + 9)} = \dfrac{1}{20}$

    Since $\lim\limits_{x \to 2} f(x) = f(2)$, the function is continuous at $x = 2$.

11. $f(36) = \dfrac{36\sqrt{36}}{(36-6)^2} = \dfrac{6}{25}$, $\lim\limits_{x \to 36} f(x) = \dfrac{\lim\limits_{x \to 36} x \lim\limits_{x \to 36}\sqrt{x}}{\lim\limits_{x \to 36}(x-6)^2} = \dfrac{6}{25}$

    Since $\lim\limits_{x \to 36} f(x) = f(36)$, the function is continuous at $x = 36$.

13. The function is undefined at $x = 3$, and therefore discontinuous there.

15. The function is undefined at $x = -1$, and therefore discontinuous there.
    $\left[\text{Note: } \lim\limits_{x \to -1} f(x) = -3/2, \text{ but this is irrelevant.}\right]$

17. $\lim\limits_{x \to 0} f(x) = \lim\limits_{x \to 0} x^2 = 0$. However, $f(0) = 1 \neq \lim\limits_{x \to 0} f(x)$, hence the function is discontinuous.

19. $\lim\limits_{x \to 2^+} f(x) = \lim\limits_{x \to 2^+}(2x-4) = 0 = f(2)$. $\lim\limits_{x \to 2^-} f(x) = \lim\limits_{x \to 2^-}(-2x+4) = 0 = f(2)$.
    Hence $\lim\limits_{x \to 2} f(x) = f(2)$; the function is continuous at $x = 2$.

21. $\lim\limits_{x \to 0^+} f(x) = \lim\limits_{x \to 0^+} 2x^2 = 0 = f(0)$. $\lim\limits_{x \to 0^-} f(x) = \lim\limits_{x \to 0^-}(x^2 - x) = 0 = f(0)$.
    Hence $\lim\limits_{x \to 0} f(x) = f(0)$; the function is continuous at $x = 0$.

23. $\lim\limits_{x\to 3^+} f(x) = \lim\limits_{x\to 3^+}(x-3) = 0 = f(3)$. $\lim\limits_{x\to 3^-} f(x) = \lim\limits_{x\to 3^-}(3-x) = 0 = f(3)$.

Hence $\lim\limits_{x\to 3} f(x) = f(3)$; the function is continuous at $x = 3$.

25. The function is continuous everywhere with the possible exception of $x = 1$ and $x = 3$, the zeros of the denominator.

At $x = 1$, $\lim\limits_{x\to 1} f(x) = \lim\limits_{x\to 1}\dfrac{x^2+x-2}{x^2-4x+3} = \lim\limits_{x\to 1}\dfrac{(x+2)(x-1)}{(x-3)(x-1)} = \lim\limits_{x\to 1}\dfrac{x+2}{x-3} = -\dfrac{3}{2} = f(1)$.

Hence the function is continuous at $x = 1$.

At $x = 3$, however, the function is undefined and is therefore not continuous.

27. The function is continuous everywhere with the possible exception of $x = 0$ and $x = 2$. In fact, since $\lim\limits_{x\to 0^-} f(x) = 1$ and $\lim\limits_{x\to 0^+} f(x) = 0$, it is discontinuous at $x = 0$.

Similarly, since $\lim\limits_{x\to 2^-} f(x) = 2$ and $\lim\limits_{x\to 2^+} f(x) = -1$, it is also discontinuous at $x = 2$.

29. The function is discontinuous on the interval because it is discontinuous at every multiple of $\frac{1}{5}$.

31. Generally, temperature is continuous, hence $g(x)$ is continuous everywhere on the interval.

33. Solve $x^2 - 7 = 0$ by choosing $x = 2$ and $x = 3$, noting that there is a sign change between 2 and 3. Then choose $x = 2.1$, $x = 2.2$, etc. until a sign change is observed between 2.6 and 2.7. Then choose $x = 2.61$, $x = 2.62$, etc. until a sign change is observed. The process can be continued until the desired decimal place accuracy is reached.

35. The function will be discontinuous on $[-3,3]$ because it is undefined at $x = 1$. Note that $f(-3) = -10$ and $f(3) = 20$. We must show that f satisfies the conclusion of the Intermediate Value Theorem: that is, for any number k between -10 and 20, there is a number c between -3 and 3 such that $f(c) = k$.

Let k be a number between -10 and 20. First, assume that $-10 \le k \le 0$ and note that $f(-2) = 0$. Since f is continuous on $[-3,-2]$, the Intermediate Value Theorem applies on that interval. Hence, there is a number c between -3 and -2 such that $f(c) = k$. Now suppose that $0 < k \le 20$ and note that $f(2) = 0$. Since f is continuous on $[2,3]$, the Intermediate Value Theorem applies on that interval. Hence, there is a number c between 2 and 3 such that $f(c) = k$. Thus in all cases, there is a number c between -3 and 3 such that $f(c) = k$.

37. The function will be continuous if $f(3) = 3b + 4 = \lim\limits_{x\to 3^+} f(x) = bx^2 - 2 = 9b - 2$.

Solving, $b = 1$ is the only possibility.

39. Since $\lim_{x\to1} f(x) = \lim_{x\to1}\dfrac{1}{x+1} = \dfrac{1}{2}$, and $f(x) = \dfrac{1}{x+1}$ if $x \neq 1$,

choose $g(x) = \dfrac{1}{x+1}$.

41. Since $\lim_{x\to4} f(x) = \lim_{x\to4}\dfrac{1}{2+\sqrt{x}} = \dfrac{1}{4}$, and $f(x) = \dfrac{1}{2+\sqrt{x}}$ if $x \neq 4$,

choose $g(x) = \dfrac{1}{2+\sqrt{x}}$.

43. Since $\lim_{x\to0^+} f(x) = 1$ and $\lim_{x\to0^-} f(x) = -1$, no choice of $f(0)$ could possibly reconcile these two facts to provide a continuous function.

13.4 Limits Involving Infinity

1. For the function $f(x) = \sqrt{x^2 + 1} - (x + 1)$, $f(100) = -.995$, $f(1000) = -.9995$, $f(10,000) = -.99995$, $f(100,000) = -.999995$. The limit appears to be -1.

3. For the function $f(x) = \dfrac{x^{2/3} - x^{4/3}}{x^3}$, $f(-100) = .0004$, $f(-1000) = .00001$, $f(-10,000) = .0000002$, $f(-100,000) = .000000005$. The limit appears to be 0.

5. For the function $f(x) = \sin\left(\dfrac{1}{x}\right)$ some values are as shown:

    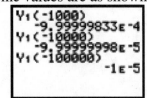

    The limit appears to be 0.

7. For the function $f(x) = \dfrac{\ln x}{x}$ some values are as shown:

    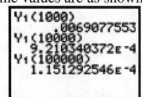

    The limit appears to be 0.

9. No vertical asymptotes. $\lim\limits_{x \to \infty} f(x) = \infty$. $\lim\limits_{x \to -\infty} f(x) = 0$.

11. No vertical asymptotes. $\lim\limits_{x \to \infty} f(x) = 2$. $\lim\limits_{x \to -\infty} f(x) = -1$.

13. Vertical asymptotes: $x = -10$. $\lim\limits_{x \to \infty} f(x) = \infty$. $\lim\limits_{x \to -\infty} f(x) = -\infty$.

15. $\lim\limits_{x \to \pm\infty} \dfrac{3x^2 + 5}{4x^2 - 6x + 2} = \lim\limits_{x \to \pm\infty} \dfrac{3 + 5/x^2}{4 - 6/x + 2/x^2} = \dfrac{3}{4}$. $y = \dfrac{3}{4}$ is the horizontal asymptote.

17. $\lim\limits_{x \to \pm\infty} \dfrac{2x^2 - 6x + 1}{2 + x - x^2} = \lim\limits_{x \to \pm\infty} \dfrac{2 - 6/x + 1/x^2}{2/x^2 + 1/x - 1} = -2$. $y = -2$ is the horizontal asymptote.

19. Since the numerator is of greater degree than the denominator, the limits as $x \to \infty$ and as $x \to -\infty$ are infinite and there is no horizontal asymptote.

21. $\lim\limits_{x \to -\infty} \dfrac{(x - 3)(x + 2)}{2x^2 + x + 1} = \lim\limits_{x \to -\infty} \dfrac{(1 - 3/x)(1 + 2/x)}{2 + 1/x + 1/x^2} = \dfrac{1}{2}$.

23. $\lim\limits_{x \to \infty}(3x - 1/x^2) = \lim\limits_{x \to \infty} 3x - \lim\limits_{x \to \infty} 1/x^2 = \infty - 0 = \infty$.

25. $\lim\limits_{x \to -\infty}\left(\dfrac{3x}{x+2}+\dfrac{2x}{x-1}\right) = \lim\limits_{x \to -\infty}\dfrac{3x}{x+2}+\lim\limits_{x \to -\infty}\dfrac{2x}{x-1}=3+2=5.$

27. $\lim\limits_{x \to \infty}\dfrac{2x}{\sqrt{x^2-2x}} = \lim\limits_{x \to \infty}\dfrac{2x}{\sqrt{x^2-2x}}\div\dfrac{x}{\sqrt{x^2}}=\lim\limits_{x \to \infty}\dfrac{2}{\sqrt{1-2/x}}=2.$

$\left(\text{since when } x \text{ is positive, } \sqrt{x^2}=x\right)$

29. $\lim\limits_{x \to -\infty}\dfrac{3x-2}{\sqrt{2x^2+1}} = \lim\limits_{x \to -\infty}\dfrac{3x-2}{\sqrt{2x^2+1}}\div\dfrac{(-x)}{\sqrt{x^2}}=\lim\limits_{x \to -\infty}\dfrac{-3+2/x}{\sqrt{2+1/x^2}}=\dfrac{-3}{\sqrt{2}}.$

$\left(\text{since when } x \text{ is negative, } \sqrt{x^2}=-x\right)$

31. $\lim\limits_{x \to \infty}\dfrac{\sqrt{2x^2+1}}{3x-5} = \lim\limits_{x \to \infty}\dfrac{\sqrt{2x^2+1}}{3x-5}\div\dfrac{\sqrt{x^2}}{x}=\lim\limits_{x \to \infty}\dfrac{\sqrt{2+1/x^2}}{3-5/x}=\dfrac{\sqrt{2}}{3}.$

$\left(\text{since when } x \text{ is positive, } \sqrt{x^2}=x\right)$

33. $\lim\limits_{x \to -\infty}\dfrac{\sqrt{3x^2+3}}{x+3} = \lim\limits_{x \to -\infty}\dfrac{\sqrt{3x^2+3}}{x+3}\div\dfrac{\sqrt{x^2}}{(-x)}=\lim\limits_{x \to -\infty}\dfrac{\sqrt{3+3/x^2}}{-1-3/x}=-\sqrt{3}.$

$\left(\text{since when } x \text{ is negative, } \sqrt{x^2}=-x\right)$

35. $\lim\limits_{x \to \infty}\dfrac{x^2+2x+1}{\sqrt{x^4+2x}} = \lim\limits_{x \to \infty}\dfrac{\left(x^2+2x+1\right)\div x^2}{\sqrt{x^4+2x}\div\sqrt{x^4}}=\lim\limits_{x \to \infty}\dfrac{1+2/x+1/x^2}{\sqrt{1+2/x^3}}=1.$

37. $\lim\limits_{x \to \infty}\dfrac{1-\sqrt{x}}{1+\sqrt{x}} = \lim\limits_{x \to \infty}\dfrac{\left(1-\sqrt{x}\right)\left(1-\sqrt{x}\right)}{\left(1+\sqrt{x}\right)\left(1-\sqrt{x}\right)}=\lim\limits_{x \to \infty}\dfrac{1-2\sqrt{x}+x}{1-x}=\lim\limits_{x \to \infty}\dfrac{1/x-2/\sqrt{x}+1}{1/x-1}=-1.$

39. $\lim\limits_{x \to \infty}\left(\sqrt{x^2+1}-x\right) = \lim\limits_{x \to \infty}\dfrac{\left(\sqrt{x^2+1}-x\right)\left(\sqrt{x^2+1}+x\right)}{1\cdot\left(\sqrt{x^2+1}+x\right)}=\lim\limits_{x \to \infty}\dfrac{1}{\sqrt{x^2+1}+x}=0.$

41. $\lim\limits_{x \to \infty}\left(\sqrt{x^2-1}-\sqrt{x^2+1}\right) = \lim\limits_{x \to \infty}\dfrac{\left(\sqrt{x^2-1}-\sqrt{x^2+1}\right)\left(\sqrt{x^2-1}+\sqrt{x^2+1}\right)}{1\cdot\left(\sqrt{x^2-1}+\sqrt{x^2+1}\right)}$

$= \lim\limits_{x \to \infty}\dfrac{-2}{\left(\sqrt{x^2-1}+\sqrt{x^2+1}\right)}=0.$

43. $\lim\limits_{x\to\infty}\dfrac{x}{|x|}=\lim\limits_{x\to\infty}\dfrac{x}{x}=\lim\limits_{x\to\infty}1=1.$

45. Since if $|x|$ is very large, $[\![x]\!]$ differs from x by at most 1,

    a. $\lim\limits_{x\to\infty}\dfrac{[\![x]\!]}{x}=\lim\limits_{x\to\infty}\dfrac{x-\alpha}{x}=\lim\limits_{x\to\infty}\dfrac{1-\alpha/x}{1}=1.$

    b. $\lim\limits_{x\to-\infty}\dfrac{[\![x]\!]}{x}=\lim\limits_{x\to-\infty}\dfrac{x-\beta}{x}=\lim\limits_{x\to-\infty}\dfrac{1-\beta/x}{1}=1.$

    Here α and β are variable, but always less than 1.

47. $\lim\limits_{x\to\infty}\dfrac{\sqrt{x+\sqrt{x+\sqrt{x}}}}{\sqrt{x+1}}=\lim\limits_{x\to\infty}\dfrac{\sqrt{x+\sqrt{x+\sqrt{x}}}\div\sqrt{x}}{\sqrt{x+1}\div\sqrt{x}}=\lim\limits_{x\to\infty}\dfrac{\sqrt{1+\sqrt{x+\sqrt{x}}/x}}{\sqrt{1+1/x}}$

                  $=\lim\limits_{x\to\infty}\dfrac{\sqrt{1+\sqrt{1/x+\sqrt{x}/x^2}}}{\sqrt{1+1/x}}=\lim\limits_{x\to\infty}\dfrac{\sqrt{1+\sqrt{1/x+\sqrt{1/x^3}}}}{\sqrt{1+1/x}}=1.$

49. The first part of the informal definition is included in the second part, which states: the values of $f(x)$ can be made arbitrarily close to L by taking large enough values of x. This means that whenever you specify how close $f(x)$ should be to L, we can tell you how large x must be to guarantee this. In other words, you specify how close you want $f(x)$ to be to L by giving a positive number ε and we tell you how large x must be to guarantee that $f(x)$ is within ε of L, that is, to guarantee that $|f(x)-L|<\varepsilon$. We do this by giving a positive number k such that $|f(x)-L|<\varepsilon$ whenever $x>k$. In other words, for every positive number ε, there is a positive number k (depending on ε) such that: If $x>k$, then $|f(x)-L|<\varepsilon$.

51. a. For the function $f(x)=(1+1/x)^x$ some values are as shown:

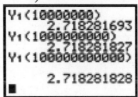

    The last two figures agree to seven decimal places: 2.7182818.

    b. $e\approx 2.718281828$ **c.** e **d.** The limits are the same. Substitute $z=\dfrac{1}{x}$ for x.

Chapter 13 Review Exercises

1. For the function $f(x) = \dfrac{3x - \sin x}{x}$ some values are as shown:

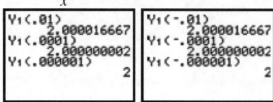

The limit appears to be 2.

3. For the function $f(x) = \dfrac{10^x - .1}{x + 1}$, some values are as shown:

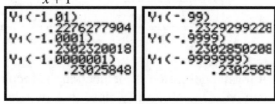

The limit appears to be approximately .23026. (This is actually an approximation of $(.1)\ln 10$.)

5. Since $\lim\limits_{x \to 2^+} f(x) = 2$, $\lim\limits_{x \to 2^-} f(x) = 2$, $\lim\limits_{x \to 2} f(x) = 2$.

7. $\lim\limits_{x \to 3}\left[3f(x) - 15\right] = 3\lim\limits_{x \to 3} f(x) - 15 = 3 \cdot 5 - 15 = 0$.

9. $\lim\limits_{x \to 3} \dfrac{f(x)g(x) - 2f(x)}{\left[g(x)\right]^2} = \dfrac{\lim\limits_{x \to 3} f(x)\lim\limits_{x \to 3} g(x) - 2\lim\limits_{x \to 3} f(x)}{\left[\lim\limits_{x \to 3} g(x)\right]^2} = \dfrac{5(-2) - 2(5)}{(-2)^2} = -5$.

11. $\lim\limits_{x \to -2}\left(x^3 - 3x + 1\right) = (-2)^3 - 3(-2) + 1 = -1$.

13. $\lim\limits_{x \to 1} \dfrac{x^2 - 1}{x^2 - 3x + 2} = \lim\limits_{x \to 1} \dfrac{x + 1}{x - 2} = \dfrac{1 + 1}{1 - 2} = -2$.

15. $\lim\limits_{x \to 0} \dfrac{\sqrt{1 + x} - 1}{x} = \lim\limits_{x \to 0} \dfrac{\left(\sqrt{1 + x} - 1\right)\left(\sqrt{1 + x} + 1\right)}{x\left(\sqrt{1 + x} + 1\right)} = \lim\limits_{x \to 0} \dfrac{x}{x\left(\sqrt{1 + x} + 1\right)} = \dfrac{1}{2}$.

17. $\lim\limits_{x \to -1^+} \sqrt{9 + 8x - x^2} = \sqrt{9 + 8(-1) - (-1)^2} = 0$.

19. $\lim\limits_{x \to -5^+} \dfrac{|x + 5|}{x + 5} = \lim\limits_{x \to -5^+} \dfrac{x + 5}{x + 5} = \lim\limits_{x \to -5^+} 1 = 1$.

21. $\lim\limits_{h\to 0}\dfrac{f(2+h)-f(2)}{h}=\lim\limits_{h\to 0}\dfrac{\left[(2+h)^2+1\right]-(2^2+1)}{h}$

$\qquad\qquad =\lim\limits_{h\to 0}\dfrac{5+4h+h^2-5}{h}=\lim\limits_{h\to 0}(4+h)=4.$

23. $\lim\limits_{h\to 0}\dfrac{f(x+h)-f(x)}{h}=\lim\limits_{h\to 0}\dfrac{\left[4(x+h)-2\right]-(4x-2)}{h}$

$\qquad\qquad =\lim\limits_{h\to 0}\dfrac{4x+4h-2-4x+2}{h}=\lim\limits_{h\to 0}4=4.$

25. We require a $\delta>0$ such that if $0<|x-3|<\delta$, then $|(2x+1)-7|<\varepsilon.$
Since $|(2x+1)-7|=|2x-6|=2|x-3|<\varepsilon$ if $|x-3|<\varepsilon/2$, choose $\delta=\varepsilon/2.$
Then if $0<|x-3|<\delta,\ 0<|x-3|<\varepsilon/2$
$$0<2|x-3|<\varepsilon$$
$$|2x-6|<\varepsilon$$
$$|(2x+1)-7|<\varepsilon \text{ as required.}$$

27. Since $\lim\limits_{x\to-3}f(x)$ appears to equals $f(-3)$, the function is continuous at $x=-3$. Since $f(2)$ is not defined, the function is not continuous at $x=2$.

29. $\lim\limits_{x\to4}\sqrt{x^2-8}=\sqrt{4^2-8}=\sqrt{8}$, hence the function is continuous at $x=4$.
Since $\lim\limits_{x\to\sqrt{8}^-}f(x)$ does not exist; however, $\lim\limits_{x\to\sqrt{8}^+}f(x)=0$, so the function is continuous at $x=\sqrt{8}.$

31. Since $h(x)$ is not defined at $x=-4$, the function is not continuous at $x=-4$. Since $\lim\limits_{x\to0}h(x)=h(0)$, the function is continuous at $x=0$.

33. a. $\lim\limits_{x\to2}f(x)=\lim\limits_{x\to2}\dfrac{x^2-x-6}{x^2-9}=\dfrac{\lim\limits_{x\to2}\left(x^2-x-6\right)}{\lim\limits_{x\to2}\left(x^2-9\right)}$

$\qquad =\dfrac{2^2-2-6}{2^2-9}=\dfrac{-4}{-5}=\dfrac{4}{5}=f(2)$

b. Since $f(x)$ is not defined at $x=3$, the function is not continuous at $x=3$.

35. Vertical asymptotes at $x=-1$ and $x=2$. The graph moves upward as x approaches -1 from the left and downward as x approaches -1 from the right. The graph moves downward as x approaches 2 from the left and upward as x approaches 2 from the left.

37. The limit does not exist. One can write $\lim\limits_{x\to 3}\dfrac{1}{(x-3)^2}=\infty$, however.

39. Since $\lim\limits_{x\to\infty}\dfrac{1}{x^2}=\lim\limits_{x\to\infty}\left(-\dfrac{1}{x^2}\right)=0$, and $-\dfrac{1}{x^2}\le\dfrac{\sin x}{x^2}\le\dfrac{1}{x^2}$ for all x, $\lim\limits_{x\to\infty}\dfrac{\sin x}{x^2}=0$.

41. $\lim\limits_{x\to\infty}\dfrac{2x^3-3x^2+5x-1}{4x^3+2x^2-x+10}=\lim\limits_{x\to\infty}\dfrac{2-3/x+5/x^2-1/x^3}{4+2/x-1/x^2+10/x^3}=\dfrac{1}{2}.$

43. $\lim\limits_{x\to-\infty}\left(\dfrac{2x+1}{x-3}+\dfrac{4x-1}{3x}\right)=\lim\limits_{x\to-\infty}\dfrac{2+1/x}{1-3/x}+\lim\limits_{x\to-\infty}\dfrac{4-1/x}{3}=2+\dfrac{4}{3}=\dfrac{10}{3}.$

45. $\lim\limits_{x\to-\infty}f(x)=\lim\limits_{x\to-\infty}\dfrac{1-1/x+7/x^2}{2+5/x+7/x^2}=\dfrac{1}{2}.$ The horizontal asymptote is $y=\dfrac{1}{2}$. Graph using the window $-20\le x\le 20, 0\le y\le 3$.

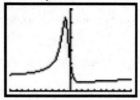

Chapter 13 Test

1. (a) 2   (b) Does not exist   (c) 1   (d) 3

2. $\lim\limits_{x\to-2}\dfrac{x^2+3x+2}{x^2-5x-14}=\lim\limits_{x\to-2}\dfrac{(x+2)(x+1)}{(x+2)(x-7)}=\lim\limits_{x\to-2}\dfrac{x+1}{x-7}=\dfrac{\lim\limits_{x\to-2}(x+1)}{\lim\limits_{x\to-2}(x-7)}=\dfrac{-2+1}{-2-7}=\dfrac{-1}{-9}=\dfrac{1}{9}$

3. For the function $f(x)=\dfrac{e^{8x}+e^x-2}{e^x-1}$ some values are as shown:

```
Y₁(.01)
          9.287132639
Y₁(.001)
          9.028070131
Y₁(.0001)
          9.002800698
■
```

```
Y₁(-.01)
          8.726871258
Y₁(-.001)
          8.97206987
Y₁(-.0001)
          8.9972007
```

The limit appears to be 9.

4. $\lim\limits_{x\to4}\dfrac{f(x)-2g(x)}{4h(x)}=\dfrac{\lim\limits_{x\to4}[f(x)-2g(x)]}{\lim\limits_{x\to4}[4h(x)]}=\dfrac{\lim\limits_{x\to4}f(x)-2\lim\limits_{x\to4}g(x)}{4\lim\limits_{x\to4}h(x)}=\dfrac{5-2(0)}{4(-8)}=-\dfrac{5}{32}$

5. a.

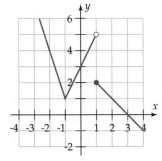

b. 1

c. 1

d. 5

e. 2

6. $\lim\limits_{x\to-2}\left(\dfrac{x^2+2x}{x+2}\right)=\lim\limits_{x\to-2}\left[\dfrac{x(x+2)}{x+2}\right]=\lim\limits_{x\to-2}x=-2$

7. $\lim\limits_{x\to3^-}\dfrac{|x-3|}{x-3}=\dfrac{1}{-1}=-1$ *but* $\lim\limits_{x\to3^+}\dfrac{|x-3|}{x-3}=\dfrac{1}{1}=1$, hence the limit does not exist.

8. $\lim\limits_{h\to0}\dfrac{f(3+h)-f(3)}{h}=\lim\limits_{h\to0}\dfrac{(3+h)^2-(3)^2}{h}=\lim\limits_{h\to0}\dfrac{9+6h+h^2-9}{h}$

$=\lim\limits_{h\to0}\dfrac{6h+h^2}{h}=\lim\limits_{h\to0}\dfrac{h(6+h)}{h}=\lim\limits_{h\to0}(6+h)=6+0=6$

9. (a) For the function $f(x) = (e^x - 1)\ln(|x|)$ some values are as shown:

```
Y₁(.01)
        -.0462827298
Y₁(.001)
        -.0069112103
Y₁(.0001)
     -9.21080091E-4
■
```
```
Y₁(-.01)
         .045822209
Y₁(-.001)
         .0069043026
Y₁(-.0001)
      9.209879871E-4
■
```

The limit appears to be 0.

(b) $f(0) = (e^0 - 1)\ln(|0|) = $ not defined

10. $f'(x) = \lim_{h \to 0} \dfrac{f(x+h) - f(x)}{h} = \lim_{h \to 0} \dfrac{\left[(x+h)^2 - 5(x+h) + 6\right] - \left[x^2 - 5x + 6\right]}{h}$

$\lim_{h \to 0} \dfrac{x^2 + 2xh + h^2 - 5x - 5h + 6 - x^2 + 5x - 6}{h} = \lim_{h \to 0} \dfrac{2xh + h^2 - 5h}{h} = \lim_{h \to 0} \dfrac{h(2x + h - 5)}{h}$

$\lim_{h \to 0} (2x + h - 5) = 2x - 5$

11. If the problem is: $\lim_{x \to \sqrt{2}} \left(\dfrac{x - 2}{x - \sqrt{2}} \right) = \dfrac{\sqrt{2} - 2}{\sqrt{2} - \sqrt{2}} = \dfrac{\sqrt{2} - 2}{0} = $ limit does not exist

If the problem is: $\lim_{x \to \sqrt{2}} \left(\dfrac{x^2 - 2}{x - \sqrt{2}} \right) = \lim_{x \to \sqrt{2}} \left(\dfrac{(x - \sqrt{2})(x + \sqrt{2})}{x - \sqrt{2}} \right)$

$= \lim_{x \to \sqrt{2}} \left(x + \sqrt{2} \right)$

$= \sqrt{2} + \sqrt{2} = 2\sqrt{2}$

12. Given $\varepsilon > 0$, we require a $\delta > 0$ such that if $0 < |x - 2| < \delta, |(3x - 1) - 5| < \varepsilon$.

Since $|(3x - 1) - 5| = |3x - 6| = 3|(x - 2)| < \varepsilon$ if $|x - 2| < \varepsilon/3$, choose $\delta = \varepsilon/3$.

Then if $0 < |x - 2| < \delta$, $0 < |x - 2| < \varepsilon/3$

$0 < 3|x - 2| < \varepsilon$

$|3x - 6| < \varepsilon$

$|3x - 1 - 5| < \varepsilon$ as required.

13. $k(-2) = \dfrac{\sqrt{8-(-2)^2}}{4(-2)^2-7} = \dfrac{\sqrt{4}}{9} = \dfrac{2}{9}$

$$\lim_{x\to-2} k(x) = \lim_{x\to-2} \frac{\sqrt{8-x^2}}{4x^2-7} = \frac{\lim\limits_{x\to-2}\sqrt{8-x^2}}{\lim\limits_{x\to-2}(4x^2-7)} = \frac{\sqrt{\lim\limits_{x\to-2}(8-x^2)}}{\lim\limits_{x\to-2}(4x^2-7)} = \frac{8-(-2)^2}{4(-2)^2-7}$$

$\dfrac{\sqrt{8-4}}{4(4)-7} = \dfrac{\sqrt{4}}{9} = \dfrac{2}{9}$

Since $\lim\limits_{x\to-2} k(x) = k(-2)$, the function is continuous at $x=-2$.

14. (a) 0   (b) 0   (c) ∞ (d) ∞

15. $g(-3) = -4(-3)-5 = 12-5 = 7$ *but* $\lim\limits_{x\to-3^-} g(x) = \lim\limits_{x\to-3^-}(4x+17) = 4(-3)+17 = 5$

Since $\lim\limits_{x\to-3^-} g(x) \neq g(-3)$, the function is not continuous at $x=-3$.

16. (a) $x=10$ (b) -1   (c) -1

17. The function is undefined for $x=0$ so there is a discontinuity at $x=0$. Need to check $x=1$.

$$g(1) = 5(1)^2 = 5(1) = 5 \textit{ but } \lim_{x\to1^-} g(x) = \lim_{x\to1^-}\frac{5}{x} = \frac{5}{1} = 5$$

Since $\lim\limits_{x\to1^-} g(x) = g(1)$, the function is continuous at $x=1$.

18. $\lim\limits_{x\to\infty} \dfrac{4x^3+3x^2-2x+1}{2x^3-3x^2+4x-5} = \lim\limits_{x\to\infty} \dfrac{(4x^3+3x^2-2x+1)\div x^3}{(2x^3-3x^2+4x-5)\div x^3} = \lim\limits_{x\to\infty} \dfrac{4+\dfrac{3}{x}-\dfrac{2}{x^2}+\dfrac{1}{x^3}}{2-\dfrac{3}{x}+\dfrac{4}{x^2}-\dfrac{5}{x^3}} = \lim\limits_{x\to\infty}\dfrac{4}{2} = 2$

19. (a) Yes   (b) No, because the discontinuity will be at .2, .4, .6, .8 before 1.

20. $\lim\limits_{x\to\infty}\left[\dfrac{x}{x^2+1} + \dfrac{2x^2}{x(x^2+1)}\right] = \lim\limits_{x\to\infty}\left[\dfrac{x^2}{x(x^2+1)} + \dfrac{2x^2}{x(x^2+1)}\right] = \lim\limits_{x\to\infty}\dfrac{3x^2}{x^3+x} = \lim\limits_{x\to\infty}\dfrac{\dfrac{3x^2}{x^3}}{\dfrac{x^3}{x^3}+\dfrac{x}{x^3}}$

$= \lim\limits_{x\to\infty}\dfrac{0}{1} = 0$

21. The function will be continuous if $f(3) = 3b + 4 = \lim_{x \to 3^+} f(x) = bx^2 - 2 = 9b^2 - 2$.

Solving, $9b^2 - 2 = 3b + 4$

$$9b^2 - 3b - 6 = 0$$

$$3b^2 - b - 2 = 0$$

$$(3b + 2)(b - 1) = 0$$

$$b = -\frac{2}{3}, 1$$

22. $\lim_{x \to \infty} \dfrac{16x^2 + 50x + 25}{4x^2 + 2x - 3} = \lim_{x \to \infty} \dfrac{16 + \dfrac{50}{x} + \dfrac{25}{x^2}}{4 + \dfrac{2}{x} - \dfrac{3}{x^2}} = \lim_{x \to \infty} \dfrac{16}{4} = 4$

23.

Function	Interval			
	$[0,1]$	$(0,1]$	$[0,1)$	$(0,1)$
$f(x) = \dfrac{1}{x}$	No	Yes	No	Yes
$g(x) = \dfrac{2x^2}{x-1}$	No	No	Yes	Yes
$H(x) = \sqrt{x}$	Yes	Yes	Yes	Yes
$K(x) = \dfrac{2x^2}{2x-1}$	No	No	No	No

24. $\lim_{x \to \pm\infty} \dfrac{5x^5 - x^3 + 2x - 1}{7 - x^5} = \lim_{x \to \pm\infty} \dfrac{5 - 1/x^2 + 2/x^4 - 1/x^5}{7/x^5 - 1} = -5.$ $y = -5$ is the horizontal asymptote.

Appendix 1
Algebra Review

1.A Integral Exponents

1. 36

3. $5+4\left(3^2+2^3\right)=5+4(9+8)$
$$=5+4\cdot17=73$$

5. $\dfrac{(-3)^2+(-2)^4}{-2^2-1}=\dfrac{9+16}{-4-1}=\dfrac{25}{-5}=-5$

7. $\left(-\dfrac{5}{4}\right)^3=-\dfrac{5^3}{4^3}=-\dfrac{125}{64}$

9. $\left(\dfrac{1}{3}\right)^3+\left(\dfrac{2}{3}\right)^3=\dfrac{1}{27}+\dfrac{8}{27}=\dfrac{9}{27}=\dfrac{1}{3}$

11. $2^4-2^7=16-128=-112$

13. $\left(2^{-2}+2\right)^2=\left(\dfrac{1}{4}+2\right)^2=\left(\dfrac{9}{4}\right)^2=\dfrac{81}{16}$

15. $2^2\cdot3^{-3}-3^2\cdot2^{-3}=4\cdot\dfrac{1}{27}-9\cdot\dfrac{1}{8}$
$$=-\dfrac{211}{216}$$

17. $\dfrac{1}{2^3}+\dfrac{1}{2^{-4}}=\dfrac{1}{8}+2^4=\dfrac{129}{8}$

19. $x^2\cdot x^3\cdot x^5=x^{2+3+5}=x^{10}$

21. $(.03)y^2\cdot y^7=.03y^{2+7}=.03y^9$

23. $\left(2x^2\right)^3 3x=2^3 x^6\cdot3x=24x^7$

25. $\left(3x^2y\right)^2=3^2\left(x^2\right)^2 y^2=9x^4y^2$

27. $\left(a^2\right)(7a)\left(-3a^3\right)=-21a^{2+1+3}=-21a^6$

29. $(2w)^3(3w)(4w)^2=8w^3\cdot3w\cdot16w^2$
$$=384w^6$$

31. ab^3

33. $(2x)^{-2}(2y)^3(4x)=2^{-2}x^{-2}\cdot2^3 y^3\cdot4x$
$$=8x^{-1}y^3=\dfrac{8y^3}{x}$$

35. $\left(-3a^4\right)^2\left(9x^3\right)^{-1}=9a^8\cdot\dfrac{1}{9x^3}=\dfrac{a^8}{x^3}$

37. $\left(2x^2y\right)^0(3xy)=1\cdot3xy=3xy$

39. $(64)^2=\left(2^6\right)^2=2^{12}$

41. $\left(2^4\cdot16^{-2}\right)^3=\left[2^4\cdot\left(2^4\right)^{-2}\right]^3=\left(2^4\cdot2^{-8}\right)^3$
$$=\left(2^{-4}\right)^3=2^{-12}$$

43. $\dfrac{x^4\left(x^2\right)^3}{x^3}=\dfrac{x^4 x^6}{x^3}=\dfrac{x^{10}}{x^3}=x^7$

45. $\left(\dfrac{e^6}{c^4}\right)^2 \cdot \left(\dfrac{c^3}{e}\right)^3 = \dfrac{e^{12}}{c^8} \cdot \dfrac{c^9}{e^3} = e^9 c$

47. $\left(\dfrac{ab^2c^3d^4}{abc^2d}\right)^2 = \left(bcd^3\right)^2 = b^2c^2d^6$

49. $\left(\dfrac{a^6}{b^{-4}}\right)^2 = \dfrac{a^{12}}{b^{-8}} = a^{12}b^8$

51. $\left(\dfrac{c^5}{d^{-3}}\right)^{-2} = \dfrac{c^{-10}}{d^6} = \dfrac{1}{c^{10}d^6}$

53. $\left(\dfrac{3x}{y^2}\right)^{-3}\left(\dfrac{-x}{2y^3}\right)^2 = \left(\dfrac{y^2}{3x}\right)^3\left(\dfrac{-x}{2y^3}\right)^2$

$= \dfrac{y^6}{27x^3}\cdot\dfrac{x^2}{2^2y^6}$

$= \dfrac{1}{108x}$

55. $\dfrac{\left(a^{-3}b^2c\right)^{-2}}{\left(ab^{-2}c^3\right)^{-1}} = \dfrac{a^6b^{-4}c^{-2}}{a^{-1}b^2c^{-3}} = \dfrac{a^7c}{b^6}$

57. c^3d^6

59. $a^2\left(a^{-1}+a^{-3}\right) = a + a^{-1} = a + \dfrac{1}{a}$

61. Since the first factor is negative after cubing and the second is positive after squaring, the product is negative.

63. Since the first factor is negative after taking the factor to the ninth power and the second is positive, the product is negative.

65. Since the first factor is negative after cubing and the second and third factors are positive, the product is negative.

67. $\dfrac{3^{-r}}{3^{-s-r}} = 3^{-r-(-s-r)} = 3^s$

69. $\left(\dfrac{a^6}{b^{-4}}\right)^t = \left(a^6b^4\right)^t = a^{6t}b^{4t}$

71. $\dfrac{\left(c^{-r}b^s\right)^t}{\left(c^tb^{-s}\right)^r} = \dfrac{c^{-rt}b^{st}}{c^{rt}b^{-sr}} = \dfrac{b^{rs+st}}{c^{2rt}}$

73. Many possible examples, including this one. If $a=3, b=4, r=2,$ then $a^2+b^2 = 3^2+4^2 = 9+16 = 25,$ but $(a+b)^2 = (3+4)^2 = 7^2 = 49.$

75. Many possible examples, including this one. If $a=2, b=3, r=1, s=2,$ then $a^rb^s = 2^1\cdot3^2 = 18$ but $(ab)^{r+s} = (2\cdot3)^{1+2} = 6^3 = 216$

77. Many possible examples, including this one. If $c=2, r=3, s=1,$ then $\dfrac{c^r}{c^s} = \dfrac{2^3}{2^1} = 2^2 = 4$ but $c^{r/s} = 2^{3/1} = 2^3 = 8$

79. Many possible examples, including this one. If $a = -3$, then
$$(-a)^2 = \left(-(-3)\right)^2 = (3)^2 = 9$$
but
$$-(-3)^2 = -(9) = -9$$

1.B Arithmetic of Algebraic Expressions

1. $8x$

3. $6a^2b + (-8b)a^2 = 6a^2b - 8a^2b = -2a^2b$

5. $(x^2 + 2x + 1) - (x^3 - 3x^2 + 4) = x^2 + 2x + 1 - x^3 + 3x^2 - 4 = -x^3 + 4x^2 + 2x - 3$

7. $\left[u^4 - (-3)u^3 + u/2 + 1\right] - \left(u^4 - 2u^3 + 5 - u/2\right) = u^4 + 3u^3 + u/2 + 1 - u^4 + 2u^3 - 5 + u/2$
$$= 5u^3 + u - 4$$

9. $\left[4z - 6z^2w - (-2)z^3w^2\right] + \left(8 - 6z^2w - zw^3 + 4z^3w^2\right)$
$$= 4z - 6z^2w + 2z^3w^2 + 8 - 6z^2w - zw^3 + 4z^3w^2$$
$$= 6z^3w^2 - 12z^2w - zw^3 + 4z + 8$$

11. $(9x - x^3 + 1) - \left[2x^3 + (-6)x + (-7)\right] = 9x - x^3 + 1 - 2x^3 + 6x + 7 = -3x^3 + 15x + 8$

13. $(x^2 - 3xy) - (x + xy) - (x^2 + xy) = x^2 - 3xy - x - xy - x^2 - xy = -5xy - x$

15. $15y^3 - 5y$

17. $12a^2x^2 - 6a^3xy + 6a^2xy$

19. $12z^4 + 30z^3$

21. $12a^2b - 18ab^2 + 6a^3b^2$

23. $(x + 1)(x - 2) = x^2 - 2x + 1x - 2 = x^2 - x - 2$

25. $(-2x + 4)(-x - 3) = 2x^2 + 6x - 4x - 12 = 2x^2 + 2x - 12$

27. $(y + 3)(y + 4) = y^2 + 4y + 3y + 12 = y^2 + 7y + 12$

29. $(3x + 7)(-2x + 5) = -6x^2 + 15x - 14x + 35 = -6x^2 + x + 35$

31. $3y^3 - 9y^2 + 4y - 12$

33. $(x + 4)(x - 4) = x^2 - 4x + 4x - 16 = x^2 - 16$

35. $(4a + 5b)(4a - 5b) = 16a^2 - 20ab + 20ab - 25b^2 = 16a^2 - 25b^2$

37. $(y - 11)^2 = (y - 11)(y - 11) = y^2 - 11y - 11y + 121 = y^2 - 22y + 121$

39. $(5x - b)^2 = (5x - b)(5x - b) = 25x^2 - 5bx - 5bx + b^2 = 25x^2 - 10bx + b^2$

41. $\left(4x^3 - y^4\right)^2 = \left(4x^3 - y^4\right)\left(4x^3 - y^4\right) = 16x^6 - 4x^3y^4 - 4x^3y^4 + y^8 = 16x^6 - 8x^3y^4 + y^8$

43. $\left(-3x^2+2y^4\right)^2=\left(-3x^2+2y^4\right)\left(-3x^2+2y^4\right)=9x^4-6x^2y^4-6x^2y^4+4y^8$
$=9x^4-12x^2y^4+4y^8$

45. $(2y+3)\left(y^2+3y-1\right)=2y\left(y^2+3y-1\right)+3\left(y^2+3y-1\right)$
$=2y^3+6y^2-2y+3y^2+9y-3=2y^3+9y^2+7y-3$

47. $(5w+6)\left(-3w^2+4w-3\right)=5w\left(-3w^2+4w-3\right)+6\left(-3w^2+4w-3\right)$
$=-15w^3+20w^2-15w-18w^2+24w-18=-15w^3+2w^2+9w-18$

49. $2x(3x+1)(4x-2)=2x\left(12x^2-6x+4x-2\right)=2x\left(12x^2-2x-2\right)=24x^3-4x^2-4x$

51. $(x-1)(x-2)(x-3)=(x-1)\left(x^2-3x-2x+6\right)=(x-1)\left(x^2-5x+6\right)$
$=x\left(x^2-5x+6\right)-1\left(x^2-5x+6\right)=x^3-5x^2+6x-x^2+5x-6=x^3-6x^2+11x-6$

53. $(x+4y)(2y-x)(3x-y)=(x+4y)\left(6xy-2y^2-3x^2+xy\right)$
$=(x+4y)\left(-3x^2+7xy-2y^2\right)=x\left(-3x^2+7xy-2y^2\right)+4y\left(-3x^2+7xy-2y^2\right)$
$=-3x^3+7x^2y-2xy^2-12x^2y+28xy^2-8y^3=-3x^3-5x^2y+26xy^2-8y^3$

55. The terms involving x^2 will be $x^2(-3)+(3x)(2x)=-3x^2+6x^2=3x^2$.   3

57. The term involving x^2 will be $-6\left(x^2\right)$. -6

59. $(x+2)^3=(x+2)(x+2)(x+2)=(x+2)\left(x^2+4x+4\right)$
The terms involving x^2 will be $2x^2+x(4x)=2x^2+4x^2=6x^2$.   6

61. The terms involving x^2 will be $x^2(1)+x(-x)+1\left(x^2\right)=x^2-x^2+x^2=x^2$.   1

63. The terms involving x^2 will be $2x(3x)-1\left(x^2\right)=6x^2-x^2=5x^2$.   5

65. $\left(\sqrt{x}+5\right)\left(\sqrt{x}-5\right)=\sqrt{x}\sqrt{x}-5\sqrt{x}+5\sqrt{x}-25=x-25$

67. $\left(3+\sqrt{y}\right)^2=\left(3+\sqrt{y}\right)\left(3+\sqrt{y}\right)=9+3\sqrt{y}+3\sqrt{y}+\sqrt{y}\sqrt{y}=9+6\sqrt{y}+y$

69. $\left(1+\sqrt{3}x\right)\left(x+\sqrt{3}\right)=x+\sqrt{3}+\sqrt{3}x^2+\sqrt{3}x\sqrt{3}=\sqrt{3}x^2+4x+\sqrt{3}$

71. $(ax+b)(3x+2)=3ax^2+2ax+3bx+2b=3ax^2+(2a+3b)x+2b$

73. $(ax+b)(bx+a)=abx^2+a^2x+b^2x+ab=abx^2+\left(a^2+b^2\right)x+ab$

75. $(x-a)(x-b)(x-c) = (x-a)(x^2 - cx - bx + bc)$
$$= x(x^2 - cx - bx + bc) - a(x^2 - cx - bx + bc)$$
$$= x^3 - cx^2 - bx^2 + bcx - ax^2 + acx + abx - abc$$
$$= x^3 - (a+b+c)x^2 + (bc + ac + ab)x - abc$$

77. $16x^{n+k}$

79. $(y^r + 1)(y^s - 4) = y^r y^s - 4y^r + y^s - 4 = y^{r+s} - 4y^r + y^s - 4$

81. The outer rectangle has dimensions $10 + 2x$ and $6 + 2x$ and area $(10 + 2x)(6 + 2x)$.
The inner rectangle has area $6 \cdot 10 = 60$.
The shaded area is $(10 + 2x)(6 + 2x) - 60 = 60 + 20x + 12x + 4x^2 - 60 = 4x^2 + 32x$.

83. The statement $3(y+2) = 3y + 2$ is false when $y = 1$, since $3(1+2) = 9$ but
$3(1) + 2 = 5$. This mistake is the value 2. The correct statement is $3(y+2) = 3y + 6$.

85. The statement $(x+y)^2 = x + y^2$ is false when $x = 1$ and $y = 2$, since $(1+2)^2 = 9$ but
$1 + 2^2 = 5$. The correct statement is $(x+y)^2 = x^2 + 2xy + y^2$.

87. The statement $(7x)(7y) = 7xy$ is false when $x = 1$ and $y = 1$, since $(7 \cdot 1)(7 \cdot 1) = 49$
but $7 \cdot 1 \cdot 1 = 7$. The mistake is dropping a 7 on the left side. The correct statement is
$(7x)(7y) = 49xy$.

89. The statement $y + y + y = y^3$ is false when $y = 1$, since $1 + 1 + 1 = 3$ but $1^3 = 1$. The
mistake is the use of 3 as an exponent. The correct statement is $y + y + y = 3y$.

91. The statement $(x-3)(x-2) = x^2 - 5x - 6$ is false when $x = 3$, since $(3-3)(3-2) = 0$
but $3^2 - 5 \cdot 3 - 6 = -12$. The mistake is the sign on the 6. The correct statement is
$(x-3)(x-2) = x^2 - 5x + 6$.

93. This is represented algebraically by $\dfrac{(x+1)^2 - (x-1)^2}{x}$. This simplifies to
$$\frac{(x^2 + 2x + 1) - (x^2 - 2x + 1)}{x} = \frac{4x}{x} = 4. \text{ Thus, the result is always 4.}$$

95. Answers will vary. One example will be to take a number. Subtract 3 from it to get a
first number. Now add three to the original number to get a second number. Multiply
the first number and the second number. Add 9 to the result. Divide by the original
number, and the result is the number that you started with.

1.C Factoring

1. $(x+2)(x-2)$

3. $9y^2 - 25 = (3y)^2 - 5^2 = (3y+5)(3y-5)$

5. $81x^2 + 36x + 4 = (9x)^2 + 2(9x)2 + 2^2 = (9x+2)^2$

7. $5 - x^2 = \left(\sqrt{5}\right)^2 - x^2 = \left(\sqrt{5}+x\right)\left(\sqrt{5}-x\right)$

9. $49 + 28z + 4z^2 = 7^2 + 2(7)(2z) + (2z)^2 = (7+2z)^2$

11. $x^4 - y^4 = \left(x^2\right)^2 - \left(y^2\right)^2 = \left(x^2 + y^2\right)\left(x^2 - y^2\right) = \left(x^2 + y^2\right)(x+y)(x-y)$

13. $(x+3)(x-2)$

15. $(z+1)(z+3)$

17. $(y+9)(y-4)$

19. $x^2 - 6x + 9 = x^2 - 2(3x) + 3^2 = (x-3)^2$

21. $(x+5)(x+2)$

23. $(x+9)(x+2)$

25. $(3x+1)(x+1)$

27. $(2z+3)(z+4)$

29. $9x(x-8)$

31. $10x^2 - 8x - 2 = 2\left(5x^2 - 4x - 1\right) = 2(5x+1)(x-1)$

33. $(4u-3)(2u+3)$

35. $4x^2 + 20xy + 25y^2 = (2x)^2 + 2(2x)(5y) + (5y)^2 = (2x+5y)^2$

37. $x^3 - 125 = x^3 - 5^3 = (x-5)\left(x^2 + x \cdot 5 + 5^2\right) = (x-5)\left(x^2 + 5x + 25\right)$

39. $x^3 + 6x^2 + 12x + 8 = x^3 + 3 \cdot x^2 \cdot 2 + 3 \cdot x \cdot 2^2 + 2^3 = (x+2)^3$

41. $8 + x^3 = 2^3 + x^3 = (2+x)\left(2^2 - 2x + x^2\right) = (2+x)\left(4 - 2x + x^2\right)$

43. $-x^3 + 15x^2 - 75x + 125 = (-x)^3 + 3 \cdot 5(-x)^2 + 3 \cdot 5^2(-x) + 5^3 = (-x+5)^3$ or $(5-x)^3$

45. $x^3 + 1 = x^3 + 1^3 = (x+1)\left(x^2 - x \cdot 1 + 1^2\right) = (x+1)\left(x^2 - x + 1\right)$

47. $8x^3 - y^3 = (2x)^3 - y^3 = (2x-y)\left[(2x)^2 + (2x)y + y^2\right] = (2x-y)\left(4x^2 + 2xy + y^2\right)$

49. $x^6 - 64 = \left(x^3\right)^2 - 8^2 = \left(x^3 + 8\right)\left(x^3 - 8\right) = (x+2)\left(x^2 - 2x + 4\right)(x-2)\left(x^2 + 2x + 4\right)$

51. $\left(y^2 + 5\right)\left(y^2 + 2\right)$

53. $81 - y^4 = 9^2 - \left(y^2\right)^2 = \left(9 + y^2\right)\left(9 - y^2\right) = \left(9 + y^2\right)(3 + y)(3 - y)$

55. $z^6 - 1 = \left(z^3\right)^2 - 1^2 = \left(z^3 + 1\right)\left(z^3 - 1\right) = (z + 1)\left(z^2 - z + 1\right)(z - 1)\left(z^2 + z + 1\right)$

57. $\left(x^2 + 3y\right)\left(x^2 - y\right)$

59. $x^2 - yz + xz - xy = x^2 + xz - yz - xy = x(x + z) - y(z + x) = (x + z)(x - y)$

61. $a^3 - 2b^2 + 2a^2b - ab = a^3 + 2a^2b - 2b^2 - ab$
$$= a^2(a + 2b) - b(2b + a) = (a + 2b)\left(a^2 - b\right)$$

63. $x^3 + 4x^2 - 8x - 32 = x^2(x + 4) - 8(x + 4)$
$$= (x + 4)\left(x^2 - 8\right) \ or \left(x + \sqrt{8}\right)\left(x - \sqrt{8}\right)(x + 4)$$

65. If $x^2 + 1 = (x + c)(x + d) = x^2 + (c + d)x + cd$, then $c + d = 0$ and $cd = 1$. But $c + d = 0$ implies that $c = -d$ and hence that $1 = cd = (-d)d = -d^2$, or equivalently, that $d^2 = -1$. Since there is no real number with this property, $x^2 + 1$ cannot possibly factor in this way.

1.D Fractional Expressions

1. $\dfrac{63}{49} = \dfrac{7 \cdot 9}{7 \cdot 7} = \dfrac{9}{7}$

3. $\dfrac{13 \cdot 27 \cdot 22 \cdot 10}{6 \cdot 4 \cdot 11 \cdot 12} = \dfrac{13 \cdot 3 \cdot 3 \cdot 3 \cdot 2 \cdot 11 \cdot 2 \cdot 5}{3 \cdot 2 \cdot 2 \cdot 2 \cdot 11 \cdot 2 \cdot 2 \cdot 3} = \dfrac{13 \cdot 3 \cdot 5}{2 \cdot 2 \cdot 2} = \dfrac{195}{8}$

5. $\dfrac{x^2 - x - 2}{x^2 + 2x + 1} = \dfrac{(x+1)(x-2)}{(x+1)(x+1)} = \dfrac{x-2}{x+1}$

7. $\dfrac{a^2 - b^2}{a^3 - b^3} = \dfrac{(a-b)(a+b)}{(a-b)(a^2 + ab + b^2)} = \dfrac{a+b}{a^2 + ab + b^2}$

9. $\dfrac{(x+c)(x^2 - cx + c^2)}{x^4 + c^3 x} = \dfrac{x^3 + c^3}{x(x^3 + c^3)} = \dfrac{1}{x}$

11. $\dfrac{3}{7} + \dfrac{2}{5} = \dfrac{15}{35} + \dfrac{14}{35} = \dfrac{29}{35}$

13. $\left(\dfrac{19}{7} + \dfrac{1}{2}\right) - \dfrac{1}{3} = \dfrac{114}{42} + \dfrac{21}{42} - \dfrac{14}{42} = \dfrac{121}{42}$

15. $\dfrac{c}{d} + \dfrac{3c}{e} = \dfrac{ce}{de} + \dfrac{3cd}{de} = \dfrac{ce + 3cd}{de}$

17. $\dfrac{b}{c} - \dfrac{c}{b} = \dfrac{b^2}{bc} - \dfrac{c^2}{bc} = \dfrac{b^2 - c^2}{bc}$

19. $\dfrac{1}{x+1} - \dfrac{1}{x} = \dfrac{x}{x(x+1)} - \dfrac{1(x+1)}{x(x+1)} = \dfrac{x - (x+1)}{x(x+1)} = \dfrac{x - x - 1}{x(x+1)} = -\dfrac{1}{x(x+1)}$

21. $\dfrac{1}{x+4} + \dfrac{2}{(x+4)^2} - \dfrac{3}{x^2 + 8x + 16} = \dfrac{x+4}{(x+4)^2} + \dfrac{2}{(x+4)^2} - \dfrac{3}{(x+4)^2}$

$= \dfrac{x + 4 + 2 - 3}{(x+4)^2} = \dfrac{x+3}{(x+4)^2}$

23. $\dfrac{1}{x} - \dfrac{1}{3x - 4} = \dfrac{3x - 4}{x(3x-4)} - \dfrac{x}{x(3x-4)} = \dfrac{(3x-4) - x}{x(3x-4)} = \dfrac{2x - 4}{x(3x-4)}$

25. $\dfrac{1}{x+y} + \dfrac{x+y}{x^3 + y^3} = \dfrac{(x^2 - xy + y^2)}{(x+y)(x^2 - xy + y^2)} + \dfrac{x+y}{x^3 + y^3}$

$= \dfrac{(x^2 - xy + y^2) + (x+y)}{x^3 + y^3} = \dfrac{x^2 - xy + y^2 + x + y}{x^3 + y^3}$

27. $\dfrac{1}{4x(x+1)(x+2)^3} - \dfrac{6x+2}{4(x+1)^3} = \dfrac{(x+1)^2}{4x(x+1)^3(x+2)^3} - \dfrac{x(6x+2)(x+2)^3}{4x(x+1)^3(x+2)^3}$

$$= \dfrac{x^2+2x+1-(6x^2+2x)(x^3+6x^2+12x+8)}{4x(x+1)^3(x+2)^3}$$

$$= \dfrac{x^2+2x+1-(6x^5+38x^4+84x^3+72x^2+16x)}{4x(x+1)^3(x+2)^3}$$

$$= \dfrac{-6x^5-38x^4-84x^3-71x^2-14x+1}{4x(x+1)^3(x+2)^3}$$

29. $\dfrac{3}{4} \cdot \dfrac{12}{5} \cdot \dfrac{10}{9} = \dfrac{3 \cdot 2 \cdot 2 \cdot 3 \cdot 2 \cdot 5}{2 \cdot 2 \cdot 5 \cdot 3 \cdot 3} = 2$ **31.** $\dfrac{3a^2c}{4ac} \cdot \dfrac{8ac^3}{9a^2c^4} = \dfrac{24a^3c^4}{36a^3c^5} = \dfrac{2}{3c}$

33. $\dfrac{7x}{11y} \cdot \dfrac{66y^2}{14x^3} = \dfrac{1}{1} \cdot \dfrac{6y}{2x^2} = \dfrac{3y}{x^2}$

35. $\dfrac{3x+9}{2x} \cdot \dfrac{8x^2}{x^2-9} = \dfrac{3(x+3)}{2x} \cdot \dfrac{2x \cdot 4x}{(x+3)(x-3)} = \dfrac{3}{1} \cdot \dfrac{4x}{x-3} = \dfrac{12x}{x-3}$

37. $\dfrac{5y-25}{3} \cdot \dfrac{y^2}{y^2-25} = \dfrac{5(y-5)}{3} \cdot \dfrac{y^2}{(y+5)(y-5)} = \dfrac{5}{3} \cdot \dfrac{y^2}{y+5} = \dfrac{5y^2}{3(y+5)}$

39. $\dfrac{u}{u-1} \cdot \dfrac{u^2-1}{u^2} = \dfrac{u}{u-1} \cdot \dfrac{(u-1)(u+1)}{u \cdot u} = \dfrac{1}{1} \cdot \dfrac{u+1}{u} = \dfrac{u+1}{u}$

41. $\dfrac{2u^2+uv-v^2}{4u^2-4uv+v^2} \cdot \dfrac{8u^2+6uv-9v^2}{4u^2-9v^2} = \dfrac{(2u-v)(u+v)}{(2u-v)(2u-v)} \cdot \dfrac{(4u-3v)(2u+3v)}{(2u-3v)(2u+3v)}$

$$= \dfrac{(u+v)(4u-3v)}{(2u-v)(2u-3v)}$$

43. $\dfrac{5}{12} \div \dfrac{4}{14} = \dfrac{5}{12} \cdot \dfrac{14}{4} = \dfrac{5}{12} \cdot \dfrac{7}{2} = \dfrac{35}{24}$ **45.** $\dfrac{uv}{v^2w} \div \dfrac{uv}{u^2v} = \dfrac{uv}{v^2w} \cdot \dfrac{u^2v}{uv} = \dfrac{u}{vw} \cdot \dfrac{u}{1} = \dfrac{u^2}{vw}$

47. $\dfrac{\dfrac{x+3}{x+4}}{\dfrac{2x}{x+4}} = \dfrac{x+3}{x+4} \div \dfrac{2x}{x+4} = \dfrac{x+3}{x+4} \cdot \dfrac{x+4}{2x} = \dfrac{x+3}{2x}$

49. $\dfrac{x+y}{x+2y} \div \left(\dfrac{x+y}{xy}\right)^2 = \dfrac{x+y}{x+2y} \cdot \dfrac{x^2y^2}{(x+y)^2} = \dfrac{x^2y^2}{(x+y)(x+2y)}$

51. $\dfrac{\dfrac{(c+d)^2}{c^2-d^2}}{cd} = \dfrac{(c+d)^2}{1} \div \dfrac{c^2-d^2}{cd} = \dfrac{(c+d)^2}{1} \cdot \dfrac{cd}{(c+d)(c-d)} = \dfrac{cd(c+d)}{c-d}$

53. $\dfrac{\dfrac{1}{x^2}-\dfrac{1}{y^2}}{\dfrac{1}{x}+\dfrac{1}{y}} = \dfrac{\dfrac{1}{x^2}-\dfrac{1}{y^2}}{\dfrac{1}{x}+\dfrac{1}{y}} \cdot \dfrac{x^2y^2}{x^2y^2} = \dfrac{y^2-x^2}{xy^2+x^2y} = \dfrac{(y+x)(y-x)}{xy(y+x)} = \dfrac{y-x}{xy}$

55. $\dfrac{\dfrac{6}{y}-3}{1-\dfrac{1}{y-1}} = \dfrac{\dfrac{6-3y}{y}}{\dfrac{(y-1)-1}{y-1}} = \dfrac{6-3y}{y} \div \dfrac{y-2}{y-1} = \dfrac{3(2-y)}{y} \cdot \dfrac{y-1}{y-2} = \dfrac{-3(y-1)}{y}$

57. $\dfrac{\dfrac{1}{2x+2h}-\dfrac{1}{2x}}{h} = \dfrac{\dfrac{1}{2x+2h}-\dfrac{1}{2x}}{h} \cdot \dfrac{2x(x+h)}{2x(x+h)} = \dfrac{x-(x+h)}{2hx(x+h)} = \dfrac{-h}{2hx(x+h)} = \dfrac{-1}{2x(x+h)}$

59. $\left(x^{-1}+y^{-1}\right) = \left(\dfrac{1}{x}+\dfrac{1}{y}\right)^{-1} = \dfrac{1}{\left(\dfrac{1}{x}+\dfrac{1}{y}\right)} = \dfrac{1}{\left(\dfrac{1}{x}+\dfrac{1}{y}\right)} \cdot \dfrac{xy}{xy} = \dfrac{xy}{y+x} \text{ or } \dfrac{xy}{x+y}$

61. $\dfrac{\text{Area of shaded region}}{\text{Area of dartboard}} = \dfrac{\pi x^2}{(2x)^2} = \dfrac{\pi}{4}$

63. $\dfrac{\text{Area of shaded region}}{\text{Area of dartboard}} = \dfrac{x^2}{(5x)^2} = \dfrac{1}{25}$

65. Let $a = b = 1$.

$$\frac{1}{a} + \frac{1}{b} = \frac{1}{1} + \frac{1}{1} = 2$$

but

$$\frac{1}{a+b} = \frac{1}{1+1} = \frac{1}{2}$$

The correct statement is:

$$\frac{1}{a} + \frac{1}{b} = \frac{a+b}{ab}$$

67. Let $a = b = 1$.

$$\left(\frac{1}{\sqrt{a}+\sqrt{b}}\right)^2 = \left(\frac{1}{\sqrt{1}+\sqrt{1}}\right)^2 = \frac{1}{4}$$

but

$$\frac{1}{a+b} = \frac{1}{1+1} = \frac{1}{2}$$

The correct statement is:

$$\left(\frac{1}{\sqrt{a}+\sqrt{b}}\right)^2 = \frac{1}{a+2\sqrt{ab}+b}$$

69. Let $u = 1$ and $v = 2$.

$$\frac{u}{v} + \frac{v}{u} = \frac{1}{2} + \frac{2}{1} = 2\frac{1}{2} \neq 1$$

The correct statement is:

$$\frac{u}{v} + \frac{v}{u} = \frac{u^2 + v^2}{uv}$$

71. Let $x = y = 1$.

$$\left(\sqrt{x}+\sqrt{y}\right)\frac{1}{\sqrt{x}+\sqrt{y}} = \left(\sqrt{1}+\sqrt{1}\right)\frac{1}{\sqrt{1}+\sqrt{1}} = 1$$

but

$$x + y = 1 + 1 = 2$$

The correct statement is:

$$\left(\sqrt{x}+\sqrt{y}\right)\frac{1}{\sqrt{x}+\sqrt{y}} = 1$$

Appendix 2
Geometry Review

1. By the Pythagorean Theorem,
$$2^2 + 4^2 = b^2$$
$$b^2 = 20$$
$$b = \sqrt{20} \text{ or } 2\sqrt{5}$$

3. By the Pythagorean Theorem,
$$c^2 + 9^2 = 12^2$$
$$c^2 = 63$$
$$c = \sqrt{63} \text{ or } 3\sqrt{7}$$

5. Since this is a isosceles right triangle, $c = 5$. By the Pythagorean Theorem,
$$5^2 + 5^2 = b^2$$
$$b^2 = 50$$
$$b = \sqrt{50} \text{ or } 5\sqrt{2}$$

7. Since this is a $30° - 60°$ right triangle $2c = 10$, $c = 5$. By the Pythagorean Theorem, $b^2 + 5^2 = 10^2$
$$b^2 = 75$$
$$b = \sqrt{75} \text{ or } 5\sqrt{3}$$

9. Since corresponding parts are congruent, Angle $A = 53.13°$.
$$\text{Angle } R = 36.87°.$$
$$r = 3, s = 4, t = 5$$

11. Angle $A = 180° - (90° + 25°) = 65°$
Angle $R = 180° - (90° + 65°) = 25°$
Thus $\angle A \cong \angle T$, $\angle C \cong \angle R$, $AC \cong TR$.
By ASA, the triangles are congruent.

13. $\angle u \cong \angle v$, that is, $\angle POD \cong \angle QOD$; $OD \cong OD$. By ASA, $PDO \cong QDO$. Since corresponding parts are congruent, $PD \cong QD$, and $\angle PDO \cong \angle QDO$. Since $\angle PDQ = 180°$, $\angle PDO = \angle QDD = 90°$. Hence PQ is perpendicular to OD.

15. $\angle u \cong \angle v$, that is, $\angle BAC \cong \angle DCA$. $AB \cong DC$ and $AC \cong AC$.
By SAS, the triangles ABC and CDA are congruent.

17. $\angle x \cong \angle y$, that is, $\angle BAC \cong \angle DAC$. $\angle u \cong \angle v$, that is, $\angle BCA \cong \angle DCA$.
$AC \cong AC$. By ASA, the triangles ABC and ADC are congruent.

19. By the Pythagorean Theorem,
$$a^2 + b^2 = c^2, c = \sqrt{6^2 + 4.5^2} = 7.5.$$
Since the triangles are similar,
$$\frac{r}{a} = \frac{s}{b} \qquad \frac{t}{c} = \frac{s}{b}$$
$$\frac{r}{6} = \frac{21}{4.5} \qquad \frac{t}{7.5} = \frac{21}{4.5}$$
$$r = 28 \qquad t = 35$$

21. Since the triangles are similar,
$$\frac{a}{b} = \frac{t}{s} \qquad \frac{c}{b} = \frac{r}{s}$$
$$\frac{7}{5} = \frac{t}{2.5} \qquad \frac{6}{5} = \frac{r}{2.5}$$
$$t = 3.5 \qquad r = 3$$

23. Since $\dfrac{AB}{RS} = \dfrac{15}{10} = \dfrac{AC}{RT}$ and $\angle A \cong \angle R$, the triangles are similar by S / A / S.

$a = 15$, $\angle S \cong \angle T$ and both have measure $60°$, since the triangles are equilateral.

25. Since $\dfrac{AC}{RS} = \dfrac{6.885}{4.59} = \dfrac{5.25}{3.5} = \dfrac{BC}{TS}$ and $\angle C \cong \angle S$, the triangles are similar by S/A/S.

(S/S/S can also be used.) $\angle T \cong \angle B$, and both have measure $62.46°$.

$\angle A \cong \angle R$, and both have measure $180° - (75° + 62.46°) = 42.54°$.

27. Since $\dfrac{a}{r} = \dfrac{5}{17.5} = \dfrac{b}{s}$ and $\angle C \cong \angle T$, the triangles are similar by S/A/S.

Since the triangles are isosceles right triangles, $\angle A$, $\angle B$, $\angle R$, $\angle S$ all have measure $45°$. By the Pythagorean Theorem, $c^2 = a^2 + b^2$, hence $c = \sqrt{5^2 + 5^2} = 5\sqrt{2}$ and

$t^2 = r^2 + s^2$, hence $t = \sqrt{17.5^2 + 17.5^2} = 17.5\sqrt{2}$.

29. Since $\angle A \cong \angle R$ and $\angle B \cong \angle T$, the triangles are similar by AA.

$\angle C \cong \angle S$ and both have measures $180° - (105° + 33.5°) = 41.5°$.

Since the triangles are similar, $\dfrac{r}{a} = \dfrac{s}{b}$; $\dfrac{3.5}{7} = \dfrac{s}{4.8}$; $s = 2.4$.

31. Let h = height of flagpole. Similar triangle considerations yield
$h/5 = 99/9$, $h = 55$ ft.

33. Let h = height of pyramid. Similar triangle considerations yield
$4/h = 1/20$, $h = 80$ ft.

35. Let x = distance from Marietta to Columbus, y = distance from Marietta to Cincinnati. Similar triangle considerations yield
$x/8.5 = 100/9.2$ and $y/15.75 = 100/9.2$; $x = 92.4$ mi; $y = 171.2$ mi.

37. In the figure, $AD \cong CR$, $CD \cong AR$, $AC \cong CA$, Hence by SSS,
triangle $ADC \cong$ triangle CRA, and so $\angle x \cong \angle u$. Similarly $\angle v \cong \angle y$. Since
$\angle x + \angle w + \angle y = 180°$, it follows that $\angle u + \angle w + \angle y = 180°$.